Dear Dave

Best Wishes,

Hope you enjoy it

Dom

Sediment Flux Modeling

Wellfleet, MA, 1999

Sediment Flux Modeling

Dominic M. DiToro

A John Wiley & Sons, Inc., Publication

New York • Chichester • Weinheim • Brisbane • Singapore • Toronto

For ordering and customer service, call 1-800-CALL-WILEY.

Library of Congress Cataloging-in-Publication Data:

Library of Congress Cataloging-in-Publication Data is available.
ISBN 0-471-13535-6

Printed in the United States of America.

10 9 8 7 6 5 4 3 2 1

Contents

Preface

Sediments are an important feature of natural water bodies. The physical, chemical, and biological processes that take place are critically influenced by their presence. A primary interaction is the exchange of solutes between the sediment and the overlying water. The flux – the transport of mass – of dissolved and particulate chemical species to and from the sediments are important components of the chemical and biological cycles that take place. For example, the consumption of dissolved oxygen by organic matter that settles to the sediment in the spring is usually the primary cause, or at least an important cause, of summertime oxygen depletion in the bottom waters of lakes and estuaries. The cycling of nutrients and metals are controlled to a significant extent by processes in sediments. In particular the degree to which they are either trapped or transformed in sediments determines the extent to which they continue to interact with the water column. Dissolved nutrients are returned to the water column where they can be reused by the biological community. Dissolved toxic metals that recycle to the water column are almost always more toxic than their particulate counterparts in the sediment. It is the degree of remobilization that determines the extent to which materials stored in sediments: e.g., organic matter, inorganic phosphorus, toxic and non-toxic metals, become available to interact chemically and biologically in the water column. It determines the extent to which the consequences of historical discharges to water bodies – the materials stored in the sediments – can still exert an influence.

This book addresses the problem of mathematically modeling the processes in sediments that determine the extent to which materials that settle to the sediment are recycled to the water column. It is the flux of materials to and from the water column that is the primary focus of this book. The substances considered are the nutrients: ammonia, nitrate, phosphate, and silica; dissolved oxygen; and the metals: calcium, manganese, iron, and cadmium. They were chosen primarily because of their importance in water quality problems: eutrophication and its consequences (N, P, Si), low dissolved oxygen with its attendant severe biological impacts; excessive concentrations of toxic metals (Mn, Cd), and for their importance to other processes (Ca, Fe).

The modeling philosophy employed is one of parsimony,* sometimes referred to as Occam's razor.[†] Thus, for the most part, we adopt a two-layer rather than a many-layer representation of the sediment. Also we approximate some complex processes with empirical formulations in order to proceed, and then evaluate the consequences. An important component of the model building exercise is the constant reliance on high quality, large data sets, the sources of which are specifically acknowledged below. In fact, most of the effort is spent in an interplay of model formulations and data comparisons.

The mathematical methods employed are algebraic equations for the steady state models, and ordinary differential equations for the time variable models, for which an appendix is provided. With the exception of a few illustrative solutions, the differential equations are approximated by finite differences and solved numerically. A unique feature is the use of MACSYMA, a symbolic computational program to produce analytical solutions of the algebraic equations. This is such a useful, powerful, and necessary technique that a brief introduction to MACSYMA programming is included in an appendix. In addition, the actual MACSYMA inputs and outputs are listed throughout the book. This should provide an excellent starting point for developing a facility with this most useful of tools. The majority of the more complex models, especially the metals models, could not have been developed without this aid.

Also included in the appendices are the data used for the model calibrations and applications. It is likely that the data will ultimately prove to be more enduring, as the models are superseded. No small effort is required to assemble large data sets – not to mention the effort to collect them – so these tables should prove to be useful in themselves. The FORTRAN code for the steady state model is also provided. It can be used for applications of the model to other data sets and to provide the basis for the development of improved and more general models.

The book is divided into six parts and two appendices, listed below. Part I, comprising the first two chapters, presents the necessary background information: the physical, chemical and biological facts that are directly relevant to sediment processes. Mass balance models are introduced and applied to simplified situations. The mathematical and MACSYMA appendices conclude the presentation.

---

*"Everything should be made as simple as possible, but not simpler." Albert Einstein (1879–1955)
[†]"Entities should not be multiplied unnecessarily." William of Occam (1300–1439)

Part II, comprising five chapters, presents the steady state models for nutrients. Ammonia and nitrate flux models are considered in Chapters 3 and 4. A general steady state model is formulated and analyzed in Chapter 5. Phosphate and silica flux models are considered in Chapters 6 and 7. Of particular interest is the failure of the phosphate model to reproduce observed anaerobic flux rates, which points directly to a time variable effect.

The four chapters of Part III are devoted to modeling oxygen fluxes. Modeling sediment oxygen demand (SOD) has been a preoccupation of mine for many years. The chapters chronicle the progression of models from a formulation that focuses solely on oxygen equivalents (Chapter 8), to the explicit consideration of the oxidation of reactive intermediates: sulfide (Chapter 9) which predominates in marine sediments, and methane (Chapter 10) in freshwater sediments. Chapter 11, which considers both by explicitly modeling sulfate uptake, concludes the presentation.

All the models in Parts II and III are steady state models. They are very instructive and quite successful in some cases. However, modeling the seasonal variation of fluxes requires a time variable formulation. Part IV begins with Chapter 12 that presents the kinetic model for the organic matter diagenesis. The "three G" model of Westrich and Berner[*] is used, which divides the organic matter into three reactivity classes in order to model properly the differing time scales of organic matter mineralization. Chapter 13 presents the structure of the time variable version of the

[*]Berner (1980a), Westrich and Berner (1984)

model. Included are the formulation for particle mixing due to benthic organisms (bioturbation), and the use of a benthic stress surrogate in order to include the effect of low (or zero) dissolved oxygen on the benthos.

Part V presents the calibration and applications of the model. The calibration of the time variable model to the Chesapeake Bay data set is presented in Chapter 14. In Chapter 15 the model is applied to three additional data sets, initially with no change in any of the model parameters. For the MERL data set, the model succeeds and fails in interesting ways, pointing to the strengths and weaknesses of the model and the price that is paid for simplicity. Two smaller but still instructive data sets are examined, Long Island Sound and Lake Champlain. The latter is an application of the model to a freshwater lake. The chapter concludes with a summary table of all the model parameters used in each application, together with sets from other applications. There is almost no variation with the exception of the phosphate partitioning coefficients, which points out the need for a more general formulation. Part V ends with Chapter 16 that revisits the steady state model, applying it to all the data sets, and using it to perform a number of sensitivity investigations. It ends by examining what controls the time to steady state for the time variable model, a question of great practical importance when the model is being used to examine remedial alternatives as part of a coupled water column-sediment model.

Part VI presents the models for metal fluxes. These are considerably more complex than the nutrient flux models because of the changing redox states for manganese and iron. Both oxidized and reduced manganese and iron need to be modeled explicitly. As a prelude, a simple model of calcium and alkalinity flux is considered in Chapter 17. The novel feature is explicitly considering calcium carbonate formation. This produces the first nonlinear set of algebraic equations for which MACSYMA is required. Manganese fluxes were also measured during the MERL experiments, which were analyzed in Chapter 15, and a model is formulated and applied to these observations in Chapters 18 and 19. In Chapter 18 the normal progression is followed, a steady state model followed by a time variable analysis. The influence of pH on the oxidation rate of Mn(II) is examined. In Chapter 19, an explicit model of the overlying water column is included with the same oxidation kinetics. This is an interesting model, since the coupled behavior is most instructive. This is followed by an iron flux model in Chapter 20, which is calibrated using field data from a lake and a reservoir.

Chapter 21 presents a model for cadmium in sediments. Unlike the preceding models, it is no longer two-layer, but rather examines the vertical distribution of sediment and pore water constituents. It focuses on the oxidation of cadmium sulfide and the liberation of dissolved cadmium to the pore water as well as the overlying water. The motivation for its construction is to aid in understanding the toxicity of metals in sediments. It uses a multilayered structure and it is the most complex of the models discussed.

The book concludes with two appendices that contain tabulations of the calibration data, a listing of the sediment flux model, and a listing of the mathematical nomenclature. The flux data from the Chesapeake Bay stations are tabulated. Pore water data from a much earlier study are also included. The MERL nutrient addition study data including the manganese data, and the Lake Champlain data complete the presentation.

I hope this book will be of use to anyone with an interest in sediments. It provides an exposition of sediment flux models for the water quality modeling community. It can be thought of as the next step to books that discuss surface water quality modeling: Thomann and Mueller (1987), Schnoor (1996), and Chapra (1997).

DOMINIC M. DI TORO

Englewood, NJ and Wellfleet, MA

Acknowledgments

Many colleagues and friends have been involved with the work presented in this book. The first sediment flux model was constructed as part of the Lake Erie Eutrophication model (Di Toro, 1980) in response to a question by our EPA project officer Nelson Thomas, an old sediment oxygen demand (SOD) hand himself: "And when the phosphorus loading to the lake changes, what are you going to do about the SOD?" The result was the oxygen equivalents model for SOD.

The close relationships with my colleagues and friends at the Department of Environmental Engineering at Manhattan College contributed greatly to my development and this work. My mentor and dear friend, the late Donald J. O'Connor, who trained and inspired a generation of environmental engineers and water quality modelers, suggested that we embark on building what are now called eutrophication models. His wide-ranging interests, rigorous approach to problems, and the primacy of data in model building, were lessons we all learned by his example. My long-time friend and colleague, Robert V. Thomann, played a critical role as a careful and insightful critic and collaborator. His scholarship and authorship of two books were an accomplishment to be emulated. I had the good fortune and privilege of being the youngest member of this group. I owe a debt of gratitude to Don and Bob. I hope this book is a small payment.

My recent appointment to the Donald J. O'Connor Chair of Environmental Engineering has enabled me to complete this project. I thank the contributors, the college administration, and the faculty of the department: Kevin Farley, Scott Lowe, John Mahony, Jim Mueller, and Rob Sharp, for their support.

Part of my professional life has been spent at the consulting firm HydroQual, and its predecessor firm, Hydroscience. My former and present colleagues Alan Blumberg, Charlie Dujardin, Jim Fitzpatrick, Tom Gallagher, Ed Garland, Jim Hallden, Bill Leo, John Mancini, Tom Mulligan, Paul Paquin, Karl Scheible, John St. John, and the rest of the members of the firm have helped in many ways.

The models that are the subject of this book have been developed during the conduct of various projects. The methane oxidation model was constructed as part of the Milwaukee River Comprehensive Study (SWRPC, 1987) in response to measurements of gas fluxes that indicated the importance of methane as the endproduct of diagenesis and the inadequacy of the oxygen equivalents assumption for this situation. My colleagues Paul Paquin and Thomas Gallagher at HydroQual, Inc. and I struggled with this model for some time. The data collected by David A. Gruber (Milwaukee Metropolitan Sewerage District) was central to its development (Di Toro et al., 1990).

The Chesapeake Bay project, sponsored by the Chesapeake Bay Program Office of the US Environmental Protection Agency and the US Army Corps of Engineers, provided the opportunity for the next developments. Much of contents of Chapters 3–7, 9, and 12–14 are from that effort. The importance of sulfide as an intermediate in oxygen consumption reaction was apparent. The presence of sulfide as a solid phase as well as in pore water clearly invalidated the oxygen equivalents idea. Perhaps the biggest impetus was the failure of a summer steady state water column model, which had no interactive sediment model, to respond to loading reductions in a sensible way – that is, to respond at all. The coupled water column-sediment eutrophication model was then developed (Cerco and Cole, 1993) with the two layer sediment flux model providing the sediment fluxes (Di Toro and Fitzpatrick, 1993). My colleagues James Fitzpatrick at HydroQual and Carl Cerco at the US Army Engineer Waterways Experiment Station (WES) were of great help. Kai-Yuan Yang (HydroQual), Mark Dortch, and Don Robey (WES) also helped with moral and program support. Ongoing commentary and advice was provided by Michael Kemp and Walter Boynton (University of Maryland). The members of the technical review committee were: Robert Thomann (Manhattan College), Donald Harleman (MIT), and Jay Taft (Harvard University), who also aided in the development.

The success of this modeling effort is due in no small measure to the excellent and comprehensive experimental data sets used in the development and calibration. They are the result of the efforts of scientists who developed the methods for reliably measuring sediment fluxes and applied these techniques in a systematic and comprehensive way. Their efforts are specifically acknowledged and appreciated: Walter Boynton, Michael Kemp, Jeffery Cornwell, and Peter Sampou (University of Maryland's Center for Estuarine and Ecosystem Studies), Jonathan Garber (EPA Research Laboratory, Narragansett, RI), and David Burdige (Old Dominion University). An older pore water data set collected under the supervision of Owen Bricker (then with the Maryland Geological Survey) was very useful.

The application of the sediment flux model to the MERL data set was sponsored by an NSF research project at Manhattan College. The support, generous cooperation, and insights of Candace Oviatt and Scott Nixon (University of Rhode Island, Graduate School of Oceanography) are gratefully acknowledged.

The manganese and iron models were developed as part of the Water Quality Research Program at WES (Di Toro et al., 1998). Chapters 17–20 are based on this work. Carl Cerco was the project officer and provided his usual generous and insightful support. My colleagues Jim Fitzpatrick and Richard Isleib (HydroQual) and Scott Lowe (Manhattan College) participated in this study and provided their usual high level of professional help. The MERL manganese data were collected and provided to the author by Carlton D. Hunt of Battelle Ocean Services, Duxbury, MA. These were unpublished data, generously provided from his personal notebooks. A special thanks for a wonderful data set.

The cadmium and iron model (Chapter 21) is based on research sponsored by a US EPA Office of Water Cooperative Agreement and by a National Institutes of Environmental Health Sciences Superfund Hazardous Substances Basic Research Program project. My coworkers John D. Mahony (Manhattan College), David J. Hansen and Walter J. Berry (EPA Research Laboratory, Narragansett, RI) and project officers Chris Zarba and Mary Reiley (EPA) provided necessary support and data with which to build the model. Additional data were kindly provided by Ed Leonard (EPA Research Lab, Duluth MN), Karsten Liber (University of Saskatchewan), Landis Hare, and Andre Tessier (University de Quebec), all of whom are thanked.

A number of colleagues read various portions of the manuscript and provided useful suggestions: Kevin Farley and John Mahony (Manhattan College); Ferdi Hellweger, Richard Isleib, and Paul Paquin (HydroQual); and Steve Chapra (Tufts University). A special note of thanks to John Mueller who read the entire manuscript with great care and devotion to detail.

The final acknowledgments are to my parents: my father, Michael, who was my first inspiration, and my mother Josephine. And, lastly, my deepest gratitude to my wife Marilyn and our children Jennifer and Joseph, who have contributed to this effort in more ways than they know.

D.M.D.

Sediment Flux Modeling

One cannot escape the feeling that these mathematical formulas have an independent existence and an intelligence of their own, that they are wiser than we are, wiser even than their discoverers, that we get more out of them than was originally put into them.

—Heinrich Hertz

Part I

Preliminaries

All models are wrong. Some models are useful.

—G.E.P. Box

Part I of the book, the first two chapters, presents the introductory material that sets the stage for the subsequent development of the models. The subject of sediments, their physical and chemical properties, and the plants and animals that are resident, is a very large and diverse one. Therefore, it is difficult to choose just those topics, and no others, that are necessary for an appreciation of the material in the following parts of the book. The following chapter is a brief survey with references to more complete expositions.

The second chapter is an introduction to the mathematical models that are the subject of the book. Elementary examples are presented in detail. These are followed by a two-layer model, the basic framework for what follows. The appendix contains the mathematical derivations of the solutions for the two differential equations employed. In addition, a primer is included for the symbolic computational program MACSYMA, which is used to solve all the difficult algebraic equations we will encounter. MACSYMA provides the means to bypass what would be, at best, long and error-prone algebraic manipulations.

1
Properties of Sediments

Sediments are formed at the bottom of water bodies by the deposition of particles from the overlying water. The particles either originate from soil and other suspended matter that are carried in with the inflowing water or from direct discharges (allochthonous particles) or they are formed within the water body itself (autocthonous particles) as a result of the growth, metabolism, and death of plants and animals. These particles settle and eventually become part of the consolidated material at the bottom of the water body. This mix of particles and water makes up the sediment.

1.1 PHYSICAL CHARACTERISTICS

Sediments have certain physical characteristics that influence the magnitude and extent of chemical fluxes to the overlying water. These are reviewed below.

1.1 A   Porosity, Density, and Solids Concentration

When particles accumulate at the bottom of a water body, they do not pack together very efficiently. As a consequence sediment is mostly water. The quantitative measure of the water content is the porosity ϕ, which is the volume fraction of the sediment that is water. The units are cm^3/cm^3 (i.e., unitless):

$$\phi = \frac{\text{volume of water}}{\text{volume of water + volume of solids}} \qquad (1.1)$$

The porosity of a sediment sample cannot be measured directly. Rather the bulk density ρ_b – the wet weight of a known volume – and the density of the solids ρ_s are measured – the latter by drying the sample to remove the water. Then the porosity is calculated from the relationship between the bulk density, the porosity, and the density of the constituents

$$\rho_b = (1 - \phi)\,\rho_s + \phi\rho_w \tag{1.2}$$

where ρ_w is the density of water (1 gm/cm^3). This equation is solved for porosity

$$\phi = \frac{\rho_s - \rho_b}{\rho_s - \rho_w} \tag{1.3}$$

If no solid phase density is measured, it is not unusual to assume that $\rho_s = 2.6$ gm/cm^3 which is also the solid density for quartz. Carbon-occluded minerals are in the range $\rho_s = 1.5$–2.2 gm/cm^3.

The porosity of sediments is influenced by a number of factors. The size of the particles has an effect (Fig. 1.1A). The porosity decreases as the median particle size

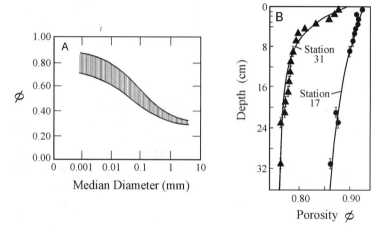

Fig. 1.1 (A) Porosity ϕ versus median particle diameter, redrawn from Berner (1980a). (B) Vertical profiles of porosity. Two stations from Lake Michigan, redrawn from Robbins and Edgington (1975).

increases, a somewhat surprising relationship. Presumably the smaller particles are clay or clay aggregates, which are usually negatively charged and less spherically shaped and thus resist close packing.

Porosity also typically decreases with depth in a sediment. Typical examples from Great Lakes sediments are shown in Fig. 1.1B (Robbins and Edgington, 1975). Although the change is regular, it is usually of a relatively small magnitude – note the abscissa scale. The decrease is due to compaction of the particles as water is squeezed out. Typical values are $\phi = 0.9$–0.7 in the top few centimeters and $\phi = 0.7$–0.4 in the deeper portions (Berner, 1980a, Robbins and Edgington, 1975).

The solids concentration of a sediment m is defined as the concentration of solids per unit volume of sediment:

$$m = \rho_s\,(1 - \phi) \tag{1.4}$$

Typical values for m using $\rho_s = 2.6$ gm/cm^3 are as follows

Porosity	ϕ	cm^3/cm^3	0.9	0.8	0.7	0.6	0.5
Solids concentration	m	gm/cm^3	0.26	0.52	0.78	1.04	1.30

1.1 B Particle Size Distribution

Sediments are usually composed of a heterogeneous collection of particle types. In particular, the particle sizes vary considerably. A standard classification system is employed to describe the size classes (Salomons and Forstner, 1984).

Clay	Silt	Sand	Gravel
$< 2\mu$m	2–60μm	60–1000μm	$> 1000\mu$m

It is important to realize that the names "clay" and "sand" refer to "clay-sized" and "sand-sized" particles, not actual pure clay or sand particles. This is a common misunderstanding of the terminology.

The particle size distribution can vary markedly, depending on the nature of the depositional environment. Fine-grained particles deposit in deeper more quiescent regions. The particle size distributions for three longitudinal transects from Chesapeake Bay are shown in Fig. 1.2. The central bay transect runs along the deepest portion of the bay and it has the largest proportion of clay-sized particles. By contrast, the sediments in the shallower western and eastern regions are composed mainly of sand-sized particles.

1.1 C Sedimentation

As particles accumulate at the sediment-water interface, the sediment depth – measured from some fixed datum such as the level of bedrock below – increases. This process is termed sedimentation, and the rate at which sediment accumulates is the sedimentation rate. The sedimentation rate can be expressed as the rate at which sediment solids accumulate J_{SS} (gm/cm^2-yr), or it can be expressed as the rate at which the sediment depth increases w (cm/yr). The two quantities are related via the porosity ϕ

$$J_{SS} = \rho_s (1 - \phi) w \tag{1.5}$$

or the solids concentration in the sediment m (Eq. 1.4)

$$J_{SS} = mw \tag{1.6}$$

Models of sediment transport are constructed in order to compute these quantities. Typical sedimentation rates are as follows:

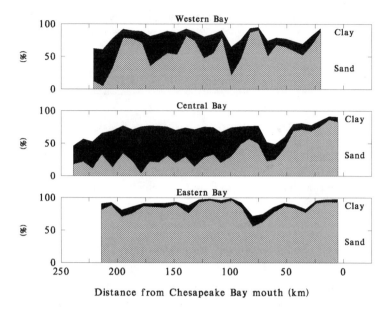

Fig. 1.2 Distribution of sand, clay, and silt-sized particles as a weight percent for three longitudinal transects. Clay ■ and sand (hatched) sized particles are shown. The remainder is silt-sized.

Location	River Deltas	Lakes and Estuaries	Oceans
w (cm/yr)	1–10	0.1–1.0	0.001–0.01

The measurement of the rate of sedimentation typically depends on the use of tracers. In some cases elevation changes in sediment bed are available from bathymetric surveys that are separated by a sufficient length of time for an appreciable change to occur. The most common method involves measuring the vertical distribution of radionuclides. The naturally occurring lead isotope ^{210}Pb forms in the atmosphere from the decay of radon and is deposited on the surface of the water body during rainfall and as dry atmospheric particle deposition. Lead has a high sorptive affinity for particles, i.e., it is very particle reactive. Therefore it sorbs to the particles in the water column and is deposited with them as they settle to the sediment. Once in the sediment ^{210}Pb decays with a known half-life of 22.3 yr. The vertical profile of ^{210}Pb activity can be analyzed to determine the sedimentation velocity. For a sediment that receives a flux of lead J_{Pb} (gm Pb/m^2-d) where burial is occurring with sedimentation velocity w (m/d) and ^{210}Pb is decaying with reaction rate k_{Pb} (d^{-1}), the vertical profile of lead concentration $c(z)$ is

$$c(z) = c(0)e^{-\frac{k_{Pb}z}{w}} \qquad (1.7)$$

where $c(0) = J_{Pb}/w$ is the concentration at the sediment surface. Thus the lead concentration decreases exponentially with depth. The derivation of this equation is presented in Chapter 2, section 2.7 B. Fig. 1.3 presents two examples with differing sedimentation rates (Robbins and Edgington, 1975). Rapid deposition causes lead to

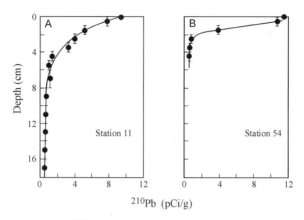

Fig. 1.3 Vertical profiles of ^{210}Pb activity for two sediments from Lake Michigan. Redrawn from Robbins and Edgington (1975).

persist deeper in the sediment (Sta. 11) versus a smaller sedimentation rate which causes ^{210}Pb to decay in the surface layers (Sta. 54).

The other common radionuclide that is used to determine sedimentation rate is cesium 137. This artificial radionuclide, with a half-life of 30.2 yr, was introduced into the atmosphere as a result of atmospheric nuclear testing during the late 1950s and '60s (Krishnaswami and Lal, 1978). It was washed out of the atmosphere as wet and dry fallout. It, too, is very particle reactive, and associated with particles as they settled to the sediment. Any sediment that was deposited before the 1950s would not be contaminated. Thus, in the absence of sediment mixing processes, the depth at which ^{137}Cs ceases to exist marks the year 1963.

Other sediment components have been used to mark a horizon in the sediment. Pollen grains, for example, persist in the sedimentary environment. In particular, ragweed pollen marks the beginning of agriculture in the drainage basin of the water body. The clearance of the forests caused the proliferation of ragweed pollen grains which found their way to the sediments via soil runoff.

Figure 1.4 presents the results of applying these three methods to the same sediment (Robbins et al., 1978). The excess or unsupported ^{210}Pb concentration is that which is above the concentration that occurs at depth and which presumably is associated with recently deposited particles. The concentration at depth – see Fig. 1.3 – originates from the continuous decay of radon that is being produced by the decay of uranium that is part of the sediment.

The concentration of ^{210}Pb decays to 1/20 of the surface concentration in 30 cm. In order to calculate the sedimentation rate we require the equation that is analogous to Eq. (1.7) in terms of half-life. For a substance with a half-life of $t_{1/2}$ the

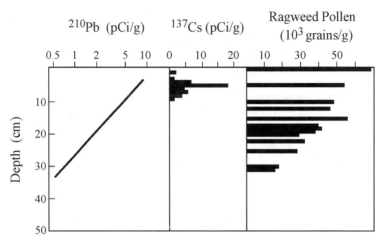

Fig. 1.4 Vertical distributions of ^{210}Pb, ^{137}Cs, and ragweed pollen. Redrawn from Robbins et al. (1978).

concentration $c(t)$ at time t is

$$c(t) = c(0)2^{-\left(\frac{t}{t_{1/2}}\right)} \tag{1.8}$$

where $c(0)$ is the concentration at $t = 0$. If $c(t) = 0.05c(0)$ corresponding to $1/20^{\text{th}}$ decrease, then since

$$\ln\left(\frac{c(t)}{c(0)}\right) = -\left(\frac{t}{t_{1/2}}\right)\ln(2) \tag{1.9}$$

$(t/t_{1/2}) = -\ln(0.05)/\ln(2) = 4.3$ half-lives or 96 yr which corresponds to the roughly 100 yr that marks the ragweed pollen horizon. This is a sedimentation rate of approximately 0.3 cm/yr. The ^{137}Cs concentration terminates at about 10 cm which corresponds to a sedimentation rate of 10 cm / 20 yr or 0.5 cm/yr which is larger than the 0.3 cm/yr deduced from the pollen and ^{210}Pb. The discrepancy is partially due to the presence of particle mixing that causes ^{137}Cs and all other tracers to be mixed downward. An analysis that does not take that mechanism into account produces erroneously large sedimentation rates. Models are available to perform the analysis properly (Robbins et al., 1978).

Nevertheless, the determination of sedimentation rates is somewhat problematical due to the spatially heterogeneous, i.e., patchy, nature of sediments. A comparison of the results of various methods applied to Chesapeake Bay sediments is shown in Fig. 1.5. In the face of such variation, it is fortunate that sediment fluxes computed from the models discussed in this book are not overly sensitive to the value of the sedimentation velocity.

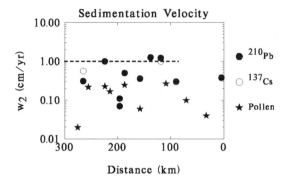

Fig. 1.5 Sedimentation velocity w_2 versus distance from the mouth of Chesapeake Bay determined by various methods.

1.2 CHEMICAL PRELIMINARIES

Before we examine some examples, we will review certain chemical concepts that are important for understanding the development of the models presented below. The first of these is stoichiometry.

1.2 A Stoichiometry

The concept of stoichiometry is fundamental to the study of chemistry and chemical reactions, wherein it is referred to as Dalton's law of constant proportion. It is defined as the numerical relationship between the relative quantities of substances in a reaction or compound. Thus in carbon dioxide CO_2 two moles of oxygen are always found combined with one mole of carbon. The oxygen to carbon stoichiometry is 2 moles O to one mole C, or on a weight basis: 32 gm O_2/12 gm C.

Certain inorganic compounds may have variable stoichiometry. An example is the mineral pyrrhotite, a form of iron sulfide that has a variable stoichiometry $Fe_{1-x}S$ where x can approach 0.15 (Vaughan and Craig, 1978). This mineral is sometimes viewed as a mixture of fixed stoichiometry minerals: FeS and Fe_7S_8, thus preserving the notion of fixed stoichiometry compounds.

Organic Matter A most useful and surprising stoichiometric relationship characterizes both freshwater and marine phytoplankton. It is called the *Redfield ratio*, named after the discoverer (Redfield et al., 1963). In fact the original idea – that these organisms are responsible for the inorganic composition of seawater – is an even more startling idea.

The organic matter produced by phytoplankton has the composition

$$C_{106}H_{263}O_{110}N_{16}P \tag{1.10}$$

which can be written more usefully as

$$(CH_2O)_{106}(NH_3)_{16}(H_3PO_4) \qquad (1.11)$$

thereby isolating the carbon, nitrogen, and phosphorus components as more readily identifiable species. This stoichiometry establishes the molar ratios C/N and C/P as well as the amount of oxygen required to oxidize the carbonaceous component of organic matter

$$CH_2O + O_2 \rightarrow CO_2 + H_2O \qquad (1.12)$$

The Redfield stoichiometric molar and weight ratios are given in Table 1.1.

Table 1.1 Redfield Ratios

Ratio	mol/mol	gm/gm
C/N	6.63	5.68
C/P	106	41.1
O_2/C	1	2.67
C/Si	6–8	2.6–3.4
C/Chl$_a$	–	25–100

There are two other constituents of the organic matter synthesized by phytoplankton: the silica content that is associated with the hard parts of diatoms, and the chlorophyll *a* content. These constituents are more variable, but ranges are given in Table 1.1.

In the absence of any more detailed information, Redfield ratios are used to specify the stoichiometry of particulate organic matter. They are usually appropriate for phytoplankton and other "fresh" particulate organic matter – the exception being nutrient-limited algae whose stoichiometry becomes depleted in the limiting nutrient (Di Toro, 1980a, Droop, 1973). The ratio is less appropriate for weathered organic matter which tends to be somewhat depleted in nitrogen as discussed in Chapter 12.

1.2 B Electron Stoichiometry

The decay of organic matter in sediments is the reaction which begins the sequence that results in fluxes to the overlying water. When particulate organic matter decays, the first step is a solubilization reaction. The soluble endproducts are then available for further reactions to take place.

The initial solubilization reaction can be idealized as

$$(CH_2O)_{106}(NH_3)_{16}(H_3PO_4)(s)$$

$$\rightarrow 106CH_2O(aq) + 16NH_3(aq) + H_3PO_4(aq) \qquad (1.13)$$

where particulate organic matter, denoted by (s) to denote a solid species, is converted to soluble intermediates, denoted by (aq) to denote dissolved species. We will concentrate on the sequence of oxidations that can take place. The phosphate is already in the most oxidized state that occurs at normal temperatures and pressures, so no further oxidation is possible. However, the soluble organic carbon and ammonia can be oxidized by various oxidants.

A convenient way of accounting for the quantities of oxidants that are required to oxidize one mole of organic carbon is to use the idea of electron equivalents. In order to oxidize organic matter, electrons must be transferred from the electron donor, the organic mater, to the electron acceptors, the oxidizers. To see this, a "half reaction" is written that displays the stoichiometries involved in the release of one electron. For CH_2O the reaction is

$$\frac{1}{4}CH_2O + \frac{1}{4}H_2O \rightarrow \frac{1}{4}CO_2 + H^+ + e^- \tag{1.14}$$

In order for the reaction to be completed an electron acceptor – the oxidant – is required. The usual oxidants present in natural waters and sediments and their half reactions are

Oxygen

$$\frac{1}{4}O_2 + H^+ + e^- \rightarrow \frac{1}{2}H_2O \tag{1.15}$$

Manganese

$$\frac{1}{2}MnO_2\,(s) + 2H^+ + e^- \rightarrow \frac{1}{2}Mn^{2+} + H_2O \tag{1.16}$$

Nitrate

$$\frac{1}{5}NO_3^- + \frac{6}{5}H^+ + e^- \rightarrow \frac{1}{10}N_2 + \frac{3}{5}H_2O \tag{1.17}$$

Iron

$$Fe(OH)_3(s) + 3H^+ + e^- \rightarrow Fe^{2+} + 3H_2O \tag{1.18}$$

Sulfate

$$\frac{1}{8}SO_4^{2-} + \frac{9}{8}H^+ + e^- \rightarrow \frac{1}{8}HS^- + \frac{1}{2}H_2O \tag{1.19}$$

Carbon dioxide

$$\frac{1}{8}CO_2 + H^+ + e^- \rightarrow \frac{1}{8}CH_4 + \frac{1}{4}H_2O \tag{1.20}$$

These reactions are written assuming that the reduced species, which occurs when the oxidants are consumed, is known. For example, the reduction of nitrate (Eq. 1.17) is to nitrogen gas, not ammonia.

To find the stoichiometry for a particular electron donor and acceptor pair, the two half reactions are added to form a complete reaction. For example the stoichiometry for the oxidation of CH_2O using O_2 as the electron acceptor is obtained by adding Eq. (1.14) to Eq. (1.15)

$$\frac{1}{4}CH_2O + \frac{1}{4}H_2O + \frac{1}{4}O_2 + H^+ + e^- \rightarrow \frac{1}{4}CO_2 + H^+ + e^- + \frac{1}{2}H_2O \quad (1.21)$$

which when simplified by treating the reaction as an algebraic equation with an "$\rightarrow$" replaced by an "=" yields

$$\frac{1}{4}CH_2O + \frac{1}{4}O_2 \rightarrow \frac{1}{4}CO_2 + \frac{1}{4}H_2O \quad (1.22)$$

as found above (Eq. 1.12).

In order to compute the oxygen required to oxidize ammonia, the half reaction for the oxidation of ammonia to nitrate is used, since this is the known pathway of nitrification

$$\frac{1}{8}NH_4^+ + \frac{3}{8}H_2O \rightarrow \frac{1}{8}NO_3^- + \frac{5}{4}H^+ + e^- \quad (1.23)$$

The overall reaction is found by adding Eq. (1.15) to the half reaction for nitrification (Eq. 1.23), i.e.,

$$\frac{1}{8}NH_4^+ + \frac{3}{8}H_2O + \frac{1}{4}O_2 + H^+ + e^- \rightarrow \frac{1}{2}H_2O + \frac{1}{8}NO_3^- + \frac{5}{4}H^+ + e^- \quad (1.24)$$

and, after simplification yields

$$\frac{1}{8}NH_4^+ + \frac{1}{4}O_2 \rightarrow \frac{1}{8}H_2O + \frac{1}{8}NO_3^- + \frac{1}{4}H^+ \quad (1.25)$$

Thus the stoichiometry for any electron donor–electron acceptor pair can easily be computed.

The utility of the idea of electron equivalents is not solely limited to the balancing of oxidation-reduction reactions. After all, the balance can be computed by a brute force application of atom and charge balances. Consider the reaction

$$a_1 NO_3^- + a_2 H^+ + e^- = a_3 N_2 + a_4 H_2O \quad (1.26)$$

The atom and charge balances are

$$
\begin{array}{ll}
H & a_2 = 2a_4 \\
N & a_1 = 2a_3 \\
O & 3a_1 = a_4 \\
e & -a_2 + a_1 + 1 = 0
\end{array}
\quad (1.27)
$$

the solution of which is the stoichiometry of Eq. (1.17). The utility is that one electron equivalent of any of the oxidants that are available can potentially oxidize one electron equivalent of any reduced substances that are present. This interchangeability of oxidants and reductants appears at first glance to lead to a hopeless number of possible combinations of oxidations and reductions that may take place. However, there is a simple principle that appears to determine which of these oxidation-reduction reactions, in fact, do take place.

1.2 C   Sequence of Redox Reactions

It is a remarkable fact that when organic matter is oxidized in water bodies and in sediments, the sequence in which the electron acceptors (oxidants) are used follows the thermodynamically predicted sequence (see Stumm and Morgan, 1996, Section 8.5) . This sequence has also been observed in sediment pore water (Luther III et al., 1998). Modifications of this idea have been suggested (Postma and Jakobsen, 1996) to account for the observation of some overlap in concentration profiles. It is not intuitively obvious that this should be the case. After all, these reactions are all bacterially mediated. Nevertheless, it is indeed the case with one notable exception. Table 1.2 presents the change in Gibbs free energy ΔG_r that occurs when one electron equivalent of organic matter (Eq. 1.14) is oxidized with the indicated oxidant. The oxidants are listed in order of largest ΔG_r – the most favorable reaction from an energetic point of view – to the smallest ΔG_r, the least favorable.

Table 1.2   Change in Gibbs Free Energy ΔG_r

Electron Acceptor	ΔG_r^*	Half Reaction
O_2	−29.9	Eq. (1.15)
NO_3^-	−19.6	Eq. (1.17)
SO_4^{2-}	−5.9	Eq. (1.19)
CO_2	−5.6	Eq. (1.20)

*kcal/electron equivalent, pH = 7

Thus aerobic oxidation of organic matter – O_2 as the electron acceptor – releases the most energy, and fermentation of organic matter to CO_2 and methane releases the least. The examples shown below verify that indeed this is the observed sequence of the use of oxidants.

The exception is that the thermodynamically predicted route of denitrification is from NO_3^- to NH_4^+, whereas most if not all of the reduced nitrogen appears as N_2. Also, in oxic environments N_2 is predicted to oxidize to NO_3^- spontaneously, whereas it actually occurs only at high temperature, for example, in a lightning discharge or in internal combustion engines. Nitrogen gas is inert in both the atmosphere and in natural waters at normal temperature and pressure.

1.3 CHEMICAL CHARACTERISTICS

The chemical characteristics of sediments can be conveniently divided into those comprising the solid and pore water phases.

1.3 A Solid Phase

The solid phase of the sediment is made up of the particles. The constituents of the particles can be further subdivided into the inorganic and organic species.

Inorganic Species The inorganic fraction of the sediment solids is usually the largest. It is made up primarily of various clays whose major components are aluminium and silica (the aluminosilicates), of quartz (SiO_2), of iron oxides, and of calcium and perhaps magnesium carbonate. Fig. 1.6 presents a mineralogical analysis of two marine sediments (Hirst, 1962). Note the preponderance of quartz, indicating a high sand content. Clay components make up another significant fraction. The calcium carbonate ($CaCO_3$) content – these samples contain a few percent – will be examined in Chapter 17 where calcium and alkalinity fluxes are modeled.

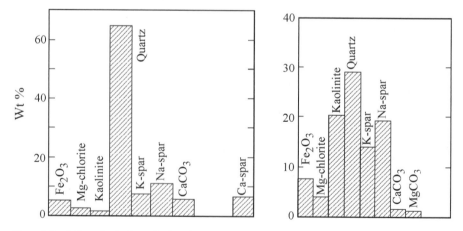

Fig. 1.6 Mineralogical analysis of two marine sediments. Nearshore sand (*left*) and nearshore marine clay (*right*). Redrawn from Hirst (1962).

Of more interest for the purposes of this book are the components of the sediments that affect various processes that are involved in generating sediment fluxes. Fig. 1.7 presents certain of these constituents as a function of particle size (Salomons and Forstner, 1984). The abscissa is a logarithmic scale. Note that the particle size distribution (Fig. 1.7A) has a bell shape, which suggests a Gaussian distribution of the logs of the particle sizes (i.e., a lognormal distribution). The specific surface area – the surface area per unit weight – of the particles (Fig. 1.7B) is a measure of their adsorption capacity. The smaller-sized particles, presumably corresponding to a predominance of clay, have the larger sorptive capacity. The iron extracted using

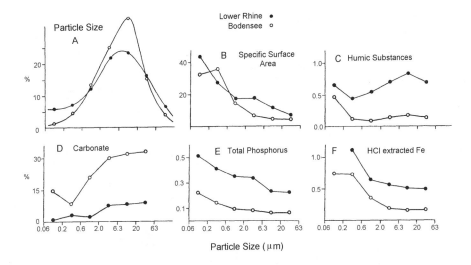

Fig. 1.7 Particulate constituent concentrations as a function of particle size. Sediments from Lake Constance and the lower Rhine River. Redrawn from Salomons and Forstner (1984).

0.5N HCl (Fig. 1.7F) is a measure of the reactive iron component of the sediment. It comprises approximately 0.1 to 1.0%. By contrast, the calcium carbonate (Fig. 1.7D) is a much larger fraction of these sediments (10–30%). The total phosphorus (Fig. 1.7E) ranges from 0.1 to 0.5%. The significance of these constituents will become apparent in subsequent chapters.

Organic Matter   The organic matter content of the sediment originates from the soil, plant, and directly discharged particles that are deposited. The most common quantification of the composition is to determine the organic carbon, nitrogen, phosphorus, and sometimes sulfur content of the particles. The measurement methods used can distinguish between the inorganic constituents, which also contain C, P, and S (e.g., calcium carbonate, sorbed PO_4, FeS). Organic carbon itself can also be fractionated by various extractions. The humic fraction, for example (Fig. 1.7C), is the organic carbon extracted by a strong base and precipitated by an acid.

The organic carbon concentration of Chesapeake Bay sediments is shown in Fig. 1.8 for the transects discussed above in Fig. 1.2. The units are percent of the dry weight (100% × gm C/gm dry weight). Note the high concentrations at the head end of the estuary and the progressive decline toward the mouth. Sediment organic carbon can vary quite a lot. Data from coastal marine and harbor sediments are shown in Fig. 1.9. Sediment organic carbon content ranges from 0.1% to over 10%. It is interesting to note that the toxicity of organic chemicals in sediments is strongly affected by the organic carbon content – see the review in Di Toro et al. (1991b)– so this parameter is important for other reasons as well. Even more variable is the acid volatile sulfide (AVS) concentration of sediments: the acid extractable component of

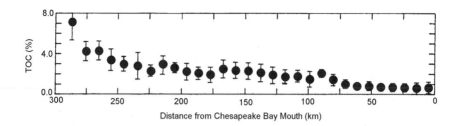

Fig. 1.8   Longitudinal profile of total organic carbon (TOC) for the sediments of the central bay transect of Chesapeake Bay.

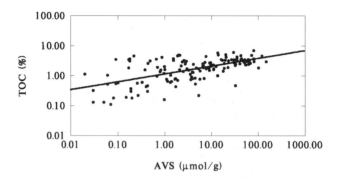

Fig. 1.9   Total organic carbon (TOC) versus acid volatile sulfide (AVS) for marine sediments.

the inorganic sulfide in sediments. It varies over four orders of magnitude. Its source is the decomposition of organic matter. Its sink is oxidation by oxygen. The result is the sediment oxygen demand of marine sediments. These topics will be discussed in Chapters 9 and 11. It is also directly implicated in regulating the toxicity of metals in sediments (Di Toro et al., 1990a). That relationship is discussed in Chapter 21.

One property of organic matter that is of fundamental importance to the generation of fluxes from sediments is the rate at which it mineralizes (Eq. 1.13). This will be examined in detail in Chapter 12, where models of the kinetics of organic matter mineralization are presented.* The usual situation is more complex than this, but if organic matter decay followed first-order kinetics and only sedimentation were occurring in a sediment, then an equation analogous to Eq. (1.7) would apply

$$\text{POC}(z) = \frac{J_{\text{POC}}}{w} e^{-\frac{k_{\text{POC}} z}{w}} \tag{1.28}$$

and one would expect an exponential decrease of POC with depth. Figure 1.10 presents an example from Chesapeake Bay (Schubel and Hirschberg, 1977). Both

---

*The term "mineralization" is used to denote the reactions that convert organic matter to inorganic (mineral) matter, as is conventional in biogeochemistry. It is not used to denote the formation of minerals.

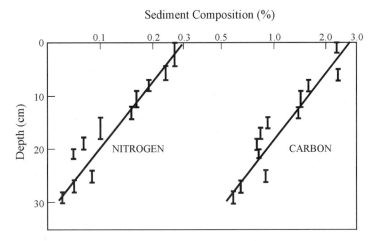

Fig. 1.10 Logarithm of the organic carbon and nitrogen concentration versus depth in a sediment core. Data from Schubel and Hirschberg (1977).

organic carbon and organic nitrogen are decreasing exponentially as shown. The line corresponds to a value of $w/k_{POC} = 18$ cm. For a sedimentation rate of $J_{SS} = 0.25$ gm/cm^2-yr , the sedimentation velocity is approximately $w_s \approx 0.25$ cm/yr (from Eq. 1.6 with $m = 1$ kg/L). This corresponds to a POC decay rate of $k_{POC} = 0.014$ yr^{-1} or a half-life of 50 years, a very slow decay rate.

The decay of organic matter nitrogen is also important. The lines in Fig. 1.10 are drawn to be parallel, indicating that the rate of decay of PON is the same as that of POC. The stoichiometric composition is approximately 10 moles of C to 1 mole of N. Thus this sample of sedimentary organic matter is depleted of nitrogen by almost twofold relative to the Redfield ratio of 6.63 (Table 1.1).

1.3 B Pore Water

The water that is contained in the sediment is called the pore water or interstitial water because it fills the pores or interstices – the intervening spaces – between the particles. It is the phase in which aqueous chemical reactions occur. It is also the phase that interacts with the overlying water via diffusion. Hence all the fluxes of dissolved constituents originate in this phase.

The vertical profiles of dissolved constituents are the result of a number of processes: the chemical reactions that have occurred or are occurring in the sediment; and the transport of these solutes within the sediment and to or from the overlying water. Models of pore water profiles of radionuclides (Lerman and Taniguchi, 1972) and nutrients (Berner, 1971, 1974) were among the first models of reactions in sediments. In principle, the modeling of the vertical distribution of solutes will produce the proper flux to/from the overlying water because the profiles that are observed reflect this loss/gain. However, as we will see, the proper vertical scale is quite small near the sediment-water interface (~1–5 mm) so that models that are designed to predict nutrient, oxygen, and metal fluxes must properly represent this surface layer.

1.3 C   Major Cations and Anions

The solutes in the overlying water that are not very reactive in the pore waters tend to have the same concentration in both phases. These include sodium, magnesium, potassium, and chloride. An example (Fig. 1.11) with measurements that extend into the overlying water illustrates the differences. Calcium and magnesium change slowly, whereas nutrients (NH_4^+, PO_4^{3-}), oxidants (NO_3^-, SO_4^{2-}), and metals (Mn(II) and Fe(II)) change more rapidly. The decay of organic matter initiates many of these reactions.

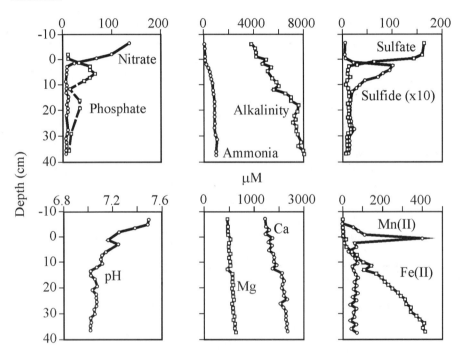

Fig. 1.11   Vertical distribution of the indicated constituents. Concentrations in μM except pH. Concentrations extend into the overlying water, indicated by the negative depths. Redrawn from Wersin et al. (1991) in Stumm and Morgan (1996).

1.3 D   Nutrients and Oxidants

When particulate organic matter decays, the endproducts are released to the pore water (Eq. 1.13) where further reactions take place. The oxidation of organic carbon occurs initially with dissolved oxygen O_2 as the oxidant. However, since oxygen is rapidly consumed, it only penetrates a few millimeters into the sediment. Figure

1.12 presents results determined using a microelectrode (Revsbach et al., 1980). The

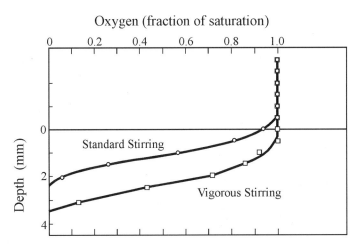

Fig. 1.12 Vertical distribution of dissolved oxygen in the pore water of a sediment core. Two different degrees of stirring. Data from Revsbach et al. (1980).

proper modeling of this aerobic layer is a key component in the flux models to be presented subsequently.

The other constituents can be sampled using more conventional pore water sampling techniques (Bufflap and Allen, 1995, Mudroch and MacKnight, 1991). Examples are shown in Fig. 1.11. The increase in ammonia NH_4^+ is a direct result of the mineralization reaction (Eq. 1.13). The organic carbon that is released $CH_2O(aq)$ is oxidized using manganese, nitrate, sulfate, and iron as the oxidants. The manganese dioxide presumably decreases – it was not measured – but the endproduct Mn(II) increases. The chemistry and a flux model for manganese is presented in Chapters 18 and 19.

The nitrate concentration in Fig. 1.11 decreases and is converted to N_2, which is difficult to measure above the background of dissolved nitrogen gas from the atmosphere. The sulfate decreases and the sulfide increases at least initially. The subsequent decrease is likely due to the formation of FeS(s). The increase in Fe(II) is presumably due to the oxidized iron that is consumed as an oxidant. The phosphate PO_4^{3-} that is released increases at first. Then it decreases, because of further reactions such as precipitation or adsorption. The protons H^+ consumed and released by the various reactions cause an alkalinity increase. The decrease in pH is the result of all the changes in pore water composition that affect the proton balance. The increase in calcium reflects the reduction in pH that causes a dissolution of calcium carbonate. Models that describe certain features of these pore water profiles in detail have been constructed* and methods for measuring pore water profiles of multiple

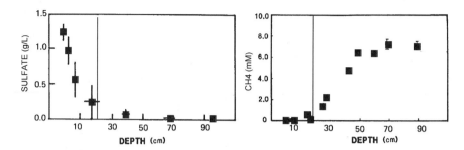

Fig. 1.13 Sulfate and methane distributions in pore water. The line at 20 cm denotes the boundary of the zone of sulfate reduction and methane formation.

solutes using microelectrodes have been developed.[†] The models tend to be quite complicated and have parameters that are specific to one set of observations.

Once the oxygen, nitrate, sulfate, and iron are completely reduced, the organic matter is converted to methane and carbon dioxide. The reaction is obtained by adding Eq. (1.14) and Eq. (1.20) yielding

$$\frac{1}{4}CH_2O \rightarrow \frac{1}{8}CO_2 + \frac{1}{8}CH_4 \qquad (1.29)$$

An example is shown in Fig. 1.13. Note that the methane increases only after the sulfate is depleted. For this to be occurring, not only can no methane be forming in the zone of sulfate reduction, the methane that is diffusing into that zone is being oxidized with sulfate as the oxidant. The reaction is obtained using the difference of two half reactions: (Eq. 1.19) minus (Eq. 1.20)

$$\frac{1}{8}CH_4 + \frac{1}{8}SO_4^{2-} + \frac{1}{8}H^+ \rightarrow \frac{1}{8}CO_2 + \frac{1}{8}HS^- + \frac{1}{4}H_2O \qquad (1.30)$$

These reactions are considered in more detail in Chapters 8 to 11 where models of the dissolved oxygen flux to the sediment are considered.

1.4 BIOLOGICAL CHARACTERISTICS

The biological community that inhabits the sediments are as diverse as the water column community. The lower levels of the sediment food chain are an important source of food for higher trophic levels such as bottom feeding crustacea and fish. Our concern here is with those aspects of the biological world that affect the production of fluxes to and from the sediment.

---

[*]Jahnke et al. (1994), Rabouille and Gaillard (1991), Soetaert et al. (1996), van Cappellen and Berner (1988), van Cappellen and Wang (1996), Wang and van Cappellen (1996). [†]Luther III et al. (1998).

1.4 A   Bacteria

Most of the reactions discussed above – the mineralizations and oxidations – are performed by bacteria and other micro-organisms (Capone and Kiene, 1988). The concentration of bacteria in the surface layers of sediments is quite large (Gachter et al., 1988, Schmidt et al., 1998). In this book, we follow the traditional approach employed in water quality modeling (Thomann and Mueller, 1987) in which no *explicit* account is taken of the bacterial biomass. Rather, the biomass concentration is implicitly part of the reaction rate constant. The justification for this procedure is twofold. First, measures of bacterial biomass are not part of the flux data sets to which the model will be applied. It is not good modeling practice to model state variables – in this case bacterial biomass – for which no measurements are available. Second, for reasons that are not clear, these empirical reaction rates appear to have a certain general applicability. As we will see in Chapter 15, the same reaction rate constants can be applied to quite different data sets with satisfactory results. Hence the effect of bacterial biomass must be similar in these cases. Nevertheless, it would be aesthetically more pleasing if the agents of the reactions were explicitly included in the formulation. This refinement awaits future developments.

1.4 B   Macrofauna

The larger animals that inhabit the sediments have a significant effect on the fluxes that result (Aller and Aller, 1998). These occur both as a direct result of their metabolic processes, for example feeding and respiration, and as a consequence of their behavior, for example, tube building. Schematic drawings (Figs. 1.14–1.15) of the major fauna from two marine sites in Long Island Sound illustrate the variety and density of these animals (Aller, 1980a). Note the penetration of the borrows to 10 cm and beyond. Data from these sites are analyzed in Chapter 15.

The feeding mode of an animal determines the role it plays in affecting sedimentary processes. Suspension feeders remove particles from the overlying water and deposit them to the sediment. Thus they directly affect the flux of suspended organic matter to the sediment. Deposit feeders ingest sediment and return it to a different location in the sediment. Thus they mix particles, as well as aid in the mineralization of organic matter. The mixing that results from the feeding and other activities of macrofauna is termed bioturbation.

1.4 C   Mixing and Bioturbation

A depiction of the influence of macrofauna on the sediment surface is shown by the X-radiograph of two sediment cores (Schaffner et al., 1999). The layering of the sediment in Fig. 1.16A indicates an absence of mixing, whereas the homogeneous texture in Fig. 1.16B, and the burrows and tubes, demonstrates the presence of the organisms.

In addition to the qualitative visual evidence, there is also evidence from tracer profiles. For marine sediments, the quantification of the extent of mixing has been

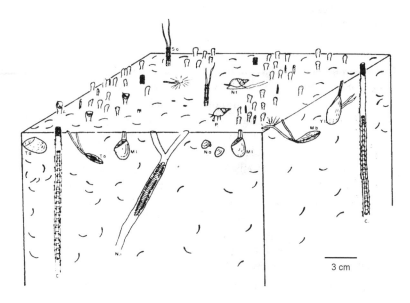

Fig. 1.14 Major fauna at the FOAM site in Long Island Sound (Aller, 1980a).

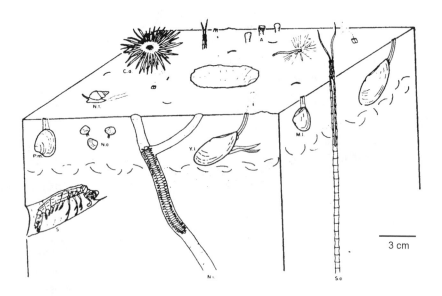

Fig. 1.15 Major fauna at the NWC site in Long Island Sound (Aller, 1980a).

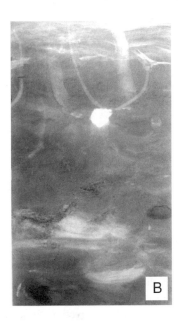

Fig. 1.16   X-radiograph of two sediment cores from Schaffner et al. (1999).

made using thorium 234. ^{234}Th is formed in seawater by the decay of ^{238}Uranium (Aller et al., 1980). It has a half-life of 24.1 days so it is quite short-lived relative to processes in the sediment. It is particle reactive, so it serves as a tracer for particles from the overlying water. The excess ^{234}Th profiles for two cores are shown in Fig. 1.17. Concentrations above the background are found to depths of 4 cm, where the concentration has decreased to 0.1 of the surface value. These particles are $-\ln(0.1)/\ln(2) = 3.3$ half-lives or 80 days old. Hence they made the transit from the surface to a depth of 4 cm in 80 days. The only possibility is that particles are mixing rapidly enough to account for the ^{234}Th at this depth. The particle diffusion coefficients that are deduced from these profiles are 0.023 to 0.050 cm^2/d. Typical diffusion coefficients for solutes are 0.5 to 2.0 cm^2/d (Table 2.1). Thus particles, which one would expect to be stationary in a sediment, mix at approximately one-tenth the rate of pore water mixing, an entirely unexpected result. Other examples of ^{234}Th profiles that demonstrate similar mixing intensities have been reported (Wheatcroft and Martin, 1996).

Bioturbation is not limited to marine sediments. In a series of elegant experiments employing freshwater organisms, the vertical migration of a particle tracer ^{137}Cs placed on the surface of the sediment was followed in time using a well-collimated detector scanning along the vertical axis of the sediment core (Robbins et al., 1979). The results are shown in Fig. 1.18 for the amphipod *Pontoporeia hoyi* and for the oligochaete *Tubifex*. The mixing produced by *P. hoyi* can be modeled using simple particle diffusion. However, the mixing produced by *Tubifex* resembles a conveyer

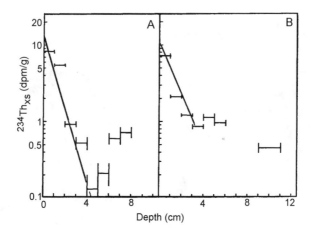

Fig. 1.17   Excess ^{234}Th activity versus depth. Redrawn from Cochran and Aller (1979).

belt mechanism (Robbins, 1986). Particles are ingested by the organism at the bottom of its tube and egested at the surface. By approximately day 50 the contaminated surface layer has moved down to the depth of ingestion. A portion is then moved to the surface where it reappears. An interesting variant of this experimental design is to start with a series of layers of ^{137}Cs tags (Matisoff et al., 1999).

Particle mixing is an important component of the mass transport of materials in sediments. Therefore for the models presented subsequently, where solid phase constituents are important, particle mixing is explicitly included.

1.5   CONCLUSION

This chapter presented a brief overview of the features of the physics, chemistry, and biology of sediments and sedimentary processes that will be utilized in the model building that follows. The models are idealizations that attempt to capture the major features of these processes. The success of this enterprise will depend on the degree to which this is achieved.

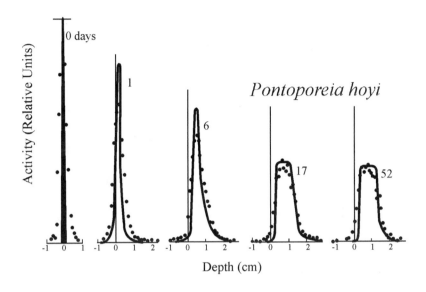

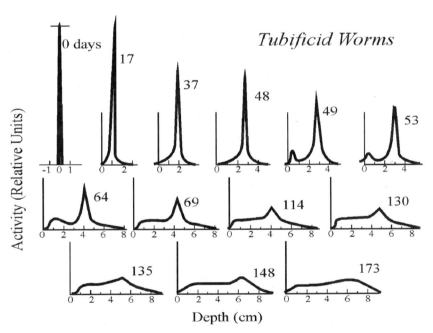

Fig. 1.18 Depth-concentration profiles of particle bound ^{137}Cs for various times after the introduction of a tagged surface layer. Redrawn from Robbins et al. (1979).

2

Model Formulation

2.1 FRAMEWORK

A number of choices need to be made at the start of any model development. First of these is what level of detail to include in the spatial representation. The compromise is between simplicity and realism. The best solution emphasizes the former without undue violence to the latter.

2.1 A Spatial Representation

The most common spatial representation for sediments is continuous and one-dimensional in the vertical direction (Berner, 1971, 1980a) although a two-dimensional analysis – radially varying as well as varying in depth – has been employed (Aller, 1980b), and three-dimensional variability has been measured (Harper et al., 1999). For simple kinetics and non-interacting species the differential equations that result from the one-dimensional mass balances, e.g., Eqs. (2.87) and (2.91) to be discussed subsequently, can be solved analytically. However, extending these solutions to include more realistic kinetic formulations, to explicitly consider soluble and particulate species, and to distinguish the aerobic and anaerobic zones, rapidly leads to intractable equations (Boudreau, 1996).

An alternate formulation results from representing the sediment as a series of homogeneous layers each characterized by the oxidation reaction occurring in that layer, e.g., Klapwijk and Snodgrass (1986), Smits and van der Molen (1993). For time variable computations, however, models with a number of layers of varying thicknesses can lead to complicated numerical problems. We will confront this problem in Chapter 11.

Finally, an entirely numerical approach can be adopted. The continuous equations are solved numerically using finite difference methods. For models that focus on a single location, or for which the point is to elaborate a complex set of kinetic and transport interactions, such an approach can be quite productive – see van Cappellen and Gaillard (1996) for a review. The model in Chapter 21 is of this type.

For the present, however, we will employ the simplest useful representation: two layers. The aerobic top layer will be allowed to vary in thickness because of its primary role in determining the fluxes of concern. Underneath is the anaerobic layer of fixed depth, in which the mineralization and other reactions occur. Therefore, the sediment is represented using two well-mixed layers that represent the aerobic and anaerobic layers of the sediment. Refinements, for example the entrainment of the lower layer as the upper layer increases in thickness (Chapter 13), will be added as they become necessary.

This choice of two layers has a number of advantages. Analytical solutions to the steady state equations are available for reasonably realistic formulations. They provide useful results that clarify which parameter groups determine the fluxes. Although numerical integrations are still required for time variable solutions to obtain the annual cycle of fluxes, the structure of the model is clarified by the steady state results. Further, comparisons of the two-layer solution and the continuous analytical solutions that are made throughout the book, indicate that little is lost by using the two-layer discretization.

The final reason is discussed in Chapter 1, namely that the sediment flux model is meant to be used as a part of comprehensive water quality models. As a consequence each bottom segment of the water quality model is underlaid with sediment segments within which the mass transport and kinetics need to be computed. If the sediment is represented by many layers to approximate the solution in one continuous dimension, the computational burden would be excessive. Hence the choice of two layers.

2.1 B   Conceptual Framework

The conceptual framework for the sediment model developed in this book is diagramed in Fig. 2.1. Four separate processes are considered. (1) Particulate organic matter (POM) from the overlying water is deposited into the aerobic and anaerobic layers of the sediment. This is referred to as the depositional flux. (2) The particulate organic matter is mineralized in the sediment. This reaction, which is termed *diagenesis*, converts POM into soluble intermediates. Other reactions occur that can convert a portion of the soluble species into particulate species, for example, via sorption. (3) These species are transported by diffusion and particle mixing into the aerobic layer, from which they are either transferred to the overlying water, further react and possibly consume oxygen, or are remixed into the anaerobic layer. (4) Finally, particulate and dissolved chemicals are buried via sedimentation. This general framework is employed for each of the chemical species considered below.

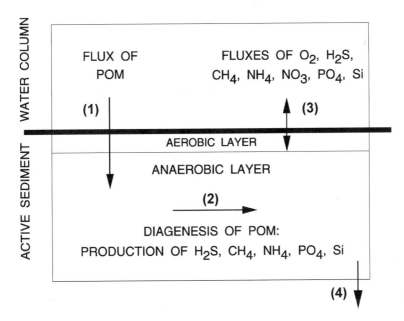

Fig. 2.1 Schematic of the nutrient and oxygen sediment flux model.

2.2 MASS BALANCE EQUATIONS

The equations for the models developed in this book are based on the principle of mass balance. The mass balance principle is important because it provides the starting point for model construction. The idea is simple enough. The change in the mass of any constituent within a volume must be the result of sources of mass to the volume, less the losses within the volume, and the export from the volume. We examine each of these components in sequence.

2.2 A Sources

Consider the situation depicted in Fig. 2.2. A volume V with units of cubic meters (m^3) receives a quantity of a substance at a rate W with units of grams per day (gm/d). The result is a mass of substance M within the volume, with units of grams, which increases as a function of time – hence the notation $M(t)$. If the rate of mass input W is constant in time, then the mass within V increases linearly in time

$$M(t) = Wt \tag{2.1}$$

The concentration that results in the volume – the mass per unit volume – is denoted by $c(t)$ with units of gm/m^3. If the volume is assumed to be completely mixed, so that the mass is distributed uniformly within the volume, then the concentration is

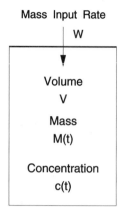

Fig. 2.2   Mass balance of sediment solids.

given by

$$c(t) = \frac{M(t)}{V} = \frac{W}{V}t \tag{2.2}$$

We now formulate this problem using a mass balance expressed as a differential equation in terms of the concentration $c(t)$. The mass within V is given as the product of the concentration and the volume itself (see Eq. 2.2)

$$M(t) = Vc(t) \tag{2.3}$$

The rate of change of this mass with respect to time is its derivative

$$\frac{dM(t)}{dt} = \frac{d(Vc(t))}{dt} = \frac{\text{change in mass } Vc(t)}{\text{change in time } t} \tag{2.4}$$

The mass balance equation is formulated by equating the time rate of change of the mass to the sum of the sources and sinks of this mass. For the situation depicted in Fig. 2.2, there is only a source W. Thus the mass balance equation becomes

$$\frac{d(Vc(t))}{dt} = W \tag{2.5}$$

Since the volume is constant in this example this equation can be written as

$$V\frac{dc(t)}{dt} = W \tag{2.6}$$

or

$$\frac{dc(t)}{dt} = \frac{W}{V} \tag{2.7}$$

which can be solved by integration since the right-hand side is independent of $c(t)$

$$\int \frac{dc(t)}{dt} dt = \int \frac{W}{V} dt \tag{2.8}$$

By using definite integrals, the initial conditions can be included

$$\int_{c(0)}^{c(t)} dc = \frac{W}{V} \int_{0}^{t} dt \tag{2.9}$$

The result is

$$c(t) = c(0) + \frac{W}{V} t \tag{2.10}$$

where $c(0)$ its the concentration at $t = 0$. This is the initial condition, the concentration that was present at the time $t = 0$, the start of the input W to the volume. This is the same result as obtained above by direct reasoning, Eq. (2.2), with $c(0) = 0$. This example is a demonstration of the utility of formulating mass balances using differential equations.

2.2 B   Reaction Sink

If mass only accumulated within the volume, as in the case discussed above, then the concentration would increase indefinitely as indicated by Eq. (2.10). However, if the substance is reactive so that chemical and/or biological processes are occurring that change the substance into something else, then there is a sink term to be considered as well.

The formulation of models of chemical and biological reactions is a subject unto itself e.g., Brezonik (1994). We will present the simplest – and what turns out to be the most useful – example: the first-order reaction. Consider a reaction for which the rate of decay R, with units gm/m^3 per day (gm/m^3-d), is proportional to the concentration $c(t)$ of the substance that is present. Thus

$$R(t) = -kc(t) \tag{2.11}$$

where k is the reaction rate constant with units 1/day (d^{-1}). If this reaction is occurring in the volume V, then it needs to be added to the mass balance equation, Eq. (2.6)

$$V \frac{dc(t)}{dt} = W - V R(t) \tag{2.12}$$

The inclusion of the volume V in the term $V R(t)$ is necessary to convert the units of $R(t)$ gm/m^3-d to the units of the mass balance equation: gm/d. Substituting Eq. (2.11) into Eq. (2.12) and dividing by the volume yields

$$\frac{dc(t)}{dt} = \frac{W}{V} - kc(t) \tag{2.13}$$

Simple integration can no longer be used because $c(t)$ is present as well as $dc(t)/dt$. The solution method is discussed in Appendix A.1. The result is

$$c(t) = c(0)e^{-kt} + \frac{W}{kV}\left(1 - e^{-kt}\right) \tag{2.14}$$

The concentration starts at $c(0)$ at $t = 0$, which is called the initial condition, and eventually approaches W/kV as time increases. The rate at which this transition occurs is controlled by k, the reaction rate constant.

An important special case is the steady state solution which corresponds to the time after which no further change in concentration is occurring. This implies that

$$\frac{dc(t)}{dt} = 0 \tag{2.15}$$

It occurs as time increases to a large number, which is denoted by $t \rightarrow \infty$, so the steady state concentration is denoted by $c(\infty)$. The equation to be solved is Eq. (2.13) with the condition Eq. (2.15), i.e.,

$$0 = \frac{W}{V} - kc(\infty) \tag{2.16}$$

so that

$$c(\infty) = \frac{W}{kV} \tag{2.17}$$

The steady state concentration is determined by two quantities: W, the rate at which mass is being inputted into the volume V, and kV, the rate at which mass is being converted by the reaction. As the reaction rate increases, the concentration $c(\infty)$ decreases since the mass of this constituent is being converted to some other form.

One final point: The time to reach steady state is theoretically infinite. But practically it is determined by comparing the magnitude of the actual solution Eq. (2.14) to the steady state solution Eq. (2.17). Note that the time t_{ss} at which

$$\frac{W}{kV} \simeq c(0)e^{-kt_{ss}} + \frac{W}{kV}(1 - e^{-kt_{ss}}) \tag{2.18}$$

is dependent on the relative magnitude of the initial concentration $c(0)$ and the steady state concentration W/kV as well as the reaction rate constant k.

A special case for which t_{ss} can be evaluated easily is to consider only the exponential terms. If t_{ss} is chosen so that

$$e^{-kt_{ss}} \simeq 0 \tag{2.19}$$

then this would define steady state. Normally, steady state is assumed if

$$e^{-kt_{ss}} = 0.05 \tag{2.20}$$

which requires that

$$kt_{ss} = 3 \tag{2.21}$$

Thus the time to steady state is estimated as

$$t_{ss} = \frac{3}{k} \tag{2.22}$$

which is the time after which the time variable solution is within 5% of the steady state solution.

2.3 SEDIMENTATION AND BURIAL

The loss of mass by burial is a sink that needs to be included in the mass balance. Burial occurs because solids accumulate at the surface of the sediment. The most convenient formulation for the effect of burial is in terms of a sedimentation velocity. It can be derived from a mass balance of the solids that comprise the sediment.

2.3 A Sediment Solids

Consider a mass balance of the solid particles that make up the sediment. For this analysis no distinction is made between differing types of particles. Only their mass is considered. The situation is illustrated in Fig. 2.3. A column of sediment with

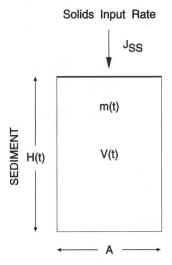

Fig. 2.3 Mass balance of sediment solids. $H(t)$ denotes the depth. A denotes the cross-sectional area.

volume $V(t)$, depth $H(t)$, and cross-sectional area A, receives a constant flux of suspended solids to the sediment surface. This source of solids from the water to the sediment column is quantified by specifying the flux of solids to the sediment water interface. The flux is defined as the mass per unit area per unit time J_{SS} (gm/m^2-d). The rate at which mass of solids is delivered to the sediment column is $J_{SS}A$ (gm/m^2-d)(m^2) = grams per day which depends on the cross-sectional area being considered since the flux is a mass per unit area. This must be balanced by the increase in solids in the volume. Thus the mass balance equation is

$$\frac{dM(t)}{dt} = \frac{d(m(t)V(t))}{dt} = J_{SS}A \tag{2.23}$$

where $M(t)$ = the mass of solids in V, and $m(t)$ = the concentration of solids in V. This equation states that the rate of change of the mass of solids $m(t)V(t)$ in the column must be equal to the rate at which solids enter the volume $J_{SS}A$.

2.3 B   Sedimentation Velocity

Note that Eq. (2.23) does not specify whether the solids concentration $m(t)$ or the volume $V(t)$ is increasing, only that either or both must be changing to accommodate the solids settling to the sediment. Two extremes span the range of possibilities as can be seen from expanding the derivative in Eq. (2.23) using the product rule

$$\frac{d(m(t)V(t))}{dt} = V(t)\frac{dm(t)}{dt} + m(t)\frac{dV(t)}{dt} = J_{SS}A \tag{2.24}$$

This expresses the fact that either $m(t)$ or $V(t)$ or both can be changing. Dividing through by the cross-sectional area A converts the volume to the depth of the sediment column

$$H(t) = \frac{V(t)}{A} \tag{2.25}$$

so that Eq. (2.24) becomes

$$H(t)\frac{dm(t)}{dt} + m(t)\frac{dH(t)}{dt} = J_{SS} \tag{2.26}$$

If we require that the depth of the sediment layer be constant so that

$$\frac{dH(t)}{dt} = 0 \tag{2.27}$$

then just the solids concentration increases

$$H\frac{dm(t)}{dt} = J_{SS} \tag{2.28}$$

Since J_{SS} is assumed to be constant in time, the result is (see Eq. 2.9)

$$m(t) = m(0) + \frac{J_{SS}}{H}t \tag{2.29}$$

a continuously increasing concentration of solids in the volume. This is an impossible result, since the concentration can only continue to increase until the volume is completely filled with solid material.

Alternately, if the solids concentration is required to be constant

$$\frac{dm(t)}{dt} = 0 \tag{2.30}$$

then Eq. (2.26) becomes

$$m\frac{dH(t)}{dt} = J_{SS} \tag{2.31}$$

and

$$H(t) = H(0) + \frac{J_{SS}}{m}t \tag{2.32}$$

so that the depth is increasing in time. This is a reasonable approximation to what is actually occurring during sedimentation. The concentration of solids m in a layer of sediment is controlled by the properties of the particles that make it up, and the local conditions where it is deposited. If these remain constant, then m is a constant and Eq. (2.32) applies.

For modeling sediment processes, however, it is inconvenient to have the layer thickness changing as a function of time, since steady state solutions are not possible. Rather we focus on a layer of constant depth that is anchored to the top of the sediment column and moves as the column changes in depth. We will label this layer H_2 – and refer to it as layer 2 – since we will be using H_1 for the thin aerobic layer 1 that overlays H_2. The situation is illustrated in Fig. 2.4. Initially, H_2 is the top portion of the sediment shown by the cross-hatching (A). As solids accumulate above the H_2 layer due to solids flux to the sediment J_{SS}, the depth of the column of solids $H(t)$ increases (B). If the layer H_2 is anchored to the top of the sediment, then it is as if sediment solids are passing through the bottom of H_2 with velocity w_2 (C) where w_2 is the rate of increase of $H(t)$.

Consider a mass balance equation around H_2. Solids are passing through the layer at a velocity

$$w_2 = \frac{dH(t)}{dt} \tag{2.33}$$

which corresponds to the rate at which the column is growing. The mass balance equation for the solids in layer 2 becomes

$$H_2\frac{dm_2(t)}{dt} = J_{SS} - w_2m_2(t) \tag{2.34}$$

This equation can now be solved for the steady state solids concentration in layer 2

$$0 = H_2\frac{dm_2(t)}{dt} = J_{SS} - w_2m_2(t) \tag{2.35}$$

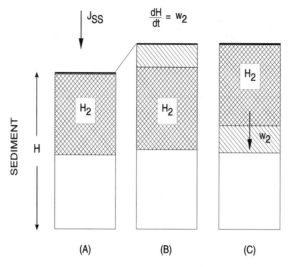

Fig. 2.4   Sedimentation velocity and the change in sediment depth.

or

$$m_2(\infty) = \frac{J_{SS}}{w_2} \qquad (2.36)$$

This suggests that the solids concentration in a layer of fixed depth H_2 is determined by the rate of solids input J_{SS} and the burial velocity w_2.

In a formal way this is true since the algebra leading to Eq. (2.36) is unassailable. However the causality is not correct. The concentration of solids m_2 in a layer of sediment is controlled by the properties of the particles that make it up, and the local conditions where it is deposited. In particular, the median particle size plays an important role (Fig. 1.1). Hence it is m_2 that is fixed, and the burial velocity adjusts to satisfy Eq. (2.35). Thus the correct interpretation of Eq. (2.35) is

$$w_2 = \frac{J_{SS}}{m_2} \qquad (2.37)$$

which determines the burial velocity as the ratio of the flux of particles to the sediment surface and the concentration of solids in the layer.

2.3 C   Inclusion in the Mass Balance Equation

The representation of the loss of mass by burial using a sedimentation velocity w in the mass balance equation is straightforward. The sedimentation velocity term is added to the mass balance equation for the constituent of interest $c(t)$ (Eq. 2.12) as a sink

$$V\frac{dc(t)}{dt} = W - Vkc(t) - wAc(t) \qquad (2.38)$$

where the cross-sectional area is included so that the units of the loss of mass by sedimentation $wAc(t)$ are $(m/d)(m^2)(gm/m^3) = gm/d$, consistent with the other terms. If Eq. (2.38) is divided through by A, the source W becomes a flux $J = W/A$. Since it is ultimately fluxes that are of interest, this is a convenient modification. Thus Eq. (2.38) becomes

$$H\frac{dc(t)}{dt} = J - kHc(t) - wc(t) \qquad (2.39)$$

where $J = W/A$ the flux of mass to the sediment.

The steady state solution for which $dc(t)/dt = 0$ is

$$c(\infty) = \frac{J}{kH + w} \qquad (2.40)$$

The concentration is determined by the balance between the flux of mass to the sediment J and the loss terms: the sum of the reaction rate–depth product kH and the sedimentation velocity w.

2.4 MIXING PROCESSES AND MASS TRANSFER COEFFICIENTS

Mixing processes are important in influencing the magnitude of sediment fluxes. Although the examples presented above illustrate the importance of the source terms, it will be seen subsequently that mixing between the anaerobic and aerobic layers of the sediment is also important.

Mixing of dissolved constituents occurs in the sediment pore waters by molecular diffusion. Transfer to and from the overlying water is also by diffusion. Both can be enhanced by biological activity – termed *bioirrigation*[*] – and by turbulence at the sediment-water interface. Mixing of solid phase constituents occurs primarily as the result of benthic animal activity, which is called *bioturbation* (Section 1.4 C).

The normal representation of mixing processes makes use of diffusion coefficients and gradients of concentrations. The relationship is called *Fick's law* (Berner, 1980a, Bird et al., 1960). An alternate parameterization uses mass transfer coefficients which are related to diffusion coefficients. The representation of fluxes of mass using mass transfer coefficients instead of diffusion coefficients is commonly used in engineering models of various sorts (Bird et al., 1960). For example, the reaeration coefficient K_L for dissolved oxygen transfer from the atmosphere to the surface of streams is formulated as a mass transfer coefficient (O'Connor, 1983, O'Connor and Dobbins, 1958).

We will make use of mass transfer coefficients in the models that follow. The derivation presented below will clarify certain of the assumptions inherent in this approach.

---

[*]See Boudreau (1996), Schluter et al. (2000).

2.4 A   Mass Transfer

Mass transport that is due to fluid or solid mixing occurs when a parcel of fluid or solid from one location exchanges position with a parcel of the same volume from another position. This is in contrast to molecular diffusion where molecular motion – the wandering of a molecule through a stationary fluid – accounts for the mass transport. The situation is illustrated in Fig. 2.5. Imagine a packet with volume v,

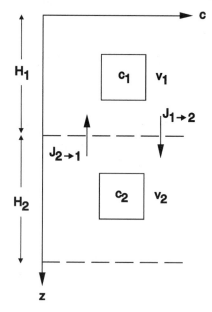

Fig. 2.5   Mass transfer between layers 1 and 2.

containing both pore water and sediment solids, which moves from H_1 to H_2 across a plane of area A. A packet with the same volume v, moves from H_2 to H_1. If the volumes that interchange position are not the same volume v, then a depletion would occur on one side of the plane and a buildup of volume would occur on the other side. This cannot persist indefinitely, so we assume that only packets of the same volume mix from either side and that the mixing has bidirectional symmetry.

The amount of mass that is transported can be related to the rate at which the volumes are exchanging position and to the quantity of the constituent of interest in the volume v. First, consider the rate. Let f be the frequency – the number of volume interchanges per unit time – with which the volumes v exchange across the plane A located at depth z (Fig. 2.5). Then the rate at which volumes of sediment and pore water are transported from H_1 to H_2 across the plane at z per unit area per unit time is

$$\frac{\text{volume exchanged} \times \text{frequency}}{\text{area}} = \frac{vf}{A} = f\ell \tag{2.41}$$

Since the quantity v/A, a volume per unit area, has the dimensions of a length, we define a characteristic length $\ell = v/A$. Thus $f\ell$ has the dimensions of a length per unit time (L/T), i.e., a velocity. In fact, it is sometimes referred to as the "piston velocity" when applied to air-water gas transfer. It is the rate f at which mixing occurs over the characteristic length ℓ. We will use the notation

$$K_L = f\ell \tag{2.42}$$

for this velocity. As we will see below, this is the mass transfer coefficient.

Consider the mass transfer that is occurring between two layers. The flux of mass from H_1 to H_2 is given by

$$J_{1\rightarrow2} = f\ell c(1) = K_L c(1) \tag{2.43}$$

where we use the notation $J_{1\rightarrow2}$ to denote the one-way flux from H_1 to H_2. K_L is the velocity with which the packets of volume are moving from H_1 to H_2. Similarly the flux of mass from H_2 to H_1 is given by

$$J_{2\rightarrow1} = K_L c(2) \tag{2.44}$$

where K_L is the velocity with which the packets of volume are moving from H_2 to H_1. The net transport of mass from H_1 to H_2 is the difference between the flux from 1 to 2 ($J_{1\rightarrow2}$) and from 2 to 1 ($J_{2\rightarrow1}$)

$$J_{12} = J_{1\rightarrow2} - J_{2\rightarrow1} = K_L[c(1) - c(2)] \tag{2.45}$$

Since the mass transport is between layer 1 and 2, the mass transfer coefficient is labeled appropriately

$$J_{12} = K_{L12}\big(c(1) - c(2)\big) \tag{2.46}$$

2.4 B Relationship to Fick's Law

Fick's law specifies the magnitude of the mass transport that occurs by molecular diffusion. The flux of mass J is proportional to the concentration gradient

$$J = -D\frac{dc(z)}{dz} \tag{2.47}$$

This type of equation is called a diffusion equation (Crank, 1975) and it is used to model other types of mixing such as turbulent mixing in a fluid or solid phase mixing by biota. It is of interest to compare the mass transfer formulation (Eq. 2.46) to diffusive mixing (Eq. 2.47).

Consider the situation illustrated in Fig. 2.6. The amount of mass of a constituent that is transported as the volumes exchange is dependent on the concentration that is contained in the volumes v. Let $c(z)$ be the concentration of a substance in the sediment at location z. Imagine that the location from which the volumes are originating are above and below the plane A a distance $\pm\Delta z$ (Fig. 2.6). Then the concentration in v at $z - \Delta z$ is $c(z - \Delta z)$. Hence the flux from H_1 to H_2, $J_{1\rightarrow2}$ is (Eq. 2.43)

$$J_{1\rightarrow2} = K_{L12}c(z - \Delta z) \tag{2.48}$$

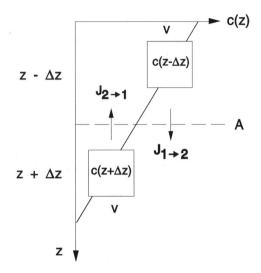

Fig. 2.6   Mass transfer and the relationship to gradients.

The transfer of mass $J_{2\to 1}$ from H_2 to H_1 (Eq. 2.44) is caused by the same mass transfer coefficient, but the concentration in the volume v is that at $z + \Delta z$, namely $c(z + \Delta z)$

$$J_{2\to 1} = K_{L12}c(z + \Delta z) \qquad (2.49)$$

The net transport of mass from H_1 to H_2 is (Eq. 2.45)

$$J_{12} = J_{1\to 2} - J_{2\to 1} = K_{L12}\big(c(z - \Delta z) - c(z + \Delta z)\big) \qquad (2.50)$$

The term in brackets multiplying K_{L12} in Eq. (2.50) is similar to the numerator of the finite difference approximation to a derivative

$$\frac{dc(z)}{dz} \simeq \frac{c(z + \Delta z) - c(z - \Delta z)}{\Delta z} \qquad (2.51)$$

To convert Eq. (2.50) to this form multiply the numerator and denominator by $-\Delta z$

$$J_{12} = -K_{L12}\Delta z \left(\frac{c(z + \Delta z) - c(z - \Delta z)}{\Delta z}\right) \qquad (2.52)$$

For $\Delta z \to 0$ the fraction approaches the derivative Eq. (2.51)

$$J_{12} \simeq -K_{L12}\Delta z \left(\frac{dc(z)}{dz}\right) \qquad (2.53)$$

Thus the relationship between the diffusion coefficient D, which for molecular diffusion is a well defined quantity, and the equivalent mass transfer coefficient is

$$K_{L12} = \frac{D}{\Delta z} \qquad (2.54)$$

In other words, for molecular diffusion the mass transfer coefficient is not a funda-mental quantity. Rather its magnitude depends on the length scale Δz of the analysis. However, for other problems such as mass or momentum transfer between air and water, or between water and sediments, it is the mass transfer coefficient that is the more useful quantity because there is no direct way to measure D and Δz directly.

The negative sign in Eq. (2.53) can be understood as follows. Consider the con-centration as drawn in Fig. 2.6. Since the concentration of mass at $z - \Delta z$ is higher than that at $z + \Delta z$, a larger amount moves from $z - \Delta z$ to $z + \Delta z$ than from $z + \Delta z$ to $z - \Delta z$. Therefore, on balance, mass moves from $z - \Delta z$ to $z + \Delta z$. This amounts to a mass flux which is in the positive z direction. Since the gradient $dc(z)/dz$ in Fig. 2.6 is negative – the concentration is decreasing as z increases – the minus sign is necessary so that the equation results in a flux with the correct sign, in this case a positive flux – a flux in the direction of positive z.

2.4 C   Relationship to Diffusion Coefficient

The relationship between the mass transfer coefficient and the diffusion coefficient for truly diffusive processes can be seen from Fick's law Eq. (2.47). The mass trans-fer approximation applies to two layers where the derivative is approximated by a finite difference

$$J = -D\frac{dc}{dz} \tag{2.55a}$$

$$\cong -D\frac{c(z + \Delta z) - c(z - \Delta z)}{\Delta z} \tag{2.55b}$$

$$= -\frac{D}{\Delta z}\left(c(z + \Delta z) - c(z - \Delta z)\right) \tag{2.55c}$$

$$= -\frac{D}{(H_1 + H_2)/2}\left(c(2) - c(1)\right) \tag{2.55d}$$

where Eq. (2.55b) is the finite difference approximation and Eq. (2.55d) is the two-layer approximation. The Δz is taken as the distance from the midpoints of the two layers. From this we see that the mass transfer coefficient is related to the diffusion coefficient via

$$K_{L12} = \frac{D}{\Delta z} = \frac{D}{(H_1 + H_2)/2} \tag{2.56}$$

For an accurate representation of actual diffusion processes, the mass transfer ap-proximation requires that (1) the concentration profile be linear so that Eq. (2.55b) is an accurate representation of Eq. (2.55a), and (2) that the layer depths be known so that $(H_1 + H_2)/2$ can be calculated. It is in situations where the shape of the profile and the actual layer depths are uncertain, for example in turbulent flows over com-plicated surfaces, that mass transfer coefficients are used. For these cases, neither D nor Δz are known whereas K_{L12} can be measured.

For the case where the depths of the layers are known, then the mass transfer coefficient is proportional to the diffusion coefficient of the species being considered (Eq. 2.56).

Table 2.1 Molecular Diffusion Coefficients[a]

Species	D	Species	D	Species	D
O_2	1.98	Ca^{2+}	0.70	OH^-	4.60
CO_2	1.66	Cd^{2+}	0.61	Cl^-	1.78
CH_4	1.44	Fe^{2+}	0.61	HCO_3^-	1.03
NH_3	1.97	Mn^{2+}	0.61	$H_2PO_4^-$	0.83
				HS^-	1.49
H^+	8.06			NO_3^-	1.66
Na^+	1.17			CO_3^{2-}	0.80
NH_4^+	1.71			HPO_4^{2-}	0.66

[a] Units: cm^2/d. $T = 10°C$ (Boudreau, 1996)

Measured values are listed in Table 2.1. With the exception of the smaller ions, the diffusion coefficients average $D = 1.22$ (± 0.52) cm^2/d. Since the mixing processes in sediments are not limited to molecular diffusion – for example bioirrigation by organisms – it is not unreasonable to assume that the mass transfer coefficients for all the dissolved species are the same. This simplifies the analysis considerably without sacrificing too much realism. Of course, there is no inherent difficulty is using the measured coefficients. It is simply convenient to have one mass transfer coefficient that applies to all dissolved species.

2.5 TWO-LAYER MASS BALANCE

Constructing a mass balance for a two-layer segmentation is a straightforward generalization of the previous equations. Consider the segmentation in Fig. 2.7. The mass balance equation for H_1 is

$$H_1 \frac{dc(1)}{dt} = J + K_{L12}\big(c(2) - c(1)\big) - w_2 c(1) \tag{2.57}$$

where the terms on the right-hand side correspond to the flux to H_1, the mass transfer between H_2 and H_1, and the loss of mass from H_1 due to sedimentation. The mass balance equation for H_2 is

$$H_2 \frac{dc(2)}{dt} = w_2 c(1) + K_{L12}\big(c(1) - c(2)\big) - w_2 c(2) \tag{2.58}$$

where the terms correspond to the burial flux from H_1, the mass transfer between H_1 and H_2, which is the negative of the term in Eq. (2.57), and the loss of mass from H_2 due to sedimentation.

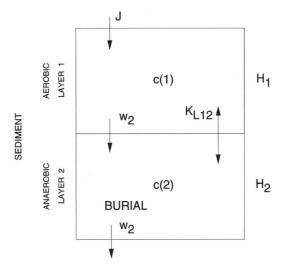

Fig. 2.7   Two-layer mass balance.

It is an interesting exercise to add Eqs. (2.57) and (2.58)

$$H_1 \frac{dc(1)}{dt} + H_2 \frac{dc(2)}{dt} = J - w_2 c(2) \tag{2.59}$$

The left-hand side is the rate of change of the total mass per unit area in the sediment

$$\frac{d\big(H_1 c(1) + H_2 c(2)\big)}{dt} = J - w_2 c(2) \tag{2.60}$$

It changes due to the input of mass J and the loss of mass by burial $-w_2 c(2)$. The other terms are internal mass transfers and they cancel out of the overall mass balance. For more complicated sets of equations, the addition verifies that the signs of the terms are correct and that the proper cancellation occurs. We will see numerous examples of this checking procedure in subsequent chapters.

2.6   PARTICULATE ORGANIC NITROGEN AND AMMONIA

The ideas that form the basis for the models of nutrient and metal fluxes to be discussed subsequently are best illustrated using an example. We will use organic nitrogen and ammonia as the constituents to be modeled since, with the exception of the nitrification reaction which is discussed in Chapter 3, ammonia is essentially unreactive in sediments. Thus the simplifications introduced in this section are not so unrealistic as to yield misleading results. Rather they illustrate a number of fundamental principles that will continue to reappear in the more complex models developed subsequently.

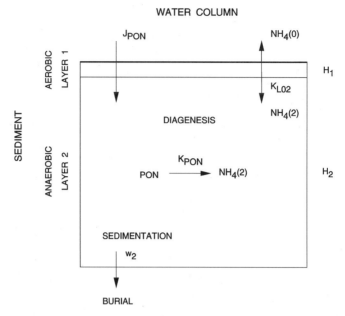

Fig. 2.8 Schematic of a simplified sediment flux model for ammonia.

2.6 A   Particulate Organic Nitrogen

The situation is illustrated in Fig. 2.8. For this example, the aerobic layer H_1 is ignored. This simplification is consistent with the assumption that no nitrification is occurring. The flux J_{PON} is the source of particulate organic nitrogen (PON) to the sediment. The concentration of PON decreases as a consequence of its mineralization to ammonia

$$PON \overset{k_{PON}}{\rightarrow} NH_4^+ \tag{2.61}$$

This reaction is termed *diagenesis*. For the sake of simplicity it is assumed to be a first-order reaction. PON is also removed from the sediment layer via sedimentation. Thus the mass balance equation for PON in layer 2, [PON(2)], is (see Eq. 2.58)

$$H_2 \frac{d[\text{PON}(2)]}{dt} = J_{PON} - k_{PON} H_2 [\text{PON}(2)] - w_2 [\text{PON}(2)] \tag{2.62}$$

The notation is as follows: (2) denotes layer 2, and the square brackets [ ] denote concentration.

The inclusion of H_2 in the reaction term can be understood by considering the units of each term in the equation. The quantity of mass per unit area being considered is $H_2[\text{PON}(2)]$ with units g N/m^2. It is the time rate of change of this mass that is being computed so that the units are g N/m^2–d. These are also the units of the flux J_{PON}. The decay rate k_{PON} has units d^{-1} which, when applied to the mass $H_2[\text{PON}(2)]$, produces the correct loss term. The sedimentation velocity w_2 has units m/d so that the product of the concentration [PON(2)] with units g N/m^3 and the velocity term w_2 results in units of a flux.

The steady state solution can be found immediately by solving

$$0 = H_2 \frac{d[PON(2)]}{dt} = J_{PON} - k_{PON} H_2 [PON(2)] - w_2 [PON(2)] \qquad (2.63)$$

to yield

$$[PON(2)] = \frac{J_{PON}}{k_{PON} H_2 + w_2} \qquad (2.64)$$

The concentration of PON that results is determined by the magnitude of the source of PON to the sediment J_{PON} and the sum of the two loss terms $k_{PON} H_2$ and w_2. Note the importance of H_2 in determining the extent to which the reaction rate influences the concentration relative to the sedimentation rate.

The time variable solution can also be easily found by solving Eq. (2.62) directly in the form

$$\frac{d[PON(2)](t)}{dt} = \frac{J_{PON}}{H_2} - \left(k_{PON} + \frac{w_2}{H_2} \right) [PON(2)](t) \qquad (2.65)$$

to yield

$$[PON(2)](t) = \frac{J_{PON}}{k_{PON} H_2 + w_2} \left\{ 1 - \exp\left(-(k_{PON} + w_2/H_2) t \right) \right\} \qquad (2.66)$$

where the initial condition – the concentration of PON in the sediment at the start of the deposition [PON(2)](0) – is assumed to be zero. The parameter group that controls the time to steady state is called the *time constant, τ*

$$\tau = \frac{1}{k_{PON} + w_2/H_2} \qquad (2.67)$$

with units of days. As shown above (Eq. 2.22), 3τ is the approximate time to steady state. In the absence of mineralization, the time to steady state is determined by the time it takes for sedimentation to replace the sediment in H_2, namely H_2/w_2. And because the volume is assumed to be completely mixed due to organism mixing, it takes three time constants to completely (95%) replace the sediment solids.

2.6 B Ammonia

The decay of PON produces ammonia at a rate $+k_{PON} H_2 [PON(2)]$ (Eq. 2.62). This source and the other processes that affect ammonia are illustrated in Fig. 2.8. Ammonia is lost to the overlying water via diffusion from the pore water $-K_{L02} [NH_4(2)]$. It may also be increased by diffusion from the overlying water to the pore water if the overlying water ammonia concentration is non-zero $+K_{L02} [NH_4(0)]$. Finally ammonia is lost via sedimentation $-w_2[NH_4(2)]$. Therefore the mass balance equation for ammonia contains all these terms

$$H_2\frac{d[NH_4(2)]}{dt} = +k_{PON}H_2[PON(2)] + K_{L02}([NH_4(0)] - [NH_4(2)])$$
$$-w_2[NH_4(2)] \tag{2.68}$$

The steady state solution can be found immediately by solving

$$0 = +k_{PON}H_2[PON(2)] + K_{L02}([NH_4(0)] - [NH_4(2)]) - w_2[NH_4(2)] \tag{2.69}$$

for $NH_4(2)$ to yield

$$[NH_4(2)] = \frac{k_{PON}H_2[PON(2)] + K_{L02}[NH_4(0)]}{K_{L02} + w_2} \tag{2.70}$$

The concentration of $NH_4(2)$ is determined by the ratio of the source terms in the numerator and the loss rate constants in the denominator. The final solution is arrived at by substituting the solution for PON(2) Eq. (2.64) into Eq. (2.70) to give

$$[NH_4(2)] = \frac{\dfrac{J_{PON}k_{PON}H_2}{H_2k_{PON} + w_2} + K_{L02}[NH_4(0)]}{K_{L02} + w_2} \tag{2.71}$$

As we will subsequently see, the sedimentation velocity is usually smaller than either of the other two loss rate constants

$$w_2 \ll K_{L02} \tag{2.72}$$

$$w_2 \ll k_{PON}H_2 \tag{2.73}$$

in which case the pore water ammonia concentration is

$$[NH_4(2)] \simeq \frac{J_{PON} + K_{L02}[NH_4(0)]}{K_{L02}} \tag{2.74}$$

It is determined by the total source of nitrogen to the sediment: the flux J_{PON} of PON and the diffusive transport of overlying water ammonia $K_{L02}[NH_4(0)]$. Note that in the absence of a flux of PON, the interstitial water concentration would equal the overlying water concentration: $[NH_4(2)] = [NH_4(0)]$. This is an entirely reasonable result since there is no source of ammonia in the sediment and the pore water concentration equilibrates with the overlying water. Note, also, the importance of the mass transfer coefficient K_{L02}. If $J_{PON} \gg K_{L02}[NH_4(0)]$, that is, the mineralization of organic matter is producing significant ammonia, then

$$[NH_4(2)] \simeq \frac{J_{PON}}{K_{L02}} \tag{2.75}$$

and the pore water ammonia concentration is inversely related to K_{L02}. Decreasing K_{L02} increases $[NH_4(2)]$. We will use this fact to estimate K_{L02} from observed pore water concentrations of ammonia.

The point of these models is to compute fluxes to or from the sediment. The flux of ammonia to the overlying water $J[NH_4]$ is given by

$$J[NH_4] = K_{L02}([NH_4(2)] - [NH_4(0)]) \tag{2.76}$$

which is the negative of the diffusive exchange term in Eq. (2.68). Substituting the solution for $[NH_4(2)]$ Eq. (2.71) into Eq. (2.76) above and simplifying yields

$$J[NH_4] = J_{PON}\frac{k_{PON}H_2 K_{L02}}{(H_2 k_{PON} + w_2)(K_{L02} + w_2)} - [NH_4(0)]\frac{K_{L02}w_2}{K_{L02} + w_2} \tag{2.77}$$

The two terms correspond to the two sources of ammonia: mineralization of PON and the diffusive exchange from the overlying water.

One of the principal uses of an analytical solution is in examining how the solution depends on the parameters. An initial simplification can be made by using the fact that w_2 is small relative to K_{L02} and $k_{PON}H_2$ (Eqs. 2.72–2.73). Setting $w_2 = 0$ is not the same as assuming w_2 is small. The proper procedure is to expand Eq. (2.77) in an ascending Taylor series in w_2 (see Appendix B). The result is

$$J[NH_4] = J_{PON}(1 - \frac{w_2}{K_{L02}} - \frac{w_2}{k_{PON}H_2} + \cdots)$$
$$- w_2[NH_4(0)](1 - \frac{w_2}{K_{L02}} + \cdots) \tag{2.78}$$

Collecting the terms that are constant and linear in w_2 yields

$$J[NH_4] \simeq J_{PON} - w_2\left(\frac{J_{PON}}{K_{L02}} + [NH_4(0)]\right) - J_{PON}\frac{w_2/H_2}{k_{PON}} \tag{2.79a}$$

$$= J_{PON} - w_2[NH_4(2)] - J_{PON}\frac{w_2/H_2}{k_{PON}} \tag{2.79b}$$

where Eq. (2.74) is used in Eq. (2.79b) for $[NH_4(2)]$. This is a very interesting result. Consider first the simple case for which $w_2 = 0$. The flux of ammonia is equal to the flux of PON to the sediment. That is, all the organic nitrogen that enters the sediment is recycled back to the overlying water as a flux of ammonia.

For small w_2, the effect of the sedimentation velocity adds two terms that slightly decrease the flux (Eq. 2.79b). The first, $w_2[NH_4(2)]$, accounts for the pore water ammonia that is buried. The second term accounts for the unreacted PON that is buried. It is inversely proportional to the product of the residence time of sediment in layer 2, H_2/w_2, and the reaction rate k_{PON}.

What is surprising is that, for small w_2, the mass transfer K_{L02} has no effect on the flux (Eq. 2.79b). One might have expected that the mass transfer coefficient K_{L02} would be controlling. After all, the flux is driven by the diffusion of pore water ammonia to the overlying water and, therefore, it is proportional to K_{L02} (Eq. 2.76). However, the pore water ammonia concentration is inversely proportional to K_{L02} (Eq. 2.75). Hence the dependency on K_{L02} cancels out, which explains mathematically Eq. (2.79).

The underlying reason for this result can be understood as follows: Ammonia is produced by the mineralization of particulate organic nitrogen which is added to the sediment at a rate J_{PON}. Where can this nitrogen go? It can be buried, or it can diffuse back to the overlying water. If burial is small, then the only alternative is for it to escape back to the overlying water. Thus $J[NH_4] \simeq J_{PON}$. And what about diffusion? If the diffusion rate – actually the mass transfer rate for this layered model – is large, then it seems reasonable that the ammonia can escape to the overlying water. But what if the mass transfer rate is lowered? Then the pore water ammonia concentration builds up (Eq. 2.75) until it is large enough to transfer all the ammonia produced by mineralization to the overlying water. This is the reason that the pore water concentration is inversely proportional to the mass transfer coefficient (Eq. 2.75).

So we see that for this model the principal driving force for the flux of ammonia from the sediment is the flux of PON to the sediment. The effects of the mineralization and mass transfer rates are secondary. As the model is made more realistic by inclusion of nitrification, for example, other processes become significant. However, the primacy of the source, the mineralization of organic matter, remains a feature throughout.

2.7   CONTINUOUS MODELS

An alternate method of modeling the fluxes from sediment is to model the concentration profiles in sediment interstitial water (Berner, 1971, 1980a). Once the concentration profile is modeled, the flux can be obtained from the slope of the profile at the sediment-water interface from Fick's law. Because of the importance of this class of models, we present an example for ammonia that parallels the single volume model presented above.

2.7 A   Mass Balance Equation

The mass balance equation for a continuous one-dimensional model can be derived from the finite volume equations. Fig. 2.9 presents the schematic. Consider the volume and concentration c_2. The mass balance for this volume is analogous to that derived above (Eq. 2.58)

$$H\frac{dc(2)}{dt} = wc(1) + K_L(c(1) - c(2)) - wc(2) + K_L(c(3) - c(2)) - k_r c(2)H$$

$$(2.80)$$

The terms represent the sedimentation flux from H_1, the diffusive exchange between H_1 and H_2, the sedimentation flux out of H_2, the diffusive exchange between H_2

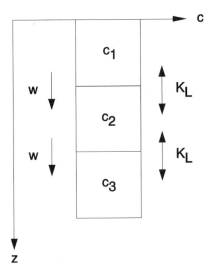

Fig. 2.9 Schematic of a continuous model. Each segment has depth H.

and H_3, and a reaction sink. We assume that w and K_L are constant in depth, so they no longer need to be indexed by segment number.

To make the model continuous, the depth of each segment H needs to shrink to zero so the differences become derivatives. Thus replacing H with Δz, dividing by Δz, and grouping terms to resemble derivatives yields

$$0 = \frac{dc(2)}{dt} = -w\left(\frac{c(2)-c(1)}{\Delta z}\right) + K_L\left(\frac{c(3)-c(2)}{\Delta z} - \frac{c(2)-c(1)}{\Delta z}\right) - k_r c(2)$$

(2.81)

where a steady state has been assumed for simplicity. As $\Delta z \to 0$ the differences become derivatives, and

$$0 = -w\left(\frac{dc}{dz}\right) + K_L\left(\frac{dc}{dz}\bigg|_{3-2} - \frac{dc}{dz}\bigg|_{2-1}\right) - k_r c$$

(2.82)

The diffusive mixing becomes the difference of two derivatives

$$0 = -w\frac{dc}{dz} + K_L \Delta z\frac{\dfrac{dc}{dz}\bigg|_{3-2} - \dfrac{dc}{dz}\bigg|_{2-1}}{\Delta z} - k_r c$$

(2.83)

which, as $\Delta z \to 0$, approaches a second derivative

$$0 = -w\frac{dc}{dz} + K_L \Delta z\frac{d^2 c}{dz^2} - k_r c$$

(2.84)

and the mass transfer coefficient becomes the diffusion coefficient (Eq. 2.56)

$$0 = -w\frac{dc}{dz} + D\frac{d^2 c}{dz^2} - k_r c$$

(2.85)

The signs of the terms can best be remembered by keeping the transport terms on the left-hand side of the equation and the sources and sinks on the right-hand side

$$-D\frac{d^2c}{dz^2} + w\frac{dc}{dz} = -k_r c \tag{2.86}$$

2.7 B   Particulate Organic Nitrogen

Applying Eq. (2.86) to particulate organic nitrogen yields

$$w\frac{d[\text{PON}]}{dz} = -k_{\text{PON}}[\text{PON}] \tag{2.87}$$

where k_{PON} is the rate constant for mineralization of PON to ammonia. We assume for the sake of simplicity that there is no particle mixing, so $D = 0$. The solution is

$$[\text{PON}(z)] = [\text{PON}(0)]\exp\left(-\frac{k_{\text{PON}}z}{w}\right) \tag{2.88}$$

The concentration of PON at the sediment water interface can be related to the flux of PON to the surface by considering a mass balance around the sediment-water interface. The result is

$$[\text{PON}(0)] = \frac{J_{\text{PON}}}{w} \tag{2.89}$$

and Eq. (2.88) becomes

$$[\text{PON}(z)] = \frac{J_{\text{PON}}}{w}\exp\left(-\frac{k_{\text{PON}}z}{w}\right) \tag{2.90}$$

The result is that the concentration of particulate organic nitrogen decays exponentially with depth, reflecting the rate at which it is being buried, and the rate at which it is mineralizing. The exponential decline with depth has been illustrated by the data in Fig. 1.10.

The decay of PON produces ammonia. If, as before, we assume that ammonia is conservative in the sediment so that no nitrification is occurring, then the mass balance equation for the ammonia concentration in pore water $[\text{NH}_4(z)]$ is (Eq. 2.86)

$$-D_{\text{NH}_4}\frac{d^2[\text{NH}_4(z)]}{dz^2} + w\frac{d[\text{NH}_4(z)]}{dz} = -k_{\text{PON}}[\text{PON}(z)] \tag{2.91}$$

where D_{NH_4} is the pore water diffusion coefficient for ammonia. The solution is

$$[\text{NH}_4(z)] = [\text{NH}_4(0)] + \frac{J_{\text{PON}}}{w}\frac{1}{1 + D_{\text{NH}_4}k_{\text{PON}}/w^2}\left\{1 - \exp\left(-\frac{k_{\text{PON}}z}{w}\right)\right\} \tag{2.92}$$

Ammonia increases from the overlying water concentration $[\text{NH}_4(0)]$ until at depth it reaches a concentration of

$$[\text{NH}_4(\infty)] = [\text{NH}_4(0)] + \Delta[\text{NH}_4] \tag{2.93}$$

where

$$\Delta[\text{NH}_4] = \frac{J_{\text{PON}}}{w} \frac{1}{1 + D_{\text{NH}_4} k_{\text{PON}}/w^2} \tag{2.94}$$

The flux of ammonia predicted by this model is

$$J[\text{NH}_4] = -D_{\text{NH}_4} \left. \frac{d[\text{NH}_4(z)]}{dz} \right|_{z=0}$$

$$= -J_{\text{PON}} \frac{D_{\text{NH}_4} k_{\text{PON}}/w^2}{1 + D_{\text{NH}_4} k_{\text{PON}}/w^2}$$

$$= -J_{\text{PON}} \frac{\xi}{1 + \xi} \tag{2.95}$$

where

$$\xi = \frac{D_{\text{NH}_4} k_{\text{PON}}}{w^2} \tag{2.96}$$

The negative sign indicates that the flux is in the negative z direction, out of the sediment toward the overlying water. Note that if the product of the diffusion coefficient D_{NH_4} and the mineralization rate constant k_{PON} is large relative to the sedimentation velocity w, then $\xi \gg 1$ and the ammonia flux is equal in magnitude to the flux of organic nitrogen to the sediment.

This solution exhibits the same property as the one-volume solution. The ammonia flux is determined first and foremost by the flux of particulate organic nitrogen to the sediment J_{PON}. The remaining mechanisms determine what fraction of the organic matter flux delivered to the sediment escapes back to the overlying water as an ammonia flux. It is determined by the relative magnitudes of the process that returns ammonia to the overlying water (mineralization and diffusion) and the processes that bury nitrogen permanently in the sediment (sedimentation).

Table 2.2 Continuous Ammonia Model Parameters

Parameter	Symbol	Jan-Mar	April-June	July-Sept	Oct-Dec	Units
Ammonia concentration at depth	$\Delta[\text{NH}_4]$	31.3	43.7	49.8	57.5	mg N/L
Apparent rate of diagenesis	K_{PON}/w	0.047	0.024	0.0424	0.0527	cm^{-1}

As an example of its utility, this equation is fit to the quarterly averaged pore water data from Chesapeake Bay Station 854 from the Bricker data set (Bricker et al., 1977) that is tabulated in Part VII of the book. The result is compared in Fig. 2.10. For the

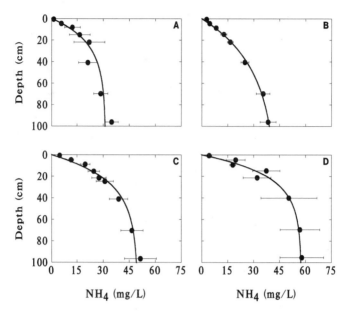

Fig. 2.10 Pore water ammonia. Comparison to Chesapeake Bay data. (A) January-March. (B) April-June. (C) July-September. (D) October-December.

sake of simplicity the overlying water concentration is assumed to be zero.

The two parameters: $\Delta[NH_4]$ (Eq. 2.94) and k_{PON}/w are listed in Table 2.2. There is an increase in the ammonia concentration at depth $\Delta[NH_4]$ whereas the reaction rate k_{PON}/w is somewhat variable. Neither of these patterns of variations seems particularly realistic, especially the variation of ammonia concentration at depth which one would expect to be constant. However, this application is presented only as an illustration. We leave the detailed development of a model for ammonia flux to subsequent chapters where much larger data sets are analyzed comprehensively.

Appendix 2A: Solution of Mass Balance Equations

2A.1   FIRST-ORDER EQUATION

Consider the differential equation

$$H\frac{dc}{dt} = -kHc + J \tag{2A.1}$$

It can be simplified to

$$\frac{dc}{dt} = -kc + R \tag{2A.2}$$

where $R = J/H$. A solution of the form

$$c(t) = A_0 + A_1 e^{\lambda t} \tag{2A.3}$$

is hypothesized where A_0, A_1, and λ are as yet undetermined. Its suitability can be determined if there are values for A_0, A_1, and λ that can make Eq. (2A.3) a solution for Eq. (2A.2). Thus substituting Eq. (2A.3) into Eq. (2A.2) yields

$$\frac{d(A_0 + A_1 e^{\lambda t})}{dt} = -k(A_0 + A_1 e^{\lambda t}) + R \tag{2A.4}$$

or

$$A_1 \lambda e^{\lambda t} = -k(A_0 + A_1 e^{\lambda t}) + R \tag{2A.5}$$

Collecting terms that multiply $e^{\lambda t}$ and the constant terms yields

$$A_1 e^{\lambda t}(\lambda + k) + A_0 k - R = 0 \tag{2A.6}$$

This equation can be satisfied if

$$\lambda = -k \tag{2A.7}$$

$$A_0 = \frac{R}{k} \tag{2A.8}$$

This leaves A_1 as an undetermined constant. It is specified by requiring that the initial condition at $t = 0$ is $c(0)$

$$c(0) = A_0 + A_1 \tag{2A.9}$$

or

$$A_1 = c(0) - A_0 \tag{2A.10}$$

Thus the solution is

$$c(t) = c(0)e^{-kt} + \frac{R}{k}(1 - e^{-kt}) \tag{2A.11}$$

2A.2 SECOND-ORDER EQUATION

The differential equation that specifies the vertical distribution of a constituent where both diffusion and sedimentation are important is

$$-D\frac{d^2c}{dz^2} + w\frac{dc}{dz} = -kc + R \tag{2A.12}$$

A solution of the form

$$c(t) = A_0 + A_1 e^{\lambda z} \tag{2A.13}$$

is hypothesized where A_0, A_1, and λ are as yet undetermined. The procedure is the same as with Eq. (2A.2). Substituting Eq. (2A.13) into Eq. (2A.12) yields

$$-D\frac{d^2(A_0 + A_1 e^{\lambda z})}{dz^2} + w\frac{d(A_0 + A_1 e^{\lambda z})}{dz} = -k(A_0 + A_1 e^{\lambda z}) + R \tag{2A.14}$$

Performing the indicated operations yields

$$-DA_1\lambda^2 e^{\lambda z} + wA_1\lambda e^{\lambda z} + k(A_0 + A_1 e^{\lambda z}) - R = 0 \tag{2A.15}$$

and collecting terms yields

$$A_1 e^{\lambda z}(-D\lambda^2 + w\lambda + k) + A_0 k - R = 0 \tag{2A.16}$$

This equation can be satisfied if

$$-D\lambda^2 + w\lambda + k = 0 \tag{2A.17}$$

and

$$A_0 = \frac{R}{k} \tag{2A.18}$$

Note that A_1 is still undetermined. The quadratic Eq. (2A.17) has two roots

$$\lambda_1 = \frac{w}{2D}\left(1 - \sqrt{1 + \frac{4Dk}{w^2}}\right) \tag{2A.19a}$$

$$\lambda_2 = \frac{w}{2D}\left(1 + \sqrt{1 + \frac{4Dk}{w^2}}\right) \tag{2A.19b}$$

which are negative λ_1 and positive λ_2. Thus the complete solution is

$$c(z) = A_1 e^{\lambda_1 z} + A_2 e^{\lambda_2 z} + \frac{R}{k} \tag{2A.20}$$

The boundary conditions determine A_1 and A_2.

For a semi-infinite model where $z \to \infty$, the condition that $c(\infty)$ is finite requires that the coefficient of the positive root λ_2 must be zero

$$A_2 = 0 \qquad (2A.21)$$

otherwise, the term $A_2 e^{\lambda_2 z} \to \infty$. Hence the solution is

$$c(z) = A_1 e^{\lambda_1 z} + \frac{R}{k} \qquad (2A.22)$$

The boundary condition at $z = 0$ determines A_1

$$c(0) = A_1 + \frac{R}{k} \qquad (2A.23)$$

and the solution is

$$c(z) = c(0)e^{-\lambda_1 z} + \frac{R}{k}\left(1 - e^{-\lambda_1 z}\right) \qquad (2A.24)$$

Other solutions come from applying different boundary conditions to Eq. (2A.20).

Appendix 2B: MACSYMA Solutions

The algebraic equations presented in this chapter can be solved using the computer program MACSYMA (Macsyma, 1993). It can perform a wide variety of symbolic mathematical computations involving algebra, trigonometry, calculus, and much more. Although it is not required for the computations presented below, it is essential for the more complicated problems analyzed in subsequent chapters. In fact, without computer algebraic assistance, many of the closed form solutions presented subsequently would not have been found.

MACSYMA is powerful and easy to use. My previous experience was with MATHEMATICA, which has many of the same features. However, it is an order of magnitude more complicated and unwieldy. For actual mathematical computations, there is really no comparison. I have had no experience with the other commonly available symbolic computation program, MAPLE.

I have found that the most efficient use of MACSYMA is as a very capable, and very accurate, mathematical assistant that carries out a plan of attack. It is much less successful if it is simply presented with a set of equations and told to "solve" them. However, nothing is lost by such a procedure and sometimes it produces useful results.

This appendix is not intended as a complete presentation of MACSYMA programming. Rather it is intended to present enough of the syntax so that the examples can be understood.

It is important to understand the distinctions among the assignment operator, the colon, and the equal sign. The colon assigns the variable or the result of a computation on the right-hand side (rhs) of the colon to the variable on the left-hand side

(lhs)

$$x : y + 1 \tag{2B.1}$$

which assigns $y + 1$ to x. Note the use of typewriter font to indicate MACSYMA code. The equal sign is a mathematical statement of equality in an equation

$$eq : x = y + 1 \tag{2B.2}$$

The variable eq is assigned the equation $x = y + 1$ which requires that x equals $y + 1$. Since MACSYMA is a symbolic computation program, it must distinguish between these two relationships.

The first example of a MACSYMA calculation is the solution of Eq. (2.63), the PON mass balance in the anaerobic layer. The code and solutions are presented in Fig. 2B.1. This figure is a direct reproduction of the actual MACSYMA output. The lines marked with "c" are the command input, the "d"s are MACSYMA's response. Both interactive responses and command batch files can be used.

The first line (c3)

$$eq1 : 0 = j[pon] - k[pon] * h[2] * pon[2] - w[2] * pon[2] \tag{2B.3}$$

assigns the mass balance equation to the variable eq1. The response (d3) is just a restatement of the equation. The next line (c4)

$$sol1 : solve([eq1], [pon[2]])[1][1] \tag{2B.4}$$

is the instruction to solve the equation eq1 for the unknown pon[2] and assign the solution to sol1. The trailing [1][1] select the first of the solutions in the set produced by solve. There is only one in this case.

Two things to note: MACSYMA reorders the equations – compare (c3) and (d3) in Fig. 2B.1. It does this to maintain a unique ordering of the terms in equations. And, it solved Eq. (2B.3) correctly, as can be seen by inspection.

The solution of the ammonia concentration in the anaerobic layer (Eq. 2.69), is also shown in Fig. 2B.1. The equation (c5) is

$$eq2 : 0 = k[pon] * h[2] * pon[2] \tag{2B.5}$$

$$+k[102] * (nh4[0] - nh4[2]) - w[2] * nh4[2] \tag{2B.6}$$

and the instruction to solve (c6)

$$sol2 : solve([eq2], [nh4[2]])[1][1] \tag{2B.7}$$

is similar to Eq. (2B.4). The solution (d6) is a function of pon[2]. The substitution for pon[2] is made (c7) using the evaluation instruction ev

$$sol3 : ev(sol2, sol1) \tag{2B.8}$$

Mass Balance Equations for PON

(c3)   eq1: 0=j[pon]-k[pon]*h[2]*pon[2]-w[2]*pon[2]

(d3)
$$0 = -h_2 \, pon_2 \, k_{pon} + j_{pon} - pon_2 \, w_2$$

Solve for pon[2]

(c4)   sol1:solve([eq1],[pon[2]])[1][1]

(d4)
$$pon_2 = \frac{j_{pon}}{h_2 \, k_{pon} + w_2}$$

Mass Balance Equation for NH4[2]

(c5)   eq2: 0=k[pon]*h[2]*pon[2]+k[l02]*(nh4[0]-nh4[2])-w[2]*nh4[2]

(d5)
$$0 = h_2 \, pon_2 \, k_{pon} + \left(nh4_0 - nh4_2\right) k_{l02} - nh4_2 \, w_2$$

Solve for NH4[2]

(c6)   sol2:solve([eq2],[nh4[2]])[1][1]

(d6)
$$nh4_2 = \frac{h_2 \, pon_2 \, k_{pon} + nh4_0 \, k_{l02}}{k_{l02} + w_2}$$

Substitute for pon[2]

(c7)   sol3:ev(sol2,sol1)

(d7)
$$nh4_2 = \frac{\dfrac{h_2 \, j_{pon} \, k_{pon}}{h_2 \, k_{pon} + w_2} + nh4_0 \, k_{l02}}{k_{l02} + w_2}$$

For small w[2]

(c8)   sol4:ev(sol3,w[2]=0)

(d8)
$$nh4_2 = \frac{j_{pon} + nh4_0 \, k_{l02}}{k_{l02}}$$

Fig. 2B.1   MACSYMA solutions for PON (Eq. 2.63)and ammonia (Eq. 2.69) mass balance equations.

It causes the expression sol2, the ammonia solution (d6), to be evaluated using the equation sol1, the PON[2] solution (d4), as a substitution for pon[2]. The result (d7) has the appropriate substitution incorporated.

A simplified version, the result for small w_2, is computed using ev (c8)

$$\text{sol4} : \text{ev}(\text{sol3}, \text{w}[2] = 0) \tag{2B.9}$$

It applies the equation w[2]=0 to the solution sol3 and stores the result in sol4.

The computation of the ammonia flux is presented in Fig. 2B.2. The mass balance equation (Eq. 2.76) is (c9)

$$\text{sol5} : \text{j}[\text{nh4}] = \text{k}[\text{l02}] * (\text{rhs}(\text{sol3}) - \text{nh4}[0]) \tag{2B.10}$$

where the rhs operator selects the right-hand side of equation sol3, the ammonia concentration (d7). The response (d9) shows that the proper selection has been made. The simplification of the solution (c10)

Ammonia flux

(c9)   sol5:j[nh4]=k[l02]*(rhs(sol3)-nh4[0])

(d9)
$$j_{nh4} = k_{l02} \left(\frac{\dfrac{h_2 j_{pon} k_{pon}}{h_2 k_{pon} + w_2} + nh4_0 k_{l02}}{k_{l02} + w_2} - nh4_0 \right)$$

(c10)   sol6:distrib(facsum(sol5,nh4[0]))

(d10)
$$j_{nh4} = \frac{h_2 k_{l02} j_{pon} k_{pon}}{\left(k_{l02} + w_2\right)\left(h_2 k_{pon} + w_2\right)} - \frac{nh4_0 w_2 k_{l02}}{k_{l02} + w_2}$$

Taylor series in w[2]
  first term

(c11)   lhs(sol6)=distrib(facsum(taylor(first(rhs(sol6)),w[2],0,1)))

(d11)
$$j_{nh4} = -\frac{w_2 j_{pon}}{h_2 k_{pon}} - \frac{w_2 j_{pon}}{k_{l02}} + j_{pon}$$

  2nd term

(c12)   lhs(sol6)=distrib(facsum(taylor(second(rhs(sol6)),w[2],0,2)))

(d12)
$$j_{nh4} = \frac{nh4_0 w_2^2}{k_{l02}} - nh4_0 w_2$$

Fig. 2B.2   MACSYMA solution for ammonia flux (Eq. 2.76) and the Taylor series expansion.

$$\texttt{sol6 : distrib(facsum(sol5, nh4[0]))} \qquad \text{(2B.11)}$$

introduces two new operators: facsum(sol5,nh4[0]) expands the expression in sol5 with respect to nh4[0] and factors its coefficients. The command distrib(eq) distributes sums over products in eq to produce the two fractions (d10). MACSYMA has many algebraic simplification and factoring commands. To some extent, producing the desired result is a trial and error proposition. Note that nesting of commands is allowed which simplifies the coding. The result (d10) is the ammonia flux equation expressed with separated terms for the two sources, J_{PON}, and $NH_4[0]$.

The equations for small w_2 are derived using a Taylor series (Kreyszig, 1972) (c11)

$$\texttt{lhs(sol6)} = \texttt{distrib(facsum} \qquad \text{(2B.12)}$$

$$\texttt{(taylor(first(rhs(sol6)), w[2], 0, 1)))} \quad \text{(2B.13)}$$

The command is taylor(eq,x,origin,n) for a Taylor series of eq in the variable x about the origin up to the n power. The series is found for the first term of the rhs of sol6, the expanded ammonia flux solution (d10). facsum and distrib are used as before. The lhs(sol6) = command reproduces the left-hand side of sol6 (c10) to provide a complete equation for display purposes. The second term (c12) is handled in a similar fashion. The reason for the unusual ordering of the solutions (d11) and (d12) has been explained above.

Part II

Nutrients

The subsequent five chapters present the model formulations for computing fluxes of nutrients to and from the sediment. The models are all formulated as steady state models. The advantages are that analytical solutions are available that illuminate the influence of the parameters, without resorting to numerical sensitivity analysis. Also, if the models fail to reproduce observations, it may be that the phenomena can only be modeled time variably. This turns out to be the case for anaerobic phosphate fluxes (Chapter 6). Of course, it is eventually necessary to use numerical methods, but with the help of MACSYMA, analytical solutions can be obtained for many interesting cases.

The nutrient models in Part II were all developed for use as part of the Chesapeake Bay Eutrophication Model (Cerco and Cole, 1994). The impetus for their construction was the failure of a summer steady state water column model, which had no interactive sediment model, to respond to loading reductions in a sensible way – that is, to respond at all. The two-layer sediment flux model was the result (Di Toro and Fitzpatrick, 1993). Its use is as part of the coupled water column-sediment model. But it was first constructed, tested, and calibrated as a stand-alone model. This strategy prevented problems from either the sediment or water column models from interacting in confusing ways. The calibration exercise was difficult enough.

The success of this modeling effort is due in no small measure to the availability of the excellent experimental data sets used in the development and calibration. The scientists who developed the methods for reliably measuring sediment fluxes and applied these techniques in a systematic and comprehensive way are listed above in the Acknowledgments. It is important to realize just how critical it is to have *large* and *comprehensive* data sets. Small data sets, say less than 30 observations, cannot possibly cover all the interactions without confounding the causes. The Chesapeake Bay data set used below comprises over 200 triplicated measurements. The models are then compared to other large, and not so large, data sets in Chapter 15.

We begin with ammonia, followed by a model for nitrate. This is followed by an analysis of a general steady state model. The equations for a chemical that partitions between sediment particles and pore water are formulated and solved. The importance of the concentrations of chemical per unit weight of solids in the aerobic r_1 and anaerobic r_2 layers becomes apparent. The model is applied to phosphate in Chapter 6 and to silica in Chapter 7. Oxygen is considered subsequently in Part III.

3

_____

Ammonia

3.1 INTRODUCTION

Models for the concentration distribution of ammonia in pore water and for the flux of ammonia from sediments have been proposed by various workers.* The original models focused on the mechanisms that generated the pore water profile: the mineralization of organic nitrogen and the mixing and adsorption processes. Subsequent models focused on the processes that occur in the aerobic layer of the sediment: primarily the nitrification reaction, and the ammonia flux that results. The model presented below is an extension of these formulations. Nitrification is formulated using Monod kinetics and the Chesapeake Bay data set is used to calibrate the model.

3.2 MODEL COMPONENTS

The model schematization for ammonia is presented in Fig. 3.1. Ammonia is produced by diagenesis in the aerobic H_1 and anaerobic H_2 layers. The production in the aerobic layer is small relative to the anaerobic layer because of the relative depths of the layers. The aerobic layer is usually a few millimeters deep (Fig. 1.12) whereas the anaerobic layer is on the order of 10 to 20 cm (Figs. 1.10 and 1.11). Nevertheless, aerobic layer diagenesis is included in this initial formulation for the sake of

*See Berner (1971, 1980a), Billen (1978, 1982, 1988), Billen et al. (1989), Blackburn (1990), Di Toro et al. (1990b), Klapwijk and Snodgrass (1982), Klump and Martens (1989), Vanderborght and Billen (1975), Vanderborght et al. (1977).

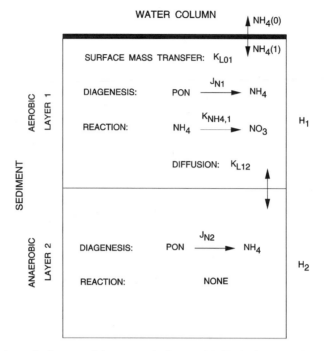

Fig. 3.1   Schematic diagram of the ammonia flux model. Typical values of the layer depths are $H_1 = 1$ mm and $H_2 = 10$ cm.

completeness. Diffusion transports ammonia from the anaerobic to the aerobic layer K_{L12} and to the overlying water K_{L01}.

If ammonia was a conservative substance, then the ammonia flux to the overlying water would be nearly equal to the diagenetically produced ammonia (see Eqs. 2.79 and the development in Chapter 2). However, ammonia can be nitrified to nitrate in the presence of oxygen. For the initial development, nitrification is formulated as a first-order reaction with respect to ammonia. Since the reaction can only occur where oxygen is present, it is restricted to the aerobic layer. Refinements to the nitrification kinetics – the use of Monod kinetics and the inclusion of the oxygen dependency of the nitrification rate – are subsequently included.

In the previous chapter the model was formulated by first considering the decay of PON and then subsequently using this as the source of ammonia in the mass balance model. For this chapter a slightly different strategy is followed. The mass balance equations are formulated with an explicit ammonia source term expressed as a flux J_N rather than the source term $+K_{PON}H_2 [PON (2)]$ used in the previous chapter (Eq. 2.68). The reason is that for the data sets that are available, there are no measurements of reactive PON from which $K_{PON}H_2 [PON (2)]$ could be estimated. In fact, measuring the anaerobic production of ammonia as part of a sediment flux data collection program is a very useful thing to do since it provides the source term for ammonia flux – see, for example, Aller et al. (1985).

3.3 MASS BALANCE EQUATIONS

The model is based on mass balance equations for the aerobic and anaerobic layers. Fig. 3.1 presents the schematization. The mass balance equations for the two layers are

$$H_1 \frac{d[NH_4(1)]}{dt} = -k_{NH_4,1}[NH_4(1)]H_1 - K_{L01}([NH_4(1)] - [NH_4(0)])$$

$$+K_{L12}([NH_4(2)] - [NH_4(1)]) + J_{N1} \qquad (3.1a)$$

$$H_2 \frac{d[NH_4(2)]}{dt} = -K_{L12}([NH_4(2)] - [NH_4(1)]) + J_{N2} \qquad (3.1b)$$

where

H_1 and H_2 are the depths of the aerobic (1) and anaerobic (2) layers

$[NH_4(0)]$, $[NH_4(1)]$, and $[NH_4(2)]$ are the ammonia concentrations in the overlying water (0) and layers (1) and (2)

$k_{NH_4,1}$ is the nitrification rate constant in the aerobic layer

K_{L01} is the mass transfer coefficient between the overlying water and the aerobic layer, which will be referred to as the surface mass transfer coefficient

K_{L12} is the mass transfer coefficient between the aerobic and anaerobic layers

J_{N1} and J_{N2} are the sources of ammonia in the two layers which result from the diagenesis of particulate organic nitrogen PON

This two-layer formulation employs mass transfer coefficients to parameterize the rate at which mass is transferred between the overlying water and the aerobic layer

$$K_{L01}([NH_4(1)] - [NH_4(0)]) \qquad (3.2)$$

and between the aerobic and anaerobic layers

$$K_{L12}([NH_4(2)] - [NH_4(1)]) \qquad (3.3)$$

The dimensions of K_{L01} and K_{L12} are length per unit time as discussed in Chapter 2.

3.3 A Solution

The solution of the mass balance equations is elementary for the steady state case for which the derivatives are zero. Adding the steady state equations yields

$$0 = -k_{NH_4,1}[NH_4(1)]H_1 - K_{L01}([NH_4(1)] - [NH_4(0)]) + J_{N1} + J_{N2} \qquad (3.4)$$

which can be solved for the aerobic layer ammonia concentration

$$[NH_4(1)] = \frac{J_N + K_{L01}[NH_4(0)]}{K_{L01} + k_{NH_4,1}H_1} \tag{3.5}$$

where $J_N = J_{N1} + J_{N2}$, the total ammonia diagenesis flux. The anaerobic layer concentration follows from Eq. (3.1b)

$$[NH_4(2)] = \frac{J_N}{K_{L12}} + [NH_4(1)] \tag{3.6}$$

The flux of ammonia from the sediment to the overlying water is

$$J[NH_4] = K_{L01}\left([NH_4(1)] - [NH_4(0)]\right) \tag{3.7}$$

Using Eq. (3.5) for $[NH_4(1)]$ yields

$$J[NH_4] = K_{L01}\frac{J_N - k_{NH_4,1}H_1[NH_4(0)]}{K_{L01} + k_{NH_4,1}H_1} \tag{3.8}$$

This solution can be written in two parts that separate the sources of ammonia

$$J[NH_4] = J_N\frac{K_{L01}}{K_{L01} + k_{NH_4,1}H_1} \tag{3.9a}$$

$$-[NH_4(0)]\left(\frac{1}{K_{L01}} + \frac{1}{k_{NH_4,1}H_1}\right)^{-1} \tag{3.9b}$$

The first term (Eq. 3.9a) quantifies the fraction of diagenetically produced ammonia J_N which escapes as an ammonia flux. If the surface mass transfer coefficient K_{L01} is large relative to the nitrification rate-aerobic depth product $k_{NH_4,1}H_1$, then all the ammonia produced escapes to the overlying water. Conversely, a large $k_{NH_4,1}H_1$ reduces the ammonia flux, since ammonia in the aerobic layer is being nitrified to nitrate faster than it can be transported to the overlying water.

The second term (Eq. 3.9b) determines the extent to which overlying ammonia is nitrified in the sediment. The form of the coefficient multiplying $[NH_4(0)]$, a reciprocal of the reciprocal sum of parameters $(1/K_{L01} + 1/k_{NH_4,1}H_1)^{-1}$, is analogous to electrical resistors in parallel.* The *smaller* of the two parameters determines the extent of nitrification. The reason is that the reciprocal of the smaller number is the larger number, and it dominates the value of the sum. For example, if the surface mass transfer coefficient K_{L01} is the larger parameter, then the nitrification rate-aerobic depth product $k_{NH_4,1}H_1$ controls the extent of nitrification. Intuitively this is

---

*This analogy is often incorrectly referred to as resistors in series. The resistance of resistors in series is the sum of the individual resistances. It is resistors in parallel for which the formula is: $\frac{1}{R_T} = \frac{1}{R_1} + \frac{1}{R_2} + \cdots + \frac{1}{R_n}$. The reason for the misstatement is that for mass transfer problems it is mass transfer resistances in series that give rise to the sum of reciprocals formula (Section 11.3 .

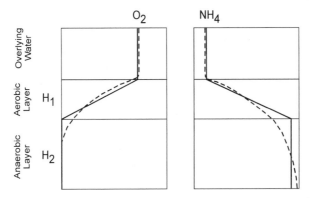

Fig. 3.2 Schematic diagram of the idealized vertical profiles of oxygen and ammonia. The dashed line are the continuous concentration profiles. The solid lines are the model idealizations.

a reasonable result. The extent to which overlying water ammonia is nitrified is controlled by which of the two necessary processes is slower: either the mass transfer from the overlying water to the aerobic layer, or the rate of nitrification. The faster process does not limit the rate of the overall reaction.

Two parameters, $k_{NH_4,1} H_1$ and K_{L01}, are required to quantify the ammonia flux. A method for estimating the latter parameter is discussed next.

3.3 B Surface Mass Transfer Coefficient

The critical observation that allows us to proceed with the analysis is that the surface water mass transfer coefficient K_{L01} can be related to the sediment oxygen demand, SOD (Di Toro et al., 1990b). The idea is that both the SOD and the ammonia flux can be expressed using the same mass transfer expressions. And, since the SOD is known, it can be used to solve for K_{L01}.

The situation is diagrammed in Fig. 3.2. We assume that the profiles of ammonia and dissolved oxygen in the aerobic layer can be approximated by straight lines. The derivation proceeds from an application of Fick's Law Eq. (2.47)

$$J[NH_4] = -D_{NH_4} \left. \frac{d[NH_4(z)]}{dz} \right|_{z=0} \tag{3.10a}$$

$$\simeq -D_{NH_4} \frac{[NH_4(0)] - [NH_4(H_1)]}{H_1} = \tag{3.10b}$$

$$-K_{L01,NH_4} \left([NH_4(0)] - [NH_4(H_1)] \right) \tag{3.10c}$$

where

$$K_{L01,NH_4} = \frac{D_{NH_4}}{H_1} \tag{3.10d}$$

Because the profile is assumed to be linear, the derivative can be approximated using the concentrations at the sediment-water interface $[NH_4(0)]$ and at the base of the aerobic layer $[NH_4(H_1)]$.

A similar argument can be made for the SOD, which is the mass flux of dissolved oxygen into the sediment. It can also be calculated from Fick's law

$$SOD = D_{O_2} \left. \frac{d[O_2(z)]}{dz} \right|_{z=0} \tag{3.11a}$$

$$\simeq D_{O_2} \frac{[O_2(0)] - [O_2(H_1)]}{H_1} \tag{3.11b}$$

$$= K_{L01,O_2} \left([O_2(0)] - [O_2(H_1)] \right) \tag{3.11c}$$

where $[O_2(z)]$ is the concentration profile of dissolved oxygen as a function of depth z, D_{O_2} is the diffusion coefficient in the aerobic layer, and

$$K_{L01,O_2} = \frac{D_{O_2}}{H_1} \tag{3.11d}$$

The minus sign is not included because SOD is defined to be positive if the flux is into the sediment. Therefore, solving Eq. (3.11c) for K_{L01,O_2} yields

$$K_{L01,O_2} = \frac{SOD}{[O_2(0)]} \tag{3.12}$$

the mass transfer coefficient for dissolved oxygen. Note that the only difference between K_{L01,O_2} (Eq. 3.10d) and K_{L01,NH_4} (Eq. 3.11d) is the respective diffusion coefficients.

We will use one surface mass transfer coefficient K_{L01} for both ammonia and oxygen for a number of reasons. First, the difference in their respective diffusion coefficients is small (Table 2.1). Second, the errors introduced by the finite difference approximations (Eqs. 3.10b and 3.11b) are also substantial. An analysis of the finite difference errors using the continuous solutions of the mass balance equations – the dashed lines in Fig. 3.2 – is given in Appendix 3B. But the most important reason is that these differences are subsumed into the kinetic parameter that is fit to the data. This is demonstrated below in Section 3.3 D. Therefore, no real error is introduced, and the model's ability to describe ammonia fluxes does not suffer as a consequence. Thus for layer 1

$$K_{L01} = \frac{D_1}{H_1} \tag{3.13}$$

where D_1 is the diffusion coefficient in layer 1. Finally, using Eq. (3.12) the surface mass transfer coefficient can be obtained from

$$K_{L01} = \frac{SOD}{[O_2(0)]} \triangleq s \tag{3.14}$$

which is the ratio of SOD and overlying water dissolved oxygen concentration $[O_2(0)]$. For notational simplicity this ratio is termed s (Eq. 3.14).

This result: $K_{L01} = s$ is of central importance. If an ammonia flux measurement is accompanied by an oxygen flux measurement and the corresponding overlying water oxygen concentration, then the surface mass transfer coefficient has essentially been measured as well via Eq. (3.14). With s known, it is possible to estimate the other model parameters.

3.3 C   Depth of the Aerobic Zone, Reaction Velocities, and the Solution

The remaining term in the equation for ammonia flux, Eq. (3.9), is the product of the reaction rate and the depth of the aerobic zone $k_{NH_4,1} H_1$. The depth of the aerobic zone H_1 can be estimated from Eq. (3.14)

$$H_1 = D_{O_2} \frac{[O_2(0)]}{SOD} = \frac{D_{O_2}}{s} \tag{3.15}$$

This relationship has been suggested on many occasions, apparently first by Grote (1934) – quoted by Hutchinson (1957), and verified by more recent measurements (Jorgensen and Revsbech, 1985, Revsbach et al., 1980). The utility of this equation is that it directly establishes the depth of the aerobic zone. Using this result, the reaction rate-depth product becomes

$$k_{NH_4,1} H_1 = \frac{D_{O_2} k_{NH_4,1}}{s} \tag{3.16}$$

The product $D_{O_2} k_{NH_4,1}$ is made up of two coefficients, neither of which is well known. The diffusion coefficient in a millimeter layer of sediment at the sediment-water interface may be larger than the diffusion coefficient in the bulk of the sediment due to the effects of overlying water shear. It is, therefore, convenient to define the parameter

$$\kappa_{NH_4,1} = \sqrt{D_{O_2} k_{NH_4,1}} \tag{3.17}$$

which can be termed a *reaction velocity* since its dimensions are length/time. The square root is used to conform to the analogous expression in the continuous form of the solution (see the next section).

The surface mass transfer coefficient and the reaction velocity can be substituted into Eq. (3.5) to obtain the ammonia concentration in the aerobic layer

$$[NH_4(1)] = \frac{s \left(J_N + s[NH_4(0)] \right)}{s^2 + \kappa_{NH_4,1}^2} \tag{3.18}$$

and into Eq. (3.9) for the ammonia flux

$$J[NH_4] = J_N \frac{s^2}{s^2 + \kappa_{NH_4,1}^2} - [NH_4(0)] \left(\frac{1}{s} + \frac{s}{\kappa_{NH_4,1}^2} \right)^{-1} \tag{3.19}$$

3.3 D Continuous Two-Layer Model

The ammonia concentrations in the two-layer model analyzed above are assumed to be constants in the layers. In fact the ammonia concentration in pore water varies smoothly with respect to depth (Fig. 2.10). A more realistic model would explicitly represent this variation. A two-layer continuous model is one such solution. The mass balance equations for interstitial water ammonia concentrations in the aerobic $n_1(z)$ and anaerobic $n_2(z)$ sediment layers are

$$-D_{NH_4}\frac{d^2 n_1(z)}{dz^2} = -K_{NH_4,1}n_1(z) + S_N \qquad 0 \leqslant z \leqslant H_1 \qquad (3.20a)$$

$$-D_{NH_4}\frac{d^2 n_2(z)}{dz^2} = S_N \qquad\qquad\qquad H_1 \leqslant z \leqslant H_2 \qquad (3.20b)$$

where $k_{NH_4,1}$ is the first-order oxidation rate of ammonia and $S_N = J_N/H_2$ is the rate of ammonia production by diagenesis, which is restricted to the depth of the active layer H_2. The surface boundary condition is

$$n_1(0) = 0 \qquad (3.21a)$$

which corresponds to negligible overlying water ammonia concentration compared with the interstitial waters. No essential complication is introduced if a nonzero concentration is included. The continuity conditions for concentration and flux at the aerobic-anaerobic layer boundary, which follow from the requirements of mass balance at the interface, yield

$$n_1(H_1) = n_2(H_1) \qquad (3.21b)$$

$$-D_{NH_4}\frac{dn_1(z)}{dz}\bigg|_{z=H_1} = -D_{NH_4}\frac{dn_2(z)}{dz}\bigg|_{z=H_1} \qquad (3.21c)$$

A zero ammonia flux condition at $z = H_2$ corresponds to no ammonia being produced below H_2, the depth of the active sediment layer

$$-D_{NH_4}\frac{dn_2(z)}{dz}\bigg|_{z=H_2} = 0 \qquad (3.21d)$$

These equations are solved for $n_1(z)$ and $n_2(z)$ in Appendix 3A. The ammonia flux to the overlying water

$$J[NH_4] = D_{NH_4}\frac{dn_1(z)}{dz}\bigg|_{z=0} \qquad (3.22)$$

is

$$J[NH_4] = S_N(H_2 - H_1)\text{sech}(\lambda_N H_1) + \frac{S_N}{\lambda_N}\tanh(\lambda_N H_1) \qquad (3.23)$$

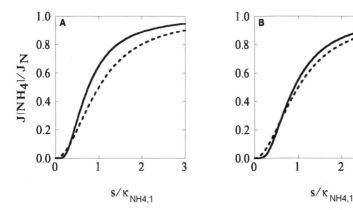

Fig. 3.3 Comparison of the continuous (solid line) and two-layer (dashed line) solutions. (A) Comparison using the same $\kappa_{NH_4,1}$. (B) Comparison using $1.2 \times \kappa_{NH_4,1}$ in the continuous solution and $1.0 \times \kappa_{NH_4,1}$ in the two-layer solution.

where

$$\lambda_N = \sqrt{\frac{k_{NH_4,1}}{D_{NH_4}}} \tag{3.24}$$

The hyperbolic functions are defined by (Abramowitz and Stegun, 1954)

$$\mathrm{sech}(x) = \frac{1}{\cosh(x)} = \frac{2}{e^x + e^{-x}} \tag{3.25}$$

$$\tanh(x) = \frac{e^x - e^{-x}}{e^x + e^{-x}} \tag{3.26}$$

An important simplification is possible because the depth of the aerobic layer H_1 is usually small (~ 1 mm) compared with the depth of the active sediment layer H_2 (~10 cm). For this case the contribution of S_N in Eq. (3.20b) can be ignored when solving for n_1. As shown in Appendix 3A, the result is

$$J[NH_4] = J_N \mathrm{sech}(\lambda_N H_1) \tag{3.27}$$

Using Eq. (3.15) for H_1 and Eq. (3.17) for $k_{NH_4,1}$ yields

$$J[NH_4] = J_N \mathrm{sech}\left(\frac{\kappa_{NH_4,1}}{s}\right) \tag{3.28}$$

A comparison of the continuous and two-layer solution is shown in Fig. 3.3A. A slightly larger $\kappa_{NH_4,1}$ in the continuous solution produces essentially the same result as the two-layer model, Fig. 3.3B. This difference is the inaccuracy introduced by the two-layer approximation that was examined in Appendix 3B.

3.3 E   Monod Kinetics

The nitrification reaction is known to follow Monod kinetics with respect to the ammonia concentration (Painter and Loveless, 1983). That is, the rate of nitrification $R_{NH_4,1}$ decreases as the ammonia concentration increases. The rate expression is

$$R_{NH_4,1} = k_{NH_4,1}[NH_4(1)]\frac{K_{M,NH_4}}{K_{M,NH_4} + [NH_4(1)]} \qquad (3.29)$$

where $k_{NH_4,1}$ is the first-order rate constant as before (Eqs. 3.1), and K_{M,NH_4} is the half saturation or Michaelis-Menton constant. For ammonia concentrations that are small relative to the half saturation concentration $[NH_4(1)] \ll K_{M,NH_4}$, the fraction in Eq. (3.29) approaches one, and the rate is first-order as before. However, as the ammonia concentration exceeds the half saturation concentration $[NH_4(1)] \gg K_{M,NH_4}$, then the rate approaches a constant $k_{NH_4,1} K_{M,NH_4}$, which is independent of the ammonia concentration $[NH_4(1)]$.

Although the first-order approximation is reasonable for small ammonia concentrations, the interstitial water ammonia concentrations can exceed the half saturation constant for ammonia oxidation, $K_{M,NH_4} \approx 1.0$ mg N/L. Therefore, it is necessary to use Monod kinetics to extend the applicable range.

In addition, the nitrification reaction rate decreases with decreasing oxygen concentrations. This can also be included using a Michaelis-Menton expression

$$R_{NH_4,1} = k_{NH_4,1}[NH_4(1)]\frac{K_{M,NH_4}}{K_{M,NH_4} + [NH_4(1)]}\frac{[O_2(1)]}{K_{O_2,NH_4} + [O_2(1)]} \qquad (3.30)$$

where K_{O_2,NH_4} is the half saturation constant for oxygen. As the oxygen concentration in the aerobic layer $[O_2(1)]$ decreases, the rate decreases to zero at zero oxygen concentration. Table 3.1 presents a summary of the information available for these parameters and their temperature coefficients.

The nitrification rate constants are not included in the table since the more modern formulations include bacterial biomass as part of the rate expression, whereas a first-order rate constant is employed above. However, the temperature coefficient is still applicable. It is applied to the square of the reaction velocity, since the square of the defining equation (3.17) is linear in the reaction rate constant $k_{NH_4,1}$

$$\kappa_{NH_4,1}^2 = D_{O_2}k_{NH_4,1}\theta_{NH_4}^{(T-20)} \qquad (3.31)$$

Solving the continuous two-layer model equations for these kinetics is very difficult if not impossible. Therefore, the discrete two-layer model is employed. The

Table 3.1 Ammonia Nitrification Parameters

Nitrification Temperature Coefficient θ_{NH_4}	NH$_4$ Half Saturation Constant (mg N/L) K_{M,NH_4}	Temperature Coefficient $\theta_{K_{M,NH_4}}$	O$_2$ Half Saturation Constant (mg O$_2$/L) K_{O_2,NH_4}	Reference
1.123	–	–	–	(a)
1.125	0.728	–	–	(b)
–	0.630	–	–	(c)
–	0.700	–	–	(d)
–	1.0	–	0.32	(e)
1.076	–	–	–	(f)
–	0.329	–	–	(g)
–	–	0.3	0.25–2.0	(h)
1.127	0.730	1.125	–	(i)
1.081	–	–	–	(j)
1.123	0.728	1.125	–	(k)
1.088	1.19	–	–	(l)
1.123	0.728	1.125	0.370	Median

[a]Antoniou et al. (1990), [b]Argaman and Miller (1979), [c]Cooke and White (1988), [d]Gee et al. (1990), [e]Hauaki (1990), [f]Painter and Loveless (1983), [g]Shuh (1979), [h] Stenstrom and Poduska (1980), [i]Stevens et al. (1989), [j] Warwick (1986), [k]Young and Thompson (1979), [l]Henriksen and Kemp (1988)

aerobic layer mass balance equation (3.1a) becomes

$$H_1 \frac{d[NH_4(1)]}{dt} = - \left(\frac{K_{M,NH_4} \theta_{K_{M,NH_4}}^{(T-20)}}{K_{M,NH_4} \theta_{K_{M,NH_4}}^{(T-20)} + [NH_4(1)]} \right) \left(\frac{[O_2(1)]}{K_{O_2,NH_4} + [O_2(1)]} \right)$$

$$\left(\frac{\kappa_{NH_4,1}^2 \theta_{NH_4}^{(T-20)}}{s} [NH_4(1)] - s \left([NH_4(1)] - [NH_4(0)] \right) \right)$$

$$+ K_{L12} \left([NH_4(2)] - [NH_4(1)] \right) + J_{N1} \qquad (3.32)$$

The ammonia concentration dependency has been formulated so that the reaction velocity, $\kappa_{NH_4,1}$, has the same meaning as in Eq. (3.17). That is, for $[NH_4(1)] \ll K_{M,NH_4}$ and $[O_2(1)] \gg 2K_{O_2,NH_4}$ this equation reduces to Eq. (3.1a).

The oxygen dependency is expressed in terms of the aerobic layer oxygen concentration $[O_2(1)]$. Since the oxygen profile is assumed to be linear in the aerobic layer, starting at $[O_2(0)]$ at the sediment-water interface, and ending at zero at the aerobic-anaerobic boundary H_1, the average aerobic layer oxygen concentration is

$$[O_2(1)] \simeq \frac{[O_2(0)] + [O_2(H_1)]}{2} = \frac{[O_2(0)]}{2} \qquad (3.33)$$

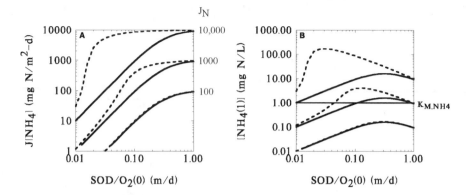

Fig. 3.4 Predicted (A) ammonia fluxes and (B) aerobic layer ammonia concentrations for the first-order (solid line) and Monod (dashed line) kinetics models. The pairs of curves represent increasing ammonia diagenesis (J_N = 100, 1000, 10,000 mg N/m^2-d)

This substitution can be used in the Michaelis-Menton expression

$$\frac{[O_2(1)]}{[O_2(1)] + K_{O_2,NH_4}} = \frac{\frac{1}{2}[O_2(0)]}{\frac{1}{2}[O_2(0)] + K_{O_2,NH_4}} = \frac{[O_2(0)]}{[O_2(0)] + 2K_{O_2,NH_4}} \quad (3.34)$$

The solution is obtained by assuming steady state and adding this equation to the layer 2 mass balance equation (Eq. 3.1b)

$$0 = -\left(\frac{K_{M,NH_4}\theta_{K_{M,NH_4}}^{(T-20)}}{K_{M,NH_4}\theta_{K_{M,NH_4}}^{(T-20)} + [NH_4(1)]}\right)\left(\frac{[O_2(0)]}{2K_{O_2,NH_4} + [O_2(0)]}\right)$$

$$\frac{\kappa_{NH_4,1}^2\theta_{NH_4}^{(T-20)}}{s}[NH_4(1)] - s([NH_4(1)] - [NH_4(0)]) + J_N \quad (3.35)$$

which is a quadratic equation in $[NH_4(1)]$ and can easily be solved, as shown in Section 3.4 B.

The predicted ammonia fluxes and aerobic layer ammonia concentrations for the first-order and Monod kinetics models are compared in Fig. 3.4. The pairs of curves represent increasing ammonia diagenesis (J_N = 100, 1000, 10,000 mg N/m^2-d). When the diagenesis flux is small, there is no difference between the two solutions because the aerobic layer ammonia concentrations are well below the half saturation constant K_{M,NH_4} (Fig. 3.4B). However, for larger diagenesis fluxes, the difference increases because the aerobic layer ammonia concentration starts to exceed the half saturation constant. This causes the rate of nitrification to decrease relative to the first-order kinetic formulation (Eq. 3.29). As a consequence, less ammonia is nitrified, the aerobic layer ammonia concentration increases (Fig. 3.4B), and more escapes to the overlying water (Fig. 3.4A).

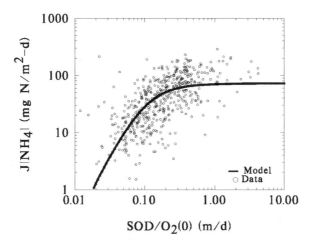

Fig. 3.5 Ammonia flux versus $s = \text{SOD}/[\text{O}_2(0)]$ for all stations and times in the Chesapeake Bay data set.

3.4 DATA ANALYSIS

Two approaches are employed to estimate the remaining parameters in the ammonia flux model. The first is a graphical analysis that provides an average estimate of the reaction velocity. The second is based on regression analysis which provides more detailed results.

3.4 A Graphical Analysis

The ammonia flux, Eq. (3.19), is determined by the two sources of ammonia: diagenesis J_N and overlying water ammonia $s[\text{NH}_4(0)]$. If the latter is a small contribution, then only the diagenesis term is significant and

$$J[\text{NH}_4] = J_N \frac{s^2}{s^2 + \kappa_{\text{NH}_4,1}^2} \tag{3.36}$$

The model predicts that $J[\text{NH}_4]$ should vary as s^2 for small s. For large s, the ammonia flux equals the ammonia diagenesis flux J_N. Fig. 3.5 is a plot of ammonia flux versus $s = \text{SOD}/[\text{O}_2(0)]$ for all stations and times in the Chesapeake Bay data sets. The triplicates are plotted separately. The line is a least squares fit of Eq. (3.36) to the data.

The data appear to roughly conform to the expected relationship: smaller ammonia fluxes are associated with smaller s. However, there is substantial scatter about the fitted line. This is not unexpected, since this comparison assumes that J_N is the same for every station at every sampling time. Because this is clearly not the case,

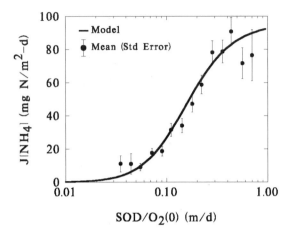

Fig. 3.6 Comparison of model calculation to data grouped into 0.1 $\log_{10}$ intervals of s.

one would expect considerable scatter in a pointwise comparison using data from different locations in the bay and from different seasons of the year.

In order to compensate for this variation, some data averaging is appropriate. The following has been found to be useful: The data are averaged within intervals of the independent variable, in this case s. Fig. 3.6 compares the model calculation to the data that have been grouped into 0.1 $\log_{10}$ intervals of s. The average and the standard error of the mean for $J[NH_4]$ are shown for intervals which contain more than five data points. The fit is quite remarkable. The estimated parameter values are listed in Table 3.2 below. The relationship to s^2 is clear, as is the flattening out of the profile at larger s. This graphical analysis should be viewed as only a first step. A more rigorous approach is to use regression methods to estimate the parameters of the model.

3.4 B Nonlinear Regression

The graphical analysis presented above assumes that the ammonia diagenesis flux J_N is a constant in time and space. This assumption can be removed by letting J_N be a function of space and time. The spatial variation can be accommodated by defining station specific diagenesis fluxes $J_N(i)$ where i indexes the stations. The temporal variation can be included by relating ammonia diagenesis to the temperature $T_{i,j}$, at location i and time t_j via the exponential approximation to the Arrhenius relationship. The result is that the diagenesis flux $J_N(i, t_j)$ is parameterized as

$$J_N(i, t_j) = J_N(i)\theta_N^{(T_{i,j}-20)} \tag{3.37}$$

The unknown parameters are the station-specific diagenesis fluxes: $J_N(i)$, the temperature coefficient for diagenesis θ_N and the nitrification reaction velocity $\kappa_{NH_4,1}$.

The median of the reported values in Table 3.1 is used for the nitrification temperature coefficient. The equation for ammonia flux that would be used in the regression analysis if linear nitrification kinetics are employed is

$$J[NH_4(i,t_j)] = J_N(i)\theta_N^{(T_{i,j}-20)} \frac{s_{i,j}^2}{s_{i,j}^2 + \kappa_{NH_4,1}^2 \theta_{NH_4}^{(T_{i,j}-20)}}$$

$$-[NH_4(0)]_{i,j} \left(\frac{1}{s_{i,j}} + \frac{s_{i,j}}{\kappa_{NH_4,1}^2 \theta_{NH_4}^{(T_{i,j}-20)}} \right)^{-1} \qquad (3.38)$$

where the subscripts i, j indicate that the temperature $T_{i,j}$ the surface mass transfer coefficient $s_{i,j} = SOD_{i,j}/[O_2(0)]_{i,j}$ and the overlying water ammonia concentrations $[NH_4(0)]_{i,j}$ are for station i at time t_j.

The regression equation using Monod kinetics is computed as follows. The aerobic layer mass balance equation (3.32) for temporal steady state is

$$0 = -\left(\frac{K_{M,NH_4}\theta_{NH_4}^{(T_{i,j}-20)}}{K_{M,NH_4}\theta_{NH_4}^{(T_{i,j}-20)} + [NH_4(1)]_{i,j}} \right) \left(\frac{[O_2(0)]}{2K_{O_2,NH_4} + [O_2(0)]} \right)$$

$$\left(\frac{\kappa_{NH_4,1}^2 \theta_{NH_4}^{(T_{i,j}-20)}}{s_{i,j}} [NH_4(1)]_{i,j} - s_{i,j} ([NH_4(1)]_{i,j} - [NH_4(0)_{i,j}]) \right)$$

$$+ J_N(i)\theta_N^{(T_{i,j}-20)} \qquad (3.39)$$

which is a quadratic equation in the unknown concentration $[NH_4(1)]_{i,j}$. The solution is

$$[NH_4(1)]_{i,j} = -\frac{b}{2a} \left(1 \pm \sqrt{1 - \frac{4ac}{b^2}} \right) \qquad (3.40)$$

where

$$a = -s_{i,j}^2 \qquad (3.41a)$$

$$b = s_{i,j} J_N(i)\theta_N^{(T_{i,j}-20)} - s_{i,j}^2 \left(K_{M,NH_4}\theta_{K_{M,NH_4}}^{(T_{i,j}-20)} - [NH_4(0)]_{i,j} \right) \qquad (3.41b)$$

$$-\kappa_{NH_4,1}^2 \theta_{NH_4}^{(T_{i,j}-20)} K_{M,NH_4}\theta_{K_{M,NH_4}}^{(T_{i,j}-20)} \frac{[O_2(0)]_{i,j}}{2K_{O_2,NH_4} + [O_2(0)]_{i,j}}$$

$$c = s_{i,j} K_{M,NH_4}\theta_{K_{M,NH_4}}^{(T_{i,j}-20)} \left(J_N(i)\theta_N^{(T_{i,j}-20)} + s_{i,j}[NH_4(0)]_{i,j} \right) \qquad (3.41c)$$

The sign of the root in Eq. (3.40) is chosen so that $[NH_4(1)]_{i,j}$ is positive. The ammonia flux is computed using

$$J[NH_4(i,t_j)] = s_{i,j} ([NH_4(1)]_{i,j} - [NH_4(0)]_{i,j}) \qquad (3.42)$$

The data used in the regression analysis are restricted to the ten Chesapeake Bay stations for the years 1985 through 1988. The regression is performed using Monod kinetics, Eqs. (3.40 to 3.42). Table 3.1 lists the reported values for nitrification kinetic coefficients. The median values are used in the regression. The data are analyzed in two ways: replicate flux measurements are treated as individual measurements, and the average of the replicates are used.

The initial regression results indicated that it is not possible to estimate both $J_N(i)$ and $\kappa_{NH_4,1}$ simultaneously. The results are too unstable to be reliable. The cause of the problem can be understood using the simplest version of the ammonia flux model, Eq. (3.36). Consider what occurs when the surface mass transfer coefficient is much less than the nitrification reaction velocity $s^2 \ll \kappa^2_{NH_4,1}$. In this case

$$J[NH_4] = J_N \frac{s^2}{s^2 + \kappa^2_{NH_4,1}} \approx \frac{J_N}{\kappa^2_{NH_4,1}} s^2 \qquad (3.43)$$

and the two parameters to be estimated, J_N and $\kappa_{NH_4,1}$, are indistinguishable in the quotient. A larger J_N can be compensated for with a larger $\kappa_{NH_4,1}$. Therefore, the ability to make independent estimates depends on the existence of a significant fraction of data for which $s^2 \gg \kappa^2_{NH_4,1}$ so that J_N can be estimated independently. Since the regression is unstable, additional data must be added.

3.4 C   Estimates of J_N

The diagenesis of organic matter releases both organic carbon and ammonia to the sediment interstitial water. As shown in Chapter 9, the organic carbon is oxidized using sulfate as the electron acceptor in marine sediments. The sulfide that results is buried, or oxidized using oxygen as the electron acceptor, or escapes as a sulfide flux. If all the sulfide were oxidized, then the oxygen flux to the sediment would be related to the carbon diagenesis at that time. This information could be used to make an estimate of ammonia diagenesis which could provide the necessary additional information for the regression analysis.

However, there are a number of intermediate steps between carbon diagenesis and eventual oxidation. Therefore, it is not true that the oxygen flux to the sediment (SOD) at any instant in time is equal to the carbon diagenesis flux (in oxygen equivalents) at that time. Nevertheless, if most of the carbon diagenesis is eventually oxidized, then the long-term average SOD could be used to make a reasonable estimate of the long-term average ammonia diagenesis using suitable stoichiometric relationships. The relationship between the long-term average J_N and SOD is

$$\overline{J_N(i)} = a_{C,N} a_{O_2,C} \overline{SOD(i)} \qquad (3.44)$$

where $\overline{J_N(i)}$ is the estimate of the long-term average ammonia diagenesis flux for station i, and $\overline{SOD(i)}$ is the observed long-term average SOD at station i. The over-bars denote averages. The Redfield stoichiometry is: $a_{O_2,C} = 2.67$ g O_2/g C and $a_{C,N} = 5.68$ g C/g N (Table 1.1). As shown in Chapter 12, these ratios are consistent with the stoichiometry of decaying sediment organic matter in Chesapeake Bay.

Table 3.2 Ammonia Model Parameters

Parameter	Units	Estimation Method				
		(a)	(b)	(c)	(d)	(e)
J_N	mg N/m^2–d	92.2	–	–	–	–
$\kappa_{NH_4,1}$	m/d	0.166	0.0722	0.116	0.151	0.148
θ_N	–	–	–	1.112	1.142	1.141	1.153

Parameter J_N	Average*	Estimation Method				
Station		(a)	(b)	(c)	(d)	(e)
Point No Pt.[#]	56.9	–	39.3	43.1	62.6	54.6
R-64[#]	44.5	–	79.2	90.6	95.8	93.8
R-78[#]	40.8	–	41.6	38.5	50.7	49.6
Still Pond	72.4	–	53.0	63.7	76.6	72.4
St. Leo	92.7	–	49.6	60.3	93.4	92.7
Buena Vista	101.6	–	67.5	73.6	105.6	101.6
Horn Pt.	88.3	–	62.5	71.1	90.1	88.3
Windy Hill[#]	118.4	–	39.6	44.0	56.1	56.5
Ragged Pt.[#]	72.7	–	109.6	98.7	109.4	100.0
Maryland Pt.	73.7	–	66.0	78.8	71.6	73.9
Mean				66.2		

*Four-year average computed from arithmetic average SOD and Redfield stoichiometry. The average temperatures for the data are very nearly 20 °C. [#]Stations with significant anoxic periods. These are not used in the regression. [a]Nonlinear regression analysis, Fig.3.6. [b]Individual replicates, least squares. [c]Averaged replicates, least squares. [d]Individual replicates, least absolute value. [e]Averaged replicates, least absolute value.

The relationship between SOD and ammonia diagenesis, Eq. (3.44), only applies for stations where no significant sulfide flux occurs. These are stations where the overlying water DO concentration does not approach zero. For the remaining stations with significant periods of anoxia, a significant fraction of the oxygen equivalents escapes as a sulfide flux, so using the long-term average SOD underestimates the diagenesis flux. For this reason, this relationship is used only for those stations for which the minimum DO is always greater than 1 mg O_2/L.

The details of the regression analysis are presented in Appendix 3C. The results are listed in Table 3.2. The nitrification reaction velocity is estimated to be in the range of $\kappa_{NH_4,1} = 0.072$ to 0.151 (m/d) depending on the details of the fitting criteria used in the regression. This is reasonably stable behavior. The estimates of ammonia diagenesis for each station are reasonably close to the estimates derived from the average SODs for those stations without significant anoxia, if least squares is used, or are essentially equal to them, if the absolute value criterion is used. Note that if ammonia diagenesis is estimated to be smaller than the SOD derived estimates (method b), then the reaction velocity is also estimated to be smaller, consistent with Eq. (3.43). The results for the least squares criterion (method c) are illustrated in

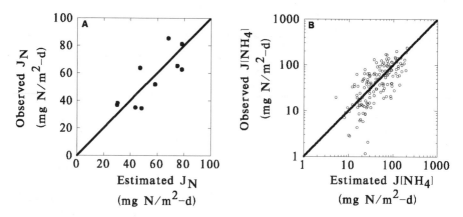

Fig. 3.7 (A) Comparison of station average diagenesis $\overline{J_N(i)}$ to model estimates $\overline{J_N(i)}^{est}$. (B) Comparison of ammonia fluxes (average of the replicates) $J[NH_4]_{i,j}^{obs}$ to model estimates $J[NH_4]_{i,j}^{model}$.

Fig. 3.7. Both the observed individual fluxes (average of the replicates) $J[NH_4]_{i,j}^{obs}$ and the station averages $\overline{J_N(i)}^{est}$ estimated from the SOD via Eq. (3.44), are compared to the model estimates $J[NH_4]_{i,j}^{model}$ and $J_N(i)$, for station i and time t_j. There is a significant scatter if the individual fluxes are compared. However, the model can reproduce the station average diagenesis fluxes reasonably well. This is not too surprising since these are part of the regression parameters. Nevertheless, their estimates are constrained by the long-term average SOD estimates for the oxic stations.

The final parameter values to be used subsequently are those estimated using the least squares criterion and the averaged data set (method c). This criterion corresponds to the maximum likelihood estimate for a lognormal distribution of the errors (Aitchison and Brown, 1957), and the replicate averages stabilize the estimates of $s_{i,j}$ which are used in the regression. This appears to be the optimal estimation procedure.

3.4 D Extent of Nitrification

The model behavior is examined in Fig. 3.8 which presents estimates of average ammonia diagenesis J_N, ammonia flux $J[NH_4]$, and by difference, the source of nitrate to the aerobic layer, $S[NO_3]$ which is the rate of nitrate production due to nitrification. The extent of nitrification at the main stem stations varies from almost none at station R-64 to almost 40% of J_N for Still Pond. This is controlled by the magnitude of the surface mass transfer coefficient and the depth of the aerobic zone, both of which are quantified using s.

The nitrate produced in the aerobic layer can either be transferred to the overlying water or be denitrified. This is examined in the next chapter.

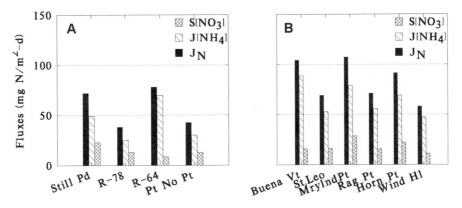

Fig. 3.8 Estimates of average ammonia diagenesis J_N ammonia flux $J[NH_4]$, and the source of nitrate to the aerobic layer $S[NO_3]$ which is the rate of nitrate production due to nitrification. (A) Main stem, (B) tributaries.

3.5 OBSERVATIONS OF CHESAPEAKE BAY NITRIFICATION

Direct measurements of the rate of nitrification in Chesapeake Bay sediments have been made during 1988 (Kemp et al., 1990, Sampou et al., 1989). These are compared to model predictions in two ways. For the stations where measurements over a season have been made (Still Pond and R-64), the station average nitrification flux is calculated and compared to the observations. The procedure is to evaluate the model, Eq. (3.40), to compute the aerobic layer ammonia concentrations $[NH_4(1)]_{i,j}$ using the observed surface mass transfer coefficients $s_{i,j}$ and temperatures $T_{i,j}$. The model parameters are the medians in Table 3.1 and the method (c) estimates in Table 3.2. The nitrification flux, denoted by $S[NO_3]$, is computed by evaluating the nitrification kinetic expression in the mass balance equation (3.39)

$$S[NO_3] = \left(\frac{K_{M,NH_4}\theta_{K_{M,NH_4}}^{(T_{i,j}-20)}}{K_{M,NH_4}\theta_{K_{M,NH_4}}^{(T_{i,j}-20)} + [NH_4(1)]_{i,j}} \right) \left(\frac{[O_2(0)]_{i,j}}{2K_{O_2,NH_4} + [O_2(0)]_{i,j}} \right)$$

$$\frac{\kappa_{NH_4,1}^2 \theta_{NH_4}^{(T_{i,j}-20)}}{s_{i,j}} [NH_4(1)]_{i,j} \tag{3.45}$$

The station averages are computed from the individual estimates. The comparison is made in Fig. 3.9. The results are in reasonable agreement considering the difficulty in measuring nitrification fluxes (Kemp et al., 1990, Rudolph et al., 1991).

An alternate method of computing the nitrification flux is to estimate the aerobic layer ammonia concentration using the observed ammonia flux, surface mass transfer coefficient, and overlying water ammonia concentration. This obviates the need for an estimate of the ammonia diagenesis flux which is required if the ammonia flux

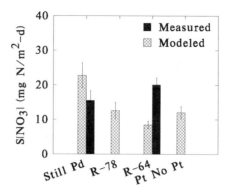

Fig. 3.9 Comparison of observed and modeled nitrification flux $S[NO_3]$ for the main stem Chesapeake bay stations.

model is used. Instead, the estimate is made from the flux equation

$$J[NH_4] = s\left([NH_4(1)] - [NH_4(0)]\right) \qquad (3.46)$$

so that

$$[NH_4(1)] = \frac{J[NH_4]}{s} + [NH_4(0)] \qquad (3.47)$$

With $[NH_4(1)]$ determined, the kinetic expression, Eq. (3.45) is used to compute the nitrification flux. Note that all the model nitrification parameters are used to compute $S[NO_3]$ so this is still a test of the model formulation. The results are compared to the observations in Fig. 3.10. There is considerable scatter in the model estimates, since they are based on observed ammonia fluxes. Nevertheless, the comparison to the observations is reasonable. In particular, the temporal variation in nitrification appears to be reproduced.

3.6 NONSTEADY STATE FEATURES

It has been pointed out by Boynton et al. (1990) that ammonia fluxes in Chesapeake Bay are not a single function of temperature. Rather they display hysteresis. The average monthly fluxes for the main stem stations and the model fluxes are plotted versus temperature in Fig. 3.11. Note the circular paths that are traversed by the data. The ammonia fluxes are generally higher in the spring months than in the fall months at the same temperature. This effect is not reproduced very well by the steady state ammonia flux model. As can be seen from the dashed lines representing the model computations, there is some hysteresis, but it is not as large as observed at most of the stations. A similar analysis using the time variable model, Chapter 14, indicates that ammonia flux hysteresis is a time variable effect that can be reproduced by the time variable model.

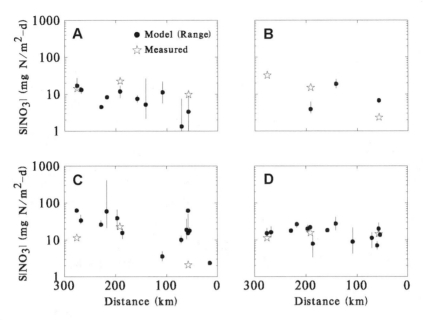

Fig. 3.10 Estimated nitrification flux using the observed ammonia flux, surface mass transfer coefficient, and overlying water ammonia concentration. (A) April, (B) June, (C) August, (D) November 1988.

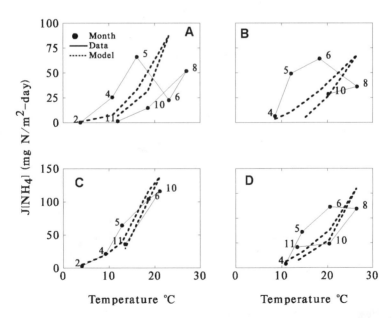

Fig. 3.11 Average monthly fluxes for main stem stations (-●-) and model fluxes (- -) versus temperature. Numbers refer to the month. (A) Still Pond, (B) R-78, (C) R-64, (D) Point No Point.

3.7 CONCLUSIONS

The steady state ammonia flux model can reproduce major features of the observed ammonia flux data. The variation with surface mass transfer coefficient s determines the extent to which nitrification takes place. A regression analysis is used to estimate the nitrification reaction velocity and the station-specific ammonia diagenesis fluxes. These are of critical importance for the analysis of the other fluxes, as will be clear in the subsequent chapters. A comparison to independently measured nitrification fluxes indicates that the model is consistent with these observations as well. However, the steady state model is not able to reproduce the hysteresis that is observed during the seasonal progression of ammonia fluxes. This limitation is directly related to the steady state assumption employed in solutions of the model equations in this chapter.

Appendix 3A: Solution of Ammonia Mass Balance Equations

The solutions for the ammonia mass balance equations, Eqs. (3.20b) and (3.20b), are

$$n_1(z) = B_{1N} \exp(\lambda_N z) + B_{2N} \exp(-\lambda_N z) + \frac{S_N}{k_{NH_4,1}} \tag{3A.1}$$

$$n_2(z) = A_{1N} + A_{2N} z - \frac{S_N z^2}{2 D_{NH_4}} \tag{3A.2}$$

Applying the boundary conditions, Eqs. (3.21a–3.21d) sequentially yields

$$B_{1N} + B_{2N} + \frac{S_N}{k_{NH_4,1}} = 0 \tag{3A.3}$$

$$B_{1N} \exp(\lambda_N H_1) + B_{2N} \exp(-\lambda_N H_1) + \frac{S_N}{k_{NH_4,1}} = A_{1N} + A_{2N} H_1 - \frac{S_N H_1^2}{2 D_{NH_4}} \tag{3A.4}$$

$$\lambda_N B_{1N} \exp(\lambda_N H_1) - \lambda_N B_{2N} \exp(-\lambda_N H_1) = A_{2N} - \frac{S_N H_1}{D_{NH_4}} \tag{3A.5}$$

$$A_{2N} - \frac{S_N H_2}{D_{NH_4}} = 0 \tag{3A.6}$$

Using Eqs. (3A.3–3A.5) to solve for B_{1N} yields

$$B_{1N} = \frac{S_N(H_2 - H_1)}{2\lambda_N D_{NH_4} \cosh(\lambda_N H_1)} - \frac{S_N \exp(-\lambda_N H_1)}{2 k_{NH_4,1} \cosh(\lambda_N H_1)} \tag{3A.7}$$

B_{2N} follows from Eq. (3A.3)

$$B_{2N} = \frac{-S_N(H_2 - H_1)}{2\lambda_N D_{NH_4} \cosh(\lambda_N H_1)} - \frac{S_N \exp(\lambda_N H_1)}{2k_{NH_4,1} \cosh(\lambda_N H_1)} \tag{3A.8}$$

The constants for the anaerobic layer solution follow from Eq. (3A.6)

$$A_{2N} = S_N \frac{H_2}{D_{NH_4}} \tag{3A.9}$$

and Eq. (3A.4)

$$A_{1N} = B_{1N} \exp(\lambda_N H_1) + B_{2N} \exp(-\lambda_N H_1) - A_{2N} H_1 + \frac{S_N H_1^2}{2D_{NH_4}} + \frac{S_N}{k_{NH_4,1}} \tag{3A.10}$$

The ammonia flux, Eq. (3C.2), follows from the relationship

$$J[NH_4] = D_{NH_4} \left. \frac{dn_1(z)}{dz} \right|_{z=0} = D_{NH_4} \lambda_N (B_{1N} - B_{2N}) \tag{3A.11}$$

where B_{1N} and B_{2N} are given by Eqs. (3A.7–3A.8).

3A.1 Simplification

The simplified solution comes from replacing Eq. (3A.1) with

$$n_1(z) = B_{1N} \exp(\lambda_N z) + B_{2N} \exp(-\lambda_N z) \tag{3A.12}$$

where the term due to the source in the aerobic layer: $S_N/k_{NH_4,1}$ is neglected. Now the first boundary condition, Eq. (3A.3), becomes

$$B_{1N} + B_{2N} = 0 \tag{3A.13}$$

Following the same derivation as above yields

$$B_{1N} = \frac{S_N(H_2 - H_1)}{2\lambda_N D_{NH_4} \cosh(\lambda_N H_1)} \tag{3A.14}$$

The ammonia flux, Eq. (3C.5), follows from the relationship

$$J[NH_4] = 2D_{NH_4} \lambda_N B_{1N} = \frac{S_N(H_2 - H_1)}{\cosh(\lambda_N H_1)} \approx \frac{S_N H_2}{\cosh(\lambda_N H_1)} \tag{3A.15}$$

Appendix 3B: Ammonia and Dissolved Oxygen Surface Mass Transfer Coefficients

For a two-layer model, the mass transfer coefficients are employed together with a difference approximation of the derivative to obtain the resulting fluxes (Eqs. 3.10

and 3.11). However if the continuous solutions were known for the ammonia and oxygen profiles, then a direct application of Fick's law could be made. We have a continuous solution for the ammonia profile for a two-layer model (Eq. 3A.12). And we analyze a similar situation for dissolved oxygen (Eq. 8.8)

$$O_2(z) = O_2(0) + \frac{R_0 z}{D_{O_2}} \left(\frac{z}{2} - H_1 \right) \tag{3B.1}$$

where R_0 is the rate of oxygen consumption in the aerobic layer. These equations can be used to obtain the equivalent mass transfer coefficients for ammonia and oxygen.

The procedure is as follows.* The flux from a two-layer model is

$$J[NH_4] = \frac{D_{NH_4}}{\Delta z} ([NH_4(1)]) \tag{3B.2}$$

where $NH_4(1)$ is the ammonia concentration in layer 1, $[NH_4(0)]$ is assumed to be zero, and $\Delta z = H_1$. The flux from the continuous model is

$$J[NH_4] = D_{NH_4} \left. \frac{dn_1(z)}{dz} \right|_{z=0} \tag{3B.3}$$

In order to compare the two flux expressions, we need the average concentration in layer 1 $\overline{n_1}$ from the continuous solution

$$\overline{n_1} = \frac{1}{H_1} \int_0^{H_1} n_1(z) dz \tag{3B.4}$$

Equating Eqs. (3B.2) and (3B.3) and solving for Δz yields

$$\Delta z = \frac{[NH_4(1)]}{\left. \dfrac{dn_1(z)}{dz} \right|_{z=0}} = \frac{\overline{n_1}}{\left. \dfrac{dn_1(z)}{dz} \right|_{z=0}} \tag{3B.5}$$

Fig. 3B.1 presents the initial portion of the MACSYMA solution for this problem, the evaluation of the arbitrary coefficient B_1 and B_2. n1 and n2 are the solutions to the differential equations. The four boundary conditions are e1-e4. The solution for the four arbitrary constants sol1 is substituted into n1 to produce the solution n1f stored in eq1. Fig. 3B.2 performs the calculations necessary to solve for Δz (Eq. 3B.5): integrate eq1 from 0 to H_1 (c13), differentiate (c14), and evaluate at $z = 0$ (c15). Finally compute Δz (c16). The result for small H_1 is computed using a Taylor expansion about $H_1 = 0$ (c17). The result is

$$\Delta z_{NH_4, 0-1} = \frac{H_1}{2} + \cdots \tag{3B.6}$$

rather than H_1 as assumed in Eq. (3.10).

The dissolved oxygen solution (Eq. 3B.1) is used to compute $\Delta z_{O_2, 0-1}$ in the same way in Fig. 3B.3. The result is

---

*This method of analysis was suggested to the author by Professor Steven C. Chapra.

Ammonia equations

(c3)   n1:b1*exp(lambda*z)+b2*exp(-lambda*z)

(d3)
$$b1\,e^{z\lambda} + b2\,e^{-z\lambda}$$

(c4)   n2:a1+a2*z-s[nh4]*z^2/2/d[nh4]

(d4)
$$-\frac{s_{nh4}\,z^2}{2\,d_{nh4}} + a2\,z + a1$$

Boundary Conditions

(c5)   e1:0=ev(n1,z=0)

(d5)
$$0 = b2 + b1$$

(c6)   e2:ev(n1,z=h[1])=ev(n2,z=h[1])

(d6)
$$b1\,e^{h_1\lambda} + b2\,e^{-h_1\lambda} = -\frac{h_1^2\,s_{nh4}}{2\,d_{nh4}} + h_1\,a2 + a1$$

(c7)   e3:ev(diff(n1,z),z=h[1])=ev(diff(n2,z),z=h[1])

(d7)
$$b1\,\lambda\,e^{h_1\lambda} - b2\,\lambda\,e^{-h_1\lambda} = a2 - \frac{h_1\,s_{nh4}}{d_{nh4}}$$

(c8)   e4:0=ev(diff(n2,z),z=h[2])

(d8)
$$0 = a2 - \frac{h_2\,s_{nh4}}{d_{nh4}}$$

(c9)   sol1:solve([e1,e2,e3,e4],[a1,a2,b1,b2])$

(c10)   n1f:ratsimp(ev(n1,sol1))

(d10)
$$\frac{e^{-z\lambda}\left((h_2 - h_1)\,s_{nh4}\,e^{2z\lambda + h_1\lambda} + (h_1 - h_2)\,s_{nh4}\,e^{h_1\lambda} \right)}{d_{nh4}\,\lambda\,e^{2h_1\lambda} + d_{nh4}\,\lambda}$$

(c11)   n2f:ratsimp(ev(n2,sol1))

(d11)
$$-\frac{\left(\left(\left(s_{nh4}\,z^2 - 2\,h_2\,s_{nh4}\,z + \left(2\,h_1\,h_2 - h_1^2 \right) s_{nh4} \right) \lambda + \left(2\,h_1 - 2\,h_2 \right) s_{nh4} \right) e^{2h_1\lambda} + \left(s_{nh4}\,z^2 - 2\,h_2\,s_{nh4}\,z + \left(2\,h_1\,h_2 - h_1^2 \right) s_{nh4} \right) \lambda + \left(2\,h_2 - 2\,h_1 \right) s_{nh4} \right)}{2\,d_{nh4}\,\lambda\,e^{2h_1\lambda} + 2\,d_{nh4}\,\lambda}$$

Fig. 3B.1 MACSYMA solution for the ammonia $\Delta z_{NH_4,0-1}$. Evaluating the arbitrary constants.

Integrate to find average in h[1]

(c13)   ih[1]:ratsimp(integrate(eq1,z,0,h[1])/h[1])

(d13)
$$\frac{\left(h_2 - h_1\right) s_{nh4}\, e^{2 h_1 \lambda} + \left(2 h_1 - 2 h_2\right) s_{nh4}\, e^{h_1 \lambda} + \left(h_2 - h_1\right) s_{nh4}}{h_1\, d_{nh4}\, \lambda^2\, e^{2 h_1 \lambda} + h_1\, d_{nh4}\, \lambda^2}$$

Compute the derivative at 0

(c14)   ratsimp(diff(eq1,z))

(d14)
$$\frac{e^{-z\lambda}\left(\left(h_2 - h_1\right) s_{nh4}\, e^{2 z \lambda + h_1 \lambda} + \left(h_2 - h_1\right) s_{nh4}\, e^{h_1 \lambda}\right)}{d_{nh4}\, e^{2 h_1 \lambda} + d_{nh4}}$$

(c15)   der:ratsimp(ev(%,z=0))

(d15)
$$\frac{\left(2 h_2 - 2 h_1\right) s_{nh4}\, e^{h_1 \lambda}}{d_{nh4}\, e^{2 h_1 \lambda} + d_{nh4}}$$

Compute the characteristic length

(c16)   dz[1,2]:ratsimp(ih[1]/der)

(d16)
$$\frac{e^{-h_1 \lambda}\left(e^{2 h_1 \lambda} - 2\, e^{h_1 \lambda} + 1\right)}{2 h_1\, \lambda^2}$$

Expand in Taylor series for small h[1]

(c17)   taylor(dz[1,2],h[1],0,2)

(d17)
$$\frac{h_1}{2} + \ldots$$

Fig. 3B.2   MACSYMA solution for the ammonia $\Delta z_{NH4,0-1}$.

Oxygen Concentration Profile

(c3)   eq1:o2(z)=o2[0]+r[0]*z/d[o2]*(z/2-h[1])

(d3)

$$o2(z) = \frac{r_0\left(\dfrac{z}{2} - h_1\right)z}{d_{o2}} + o2_0$$

Integrate to find average in H1

(c4)   ih[1]:ratsimp(integrate(eq1,z,0,h[1])/h[1])

(d4)

$$\frac{\displaystyle\int_0^{h_1} o2(z)\, dz}{h_1} = \frac{3\,o2_0\,d_{o2} - r_0\,h_1^{\,2}}{3\,d_{o2}}$$

Compute the derivative at z = 0

(c5)   der:ratsimp(diff(eq1,z))

(d5)

$$\frac{d}{dz}\big(o2(z)\big) = \frac{r_0\,z - r_0\,h_1}{d_{o2}}$$

(c6)   rhs(der)$

(c7)   der0:ev(%,z=0)

(d7)

$$-\frac{r_0\,h_1}{d_{o2}}$$

Compute the characteristic length

(c8)   dz:ratsimp(-(o2[0]-ih[1])/der0)

(d8)

$$-\frac{d_{o2}\displaystyle\int_0^{h_1} o2(z)\, dz - o2_0\,h_1\,d_{o2}}{r_0\,h_1^{\,2}} = \frac{h_1}{3}$$

Fig. 3B.3   MACSYMA solution for dissolved oxygen $\Delta z_{O_2,0-1}$.

$$\Delta z_{O_2,0-1} = \frac{H_1}{3} \qquad (3B.7)$$

exactly. The point is that it is different from $\Delta z_{NH_4,0-1}$ Eq. (3B.6). Hence an unavoidable approximation is introduced by using the two-layer approximation, which is different for each species considered, and which cannot always be quantified, since the calculation requires the continuous analytical solution.

To complete the analysis, the ammonia solution can also be used to calculate $\Delta z_{NH_4,1-2}$ that is appropriate for the layer 1-layer 2 mass transfer coefficient K_{L12}. Fig. 3B.4 continues the calculation from Fig. 3B.2. The layer 2 solution is listed, averaged over H_1 to H_2, the derivative is computed, evaluated at H_1, and Δz computed. The result is

$$\Delta z_{NH_4,1-2} = \frac{H_1}{6} + \frac{H_2}{3} + \cdots \qquad (3B.8)$$

rather than the linear estimate (Eq. 2.56)

$$\Delta z = \frac{H_1}{2} + \frac{H_2}{2} \qquad (3B.9)$$

Appendix 3C: Regression Analysis

A regression analysis is used to estimate both diagenesis and reaction rates. However, these quantities cannot be estimated only from the ammonia flux (see Eq. 3.43). In order to obtain stable estimates, it is necessary to include additional information. The idea is to use estimates of $J_N(i)$ as part of the regression criterion used to fit the ammonia flux. Ammonia diagenesis can estimated using Eq. (3.44)

$$\overline{J_N(i)}^{est} = a_{C,N} a_{O_2,C} \overline{SOD(i)}^{obs} \qquad (3C.1)$$

where $\overline{J_N(i)}^{est}$ is the estimate of the long-term average ammonia diagenesis flux for station i, and $\overline{SOD(i)}^{obs}$ is the observed long-term average SOD at station i. The overbars denote averages. Eq. (3C.1) can be incorporated into the regression as follows. The criterion to be minimized in ordinary least squares is

$$\min_{J_N(i),\theta_N,\kappa_{NH_4,1}} \left\{ \frac{1}{N_{obs}} \sum_{i,j}^{N_{obs}} (J[NH_4]_{i,j}^{obs} - J[NH_4]_{i,j}^{model})^2 \right\} \qquad (3C.2)$$

for station i and time t_j, giving a total of N_{obs} observations. A mixed criterion requires that $J_N(i)$ be close to the estimate $\overline{J_N(i)}^{est}$ as well as $J[NH_4]_{i,j}^{model}$ be close to $J[NH_4]_{i,j}^{obs}$. This is accomplished by properly augmenting the fitting criterion. In addition, each part of the criterion needs to be properly weighted. The augmented

Steady state equation

(c13)   eq2:n2f

$$(\text{d}13) \quad -\frac{\left(\begin{array}{l}\left(\left(s_{nh4}\,z^2 - 2\,h_2\,s_{nh4}\,z + \left(2\,h_1\,h_2 - h_1^{\,2}\right)s_{nh4}\right)\lambda + \left(2\,h_1 - 2\,h_2\right)s_{nh4}\right)\\[4pt] *\,e^{2\,h_1\,\lambda} + \left(s_{nh4}\,z^2 - 2\,h_2\,s_{nh4}\,z + \left(2\,h_1\,h_2 - h_1^{\,2}\right)s_{nh4}\right)\lambda + \left(2\,h_2 - 2\,h_1\right)s_{nh4}\end{array}\right)}{2\,d_{nh4}\,\lambda\,e^{2\,h_1\,\lambda} + 2\,d_{nh4}\,\lambda}$$

Integrate to find average in h[2]

(c15)   ih[2]:ratsimp(integrate(eq2,z,h[1],h[2])/(h[2]-h[1]))

$$(\text{d}15) \quad \frac{\left(\begin{array}{l}\left(\left(h_2^{\,2} - 2\,h_1\,h_2 + h_1^{\,2}\right)s_{nh4}\,\lambda + \left(3\,h_2 - 3\,h_1\right)s_{nh4}\right)\\[4pt] *\,e^{2\,h_1\,\lambda} + \left(h_2^{\,2} - 2\,h_1\,h_2 + h_1^{\,2}\right)s_{nh4}\,\lambda + \left(3\,h_1 - 3\,h_2\right)s_{nh4}\end{array}\right)}{3\,d_{nh4}\,\lambda\,e^{2\,h_1\,\lambda} + 3\,d_{nh4}\,\lambda}$$

Compute the derivative at h1

(c16)   ratsimp(diff(eq1,z))

$$(\text{d}16) \quad \frac{e^{-z\,\lambda}\left(\left(h_2 - h_1\right)s_{nh4}\,e^{2\,z\,\lambda + h_1\,\lambda} + \left(h_2 - h_1\right)s_{nh4}\,e^{h_1\,\lambda}\right)}{d_{nh4}\,e^{2\,h_1\,\lambda} + d_{nh4}}$$

(c17)   der:ratsimp(ev(%,z=h[1]))

$$(\text{d}17) \quad \frac{\left(h_2 - h_1\right)s_{nh4}}{d_{nh4}}$$

Compute the characteristic length

(c18)   dz[1,2]:ratsimp((ih[2]-ih[1])/der)

$$(\text{d}18) \quad \frac{\left(\left(h_1\,h_2 - h_1^{\,2}\right)\lambda^2 + 3\,h_1\,\lambda - 3\right)e^{2\,h_1\,\lambda} + 6\,e^{h_1\,\lambda} + \left(h_1\,h_2 - h_1^{\,2}\right)\lambda^2 - 3\,h_1\,\lambda - 3}{3\,h_1\,\lambda^2\,e^{2\,h_1\,\lambda} + 3\,h_1\,\lambda^2}$$

Expand in Taylor series for small h[1]

(c19)   taylor(dz[1,2],h[1],0,2)

$$(\text{d}19) \quad \frac{h_2}{3} + \frac{h_1}{6} + \cdots$$

Fig. 3B.4   MACSYMA solution for $\Delta z_{NH_4, 1-2}$.

criterion without weighting has the form

$$\min_{J_N(i),\theta_N,\kappa_{NH_4,1}} \left\{ \frac{1}{N_{obs}} \sum_{i,j}^{N_{obs}} (J[NH_4]_{i,j}^{obs} - J[NH_4]_{i,j}^{model})^2 \right.$$

$$\left. + \frac{1}{N_{sta}} \sum_{i}^{N_{sta}} \left(\overline{J_N(i)}^{est} - J_N(i) \right)^2 \right\} \tag{3C.3}$$

The natural choice for weights are the standard deviations of the ammonia fluxes, $\sigma_{J[NH_4(i)]}$, which can be computed from the replicates, and the standard deviation of the estimates of the diagenesis fluxes $\sigma_{\overline{J_N(i)}^{est}}$. However, it is not clear how to compute the latter standard deviations. Instead, the average itself is used as the weight for each station. This amounts to assuming that

$$\sigma_{\overline{J_N(i)}^{est}} = \overline{J_N(i)}^{est} \tag{3C.4}$$

which means that the coefficient of variation for $\overline{J_N(i)}^{est}$ is assumed to be one. The criterion that results is

$$\min_{J_N(i),\theta_N,\kappa_{NH_4,1}} \left\{ \frac{1}{N_{obs}} \sum_{i,j}^{N_{obs}} \left(\frac{J[NH_4]_{i,j}^{obs} - J[NH_4]_{i,j}^{model}}{\sigma_{J[NH_4(i)]}} \right)^2 \right.$$

$$\left. + \frac{1}{N_{sta}} \sum_{i}^{N_{sta}} \left(\frac{\overline{J_N(i)}^{est} - J_N(i)}{\overline{J_N(i)}^{est}} \right)^2 \right\} \tag{3C.5}$$

where the sum over N_{sta} includes only the oxic stations. The magnitudes of $\sigma_{[J[NH_4(i)]}$ and $\overline{J_N(i)}^{est}$ are approximately equal to the magnitudes of the numerator terms, respectively. Thus each term measures the deviation of the numerator relative to an approximately equal magnitude in the denominator. This gives approximately equal weight to each term.

A numerical procedure (Press et al., 1989) is used to minimize the criterion, Eq. (3C.5). A second criterion, using absolute values instead of squares as the measure of the deviations

$$\min_{J_N(i),\theta_N,\kappa_{NH_4,1}} \left\{ \frac{1}{N_{obs}} \sum_{i,j}^{N_{obs}} \left| \frac{J[NH_4]_{i,j}^{obs} - J[NH_4]_{i,j}^{model}}{\sigma_{J[NH_4(i)]}} \right| \right.$$

$$\left. + \frac{1}{N_{sta}} \sum_{i}^{N_{sta}} \left| \frac{\overline{J_N(i)}^{est} - J_N(i)}{\overline{J_N(i)}^{est}} \right| \right\} \tag{3C.6}$$

is also employed. The individual ammonia fluxes are log transformed if the fluxes are positive, or they are used as is if the flux is negative. The appropriate logarithmic or arithmetic standard deviations are used in the sum. The results are listed in Table 3.2.

4

Nitrate

4.1 INTRODUCTION

The model presented in the previous chapter quantifies the fraction of ammonia that is oxidized in the aerobic zone. The ammonia can be either produced by diagenesis of organic matter or transferred to the sediment from the overlying water. The result is a source of nitrate which, in some cases can be substantial (Figs. 3.9 and 3.10). The nitrate may either escape as a flux to the overlying water, or it may be denitrified to nitrogen gas. In addition, the flux of nitrate from the overlying water to the sediment adds to the nitrate that is available for denitrification. Since denitrification is a terminal sink for nitrate, it is important that the nitrate and nitrogen gas fluxes be properly computed. This chapter presents a model for the sediment nitrate and nitrogen gas fluxes.

4.2 MODEL FORMULATION AND SOLUTION

The model schematic is shown in Fig. 4.1. Denitrification can occur in the both the aerobic and anaerobic layers. The conventional formulation is to have denitrification occur only in a layer below the aerobic layer.[*] However a close coupling of the two zones has been suggested (Blackburn, 1994) and some oxic layer denitrification is estimated to occur using a numerical inversion technique applied to vertical

---

[*]Billen (1978, 1982, 1988), Billen et al. (1989), Blackburn (1990), Goloway and Bender (1982), Jahnke et al. (1982), Klapwijk and Snodgrass (1982), Vanderborght et al. (1977).

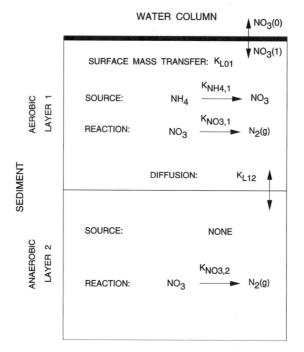

Fig. 4.1   Schematic of the nitrate flux model.

profiles (Berg et al., 1998). Aerobic layer denitrification was also employed in a one-dimensional model for oxygen and nitrate pore water profiles in a marine sediment (Brandes and Devol, 1995). A two-dimensional model with discrete sites of diagenesis was also proposed as a rationalization for the observations.

For the model formulated in this chapter, denitrification can take place in the aerobic zone as well as the anaerobic zone. Three sources of evidence are offered to support the existence of aerobic layer denitrification.

4.2 A   Aerobic Layer Denitrification

The first is the experimental results of Jenkins and Kemp (1984). An ammonia tracer, $^{15}NH_4$, was added to the water overlying sediment cores taken from two stations in the Patuxent River estuary. After 48 hours of incubation the distribution of ^{15}N among the nitrogen species was determined for the interstitial and overlying water. Approximately 10% to 20% remained as $^{15}NH_4$, the rest appeared either as $^{15}N_2$ or ^{15}PON. No appreciable $^{15}NO_3$ was observed, see Fig. 3 in Jenkins and Kemp (1984). Their interpretation is that nitrification and denitrification occur in close spatial proximity. Brezonik (1977) and Jorgensen (1977a) suggest the existence of "anoxic microsites." For example, the interior of aggregate organic particles may be anaerobic even if the exterior is aerobic. Thus both nitrification and denitrification can coexist in the aerobic layer.

The second source of evidence is the consequence of assuming that no denitrification occurs in the aerobic layer. The result is that one-half of the nitrate formed by the nitrification of diagenetically produced ammonia escapes as a nitrate flux to the overlying water. This can be seen by considering a simple continuous model of nitrate in the aerobic layer. Let $S[NO_3]$ be the source of nitrate due to ammonia oxidation. The mass balance equation for interstitial water nitrate concentration, $NO_3(z)$, is

$$-D_{NO_3} \frac{d^2[NO_3(z)]}{dz^2} = S[NO_3] \qquad 0 \leqslant z \leqslant H_1 \qquad (4.1)$$

The appropriate boundary conditions are $NO_3(0) = 0$, since the concern is the fate of internally produced nitrate and not the flux of nitrate from the overlying water to the sediment. The simplest boundary condition at the aerobic-anaerobic boundary is

$$[NO_3(H_1)] = 0 \qquad (4.2)$$

which corresponds to sufficiently rapid denitrification in the anaerobic zone to deplete the nitrate below $z = H_1$. The result is

$$[NO_3(z)] = \frac{J_{NO_3} z(z - H_1)}{2 D_{NO_3} H_1} \qquad (4.3)$$

where $J_{NO_3} = S[NO_3]H_1$, the flux of nitrate produced by ammonia oxidation. The flux of nitrate to the overlying water is

$$J[NO_3] = D_{NO_3} \frac{d[NO_3(z)]}{dz}\Big|_{z=0} = \frac{J_{NO_3}}{2} \qquad (4.4)$$

so half of the nitrate flux produced by ammonia oxidation escapes to the overlying water. If the boundary condition at the aerobic-anaerobic boundary is $[NO_3(H_1) > 0$ corresponding to a finite rate of denitrification in the anaerobic zone, more than half of J_{NO_3} escapes as $J[NO_3]$. No nitrate flux of this magnitude is observed in any data set analyzed below and in subsequent chapters.

The third source of evidence is the analysis of the composition and quantity of the measured gas flux data from the Milwaukee River sediments (Chapter 10). These data suggest that most, if not all, the nitrate produced by sediment nitrification must be denitrified to nitrogen gas. Otherwise, the magnitude of nitrogen gas flux measured from the sediments cannot be explained. If aerobic layer denitrification were not occurring, then at least half of the nitrate produced would escape as a nitrate flux to the overlying water (Eq. 4.4).

4.2 B   Model Equations and Solutions

It is assumed, therefore, that the nitrate produced in the aerobic zone of the sediment can be denitrified to nitrogen gas with a first-order rate constant $k_{NO_3,1}$. In addition, nitrate that is transported to the anaerobic layer can be denitrified as well with a first-order rate constant $k_{NO_3,2}$. The remainder of the formulation parallels the ammonia flux model (Chapter 3).

The mass balance equations for the two layers are

$$H_1 \frac{d[NO_3(1)]}{dt} = -k_{NO_3,1}[NO_3(1)]H_1 - K_{L01}([NO_3(1)] - [NO_3(0)])$$

$$+ K_{L12}([NO_3(2)] - [NO_3(1)]) + S[NO_3] \quad (4.5a)$$

$$H_2 \frac{d[NO_3(2)]}{dt} = -k_{NO_3,2}[NO_3(2)]H_2 - K_{L12}([NO_3(2)] - [NO_3(1)]) \quad (4.5b)$$

where $S[NO_3]$ is the source of nitrate from ammonia nitrification in the aerobic layer (Eq. 3.45). The solutions to these mass balance equations are slightly more complex than in the case of ammonia oxidation for which there is only an aerobic layer reaction rate. Steady state is assumed and Eqs. (4.5) are solved simultaneously for $[NO_3(1)]$ and $[NO_3(2)]$. The resulting nitrate concentrations in the aerobic and anaerobic layers are

$$[NO_3(1)] = \frac{S[NO_3] + K_{L01}[NO_3(0)]}{k_{NO_3,1}H_1 + K_{L01} + \left(\dfrac{1}{k_{NO_3,2}H_2} + \dfrac{1}{K_{L12}}\right)^{-1}} \quad (4.6)$$

$$[NO_3(2)] = [NO_3(1)]\frac{K_{L12}}{k_{NO_3,2} + K_{L12}} \quad (4.7)$$

The equality $s = K_{L01}$ (Eq. 3.14) is used for the surface mass transfer coefficient where $s = SOD/O_2(0)$. The aerobic denitrification reaction velocity is defined as

$$\kappa_{NO_3,1} = \sqrt{D_{NO_3}k_{NO_3,1}} \quad (4.8)$$

The rationale for using reaction velocities is presented in Chapter 3. The anaerobic denitrification parameter group $k_{NO_3,2}H_2$ has units of length/time and therefore

$$\kappa_{NO_3,2} = k_{NO_3,2}H_2 \quad (4.9)$$

formally qualifies as a reaction velocity. This parameter is defined for convenience of nomenclature only. It is not equivalent to the aerobic layer reaction velocities which include a diffusion coefficient as well as a reaction rate constant.

The reciprocal of the sum of the reciprocals of $\kappa_{NO_3,2}$ and K_{L12} in Eq. (4.6) can be replaced by an overall layer 2 denitrification reaction velocity

$$\kappa_{NO_3,2}^{*} = \left(\frac{1}{\kappa_{NO_3,2}} + \frac{1}{K_{L12}}\right)^{-1} \quad (4.10)$$

Using this notation, Eqs. (4.6 and 4.7) become

$$[NO_3(1)] = \frac{S[NO_3] + s[NO_3(0)]}{\kappa_{NO_3,1}^2/s + s + \kappa_{NO_3,2}^*} \qquad (4.11)$$

$$[NO_3(2)] = [NO_3(1)]\frac{K_{L12}}{\kappa_{NO_3,2} + K_{L12}} \qquad (4.12)$$

4.2 C   Nitrate Source

The source of nitrate to the aerobic layer $S[NO_3]$, which is the result of ammonia oxidation, can be quantified in a number of ways. For example, the rate of nitrification for first-order nitrification kinetics can be evaluated directly using Eq. (3.5) for $[NH_4(1)]$

$$S[NO_3] = k_{NH_4,1} H_1[NH_4(1)]$$

$$= \frac{D_{NO_3} k_{NH_4,1}}{s}[NH_4(1)]$$

$$= \frac{\kappa_{NH_4,1}^2}{s}[NH_4(1)] \qquad (4.13)$$

However, a simple mass balance argument is more instructive. Since all the ammonia sources must balance all the ammonia sinks, the nitrification sink can be found by difference. The sources of ammonia are ammonia diagenesis J_N, and ammonia transferred from the overlying water $s[NH_4(0)]$. The sinks of ammonia are the flux to the overlying water $s[NH_4(1)]$, and the loss via nitrification $S[NO_3]$. Hence the nitrate source from nitrification can be found as the difference between the sum of the ammonia sources and the ammonia loss to the overlying water

$$S[NO_3] = J_N + s[NH_4(0)] - s[NH_4(1)]$$

$$= J_N - s([NH_4(1)] - [NH_4(0)])$$

$$= J_N - J[NH_4] \qquad (4.14)$$

where the third equality follows from the mass transfer equation for ammonia flux (Eq. 3.7)

$$J[NH_4] = s([NH_4(1)] - [NH_4(0)]) \qquad (4.15)$$

Therefore, Eq. (4.14) is the required expression for $S[NO_3]$.

The nitrate flux, with the convention that positive fluxes are from the sediment, is

$$J[NO_3] = s([NO_3(1)] - [NO_3(0)]) \qquad (4.16)$$

Substituting Eq. (4.11) into Eq. (4.16) and using Eq. (4.14) yields the final expression for the nitrate flux

$$J[NO_3] = s\left(\frac{s[NO_3(0)] + J_N - J[NH_4]}{\kappa_{NO_3,1}^2/s + s + \kappa_{NO_3,2}^*} - [NO_3(0)]\right) \quad (4.17)$$

It is important to note that the nitrate flux is a linear function of the overlying water nitrate concentration, $[NO_3(0)]$. This can be seen by rearranging Eq. (4.17)

$$J[NO_3] = \left(\frac{s^2}{\kappa_{NO_3,1}^2/s + s + \kappa_{NO_3,2}^*} - s\right)[NO_3(0)] + \left(\frac{s(J_N - J[NH_4])}{\kappa_{NO_3,1}^2/s + s + \kappa_{NO_3,2}^*}\right)$$
$$(4.18)$$

The model's behavior can be examined from this point of view. The intercept – the second term in Eq. (4.18) – quantifies the extent to which nitrate produced by nitrification in the sediment appears as a nitrate flux from the sediment to the overlying water. The slope – the coefficient of $[NO_3(0)]$ in Eq. (4.18) – quantifies the extent to which overlying water nitrate is denitrified in the sediment.

4.3 NITRATE SOURCE FROM THE OVERLYING WATER

If the internal production of nitrate $J_N - J[NH_4]$ is small relative to the nitrate delivered from the overlying water $s[NO_3(0)]$, then the constant term in Eq. (4.18) is small, and the slope term dominates

$$J[NO_3] = \left(\frac{s^2}{\kappa_{NO_3,1}^2/s + s + \kappa_{NO_3,2}^*} - s\right)[NO_3(0)]$$
$$= -s\left(1 - \frac{s}{\kappa_{NO_3,1}^2/s + s + \kappa_{NO_3,2}^*}\right)[NO_3(0)] \quad (4.19)$$

This equation suggests that if the nitrate flux is normalized using the overlying water nitrate concentration, then a one-to-one relationship exists between the normalized nitrate flux to the sediment and the surface mass transfer coefficient s

$$\frac{J[NO_3]}{[NO_3(0)]} = -s\left(1 - \frac{s}{\kappa_{NO_3,1}^2/s + s + \kappa_{NO_3,2}^*}\right) \quad (4.20)$$

This result is used below in the data analysis of nitrate fluxes.

Two limiting forms of this equation exist that depend on the magnitude of s. These can be found by examining Eq. (4.20) in the following form

$$\frac{J[NO_3]}{[NO_3(0)]} = -s\left(\frac{\kappa_{NO_3,1}^2/s + \kappa_{NO_3,2}^*}{\kappa_{NO_3,1}^2/s + s + \kappa_{NO_3,2}^*}\right) \quad (4.21)$$

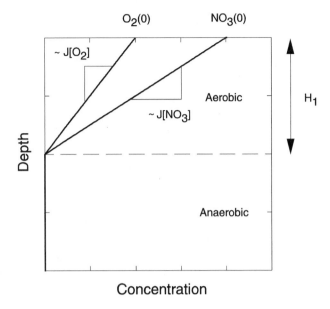

Fig. 4.2 Vertical profiles of oxygen and nitrate. The slopes are proportional to the fluxes of oxygen and nitrate.

For small s the bracketed term approaches one and Eq. (4.21) becomes

$$\frac{J[NO_3]}{[NO_3(0)]} = -s \qquad s \to 0 \qquad (4.22)$$

This result can be understood as follows: The surface mass transfer coefficient s is the ratio of SOD to $O_2(0)$, that is, the ratio of the oxygen flux into the sediment $J[O_2]$ and the overlying water oxygen concentration. The left-hand side of Eq. (4.22) is the ratio of the flux of nitrate to the sediment to the overlying water nitrate concentration. Hence Eq. (4.22) is

$$\frac{J[NO_3]}{[NO_3(0)]} = -s = -\frac{SOD}{[O_2(0)]} = \frac{J[O_2]}{[O_2(0)]} \qquad (4.23)$$

The reason for the symmetry between the equations for nitrate and oxygen fluxes is illustrated in Fig. 4.2. For small s the depth of the aerobic layer H_1 is large enough and the residence time is long enough so that all of the nitrate is denitrified either in the aerobic layer or in the anaerobic layer. Note that the bracketed term in Eq. (4.21) approaches one even if there were no aerobic layer denitrification and $\kappa_{NO_3,1} = 0$. Hence the nitrate concentration at the aerobic-anaerobic layer boundary is zero. By definition the concentration of oxygen at the aerobic-anaerobic layer boundary is also zero. Hence both profiles connect the overlying water concentration to a zero concentration. Thus the normalized fluxes are equal.

The other limiting case is for large s. The limiting form can be found from Eq. (4.21), but it is instructive to derive it directly. For large s the aerobic zone is quite small and there is no significant denitrification or mass transfer resistance in this layer. Hence $[NO_3(1)] \approx [NO_3(0)]$, and the layer 2 nitrate mass balance equation can be written as

$$0 = H_2 \frac{d[NO_3(2)]}{dt} = -\kappa_{NO_3,2}[NO_3(2)] - K_{L12}([NO_3(2)] - [NO_3(0)]) \quad (4.24)$$

which can be solved for $[NO_3(2)]$

$$[NO_3(2)] = \frac{K_{L12}}{\kappa_{NO_3,2} + K_{L12}}[NO_3(0)] \quad (4.25)$$

The nitrate flux is

$$J[NO_3] = K_{L12}([NO_3(2)] - [NO_3(0)]) \quad (4.26)$$

where the mass transfer coefficient that governs is now K_{L12} since the aerobic layer mass transfer resistance is negligibly small. The solution follows from substituting Eq. (4.25) into Eq. (4.26)

$$\frac{J[NO_3]}{[NO_3(0)]} = -\kappa^*_{NO_3,2} = -\left(\frac{1}{\kappa_{NO_3,2}} + \frac{1}{K_{L12}} \right)^{-1} \quad (4.27)$$

The result is a constant normalized flux. The nitrate flux is determined by the reciprocal of the sum of the reciprocals of the two parameters that determine the extent of denitrification: the denitrification reaction velocity and the aerobic-anaerobic layer mass transfer coefficient. The magnitude of the smaller parameter determines the extent of denitrification. This is similar to that portion of the ammonia flux expression associated with the overlying water ammonia concentration, Eq. (3.9) where the analogy to electrical resistors in parallel is explained.

To summarize the results, if the internal production of nitrate is small relative to the flux of nitrate from the overlying water, then the normalized nitrate flux to the sediment is linear in s for small s and constant for large s

$$\frac{J[NO_3]}{[NO_3(0)]} = -s \qquad s \to 0$$

$$\frac{J[NO_3]}{[NO_3(0)]} = -\kappa^*_{NO_3,2} \qquad s \to \infty \quad (4.28)$$

The $J[NO_3]$ to $[NO_3(0)]$ ratio has been used as a data analysis method on a number of occasions, e.g., Kana et al. (1998).

4.3 A   Application to Hunting Creek

The relationship between nitrate flux and the surface mass transfer coefficient can be investigated using a data set collected for Hunting Creek sediments (Cerco, 1988).

As part of an investigation of the variation of SOD as a function of overlying water DO, the nitrate flux was measured as well. Because the overlying water nitrate concentrations used in the experiments were large, it is reasonable to ignore the internal production of nitrate. Hence Eq. (4.20) applies and an analysis of $J[NO_3]/[NO_3(0)]$ versus s is appropriate.

The nitrate flux data are presented in Fig. 4.3 as a function of overlying water nitrate concentration (Fig. 4.3A), oxygen concentration (Fig. 4.3B), and SOD (Fig. 4.3C). The unfilled circles represent experiments with more rapid mixing of the over-

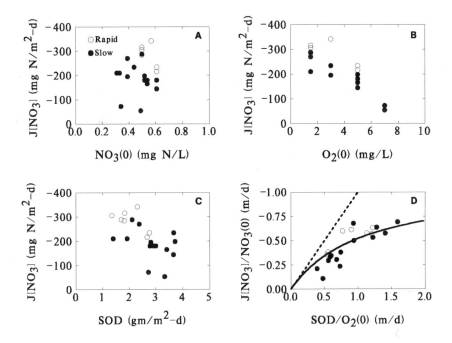

Fig. 4.3 Hunting Creek nitrate flux (Cerco, 1988). Nitrate flux versus (A) overlying water nitrate, and (B) oxygen concentrations, and (C) SOD. (D) Normalized nitrate flux versus $SOD/O_2(0)$. Solid line is Eq. (4.21) with $\kappa_{NO_3,1} = 0.05$ m/d and $\kappa^*_{NO_3,2} = 0.1$ m/d. The dotted straight line is Eq. (4.22).

lying water. The nitrate flux exhibits a decreasing dependency to overlying water DO (Fig. 4.3B), a weaker relationship to SOD (Fig. 4.3C), and almost no relationship to overlying water nitrate concentration (Fig. 4.3A). The analysis presented above suggests that the proper analysis is to examine the relationship between normalized nitrate flux, $J[NO_3]/[NO_3(0)]$, and the surface mass transfer coefficient, s. The result is presented in Fig. 4.3D, together with a model (Eq. 4.21) fit to the data. The dotted straight line is $J[NO_3]/[NO_3(0)] = s$, the small s limit (Eq. 4.22). The parameter values are listed in Fig. (4.3).

4.4   NITRATE SOURCE FROM NITRIFICATION

The intercept of the nitrate flux versus overlying water nitrate concentration is controlled by the quantity of ammonia that is nitrified but not denitrified. This is clear from the form of the constant term in Eq. (4.18)

$$J[NO_3] = (J_N - J[NH_4]) \frac{s}{\kappa_{NO_3,1}^2/s + s + \kappa_{NO_3,2}^*} \tag{4.29}$$

If the source of ammonia from the overlying water is small relative to that produced by diagenesis, then Eq. (3.36) can be substituted for $J[NH_4]$ and Eq. (4.29) becomes

$$J[NO_3] = J_N \left(\frac{\kappa_{NH_4,1}^2}{\kappa_{NH_4,1}^2 + s^2} \right) \left(\frac{s}{\kappa_{NO_3,1}^2/s + s + \kappa_{NO_3,2}^*} \right) \tag{4.30}$$

which delineates the various contributory factors. The flux is linear in ammonia diagenesis J_N, since this is the only source that is assumed to be significant. The next term is the fraction of ammonia that is nitrified to nitrate. The last term is the ratio of the rate of mass transfer to the overlying water to the sum of the rates – in mass transfer terms – of the three sinks of nitrate. It is the fraction of nitrate that escapes denitrification in either the aerobic or anaerobic layers and, therefore, escapes to the overlying water.

4.5   MODEL APPLICATIONS

The sensitivity of the model to various parameters is examined next. Then two data sets are fitted to the model.

4.5 A   Sensitivity

The behavior of the nitrate flux model (Eq. 4.17) with respect to the controlling variables is illustrated in Fig. 4.4. The reaction velocities used in the computations, which are obtained from a calibration to Chesapeake Bay data discussed below, are listed in Table 4.1. Eq. (3.36) is used for $J[NH_4]$ as discussed above in Section 4.4.

For a fixed s, the effect of increasing J_N is to increase the nitrate flux uniformly, that is, to increase the intercept of the linear relationship (Fig. 4.4A). This additional nitrate flux is that portion of the nitrate produced by the nitrification of ammonia that is not denitrified. The slope of the relationship, which is determined by s and the κ's, is unaffected.

For a fixed J_N, varying s affects both the slope and intercept (Fig. 4.4B). For small $s = 0.01$ (m/d), the nitrate flux is essentially zero, independent of overlying water nitrate concentration. The reason is that the amount of overlying water nitrate that is transferred to the aerobic zone, $s[NO_3(0)]$, is small enough, and the aerobic zone is deep enough so that denitrification is essentially complete. For the nitrate

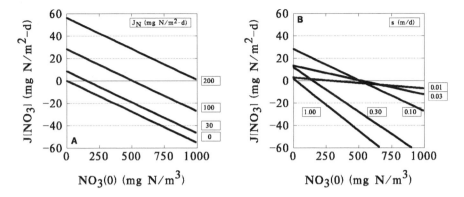

Fig. 4.4 Sensitivity analysis. Nitrate flux versus overlying water nitrate concentration for various ammonia diagenesis rates (A) and surface mass transfer rates (B). The boxed quantities are the values (A) J_N (mg N/m²-d) and (B) s (m/d) used to compute the lines. The kinetic parameter values are $\kappa_{NH_4,1} = 0.131$ m/d, $\kappa_{NO_3,1} = 0.05$ m/d, and $\kappa_{NO_3,2} = 0.10$ m/d.

produced by nitrification, the same reasoning applies, namely that the residence time in the sediment is sufficient so that denitrification is essentially complete. The result is that the nitrate produced by both sources is completely denitrified.

As s increases to 0.1 (m/d), the slope of the nitrate flux-nitrate concentration relationship increases. Surface mass transfer is increasing and aerobic layer depth is decreasing so that more nitrate is transferred to the sediment where it denitrifies. The intercept starts to increase as well, reflecting the increasing nitrate flux due to nitrate produced by ammonia nitrification. However, as s continues to increase to 1.0 (m/d), the intercept starts to decrease. The aerobic layer depth is now getting so small that less ammonia is nitrified, producing less nitrate that is available for transfer to the overlying water.

4.5 B   Application to Chesapeake Bay

The straight line relationship between nitrate flux and overlying water nitrate concentration can be used to examine data in a straightforward fashion. The aggregated Chesapeake Bay data set is shown in Fig. 4.5. The data have been divided into two classes with respect to concentration of overlying water dissolved oxygen. The observed surface mass transfer coefficient and nitrate flux are averaged in bins of width 0.2 $\log_{10}$ [$NO_3(0)$]. The number of observations in each bin is shown by the histogram.

Since there appears to be no trend in the relationship between s and [$NO_3(0)$], a constant is used in the calculation as indicated in Fig. 4.5A,B. The reaction velocities are obtained from an analysis described below in Section (4.5 C). The computed nitrate fluxes are compared to the observations in Fig. 4.5C,D. For aerobic conditions, [$O_2(0)$] > 2, the data suggest a positive nitrate flux for small [$NO_3(0)$] and exhibit the expected linear behavior as overlying water nitrate concentration increases. For

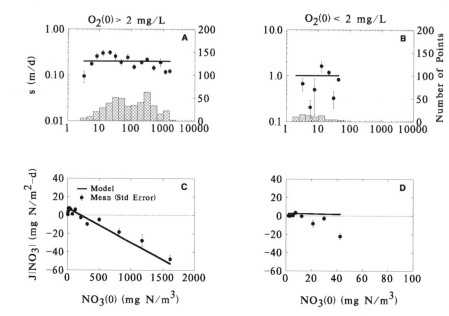

Fig. 4.5 Chesapeake Bay data for overlying water dissolved oxygen $[O_2(0)] > 2$ mg/L (A, C). Data for $[O_2(0)] < 2$ mg/L (B, D). Overlying water oxygen data and fitted line (A, B). Nitrate flux (C, D) data (•) and model (-).

hypoxic conditions, $[O_2(0)] < 2$, an almost constant relationship is predicted. Note that the abscissa scale has been changed in Fig. 4.5D. The data appear to support the absence of positive nitrate fluxes for small $[NO_3(0)]$. As $[NO_3(0)]$ increases, the model predicts a slight decrease over the range of concentrations (0–50 mg N/m^3), whereas the data suggest a larger decrease although the number of data points in the averages are small as indicated by the histogram in Fig. 4.5B.

4.5 C Application to Gunston Cove

A set of nutrient and oxygen flux measurements has been made by Cerco (1988, 1985) in Gunston Cove, a small tidal freshwater embayment of the Potomac river. Both in situ and laboratory measurements are reported. The temperature, overlying water DO, and nitrate concentrations span a reasonably wide range so their effects can be seen. In order to analyze these data within the framework of the model presented above, it is necessary to specify the variables: s and J_N.

The surface mass transfer coefficient is available from measurements of SOD and $O_2(0)$, Fig. 4.6B. The observations versus temperature and a comparison to the expression

$$s = s_{20}\theta_s^{(T-20)}$$

(4.31)

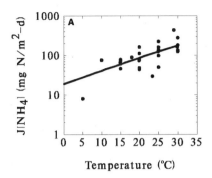
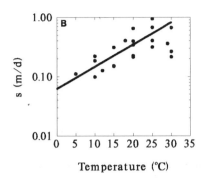

Fig. 4.6 Gunston Cove forcing functions. (A) Ammonia flux and (B) surface mass transfer versus temperature: Data and fitted line.

Table 4.1   Nitrate Model Parameters

Parameter	(a)	(b)	(c)	(d)	Units
$\kappa_{NO_3,1}$	0.0*	0.10	0.05	0.01	m/d
$\kappa^*_{NO_3,2}$	1.09	0.0096	0.075	0.0096	m/d
$\kappa_{NO_3,2}$		0.25		0.25	m/d
K_{L12}		0.01		0.01	m/d
s_{20}			0.323	(e)	m/d
θ_s			1.120	(e)	–
$J_N(20)$	0.0*			(f)	mg N/m²–d
θ_N		1.142	1.100	(f)	–

*Assigned. [a]Hunting Creek. [b]Chesapeake Bay linear analysis. [c]Gunston Cove. [d]Chesapeake Bay normalized analysis. [e]Observations are used. [f]From Table 3.2, case (c)

are shown where the parameters, s_{20} and θ_s, are estimated by regression. This formula is used strictly as a convenient interpolation for s versus temperature.

Ammonia diagenesis can be inferred from the measured ammonia fluxes using the model presented in the previous chapter, Eq. (3.36):

$$J[NH_4] = J_N \frac{s^2}{s^2 + \kappa^2_{NH_4,1}} \qquad (4.32)$$

Ammonia diagenesis is assumed to be given by an exponential function of temperature

$$J_N(T) = J_N(20)\theta_N^{(T-20)} \qquad (4.33)$$

The two parameters, $J_N(20)$ and θ_N, are found by fitting the ammonia flux model, Eq. (4.32), to the observations. The result is shown in Fig. 4.6A, and the parameters are listed in Table 4.1.

With these parameters established as a function of temperature, the nitrate flux can be predicted as a function of overlying water nitrate concentration. The reaction velocities for denitrification are found from a nonlinear least squares fit of the model, Eq. (4.17), to the data using Eqs. (4.31 and 4.33) for s and J_N at the temperature of the observation. Table 4.1 presents the results. Fig. 4.7F compares observations and predictions of the nitrate flux.

A more informative presentation can be made if the data are grouped into temperature classes. Ammonia diagenesis and surface mass transfer are calculated for the temperature indicated in each panel in Fig. 4.7. The model prediction is a straight-line relationship between nitrate flux and overlying water nitrate concentration. The slopes progressively increase as s increases with temperature. The intercept also increases as diagenesis increases with temperature. In general, the model appears to conform to the major features of these data: the linear relationship between nitrate flux and nitrate concentration, and the relationships of the slope and intercept to the surface mass transfer coefficient and the endogenous production of nitrate. However, there is considerable scatter when individual fluxes are compared to model predictions (Fig. 4.7F). As we will see, this scatter is not unique to nitrate fluxes.

4.6   FLUX NORMALIZATION AND PARAMETER ESTIMATION

A comprehensive method for the analysis of the nitrate flux data employs a normalization of the nitrate fluxes suggested by the structure of the model. Eq. (4.17) for nitrate flux can be written in the form

$$\frac{J[NO_3]}{s} = \frac{s[NO_3(0)] + J_N - J[NH_4]}{\kappa^2_{NO_3,1}/s + s + \kappa^*_{NO_3,2}} - [NO_3(0)] \tag{4.34}$$

Note that the unknown reaction velocities are in the denominator of the first term. Solving for this term yields

$$\frac{\kappa^2_{NO_3,1}}{s} + s + \kappa^*_{NO_3,2} = \frac{s[NO_3(0)] + J_N - J[NH_4]}{\dfrac{J[NO_3]}{s} + [NO_3(0)]} \tag{4.35}$$

The numerator of the right-hand side of this equation is the total source of nitrate in the aerobic layer from both the overlying water and aerobic layer nitrification. The denominator is the aerobic layer nitrate concentration, $[NO_3(1)]$. This can be seen if the nitrate mass transfer equation (4.16) is expressed as

$$[NO_3(1)] = \frac{J[NO_3]}{s} + [NO_3(0)] \tag{4.36}$$

Hence Eq. (4.35) becomes

$$\frac{\kappa^2_{NO_3,1}}{s} + s + \kappa^*_{NO_3,2} = \frac{S[NO_3]_T}{[NO_3(1)]} \tag{4.37}$$

where $S[NO_3]_T = S[NO_3] + s[NO_3(0)]$, the total nitrate source to the aerobic layer.

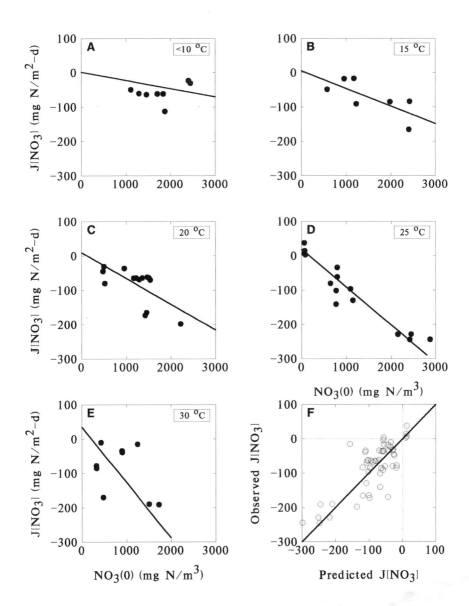

Fig. 4.7 Gunston Cove (A) to (E) nitrate flux versus overlying water nitrate concentration for various temperature ranges. (F) Observed versus predicted nitrate flux.

4.6 A   Mechanisms

The left-hand side of equation (4.37) is made up of three terms that represent the mechanisms by which nitrate is lost from the aerobic layer: (1) aerobic layer denitrification, (2) mass transfer to the overlying water, and (3) diffusion and denitrification in the anaerobic layer. For small s, so that the aerobic layer depth is large, aerobic layer denitrification predominates. For intermediate s, diffusive transport to the anaerobic layer followed by denitrification dominates. Finally, for large s, surface mass transfer dominates.

The presence of the aerobic and anaerobic denitrification terms in this equation is expected. However, the presence of the mass transfer term, s, requires clarification. The question is: Under what circumstances does

$$\frac{S[NO_3]_T}{[NO_3(1)]} \to s \qquad (4.38)$$

This can be seen by examining the normalized flux expression, Eq. (4.35)

$$\frac{S[NO_3]_T}{[NO_3(1)]} = \frac{s[NO_3(0)] + J_N - J[NH_4]}{J[NO_3]/s + [NO_3(0)]} \qquad (4.39)$$

Two cases produce the limiting behavior. The first corresponds to the case where both the nitrate source due to nitrification, $J_N - J[NH_4]$, and the nitrate flux to or from the sediment, $J[NO_3]$, are small relative to the mass transfer flux to the sediment, $s[NO_3(0)]$. This occurs for large s and/or large $[NO_3(0)]$. It is the usual situation.

The second case occurs if nitrate is behaving conservatively in the sediment and no denitrification is occurring. For this case nitrate flux is equal to the production of nitrate

$$J[NO_3] = J_N - J[NH_4] \qquad (4.40)$$

Thus Eq. (4.39) becomes

$$\frac{S[NO_3]_T}{[NO_3(1)]} = \frac{s[NO_3(1)]}{[NO_3(1)]} = s \qquad (4.41)$$

because

$$s[NO_3(0)] + J_N - J[NH_4] = s[NO_3(1)] + J[NO_3] = s[NO_3(1)] \qquad (4.42)$$

This situation corresponds to the low temperature periods when $\kappa_{NO_3,1}$ and $\kappa^*_{NO_3,2}$ are small and nitrate is behaving conservatively.

4.6 B   Sensitivity Analysis

A sensitivity analysis for the normalized flux equation is presented in Fig. 4.8. The straight line corresponds to both $\kappa_{NO_3,1}$, and $\kappa^*_{NO_3,2}$ equaling zero. As the κ's increase, the normalized flux increases as s becomes smaller. What distinguishes aerobic and anaerobic layer denitrification is that aerobic layer denitrification increases

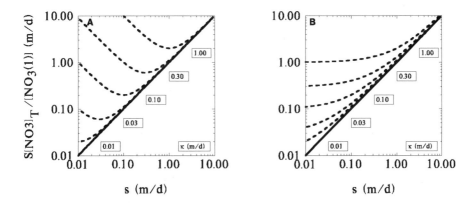

Fig. 4.8 Sensitivity to reaction velocity. Boxed quantities are (A) κ_{NO_3} and (B) $\kappa^*_{NO_3,2}$. Straight lines correspond to $\kappa_{NO_3} = 0$ and $\kappa^*_{NO_3,2} = 0$. Dashed lines are Eq. (4.37) for the indicated values of κ.

sharply as s decreases (Fig. 4.8A) whereas anaerobic layer denitrification reaches a plateau (Fig. 4.8B). Unless the normalized data have a distinctive upward curvature at small s's, it would be difficult to identify whether the denitrification was occurring in the aerobic or anaerobic layer. As shown below (Fig. 4.10) the data are equivocal.

4.6 C   Diffusive Mass Transfer Coefficient

There is an additional constraint that limits the extent of anaerobic layer denitrification. It can be limited by the rate at which nitrate is transported from the aerobic to the anaerobic layer. This is controlled by the diffusive mass transfer coefficient between the two layers K_{L12}. An independent estimate of this parameter is necessary to evaluate the extent of anaerobic layer denitrification.

A direct estimate of K_{L12} is available using a result from the ammonia flux model. The anaerobic layer ammonia concentration is given by Eq. (3.6) where $J_{N2} = J_N$ since diagenesis in layer 1 is assumed to be small

$$[NH_4(2)] = \frac{J_N}{K_{L12}} + [NH_4(1)] \tag{4.43}$$

or

$$\frac{J_N}{K_{L12}} = [NH_4(2)] - [NH_4(1)] \tag{4.44}$$

The aerobic layer ammonia concentration can be estimated from the flux equation (3.7)

$$[NH_4(1)] = \frac{J[NH_4]}{s} + [NH_4(0)] \tag{4.45}$$

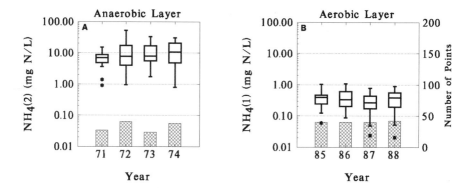

Fig. 4.9 Box plots of the anaerobic and aerobic layer ammonia concentrations. (A) Average 0 to 10 cm pore water concentrations from the Bricker et al. (1977) data set; (B) Estimated $NH_4(1)$ using Eq. (4.45) and the Chesapeake Bay data set. The outliers are denoted by $*$. The shaded bars (rhs scale) indicate the number of data points in each year.

Substituting Eq. (4.45) into Eq. (4.44) yields

$$\frac{J_N}{K_{L12}} = [NH_4(2)] - \frac{J[NH_4]}{s} - [NH_4(0)] \qquad (4.46)$$

Therefore K_{L12} can be estimated using an estimate of J_N, and measurements of ammonia flux, overlying ammonia concentration, observed $[NH_4(2)]$ concentration, and s. This result is applied in the next section.

4.7 APPLICATION TO CHESAPEAKE BAY

The anaerobic layer ammonia concentration data from the Bricker et al. (1977) data set is summarized in Fig. 4.9A. The box symbols represent the median (the horizontal line), the 25th and 75th percentiles (the lower and upper limits of the box), and the ranges, excluding outliers. The data include all the stations analyzed from 1971 to 1974. The histograms specify the number of data points in each box. An anaerobic layer ammonia concentration of $[NH_4(2)] = 10.0$ mg N/L is representative (Fig. 4.9A).

The aerobic layer ammonia concentration $[NH_4(1)]$ is estimated from the Chesapeake Bay data set using Eq. (4.45). Concentrations are computed to be less than 1.0 mg N/L (Fig. 4.9B), so $[NH_4(1)] \ll [NH_4(2)]$. Thus, from Eq. (4.44)

$$K_{L12} \approx \frac{J_N}{[NH_4(2)]} \qquad (4.47)$$

For an average ammonia diagenesis of $J_N = 100$ mg N/m^2-d and $[NH_4(2)] = 10.0$ mg N/L, the diffusive exchange mass transfer coefficient is $K_{L12} = 0.01$ m/d $= 1.0$ cm/d.

Table 4.2 Denitrification Parameters

Parameter	Values	Median	Units
θ_{NO_3}	$1.200^a, 1.070^b, 1.100^c,$ $1.056^d, 1.098^d, 1.074^d$	1.086	–
K_{M,NO_3}	$0.1^a, 3.06^c, 0.98^d$	0.98	mg N/L
K_{O_2,NH_4}	0.080^d	0.080	mg O_2/L

[a] Argaman and Miller (1979), [b] Lewandoswki (1982), [c] Messer and Brezonik (1984), [d] Nakajima et al. (1984).

This result can be compared to the mass transfer coefficient estimated using the relationship (Eq. 2.56)

$$K_{L12} = \frac{D_{NH_4}}{[H_2 + H_1]/2} \simeq \frac{D_{NH_4}}{H_2/2} \qquad (4.48)$$

since $H_2 \gg H_1$. For a depth of the anaerobic layer $H_2 = 10$ cm and the diffusion coefficient for ammonia $D_{NH_4} \sim 2.0$ cm^2/d (Table 2.1) the mass transfer coefficient is estimated to be $K_{L12} = 0.4$ cm/d. This is approximately twofold lower than the value estimated using Eq. 4.47. If the estimate from the continuous solution analysis is used for the mixing length ($H_2/3$, Eq. 3B.8), then $K_{L12} = 0.6$ cm/d. This results suggests that, perhaps, some additional mixing is present, probably due to the activities of benthic organisms (bioirrigation). However, it is not likely to be a large effect. This phenomenon is examined in further detail in Chapter 14.

4.8 ESTIMATE OF THE DENITRIFICATION REACTION VELOCITIES

The flux normalization equation (4.35) has all measured or estimated quantities on the right-hand side, and a three term expression involving the unknown parameters on the left-hand side as a function of s. Therefore, a plot of $S[NO_3]_T/[NO_3(1)]$, computed from Eq. (4.39), versus s can be used to estimate $\kappa_{NO_3,1}$ and $\kappa^*_{NO_3,2}$. There is a problem, however, because temperature affects the reaction velocities. Therefore, Eq. (4.35) becomes

$$\frac{\kappa^2_{NO_3,1}\theta_{NO_3}^{(T-20)}}{s} + s + \kappa^*_{NO_3,2}\theta_{NO_3}^{(T-20)} = \frac{S[NO_3]_T}{[NO_3(1)]} \qquad (4.49)$$

where $\theta_{NO_3,2}$ is the temperature coefficient for either $\kappa^2_{NO_3,1}$ or $\kappa^*_{NO_3,2}$. Table 4.2 lists the reported values. The terms involving the κ's are most important for small s. It happens that the temperatures are low for these observations. It is for these temperatures that the temperature correction is important. For larger s the temperatures are closer to 20°C and the correction is not significant. This suggests that it may be a reasonable approximation to move the temperature correction $\theta_{NO_3}^{(T-20)}$ to the right-hand side of the equation

$$\frac{\kappa_{NO_3,1}^2}{s} + s + \kappa_{NO_3,2}^* \approx \frac{S[NO_3]_T}{[NO_3(1)]\theta_{NO_3}^{(T-20)}} = \frac{s[NO_3(0)] + J_N - J[NH_4]}{[NO_3(1)]\theta_{NO_3}^{(T-20)}} \quad (4.50)$$

and ignore the fact that it does not multiply s. The second equality follows from substituting Eq. (4.14) – see Eq. (4.37) for the definition – into Eq. (4.50). We have found that this approximation is preferable to ignoring the temperature dependence.

The equations used for evaluating Eq. (4.50) are as follows. The aerobic layer nitrate concentration, $[NO_3(1)]$, is estimated using Eq. (4.36). The numerator terms s and $[NO_3(0)]$ are measured. The nitrate produced by nitrification $J_N - J[NH_4]$ can be estimated by evaluating the kinetic expression (Eq. 3.45)

$$S[NO_3] = J_N - J[NH_4] = \left(\frac{K_{M,NH_4}\theta_{K_{M,NH_4}}^{(T-20)}}{K_{M,NH_4}\theta_{K_{M,NH_4}}^{(T-20)} + [NH_4(1)]} \right)$$

$$\left(\frac{[O_2(0)]}{2K_{O_2,NH_4} + [O_2(0)]} \right) \frac{\kappa_{NH_4,1}^2 \theta_{NH_4}^{(T-20)}}{s}[NH_4(1)] \quad (4.51)$$

where the aerobic layer ammonia concentration is estimated using (Eq. 3.47)

$$[NH_4(1)] = \frac{J[NH_4]}{s} + [NH_4(0)] \quad (4.52)$$

The results are shown in Fig. 4.10. Since all these estimates involve measured quantities, the individual estimates are quite variable, and an averaging procedure is employed. The normalized flux in binned into 0.1 $\log_{10}$ units of s. The mean and standard error of the mean are shown. The histogram indicates the number of data points in each bin. The straight dashed line is $S[NO_3]_T/[NO_3(1)] = s$. The solid fitted line corresponds to anaerobic layer denitrification limited by the aerobic-anaerobic layer diffusive mixing: $\kappa_{NO_3,2}^* \approx K_{L12} = 0.01$ (m/d) and $\kappa_{NO_3,1} = 0.1$ (m/d).

Note that the normalized flux data exhibit increasing fluxes for intermediate s and an upward curvature that indicates aerobic layer denitrification. The model is able to reproduce the aerobic layer denitrification at the lower s, but there is some deviation from the data at intermediate s (0.1 to 0.3 m/d). This could be remedied by increased anaerobic layer denitrification (see Fig. 4.8B). However, the constraint is due to the analysis of the anaerobic layer pore water ammonia concentration that limits the diffusive exchange to $K_{L12} = 0.01$ (m/d). As shown in Fig. 4.8B, this limits the possible contribution of anaerobic layer denitrification to only a small part of the overall denitrification that is taking place.

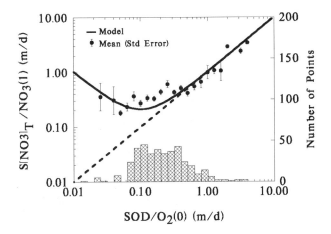

Fig. 4.10 Normalized nitrate source vs SOD/O_2 (0) . Solid line is Eq. (4.50). The shaded bars are a histogram of the number of data points in each average (rhs scale).

4.9 OBSERVATIONS OF CHESAPEAKE BAY DENITRIFICATION

Direct measurements of the rate of denitrification in Chesapeake Bay sediments have been made during 1988 (Kemp et al., 1990, Sampou et al., 1989). These are compared to model predictions in two ways, for two different approaches to making the predictions. For the main stem stations, where flux measurements over a season have been made, the ammonia flux model has been applied (Fig. 3.9) and estimates of the nitrification source of nitrate, $S[NO_3]_{i,j}$, are available. Thus the aerobic layer nitrate concentration can be computed using the model (Eq. 4.11) with $\kappa^*_{NO_3,2} \approx K_{L12}$

$$[NO_3(1)]_{i,j} = \frac{S[NO_3]_{i,j} + s_{i,j}[NO_3(0)]_{i,j}}{\kappa^2_{NO_3,1}\theta_{NO_3}^{(T_{i,j}-20)}/s_{i,j} + s_{i,j} + K_{L12}\theta_{K_{L12}}^{(T_{i,j}-20)}} \qquad (4.53)$$

with the observed surface mass transfer coefficient $s_{i,j}$ and the temperature $T_{i,j}$.

The anaerobic layer nitrate concentration follows from Eq. (4.12)

$$[NO_3(2)]_{i,j} = [NO_3(1)]_{i,j} \frac{K_{L12}\theta_{K_{L12}}^{(T_{i,j}-20)}}{\kappa_{NO_3,2}\theta_{NO_3}^{(T_{i,j}-20)} + K_{L12}\theta_{K_{L12}}^{(T_{i,j}-20)}} \qquad (4.54)$$

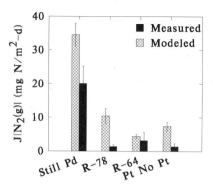

Fig. 4.11 Denitrification flux computed using Eq. (4.55). Data from Kemp et al. (1990), Sampou et al. (1989).

With the layer concentrations determined, the denitrification terms from the mass balance equations (Eq. 4.5a–4.5b) are used to compute the denitrification flux

$$J[N_2(g)]_{i,j} = k_{NO_3,1}H_1[NO_3(1)]_{i,j} + k_{NO_3,2}H_2[NO_3(2)]_{i,j}$$

$$= \frac{\kappa_{NO_3,1}^2 \theta_{NO_3}^{(T_{i,j}-20)}}{s_{i,j}}[NO_3(1)]_{i,j}$$

$$+ \kappa_{NO_3,2}\theta_{NO_3}^{(T_{i,j}-20)}[NO_3(2)]_{i,j} \qquad (4.55)$$

which is the flux of nitrogen gas to the overlying water. The model parameters that are used are the median in Table 4.2 for θ_{NO_3} – the other parameters are included for information only but are not used – and the case (d) estimates in Table 4.1.

The comparison is made in Fig. 4.11. The station averages for the main stem stations are computed from the individual estimates. The model appears to overestimate the observed $J[N_2(g)]$. Whether this is a systematic problem, or it is due to the considerable difficulty in measuring denitrification fluxes (Kemp et al., 1990), will be addressed in Chapters 14 and 15 with the time variable model.

An alternate method of computing the denitrification flux is to use the observed nitrate flux, surface mass transfer coefficient, and overlying water nitrate concentration, to estimate the aerobic layer nitrate concentration. The estimate is made from the flux equation (4.16) so that

$$[NO_3(1)] = \frac{J[NO_3]}{s} + [NO_3(0)] \qquad (4.56)$$

The anaerobic layer concentration is estimated using Eq. (4.54), and the estimate of $J[N_2(g)]$ then follows from Eq. (4.55) as before. The results are compared to the observations in Fig. 4.12. There is considerable scatter in the model estimates since they are based on observed ammonia and nitrate fluxes. Again the model seems to overestimate the observations except for August 1988.

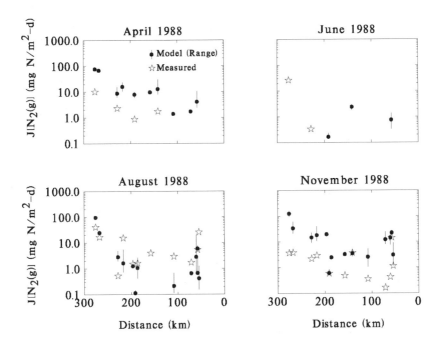

Fig. 4.12 Comparison of measured and modeled denitrification using Eqs. (4.56), (4.54), and (4.55). Data from Kemp et al. (1990), Sampou et al. (1989).

4.10   EXTENT OF DENITRIFICATION AND THE NITROGEN BALANCE

The objective of the ammonia and nitrate flux models is to compute the extent of nitrification and denitrification. The results from the model applied to the four years of data are summarized in this section for (A) main stem and (B) tributary Chesapeake Bay stations. The extent of denitrification is examined in Fig. 4.13. The source of nitrate from nitrification $S[NO_3]$ is shown and compared to the flux of nitrate to (−) or from (+) the sediment and the flux of nitrogen gas. For most stations, the nitrification source produces a small nitrate flux to the overlying water and a larger nitrogen gas flux. Where the nitrate flux is to the sediment (−), the nitrogen gas flux is considerably larger than $S[NO_3]$ since overlying water nitrate is being transported to the sediment and denitrified. These stations are characterized by high overlying water nitrate concentrations.

The nitrogen balance for the Chesapeake Bay stations are given in Fig. 4.14. The quantity of ammonia nitrogen produced by diagenesis J_N is shown. A fraction is released to the overlying water as an ammonia flux $J[NH_4]$. The remainder becomes nitrate. A portion $J[NO_3]$ either escapes to the overlying water, or additional nitrate is transported to the sediment. The quantity that remains is denitrified and a flux of nitrogen gas $J[N_2(g)]$ results. These are all shown in Fig. 4.14.

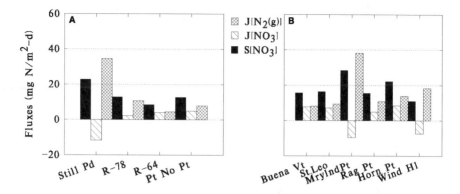

Fig. 4.13   Summary of model results. Extent of denitrification. Source of nitrate $S[NO_3]$, nitrate flux to the overlying water $J[NO_3]$, and nitrogen gas flux $J[N_2(g)]$ to the overlying water due to denitrification.

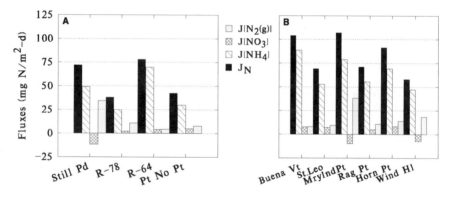

Fig. 4.14   Summary of model results: Nitrate sources and sinks. Ammonia diagenesis J_N, ammonia flux $J[NH_4]$, nitrate flux $J[NO_3]$, and nitrogen gas flux $J[N_2(g)]$.

4.11   CONCLUSIONS

The nitrate flux model reproduces the major features that relate the flux to the overlying water nitrate concentration and to the surface mass transfer coefficient. This latter relationship combines the effects of mass transport and the thickness of the aerobic layer, as it does in the ammonia flux model. The surprising result is that the primary site of denitrification is in the aerobic layer. Mass transfer of nitrate to the anaerobic layer is insufficient for significant denitrification to occur in that layer. It is possible that this result is an artifact of the two-layer segmentation. Recent measurements of vertical profiles of oxygen and nitrate in sediment pore waters indicate that the zone of nitrate reduction is below the oxic zone (Sorensen and Revsbech, 1990). However, it is difficult to reconcile this result with the results of the analysis that without aerobic layer denitrification, substantial fluxes of nitrate to the overlying water would result.

The magnitude of the denitrification flux predicted by the model is roughly comparable to independent measurements, although as with any pointwise comparison, there is considerable scatter. For the main stem and tributary stations, 76% of ammonia diagenesis is returned as ammonia flux. The rest is either denitrified or returned as a nitrate flux. The nitrogen gas flux is 22% of the ammonia diagenesis flux, but this includes the denitrification of overlying water nitrate as well.

Appendix 4A: MACSYMA

We continue the presentation of MACSYMA begun in Chapter 2 with the solution of the nitrate model equations Eqs. (4.5). The output is listed in Fig. 4A.1. The lines marked with "c" are the command input, the "d"'s are MACSYMA's response.

The first two lines (c3) and (c4) assign the mass balance equations to the variables eq1 and eq2. The response (d3) is just a restatement of the equation. The next line (c4) is the instruction to solve the equation eq2 for the unknown no3[2] and assign the solution to ss1. The trailing [1][1] selects the first of the solutions in the set produced by solve.

The solution for no3[1] proceeds in two steps. First no3[1] is eliminated from the two equation set [eq1,eq2], creating a single equation stored in ss2. That equation is solved for no3[1] and factored.

Substituting $\kappa_{2,s}$ as defined by rule1 into no3[1] is done using scsimp. The factored result is shown as d9.

Computing the flux is simply a matter of defining j[no3], simplifying the results and distributing the two terms, to produce the final answer d11.

Mass Balance Equations for NO3

(c3)   eq1: 0=s*(no3[0]-no3[1])+k[l12]*(no3[2]-no3[1])-kappa[1]^2/s*no3[1]+j[n]-j[nh4]

(d3)
$$0 = \left(no3_0 - no3_1\right) s - \frac{\kappa_1^2 \, no3_1}{s} - j_{nh4} + j_n + \left(no3_2 - no3_1\right) k_{l12}$$

(c4)   eq2: 0=-k[l12]*(no3[2]-no3[1])-kappa[2]*no3[2]

(d4)
$$0 = - \left(no3_2 - no3_1\right) k_{l12} - \kappa_2 \, no3_2$$

Solve for NO3[2]

(c5)   ss1:solve([eq2],[no3[2]])[1][1]

(d5)
$$no3_2 = \frac{no3_1 \, k_{l12}}{k_{l12} + \kappa_2}$$

Eliminate NO3[2] and solve for NO3[1]

(c6)   ss2:eliminate([eq1,eq2],[no3[2]])[1]$

(c7)   ss3:facsum(solve(ss2,no3[1]))[1]

(d7)
$$no3_1 = \frac{\left(k_{l12} + \kappa_2\right) s \left(no3_0 \, s - j_{nh4} + j_n\right)}{k_{l12} \, s^2 + \kappa_2 \, s^2 + \kappa_2 \, k_{l12} \, s + \kappa_1^2 \, k_{l12} + \kappa_1^2 \, \kappa_2}$$

Define the composite mass transport-reaction velocity coefficient

(c8)   rule1:kappa[2,s]=(1/kappa[2]+1/k[l12])^-1

(d8)
$$\kappa_{2,s} = \frac{1}{\frac{1}{k_{l12}} + \frac{1}{\kappa_2}}$$

(c9)   no3[1]:facsum(scsimp(ss3,rule1))

(d9)
$$no3_1 = \frac{s \left(no3_0 \, s - j_{nh4} + j_n\right)}{s^2 + \kappa_{2,s} \, s + \kappa_1^2}$$

Nitrate flux

(c10)   ss5:j[no3]=s*(rhs(no3[1])-no3[0])

(d10)
$$j_{no3} = s \left(\frac{s \left(no3_0 \, s - j_{nh4} + j_n\right)}{s^2 + \kappa_{2,s} \, s + \kappa_1^2} - no3_0 \right)$$

(c11)   distrib(facsum(ss5,no3[0]))

(d11)
$$j_{no3} = - \frac{\left(j_{nh4} - j_n\right) s^2}{s^2 + \kappa_{2,s} \, s + \kappa_1^2} - \frac{no3_0 \, s \left(\kappa_{2,s} \, s + \kappa_1^2\right)}{s^2 + \kappa_{2,s} \, s + \kappa_1^2}$$

Fig. 4A.1   MACSYMA solution for the nitrate mass balance (Eqs. 4.5).

5

Steady State Model

5.1 INTRODUCTION

This chapter presents the formulation for the general sediment flux model that will be applied in the succeeding chapters. The model is structured to include both dissolved and particulate species, since both are important in determining the fluxes of solutes that either adsorb to solid particles or form precipitates, or both. A model with similar mathematical structure, which describes water column-sediment interactions, has been formulated and analyzed.* It provides the basis for the analysis presented below. Analytical solutions are obtained for steady state conditions which provide valuable insights into the behavior of the model.

5.1 A Dissolved and Particulate Phases

An important feature of the chemicals produced by mineralization of organic matter in sediments is the extent to which they become particulate species. This distribution directly affects the magnitude of the chemical that is returned to the overlying water, since it determines to some extent the fraction that is buried. Therefore, any model of sediment fluxes must include this mechanism in its formulation.

For the model developed below, the distribution of a chemical between the particulate and dissolved phases in a sediment is parameterized, for the most part, using a linear partition coefficient. The choice is made for a number of reasons. First,

---

*Di Toro and Paquin (1984), Di Toro et al. (1982), Thomann and Mueller (1987).

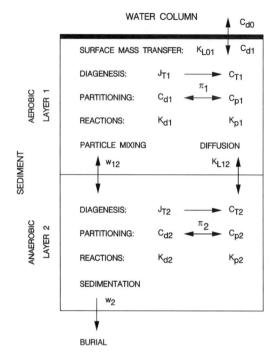

Fig. 5.1   Schematic diagram of the flux model.

the resulting equations can be solved analytically, which is an important aid to understanding the model's behavior. Second, linear partitioning can sometimes be a realistic description of the relationship between dissolved and particulate chemical. Third, the general problem of computing the chemical composition of pore water would involve using a numerical chemical equilibrium model. Mass balance equations are required for the various chemicals that affect the pore water chemistry – for example, hydrogen ion and carbon dioxide. Thermodynamic data are required for the relevant aqueous complexes, and the stable and metastable mineral phases, some of which are uncertain. Finally, sorption as well as precipitation reactions need to be considered. All this is necessary to compute the fraction of a chemical that is either dissolved or particulate.

In fact, in Parts VI and VII of this book, more realistic models of partitioning are included with the concomitant increase in mathematical complexity. Analytical solutions are still possible in some cases, but quite a lot of transparency is lost.

The equivalent partitioning model employs only a partition coefficient, the ratio of particulate to dissolved chemical concentration. If necessary, it can be varied as a function of other physical and chemical parameters in order to produce more realistic behavior. The practical question is: Does the added difficulty of including equilibrium chemistry into the model structure result in added realism? Whatever the answer, it is prudent to begin the modeling using linear partitioning and to examine the utility of the results.

5.1 B   Particle Mixing

The inclusion of particulate as well as dissolved species requires that the transport of particulate species be considered. The rate of mixing in the sediment is formulated using a particle mixing velocity. This is equivalent to representing particle mixing using a diffusion model. More elaborate models have been proposed which mimic more directly the mixing activity of benthic organisms (Boudreau (1986a,b), Robbins (1986), see Thoms et al. (1995) for a review of available models). However, as with the choice of a model for chemical partitioning, simple diffusion – in fact just particle exchange between the two layers – appears to be a reasonable first step.

5.2   MODELING FRAMEWORK

Figure 5.1 presents the framework and defines the variables used in the model. The total concentrations – the sum of the dissolved and particulate species – in layers 1 and 2 are C_{T1} and C_{T2}, respectively. The sources of chemical are denoted by J_{T1} and J_{T2}, the areal production rate in the aerobic and anaerobic layers, respectively. The fractions of the total concentration that are dissolved f_d and particulate f_p in layers 1 and 2 are specified by f_{d1}, f_{d2}, and, f_{p1}, f_{p2}, respectively. These fractions depend on the solids concentrations and partition coefficients, as shown below.

The removal reactions in layers 1 and 2 are first-order with rate constants: k_1 and k_2. The mass transport of dissolved chemical between layers 1 and 2 is via diffusion which is parameterized by a mass transfer coefficient K_{L12} as before. The mixing of particles between layers 1 and 2 due to physical and biological mechanisms is parameterized by a mixing velocity w_{12}, which has the same units as the mass transfer coefficient. Burial from layer 1 to layer 2 and out of layer 2 occurs at the sedimentation velocity w_2. Finally, the magnitude of the flux of dissolved chemical into or out of the sediment from the overlying water is determined by the surface mass transfer coefficient K_{L01}.

5.3   MASS BALANCE EQUATIONS

The mass balance equations for layers 1 and 2 are

$$H_1 \frac{dC_{T1}}{dt} = -k_1 H_1 C_{T1} + K_{L01}(f_{d0}C_{T0} - f_{d1}C_{T1})$$

$$+ w_{12}(f_{p2}C_{T2} - f_{p1}C_{T1})$$

$$+ K_{L12}(f_{d2}C_{T2} - f_{d1}C_{T1}) - w_2 C_{T1} + J_{T1} \qquad (5.1a)$$

$$H_2 \frac{dC_{T2}}{dt} = -k_2 H_2 C_{T2} - w_{12}(f_{p2}C_{T2} - f_{p1}C_{T1})$$

$$- K_{L12}(f_{d2}C_{T2} - f_{d1}C_{T1}) + w_2(C_{T1} - C_{T2}) + J_{T2} \qquad (5.1b)$$

They are formulated on the basis of the total chemical concentration. For the transport coefficients that affect only the dissolved or particulate chemical, the total concentration is multiplied by the fraction dissolved or particulate. Equations written in this form assume local equilibrium for the partitioning reaction.

The terms in Eq. (5.1a) represent, respectively, the removal of chemical by reaction, the exchange of dissolved chemical between layer 1 and the overlying water (layer 0), the exchange of particulate chemical between layers 1 and 2 via particle mixing, the exchange of dissolved chemical between layers 1 and 2 via diffusive transport, the loss of both dissolved and particulate chemical by burial into layer 2, and the source of chemical to layer 1.

The terms in Eq. (5.1b) represent, respectively, the removal of chemical by reaction, the exchange of particulate and dissolved chemical between layers 1 and 2, the gain of chemical from layer 1 and the loss of chemical from layer 2 by burial, and the source of chemical to layer 2. Note that the dissolved and particulate exchange terms have the opposite signs in layers 1 and 2. The reason is that the transport of chemical from layer 1 is a sink in that layer and a source to layer 2, and vice versa.

The dissolved f_d and particulate f_p fractions are computed from the partitioning equations

$$f_d = \frac{\phi}{1 + m_1 \pi_1 / \phi} \tag{5.2a}$$

$$f_p = \frac{m_1 \pi_1 / \phi}{1 + m_1 \pi_1 / \phi} \tag{5.2b}$$

where the solids concentrations are m_1 and m_2, and the partition coefficients are π_1 and π_2, respectively. Note that the solids concentration partition coefficient products: $m_1 \pi_1$ and $m_2 \pi_2$ determine the extent of partitioning. The concentrations of dissolved and particulate chemical are obtained as products of these fractions and the total concentrations C_{T1} and C_{T2}.

The definitions of the concentration units in Eqs. (5.1) are as follows. The concentrations C_{T1} and C_{T2} are the mass of chemical per unit bulk volume, that is the volume of the pore water and solids mixture. The dissolved concentrations $f_d C_T$ are mass of chemical per unit volume of pore water. As a consequence the dissolved fraction has the units of porosity ϕ (volume of pore water / bulk volume). The partition coefficient π is the ratio of the particulate concentration (mass of chemical per mass of solids) to the dissolved concentration (mass of chemical per volume of pore water). The solids concentration is the mass of solids per unit bulk volume. Thus the porosity ϕ is included in the solids concentration-partition coefficient product $m_1 \pi_1 / \phi$ to account for these definitions. In most cases, the porosity is usually close to one unless the particle sizes are large (see Fig. 1.1), and the corrections are usually not significant.

5.4 SOLUTION FOR ANAEROBIC LAYER SOURCE

The analytical solutions for these equations are presented in two parts. First, the solution is found for a source only in the anaerobic layer. Then, only the sources to the aerobic layer are considered. Since these equations are linear, the complete solution is the sum of the solutions for the individual sources. This procedure simplifies the derivation and the form of the solutions.

The mass balance equation for layer 1 without the source terms is

$$H_1 \frac{dC_{T1}}{dt} = -k_1 H_1 C_{T1} - K_{L01} f_{d1} C_{T1} + w_{12}(f_{p2}C_{T2} - f_{p1}C_{T1})$$

$$+ K_{L12}(f_{d2}C_{T2} - f_{d1}C_{T1}) - w_2 C_{T1} \tag{5.3}$$

The steady state solutions to Eq. (5.1b) and (5.3) are found by setting the time derivatives to zero and adding the equations

$$0 = -k_{T1} H_1 C_{T1} - (k_2 H_2 + w_2)C_{T2} + J_{T2} \tag{5.4}$$

where k_{T1} is defined as the total first-order removal rate constant in layer 1

$$k_{T1} H_1 = k_1 H_1 + K_{L01} f_{d1} \tag{5.5}$$

The overall mass balance equation (5.4) contains fewer terms because the internal mass transport terms cancel out. Only the source and removal terms remain. These terms represent sources to or removal from the entire active sediment layer. Eq. (5.4) can be solved for C_{T2} to yield

$$C_{T2} = \frac{J_{T2}}{k_2 H_2 + w_2 + k_{T1} H_1 \dfrac{C_{T1}}{C_{T2}}} \tag{5.6}$$

The ratio C_{T1}/C_{T2}, which is denoted by r_{12}, can be found by solving Eq. (5.3) at steady state

$$r_{12} \triangleq \frac{C_{T1}}{C_{T2}} = \frac{w_{12} f_{p2} + K_{L12} f_{d2}}{w_2 + w_{12} f_{p1} + K_{L12} f_{d1} + k_{T1} H_1} \tag{5.7}$$

This definition of r_{12} is slightly different from that used in the previous analysis of this model (Di Toro et al., 1982) but the idea is the same. Note that r_{12} is a function only of the reaction and transport parameters of the model. Hence, from an algebraic point of view, it is a *known* quantity. Thus the anaerobic layer concentration, Eq. (5.6), can be rewritten using this quantity

$$C_{T2} = \frac{J_{T2}}{k_2 H_2 + w_2 + k_{T1} H_1 r_{12}} \tag{5.8}$$

Finally Eq. (5.4) can be solved for the aerobic layer concentration

$$C_{T1} = \frac{J_{T2}}{k_{T1} H_1 + (k_2 H_2 + w_2)r_{21}} \tag{5.9}$$

where

$$r_{21} \triangleq r_{12}^{-1} \qquad (5.10)$$

The simplicity of this solution is due to the lack of source terms in the aerobic layer equation – compare Eq. (5.1a) to Eq. (5.3). This is the motivation for considering these source terms separately.

5.4 A   Concentration Ratio

The roles of r_{12} and r_{21} in these solutions can be explained as follows. In Eq. (5.8) for C_{T2}, the layer 2 sinks, $k_2 H_2$ and w_2, are applied directly to the denominator of the solution. The layer 1 sinks, $k_{T1} H_1$, are modified by $r_{12} = C_{T1}/C_{T2}$ so that they are applied to the equivalent layer 1 concentration. Since r_{12} is determined only by the parameters of the model, it can be viewed as a known quantity rather than the ratio of the two unknown concentrations. The inverse of this ratio, r_{21}, plays the same role in the solution for C_{T1} in Eq. (5.9).

These ratios are the only place where the layer 1–2 mixing and partitioning parameters appear. An interesting special case occurs if the particulate fractions are equal in both layers: $f_{p1} = f_{p2}$, which implies equality of the dissolved fractions as well $f_{d1} = f_{d2}$. In addition, if the layer mixing parameters are large relative to the reaction and burial terms

$$w_{12} f_{p2} + K_{L12} f_{d2} \gg w_2 + k_{T1} H_1 \qquad (5.11)$$

then

$$r_{12} = \frac{C_{T1}}{C_{T2}} = \frac{w_{12} f_{p2} + K_{L12} f_{d2}}{w_2 + w_{12} f_{p1} + K_{L12} f_{d1} + k_{T1} H_1} \rightarrow 1 \qquad (5.12)$$

Thus the mixing – either particle mixing or dissolved phase mixing – equalizes the concentrations in the two layers.

5.4 B   Final Form

For the sake of completeness the method for evaluating the aerobic layer reaction rate-depth product $k_{T1} H_1$, discussed in Chapters 3 and 4, is repeated here. The surface mass transfer coefficient K_{L01} and the depth of the aerobic zone H_1 are evaluated using the ratio of the sediment oxygen demand and the overlying water oxygen concentration

$$K_{L01} = \frac{\text{SOD}}{[O_2(0)]} \triangleq s \qquad (5.13)$$

$$H_1 = D_1 \frac{[O_2(0)]}{\text{SOD}} = \frac{D_1}{s} \qquad (5.14)$$

where D_1 is the diffusion coefficient of dissolved oxygen in layer 1. We assume that all dissolved species have the same diffusion coefficient. If the mechanism of dissolved mixing is molecular diffusion, then this assumption is an approximation, as discussed in Chapter 2. Of course it is not a necessary assumption, but it makes the analysis more convenient.

The definitions of the reaction velocities follow the convention established for ammonia and nitrate reactions

$$\kappa_1 = \sqrt{D_1 k_1} \tag{5.15}$$

$$\kappa_2 = k_2 H_2 \tag{5.16}$$

Hence, the term $k_{T1} H_1$, Eq. (5.5), becomes

$$k_{T1} H_1 = \frac{\kappa_1^2}{s} + s f_{d1} \tag{5.17}$$

The total concentrations in layers 1 and 2 become

$$C_{T1} = \frac{J_{T2}}{\dfrac{\kappa_1^2}{s} + s f_{d1} + (\kappa_2 + w_2) r_{21}} \tag{5.18}$$

$$C_{T2} = \frac{J_{T2}}{\left(\dfrac{\kappa_1^2}{s} + s f_{d1}\right) r_{12} + \kappa_2 + w_2} \tag{5.19}$$

The flux of chemical to the overlying water – not the net flux which would also include the flux from the overlying water to the sediment – is

$$J_{aq} = s f_{d1} C_{T1} = J_{T2} \frac{s f_{d1}}{\dfrac{\kappa_1^2}{s} + s f_{d1} + (\kappa_2 + w_2) r_{21}} \tag{5.20a}$$

It is convenient to define the flux reacted in layer 2, $J_{re,2} = \kappa_2 C_{T2}$, and the burial flux, $J_{br} = w_2 C_{T2}$. These can be calculated using Eq. (5.19) for the concentration C_{T2}

$$J_{re,2} = \kappa_2 C_{T2} = \kappa_2 r_{21} C_{T1} = J_{T2} \frac{\kappa_2 r_{21}}{\dfrac{\kappa_1^2}{s} + s f_{d1} + (\kappa_2 + w_2) r_{21}} \tag{5.20b}$$

$$J_{br} = w_2 C_{T2} = w_2 r_{21} C_{T1} = J_{T2} \frac{w_2 r_{21}}{\dfrac{\kappa_1^2}{s} + s f_{d1} + (\kappa_2 + w_2) r_{21}} \tag{5.20c}$$

The terms in these equations suggest the definition of the following mass transfers and equivalent reaction velocity expressions that correspond to reaction in layer 1

$$fr_{re,1} = \frac{\kappa_1^2}{s} \qquad (5.21a)$$

dissolved mixing with respect to the overlying water

$$fr_{aq} = s f_{dl} \qquad (5.21b)$$

reaction in layer 2

$$fr_{re,2} = \kappa_2 r_{21} \qquad (5.21c)$$

and to burial

$$fr_{br} = w_2 r_{21} \qquad (5.21d)$$

When compared one to another, they can be thought of as the fraction of total diagenesis that is routed to each of the pathways. Using these definitions, Eqs. (5.20) can be expressed as

$$J_{aq} = J_{T2} \frac{fr_{aq}}{fr_{re,1} + fr_{aq} + fr_{re,2} + fr_{br}} \qquad (5.22a)$$

$$J_{re,2} = J_{T2} \frac{fr_{re,2}}{fr_{re,1} + fr_{aq} + fr_{re,2} + fr_{br}} \qquad (5.22b)$$

and

$$J_{br} = J_{T2} \frac{fr_{br}}{fr_{re,1} + fr_{aq} + fr_{re,2} + fr_{br}} \qquad (5.22c)$$

The flux corresponding to the chemical that is reacted in layer 1 follows from the difference between the diagenesis flux and the losses via diffusion to the overlying water, layer 2 reaction, and burial

$$J_{re,1} = J_{T2} - J_{aq} - J_{re,2} - J_{br}$$

$$= J_{T2} \frac{fr_{re,1}}{fr_{re,1} + fr_{aq} + fr_{re,2} + fr_{br}} \qquad (5.22d)$$

5.4 C Properties

The general behavior of the steady state version of the model can be deduced from the form of these equations. The diagenesis flux is apportioned between the four removal processes: reaction in layers 1 and 2, flux to the overlying water, and burial from layer 2. The relative magnitudes of the mass transfer and reaction velocity parameters, Eqs. (5.21), determine the magnitude of each of these terminal sinks. It is important to realize that since the model is based on a mass balance, there is no other possible behavior. The chemical produced by diagenesis must exit to one of these sinks.

The parameters in the model are the reaction velocities in layers 1 and 2: κ_1 and κ_2, the mass transfer coefficients for layer 1-layer 2 dissolved and particulate mixing K_{L12} and w_{12}, the partition coefficients in the two layers π_1 and π_2, and the sedimentation velocity w_2. Estimates of the parameters and the behavior of the model are examined in the next chapters where the model is applied to phosphorus and silica.

5.5 SOLUTION FOR AEROBIC LAYER SOURCE

The previous section presented the solution for the source in the anaerobic layer. In this section, only the aerobic layer source is considered. The sources to the aerobic layer – the diffusive exchange source from the overlying water to the sediment $K_{L01} f_{d0} C_{T0}$ and the diagenesis source J_{T1} – are combined into a single source term for notational convenience

$$J_{\text{Tot},1} = J_{T1} + K_{L01} f_{d0} C_{T0} \tag{5.23}$$

The steady state solution is found as before by setting the time derivatives to zero and adding the aerobic and anaerobic layer equations (5.1a) and (5.1b) with $J_{T2} = 0$

$$0 = -k_{T1} H_1 C_{T1} - (k_2 H_2 + w_2) C_{T2} + J_{\text{Tot},1} \tag{5.24}$$

This equation can be solved for C_{T1} to yield

$$C_{T1} = \frac{J_{\text{Tot},1}}{k_{T1} H_1 + (k_2 H_2 + w_2)\dfrac{C_{T2}}{C_{T1}}} \tag{5.25}$$

The ratio C_{T2}/C_{T1}, which will be denoted by r_{21}^* to distinguish it from r_{21} (Eq. 5.10), can be found by solving Eq. (5.1b) with $J_{T2} = 0$ at steady state

$$r_{21}^* \triangleq \frac{C_{T2}}{C_{T1}} = \frac{w_{12} f_{p1} + K_{L12} f_{d1} + w_2}{w_{12} f_{p2} + K_{L12} f_{d2} + w_2 + k_2 H_2} \tag{5.26}$$

which is a known quantity. Thus the aerobic layer concentration is

$$C_{T1} = \frac{J_{\text{Tot},1}}{k_{T1} H_1 + (k_2 H_2 + w_2) r_{21}^*} \tag{5.27}$$

Eq. (5.24) can be solved for the anaerobic layer concentration

$$C_{T2} = \frac{J_{\text{Tot},1}}{(k_{T1} H_1) r_{12}^* + k_2 H_2 + w_2} \tag{5.28}$$

where

$$r_{12}^* \triangleq (r_{21}^*)^{-1} \tag{5.29}$$

5.5 A   Comparisons

The solutions for the source in the aerobic and anaerobic layer are quite similar. In fact, it might be suspected that the solutions for the source in layer 1 can be derived from the layer 2 source solutions by an interchange of the corresponding terms in layers 1 and 2. This can be checked by comparing the solutions in the layer receiving the source term: Eq. (5.27) to Eq. (5.8)

$$C_{T1} = \frac{J_{\text{Tot},1}}{k_{T1} H_1 + (k_2 H_2 + w_2) r_{21}^*} \tag{5.30}$$

$$C_{T2} = \frac{J_{T2}}{k_2 H_2 + w_2 + k_{T1} H_1 r_{12}} \tag{5.31}$$

The solutions would be identical with the replacement: $k_2 H_2 + w_2 \longleftrightarrow k_{T1} H_1$. However, it is also necessary that $r_{21}^* = r_{12}$ with $1 \longleftrightarrow 2$. That this is not the case can be seen by inspection

$$r_{21}^* = \frac{w_{12} f_{p1} + K_{L12} f_{d1} + w_2}{w_{12} f_{p2} + K_{L12} f_{d2} + w_2 + k_2 H_2} \tag{5.32}$$

$$r_{12} = \frac{w_{12} f_{p2} + K_{L12} f_{d2}}{w_{12} f_{p1} + K_{L12} f_{d1} + w_2 + k_{T1} H_1} \tag{5.33}$$

The internal mixing is completely symmetric. The loss to the overlying water in layer 1 is equivalent to loss from layer 2 by burial. However, the burial flux between layers 1 and 2 is not symmetric. It is a unidirectional advective flux from layer 1 to layer 2. This fact is reflected in the difference between r_{21}^* and r_{12}: the appearance of w_2 in the numerator of r_{21}^*.

A more practical question is: Does the location of the source term have any significant effect on the concentrations? This can be examined by comparing the layer 1 solutions for both cases

$$C_{T1} = \frac{J_{\text{Tot},1}}{k_{T1} H_1 + (k_2 H_2 + w_2) r_{21}^*} \tag{5.34}$$

$$C_{T1} = \frac{J_{T2}}{k_{T1} H_1 + (k_2 H_2 + w_2) r_{21}} \tag{5.35}$$

The concentration ratios are

$$r_{21}^* = \frac{w_{12} f_{p1} + K_{L12} f_{d1} + w_2}{w_{12} f_{p2} + K_{L12} f_{d2} + w_2 + k_2 H_2} \tag{5.36}$$

$$r_{21} = \frac{w_{12} f_{p1} + K_{L12} f_{d1} + w_2 + k_{T1} H_1}{w_{12} f_{p2} + K_{L12} f_{d2}} \tag{5.37}$$

which are not the same. However, for the case when the mixing terms are large relative to the reaction rate terms and the sedimentation velocity, $r_{21}^* \approx r_{21} \approx 1$, the solutions are identical. Therefore, for this special case, sources into either layer can be treated as though they were sources into the other layer. This simplification will be used subsequently in the application of the model to phosphorus and silica.

Appendix 5A: MACSYMA

We continue the presentation of MACSYMA with the solution to the two-layer equations. The output is listed in Fig. 5A.1. The first two lines (c2) and (c3) assign the mass balance equations to the variables eq1 and eq2. The next line (c4) checks the equations by adding them. The results are the source and sink terms: burial, reaction in both layers, and the source. (c5) is the instruction to solve both equations [eq1,eq2] for the unknowns [ct1,ct2]. The trailing $ suppresses the printing to save space. (c6) and (c7) prints the solutions.

The next instructions substitute r_{12} into the algebraically complex solutions to produce remarkably concise results (Eqs. 5.8–5.9). First r_{12} is found. The numerator (c8) and denominator (c9) are found separately, and finally used to define r_{12}. They are expressed as "rules", actually equations, that are then used to define the substitutions in scsimp. It seems almost magical that the complicated expressions (d6) and (d7) can be reduced to (d11) and (d12).

Layer 1

(c2)   eq1:-kt1*h1*ct1+kl12*(fd2*ct2-fd1*ct1)+w12*(fp2*ct2-fp1*ct1)-w2*ct1=0

(d2) $- ct1\ w2 + (ct2\ fp2 - ct1\ fp1)\ w12 - ct1\ h1\ kt1 + (ct2\ fd2 - ct1\ fd1)\ kl12 = 0$

Layer 2

(c3)   eq2:-kt2*h2*ct2-kl12*(fd2*ct2-fd1*ct1)-w12*(fp2*ct2-fp1*ct1)+w2*ct1-w2*ct2+j2=0

(d3) $- ct2\ w2 + ct1\ w2 - (ct2\ fp2 - ct1\ fp1)\ w12 - ct2\ h2\ kt2 - (ct2\ fd2 - ct1\ fd1)\ kl12 + j2 = 0$

Mass balance check

(c4)   eq1+eq2

(d4) $- ct2\ w2 - ct2\ h2\ kt2 - ct1\ h1\ kt1 + j2 = 0$

Solve the equations

(c5)   c:factorsum(solve([eq1,eq2],[ct1,ct2])[1])$

(c6)   ct[1]=ct1:rhs(c[1])

(d6) $$ct_1 = \frac{j2\ (fp2\ w12 + fd2\ kl12)}{\left(\begin{array}{l} w2^2 + fp1\ w12\ w2 + h2\ kt2\ w2 + h1\ kt1\ w2 + fd1\ kl12\ w2 + fp1\ h2\ kt2 \\ *\ w12 + fp2\ h1\ kt1\ w12 + h1\ h2\ kt1\ kt2 + fd1\ h2\ kl12\ kt2 + fd2\ h1\ kl12\ kt1 \end{array} \right)}$$

(c7)   ct[2]=ct2:rhs(c[2])

(d7) $$ct_2 = \frac{j2\ (w2 + fp1\ w12 + h1\ kt1 + fd1\ kl12)}{\left(\begin{array}{l} w2^2 + fp1\ w12\ w2 + h2\ kt2\ w2 + h1\ kt1\ w2 + fd1\ kl12\ w2 + fp1\ h2\ kt2 \\ *\ w12 + fp2\ h1\ kt1\ w12 + h1\ h2\ kt1\ kt2 + fd1\ h2\ kl12\ kt2 + fd2\ h1\ kl12\ kt1 \end{array} \right)}$$

Find r12

(c8)   rule1:r12num=ratnumer(factorsum(ct1/ct2))

(d8)
/R/ $r12num = fp2\ w12 + fd2\ kl12$

(c9)   rule2:r12den=ratdenom(factorsum(ct1/ct2))

(d9)
/R/ $r12den = w2 + fp1\ w12 + h1\ kt1 + fd1\ kl12$

(c10)   rule3:r12=r12num/r12den

(d10) $$r12 = \frac{r12num}{r12den}$$

Substitute r12

(c11)   ct[1]=facsum(scsimp(ct1,rule1,rule2,rule3),r12)

(d11) $$ct_1 = \frac{j2\ r12}{w2 + h1\ kt1\ r12 + h2\ kt2}$$

(c12)   ct[2]=facsum(scsimp(ct2,rule1,rule2,rule3),r12)

(d12) $$ct_2 = \frac{j2}{w2 + h1\ kt1\ r12 + h2\ kt2}$$

Fig. 5A.1   MACSYMA solution of the two-layer mass balance Eqs. (5.1).

6

Phosphorus

6.1 INTRODUCTION

The search for an understanding of the mechanisms that control the flux of phosphorus from sediments has a long history. For lake sediments, the classical experiments and their interpretation by Mortimer (1941, 1942) provided a framework within which to understand the profound effect of the overlying water dissolved oxygen concentration. He posited that a barrier to phosphate exists in the aerobic layer of the sediment due to the formation of iron oxyhydroxide precipitate via the oxidation of ferrous iron. The stoichiometry for this amorphous precipitate is reported to be $Fe_2O_3(H_2O)_n$ with $n = 1$ to 3 (Dzombak and Morel, 1990), which we abbreviate as FeOOH (goethite) for simplicity. Note that $2FeOOH = Fe_2O_3(H_2O)$ so the different formulas correspond to different structures with differing amounts of included water – see Section 20.2 for a more detailed discussion of iron chemistry. This particulate species strongly sorbs phosphate and prevents its escape to the overlying water via diffusion. When the overlying water oxygen concentration decreases to zero, the ferric oxyhydroxide is reduced to soluble ferrous iron, the barrier no longer exists, and phosphate escapes unimpeded. This mechanism has been invoked in many models of phosphate flux.

For marine sediments, the focus has been more on models for the interstitial water concentration distribution of phosphate.* These relate the diagenetic production of phosphate to the resulting pore water concentration distribution, usually as a one-dimensional steady state vertical model.

Models that are specifically designed to compute phosphate fluxes have been proposed. Empirical models relate phosphate flux to an extracted fraction of the phos-

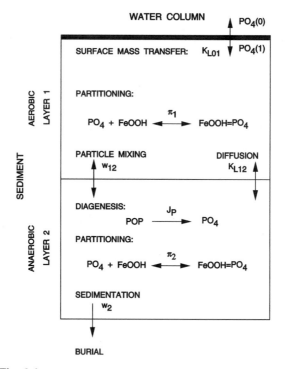

Fig. 6.1   Schematic diagram of the phosphorus flux model.

phorus concentration of the sediment.[†] More detailed, vertically segmented models have also been proposed.[‡] A continuous one-dimensional steady state model which includes partitioning to iron has been compared to flux measurements as well as pore water and sorbed concentrations.[#] The model developed below is based on both of these approaches. It incorporates the diagenetic production of phosphate, and it uses the mechanism of iron oxyhydroxide trapping. It should be pointed out that other mechanisms have been suggested that may be important: the role of sediment microorganisms[¶] and the influence of overlying water sulfate concentration[♭].

6.2   MODEL COMPONENTS

The phosphate flux model is constructed using the solutions for the steady state model equations developed in Chapter 5. The schematic is presented in Fig. 6.1. The

---

[*]Berner (1974, 1980a), van Cappellen and Berner (1988). [†]Jorgensen et al. (1975), Kamp-Nielsen (1975), Nurnberg (1988). [‡]Berner (1974, 1980a), Ishikawa and Nishimura (1989), Jorgensen et al. (1975), Kamp-Nielsen (1975), Kamp-Nielsen et al. (1982), Nurnberg (1988), van Cappellen and Berner (1988), Van der Molen (1991), Yoshida (1981). [#]Slomp et al. (1998). [¶]Gachter et al. (1988), Gachter and Meyers (1993). [♭]Caraco et al. (1993).

production of phosphate is via the diagenetic mineralization of particulate organic matter. The result is a flux of phosphate J_P to the anaerobic layer. A portion of the liberated phosphate remains in the dissolved form and a portion becomes particulate phosphate, either via precipitation of phosphate containing minerals (Troup, 1974), such as vivianite, $Fe_3(PO_4)2(s)$, or by partitioning to phosphate sorption sites.[*] The extent of particulate formation is determined by the magnitudes of the partition coefficients π_1 and π_2 in layers 1 and 2, respectively. The interaction between layers 1 and 2 is via diffusion of dissolved chemical and particle mixing. The rate of burial is determined by the sedimentation velocity w_2. Finally the flux into or out of the sediment is via diffusive exchange with the overlying water. Thus the phosphate flux model has a structure that is similar to the models discussed in the previous chapters.

6.3 SOLUTIONS

The mass balance equations for phosphate are the same as Eqs. (5.1a) and (5.1b) but without the reaction terms

$$H_1 \frac{d[PO_4(1)]_T}{dt} = s\left([PO_4(0)] - f_{d1}[PO_4(1)]_T\right)$$

$$+ w_{12}\left(f_{p2}[PO_4(2)]_T - f_{p1}[PO_4(1)]_T\right)$$

$$+ K_{L12}\left(f_{d2}[PO_4(2)]_T - f_{d1}[PO_4(1)]_T\right)$$

$$- w_2[PO_4(1)]_T \tag{6.1a}$$

$$H_2 \frac{d[PO_4(2)]_T}{dt} = -w_{12}\left(f_{p2}[PO_4(2)]_T - f_{p1}[PO_4(1)]_T\right)$$

$$- K_{L12}\left(f_{d2}[PO_4(2)]_T - f_{d1}[PO_4(1)]_T\right)$$

$$+ w_2\left([PO_4(1)]_T - [PO_4(2)]_T\right) + J_P \tag{6.1b}$$

where

H_1 and H_2 are the depths of the aerobic (1) and anaerobic (2) layers

$[PO_4(0)]$ is the dissolved phosphate concentration in the overlying water

$[PO_4(1)]_T$ and $[PO_4(2)]_T$ are the total phosphate concentrations in layers 1 and 2

f_{d1}, and f_{d2} are the dissolved fractions in layers 1 and 2

f_{p1} and f_{p2} are the particulate fractions in layers 1 and 2

---

[*]Barrow (1983), Dzombak and Morel (1990), Lijklema (1980).

s is the surface mass transfer coefficient between the overlying water and the aerobic layer

K_{L12} is the mass transfer coefficient between the aerobic and anaerobic layers

w_{12} is the particle mixing velocity between the aerobic and anaerobic layers

w_2 is the burial velocity

J_P is the source of phosphate from the diagenesis of particulate organic phosphorus POP

For simplicity, the case for zero overlying water phosphate concentration is considered. The solution is obtained from Eq. (5.18)

$$[PO_4(1)]_T = \frac{J_P}{sf_{d1} + w_2r_{21}} \tag{6.2}$$

where r_{21} is defined as

$$r_{21} = [PO_4(2)]_T/[PO_4(1)]_T \tag{6.3}$$

and is given by Eqs. (5.7) and (5.10)

$$r_{21} = \frac{w_2 + w_{12}f_{p1} + K_{L12}f_{d1} + sf_{d1}}{w_{12}f_{p2} + K_{L12}f_{d2}} \tag{6.4}$$

The phosphate flux to the overlying water is (Eq. 5.20a)

$$J[PO_4] = J_P\frac{sf_{d1}}{sf_{d1} + w_2r_{21}} \tag{6.5}$$

Because this is the steady state solution, the phosphate released by particulate organic matter diagenesis can either escape to the overlying water or be buried. The extent of partitioning in layer 1 and layer 2 affects f_{d1} and r_{21} which, in turn, control the fraction of mineralized phosphorus that is either recycled to the overlying water or buried.

6.3 A   Effect of Partitioning

It has been observed that the phosphate flux from sediments is strongly affected by the overlying water oxygen concentration $[O_2(0)]$.[*] It has been suggested that the phosphate transferred to the aerobic layer is sorbed to freshly precipitated iron oxyhydroxide which prevents it from diffusing into the overlying water. At low oxygen concentrations, the iron oxyhydroxides are reduced and dissolve, the sorption

---

[*]Bostrom et al. (1988), Mortimer (1941, 1942, 1971).

barrier is removed, and the phosphate flux escapes unimpeded.[†] This suggests that the dissolved fraction in the aerobic layer f_{d1} is changing as a function of overlying water dissolved oxygen concentration.

At first glance, it is not clear that this mechanism – a partition coefficient that is larger in the aerobic layer than in the anaerobic layer – can account for the variation of phosphate flux as a function of overlying water DO. How is it possible that, at steady state, a difference in partitioning in layer 1 and 2 can reduce the flux to the overlying water? Would not the aerobic layer barrier eventually be saturated by the diffusive transport from the anaerobic layer?

Consider the following progression in time. At any point, the sorbed phosphate equilibrates with the dissolved phosphate concentration in both layers. If a gradient of dissolved phosphate exists between layers 1 and 2, then pore water diffusion will equalize the pore water concentrations. The solid phase phosphate concentrations will adjust to accommodate the new dissolved concentrations. The process of pore water diffusion and solid phase adjustment will continue until the pore water concentrations are equal in both layers. At that point the dissolved concentrations in the aerobic and anaerobic layers are the same, the barrier is saturated, and the phosphate flux is equal to the phosphate flux from an entirely anaerobic sediment. The presence of an aerobic layer would no longer reduce the dissolved phosphate concentration at the sediment-water interface and, therefore, would no longer reduce the flux from the aerobic layer.

6.3 B   Particle Mixing

This line of reasoning depends on the assumption that the pore water concentrations will eventually equilibrate in both layers due to diffusion of dissolved phosphate. However, it is not clear that this will occur, particularly in the presence of particle mixing between layers 1 and 2. An examination of the flux equation (6.5) indicates that increasing the partitioning in layer 1 decreases f_{d1}, and thereby decreases the phosphate flux to the overlying water.

However, decreasing f_{d1} may also decrease r_{21} (see Eq. 6.4). Consider the limiting cases. Without particle mixing Eq. (6.4) becomes

$$r_{21} = \frac{w_2 + K_{L12}f_{d1} + sf_{d1}}{K_{L12}f_{d2}} \tag{6.6}$$

and the phosphate flux becomes

$$J[PO_4] = J_P \frac{sf_{d1}}{sf_{d1} + w_2\left(\dfrac{w_2 + K_{L12}f_{d1} + sf_{d1}}{K_{L12}f_{d2}}\right)} \tag{6.7a}$$

[†]Baccini (1985), Bostrom et al. (1988), Chambers and Odum (1990), Mortimer (1941, 1942, 1971), Sundby et al. (1986).

Table 6.1   Phosphate Flux Model Parameters

Parameter	Units	With Particle Mixing Fig. 6.2A-C	Without Particle Mixing Fig. 6.2D-F
w_2	cm/yr	0.25	0.25
w_{12}	m/d	0.0012	0.0
K_{L12}	m/d	0.01	0.01
m_1	kg/L	0.5	0.5
m_2	kg/L	0.5	0.5
J_p	mg P/m²-d	10	10

and

$$J[PO_4] \approx J_P \frac{sf_{d1}}{sf_{d1} + w_2 \left(\dfrac{K_{L12}f_{d1} + sf_{d1}}{K_{L12}f_{d2}} \right)} = \frac{s}{s + w_2 \left(\dfrac{K_{L12} + s}{K_{L12}f_{d2}} \right)} \qquad (6.7b)$$

The approximation in Eq. (6.7b) follows from the assumption that the burial is small relative to the other transport terms: $w_2 \ll K_{L12}f_{d1} + sf_{d1}$. For this case, the phosphate concentration is indeed independent of the layer 1 partition coefficient, and the trapping mechanism will not reduce phosphate flux at steady state since f_{d1} is not a part of Eq. (6.7b).

With intense particle mixing, Eq. (6.4) becomes

$$r_{21} = \frac{w_2 + w_{12}f_{p1}}{w_{12}f_{p2}} \approx \frac{f_{p1}}{f_{p2}} \simeq 1 \qquad (6.8)$$

since, for any realistic partition coefficient, the fraction of phosphate that is in the particulate form is essentially unity so that $f_{p1} \simeq 1$ and $f_{p2} \simeq 1$. Thus the particulate concentrations in the two layers equalize, and the phosphate flux to the overlying water becomes

$$J[PO_4] = J_P \frac{sf_{d1}}{sf_{d1} + w_2} \qquad (6.9)$$

which varies with f_{d1} and, therefore, with aerobic layer partitioning. Hence, the intensity of particle mixing determines whether the trapping mechanism can be effective in varying phosphate flux as a function of overlying water DO.

A quantitative examination is presented in Fig. 6.2. The parameters used in the calculation are given in Table 6.1. The particulate and dissolved concentrations, and the ratio of flux to the overlying water to the diagenesis flux $J[PO_4]/J_P$ are plotted versus the ratio of layer 1 to layer 2 partition coefficient π_1/π_2. The top panels (Fig. 6.2A-C) present the case with particle mixing. For equal partition coefficients, $\pi_1/\pi_2 = 1$, the layer 1 and 2 particulate and dissolved concentrations are essentially equal, and virtually all the diagenesis flux is escaping to the overlying water as indicated by the flux ratio $J[PO_4]/J_P = 1$ (Fig. 6.2C). As π_1/π_2 increases, particle

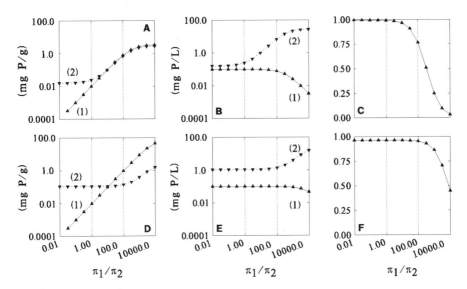

Fig. 6.2 (A and D) Particulate $f_{p2}[PO_4(2)]_T$ and (B and E) dissolved $f_{d2}[PO_4(2)]_T$ concentrations in layers 1 (▲) and 2 (▼), and (C and F) ratio of flux to the overlying water to the diagenesis flux $J[PO_4]/J_P$ versus the ratio of layer 1 to layer 2 partition coefficient π_1/π_2. (A and C) with and (D and F) without particle mixing.

mixing equilibrates the particulate concentrations (Fig. 6.2A), and the dissolved concentrations (Fig. 6.2B) become quite different in the two layers. For $\pi_1/\pi_2 > 100$, the aerobic layer concentration is sufficiently reduced (Fig. 6.2B) so that the flux ratio begins to decline significantly. Note that the particulate concentration begins to increase as well since less phosphate is being lost to the overlying water (Fig. 6.2A).

The results for no particle mixing (Fig. 6.2D-F) indicate that if the partition coefficient ratio approaches 10^4 then the phosphate flux is again reduced (Fig. 6.2F). The reason for this behavior can be seen from Eq. (6.7a) as $K_{L12}f_{d1} + sf_{d1}$ becomes small relative to w_2 (Eq. 6.10b)

$$J[PO_4] = J_P \frac{sf_{d1}}{sf_{d1} + w_2\left(\dfrac{w_2 + K_{L12}f_{d1} + sf_{d1}}{K_{L12}f_{d2}}\right)} \qquad (6.10a)$$

$$\simeq J_P \frac{sf_{d1}}{sf_{d1} + w_2\left(\dfrac{w_2}{K_{L12}f_{d2}}\right)} \qquad (6.10b)$$

for $w_2 \ll K_{L12}f_{d1} + sf_{d1}$. Now burial can become significant and the trapping mechanism operates.

The interesting question is: Why is there no equilibration of the dissolved concentrations as suggested above? This can be determined by examining the ratio of dissolved concentrations

$$\frac{[PO_4(2)]_T f_{d2}}{[PO_4(1)]_T f_{d1}} = \frac{f_{d2}}{f_{d1}} r_{21} \tag{6.11a}$$

$$= \left(\frac{f_{d2}}{f_{d1}}\right) \frac{w_2 + K_{L12} f_{d1} + s f_{d1}}{K_{L12} f_{d2}} = \frac{\dfrac{w_2}{f_{d1}} + K_{L12} + s}{K_{L12}} \tag{6.11b}$$

where Eq. (6.6) is substituted for r_{21} in Eq. (6.11b). Therefore, only if the diffusive exchange K_{L12} is much larger than $w_2/f_{d1} + s$ (Eq. 6.11b), will the dissolved concentrations equilibrate. Thus, although particle mixing is not essential for the trapping mechanism to operate, it is effective at a considerably lower partition coefficient ratio if it is present – compare Fig. 6.2C to Fig. 6.2F.

6.4   SIMPLIFIED PHOSPHATE FLUX MODEL

Since the aerobic layer trapping mechanism can reduce the aerobic layer phosphate flux, it is instructive to compare its predictions to observed phosphate fluxes. The total phosphate concentration in layer 1 is

$$[PO_4(1)]_T = \frac{J_P + s[PO_4(0)]}{s f_{d1} + w_2 r_{21}} \tag{6.12}$$

where the source due to diffusive exchange from the overlying water $s[PO_4(0)]$ has been added to the diagenesis source. Strictly speaking, this equation is not valid since diagenesis is a source to the anaerobic layer and the flux from the overlying water is a source to the aerobic layer. However, the analysis in Section 5.5 A indicates that the approximation is reasonable if $r_{12} \approx 1$. Since this is also the condition for which the trapping mechanism is most effective, the approximation is useful for this analysis.

The phosphate flux is computed in the usual way (Eqs. 3.10)

$$J[PO_4] = s([PO_4(1)] - [PO_4(0)])$$

$$= s(f_{d1}[PO_4(1)]_T - [PO_4(0)]) \tag{6.13}$$

where $s = SOD/[O_2(0)]$ is the surface mass transfer coefficient. Substituting Eq. (6.12) for the aerobic layer phosphate concentration yields

$$J[PO_4] = s\left(\frac{J_P + s[PO_4(0)]}{s + \dfrac{w_2 r_{21}}{f_{d1}}} - [PO_4(0)]\right)$$

$$= s\left(\frac{J_P + s[PO_4(0)]}{s + \Omega} - [PO_4(0)]\right) \tag{6.14}$$

where

$$\Omega_{PO_4} = \frac{w_2 r_{21}}{f_{d1}} \tag{6.15}$$

which is the parameter group that controls the extent of burial.

Since this solution is approximate – the aerobic layer source from the overlying water is treated as an anaerobic layer source – it is appropriate to check the solution's limiting behavior. With only the overlying water as a source of phosphate, $J_P = 0$, and no burial flux $\Omega_{PO_4} = 0$, it is easy to see that Eq. (6.14) predicts a zero net flux as it should.

A more interesting case occurs as $s \to \infty$ and the aerobic layer thickness approaches zero. The limit in this case can be found as follows

$$J[PO_4] = s \left(\frac{J_P + s[PO_4(0)]}{s + \Omega_{PO_4}} - [PO_4(0)] \right)$$

$$= s \left(\frac{J_P - \Omega_{PO_4}[PO_4(0)]}{s + \Omega_{PO_4}} \right)$$

$$\to J_P - \Omega_{PO_4}[PO_4(0)] \tag{6.16}$$

as $s \to \infty$. And from Eq. (6.12) it can be seen that the layer 1 dissolved concentration approaches the overlying water concentration as the aerobic layer shrinks to zero thickness $[PO_4(1)]_T f_{d1} \to [PO_4(0)]$, and this substitution can be made in Eq. (6.16). Therefore

$$J[PO_4] \to J_P - \Omega_{PO_4} f_{d1} [PO_4(1)]_T$$

$$= J_P - w_2 r_{21} [PO_4(1)]_T$$

$$= J_P - w_2 [PO_4(2)]_T \tag{6.17}$$

where the definition of $r_{21} = [PO_4(2)]_T / [PO_4(1)]_T$ (Eq. 6.3) is used to obtain the final result. This equation is a correct expression of mass balance. The flux to the overlying water is the difference between the diagenesis flux and the burial flux.

6.4 A   Numerical Analysis

The phosphate flux equation (6.14) has only one unknown parameter Ω_{PO_4}. The remaining terms are either measured: s and $[PO_4(0)]$, or for the diagenesis source J_P it can be estimated from nitrogen diagenesis J_N using the Redfield ratio $J_P/J_N = 1/7.23$ g P/g N (Table 1.1). A regression analysis yields $\Omega_{PO_4} = 0.10$ (m/d). The comparison of observed and computed fluxes is shown in Fig. 6.3A. The result is not satisfactory. The measured fluxes bear almost no relation to the modeled fluxes. The modeled fluxes vary over a range from $J[PO_4] \approx 1.0$ to 10.0 mg P/m^2-d, whereas the measured fluxes vary from $J[PO_4] \approx 0.2$ to almost 100.0 mg P/m^2-d, a nearly two orders of magnitude larger variation. This suggests that the burial fraction Ω_{PO_4} is varying, presumably as a function of overlying water dissolved oxygen $[O_2(0)]$.

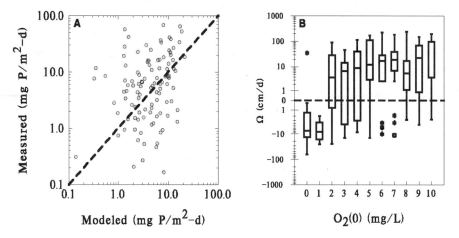

Fig. 6.3 (A) Estimated versus observed phosphate fluxes $J[PO_4]$. (B) Empirical removal parameter Ω_{PO_4} versus overlying water dissolved oxygen concentration $[O_2(0)]$.

The required variation can be estimated by computing Ω_{PO_4} directly from the data. Solving for Ω_{PO_4} in Eq. (6.14) yields

$$\Omega_{PO_4} = s \left(\frac{J_P + s[PO_4]_0}{J[PO_4] + s[PO_4]_0} - 1 \right) \tag{6.18}$$

The result is shown in Fig. 6.3B as a box plot. The ordinate scale is based on an arcsinh transformation. It is designed to accommodate both positive and negative values with a logarithmic scaling (see the Appendix of this chapter).

Note that Ω_{PO_4} is roughly constant (0.1 m/d = 10 cm/d) for $[O_2(0)] > 2$ mg/L and decreases as $[O_2(0)]$ approaches zero. In fact, negative Ω_{PO_4}'s are required to fit the observed phosphate fluxes at low dissolved oxygen concentrations. This is an indication that, in fact, this model is incapable of reproducing these observed fluxes.

6.5 STEADY STATE NUMERICAL MODEL

Perhaps the failure of the simplified model is due to the approximations introduced, namely replacing the transport and partitioning terms with the lumped parameter Ω. In this section the results of the steady state version of the full sediment model, which is presented in Chapter 16, are examined. The trapping mechanism is included by varying the aerobic layer partition coefficient as a function of the overlying water dissolved oxygen concentration.

A simple way to implement this mechanism is to make the aerobic layer partition coefficient larger than in the anaerobic layer during oxic conditions. These are defined to occur when the overlying water oxygen concentration exceeds some critical oxygen concentration, namely $[O_2(0)] > [O_2(0)]_{crit,PO_4}$. The additional sorption would be removed as $[O_2(0)]$ decreases below $[O_2(0)]_{crit,PO_4}$. Hence, if

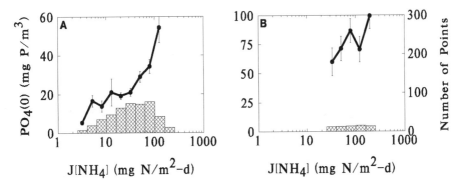

Fig. 6.4 Overlying water phosphate concentrations [$PO_4(0)$] versus ammonia flux $J[NH_4]$. (A) [$O_2(0)$] > 2 mg/L. (B) [$O_2(0)$] < 2 mg/L.

[$O_2(0)$] > [$O_2(0)$]$_{crit,PO_4}$, sorption in the aerobic layer is enhanced by an amount $\Delta\pi_{PO_4,1}$

$$\pi_1 = \pi_2(\Delta\pi_{PO_4,1}) \qquad [O_2(0)] > [O_2(0)]_{crit,PO_4} \qquad (6.19)$$

However, if oxygen falls below a critical concentration, [$O_2(0)$] < [$O_2(0)$]$_{crit,PO_4}$, then

$$\pi_1 = \pi_2(\Delta\pi_{PO_4,1})^{\beta_{PO_4}} \qquad [O_2(0)] \leqslant [O_2(0)]_{crit,PO_4} \qquad (6.20)$$

where

$$\beta_{PO_4} = \frac{[O_2(0)]}{[O_2(0)]_{crit,PO_4}} \qquad (6.21)$$

Eq. (6.20) smoothly reduces the aerobic layer partition coefficient to that in the anaerobic layer as [$O_2(0)$] goes to zero.

The steady state model is used to compare the variation of the ammonia and phosphate fluxes. The model is driven by the deposition of organic matter (Chapter 16). The appropriate depositional flux of organic matter with Redfield stoichiometry is chosen to reproduce a specific ammonia flux. Then the full model equations are solved at steady state to obtain the predicted phosphate flux. In order to complete the calculation it is necessary to specify the overlying water concentrations. The variation in overlying water phosphate concentration with respect to ammonia flux is shown in Fig. 6.4.

The results are examined in Fig. 6.5. The model successively predicts the variation in phosphate flux as ammonia flux increases for [$O_2(0)$] > 2 mg O_2/L (Fig. 6.5A). However, for [$O_2(0)$] < 2 mg O_2/L the predicted fluxes are substantially less than the observations (Fig. 6.5B).

The model behavior can be understood by examining the relationship of computed phosphate flux $J[PO_4]$ and phosphorus diagenesis J_P. For the cases where [$O_2(0)$] > 2 mg O_2/L, the model predicts a phosphate flux that is a constant fraction (0.88) of

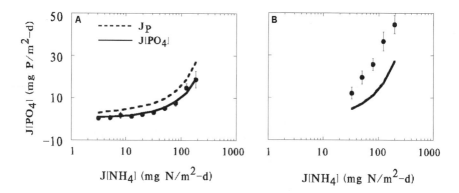

Fig. 6.5 Modeled (line) and observed (symbols) phosphate flux $J[PO_4(0)]$ versus ammonia flux $J[NH_4]$. (A) $[O_2(0)] > 2$ mg/L. (B) $[O_2(0)] < 2$ mg/L.

the phosphorus diagenesis (Fig. 6.5A). For $[O_2(0)] < 2$ mg O_2/L, however, the model predicts that $J[PO_4] = J_P$ (Fig. 6.5B). This is not unexpected since at steady state, the maximum flux possible is that generated by diagenesis. Apparently this is insufficient.

This can clearly be seen in Fig. 6.6A which compares the phosphate flux to phosphorus diagenesis, where phosphorus diagenesis is estimated from ammonia diagenesis and the Redfield ratio $a_{C,P} = 106$ mol C/mol P = 41 g C/g P (Table 1.1). A significant fraction of the phosphate fluxes are in excess of phosphorus diagenesis. This occurs when the overlying water dissolved oxygen is low as shown in Fig. 6.6B, a plot of $J[PO_4]/J_P$ versus $[O_2(0)]$. At high overlying water DO concentrations the phosphate flux is less than phosphorus diagenesis. The difference is removed by burial. However when $[O_2(0)]$ approaches zero, the phosphate flux is larger than J_P. As can be seen from Eq. (6.17), this cannot occur in a steady state model.

What is actually happening is that during the period of high overlying water DO, a portion of the phosphate produced by diagenesis is not being buried but is actually going into storage in the anaerobic layer. This corresponds to a positive derivative in the mass balance equation for phosphate in the anaerobic layer

$$H_2 \frac{d[PO_4(2)]_T}{dt} > 0 \tag{6.22}$$

In this case, storage mimics a sink in the mass balance equation. This can be seen if the derivative is included as part of the right-hand side of the mass balance equation (Eq. 6.1b)

$$0 = -w_{12}(f_{p2}[PO_4]_{T2} - f_{p1}[PO_4]_{T1}) - K_{L12}(f_{d2}[PO_4]_{T2} - f_{d1}[PO_4]_{T1})$$

$$+ w_2([PO_4]_{T1} - [PO_4]_{T2}) + J_P - H_2 \frac{d[PO_4]_{T2}}{dt} \tag{6.23}$$

The positive derivative corresponds to a loss term in the equation.

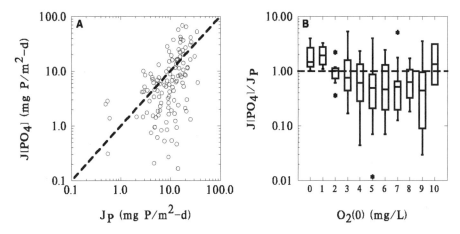

Fig. 6.6   (A) Phosphate flux $J[PO_4]$ versus phosphorus diagenesis J_P. (B) Ratio of phosphate flux to phosphorus diagenesis $J[PO_4]/J_P$ versus overlying water dissolved oxygen concentration $[O_2(0)]$.

During periods of low overlying water DO, this stored phosphate, together with the phosphate generated by diagenesis, is released to the overlying water. Then the derivative is negative and the term acts as a source to the mass balance equation. Since this fluctuating derivative is a nonsteady state phenomenon, it cannot be reproduced by a steady state model. The only solution is a time variable simulation. This is presented in Chapters 14 and 15.

6.6   CONCLUSIONS

The simplified steady state model is completely unsuccessful in simulating the range of observed phosphate fluxes. The problem is traced to the magnitude of the fluxes when the overlying water DO is less than 2 mg/L. The steady state results from the complete model reinforce this observation. The relationship between phosphate and ammonia fluxes is well reproduced for $[O_2(0)] > 2$ mg/L, but the anaerobic fluxes are underestimated. In fact, they exceed the phosphorus diagenesis flux. The source of the extra phosphate is from storage. Since this is not possible for the steady state model, only a time variable simulation will suffice.

Appendix 6A: Positive and Negative Logarithmic Scale for Plotting

For certain data analysis problems, the variable of interest may have both positive and negative values. In addition the magnitudes may have a logarithmic rather than a linear character. In order to graphically analyze these data, it is helpful to employ a transformation that has a logarithmic behavior that can be used with negative as well as positive values.

6A.1 Logarithmic Transformation

The logarithmic transformation is based on the exponential function

$$y = e^x \tag{6A.1}$$

and its inverse function

$$x = \ln(y) = \ln(e^x) = x \tag{6A.2}$$

The base 10 or common logarithmic transformation is based on the function

$$y = 10^x \tag{6A.3}$$

and its inverse function

$$x = \log_{10}(y) = \log_{10}(10^x) = x \tag{6A.4}$$

The problem is that for $y < 0$ the logarithmic function produces complex numbers, for example $\ln(-10) = \ln(10) + i\pi$, which is no help for plotting negative data on a logarithmic scale.

Consider what would be the desired answer for negative numbers. For $y = 10$, $\log_{10}(10) = 1$. Hence for $y = -10$, the appropriate answer is -1. Therefore, the appropriate transformation for negative y is

$$x = -\log(-y) \tag{6A.5}$$

or, inverting the relationship

$$y = -10^{-x} \tag{6A.6}$$

Since Eqs. (6A.1) and (6A.6) apply to positive and negative y's, their addition would define a transformation that would apply for both positive and negative y's

$$y = 10^x - 10^{-x} \tag{6A.7}$$

For large x, Eq. (6A.7) approaches either Eq. (6A.1) or (6A.6) depending on the sign of x. Thus it smoothly transitions between positive and negative y's.

The inverse function can be found by solving Eq. (6A.7) for x by using the substitution $z = 10^x$

$$y = z - \frac{1}{z} \tag{6A.8}$$

which can be solved for z

$$z = \frac{y + \sqrt{y^2 + 4}}{2} \tag{6A.9}$$

or

$$10^x = \frac{y + \sqrt{y^2 + 4}}{2} \tag{6A.10}$$

Solving for x yields

$$x = \frac{\ln\left(y + \sqrt{y^2 + 4}\right) - \ln(2)}{\ln(10)} \tag{6A.11}$$

which is the desired inverse function.

The behavior of this function can be judged using Taylor series expansions of Eq. (6A.11) about $y = 0$ and $y = \infty$.

$$x = \frac{1}{2\log(10)}\left(y - \frac{y^3}{24} + \frac{y^5}{640} - \cdots\right) \qquad y \to 0 \tag{6A.12}$$

$$x = \frac{1}{\ln(10)}\left(\ln(y) + \frac{1}{y^2} - \frac{3}{2y^4} + \cdots\right) \qquad y \to \infty \tag{6A.13}$$

For small y, x is linear in y (Eq. 6A.12). For large y, x approaches $\log_{10}(y)$ (Eq. 6A.13). One practical note: For large negative y a loss of significance occurs in evaluating Eq. (6A.11). However, since Eq. (6A.11) is symmetric in y, it can be written

$$x = \frac{y}{|y|} \frac{\ln\left(|y| + \sqrt{y^2 + 4}\right) - \ln(2)}{\ln(10)} \tag{6A.14}$$

where $y/|y|$ provides the proper sign. Table A.1 provides selected numerical values.

6A.2 Arcsinh Transformation

It is interesting to examine the result of using natural instead of common logarithms. The equation analogous to Eq. (6A.7) is

$$y = e^x - e^{-x} \tag{6A.15}$$

This function is the hyperbolic sin (or sinh) function without the leading $\frac{1}{2}$. Thus an equivalent transformation can be defined using the sinh function

$$y = \sinh(x) = \frac{1}{2}\left(e^x - e^{-x}\right) \tag{6A.16}$$

The exponential and sinh functions are compared in Fig. 6A.1A. For large positive x, the sinh function approaches an exponential (Eq. 6A.1), since the e^{-x} term approaches zero. However, unlike the exponential that results in only positive y values,

Table 6A.1   Evaluation of Eq. (6A.14)

y	$\log_{10}(y)$	Eq. (6A.14)
1.0e-6	- 6	2.17148E-7
1.0e-5	- 5	2.17148E-6
1.0e-4	- 4	2.17148E-5
0.001	- 3	2.17147E-4
0.01	- 2	0.00217
0.1	- 1	0.02171
1	0	0.20899
10	1	1.00428
100	2	2.00004
1000	3	3.0
10000	4	4.0
100000	5	5.0
1000000	6	6.0

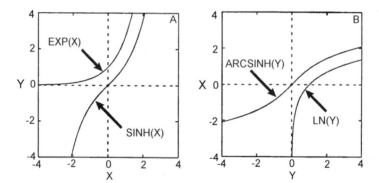

Fig. 6A.1   (A) Comparison of exponential and sinh functions. (B) Comparison of logarithmic and arcsinh functions.

the sinh function yields positive y's for positive x's and negative y's for negative x's (see Fig. 6A.1A). In fact the sinh function is antisymmetric

$$\sinh(-x) = -\sinh(x) \qquad (6A.17)$$

and it has logarithmic behavior for both positive and negative x's.

The inverse function – corresponding to the $\ln(y)$ function which is the inverse of the exponential function – is called the *arcsinh* function

$$x = \sinh^{-1}(y) = \operatorname{arcsinh}(y) \qquad (6A.18)$$

and it is the inverse of the sinh function, Eq. (6A.16). Using a derivation similar to Eqs. (6A.8–6A.11) the inverse function is

$$x = \text{arcsinh}(y) = \ln\left(y + \sqrt{y^2 + 1}\right) \qquad (6A.19)$$

The arcsinh and ln functions are compared in Fig. 6A.1B. Unlike the $\ln(y)$ function, the arcsinh(y) can be applied to both positive and negative values of y, and returns positive and negative values.

For large x, the sinh function approaches an exponential. For small x, the function is approximately linear. This can be seen by substituting the Taylor expansion for the exponentials in Eq. (6A.16)

$$\sinh(x) = \frac{1}{2}\left[\left(1 + \frac{x}{1!} + \frac{x^2}{2!} + \cdots\right) - \left(1 - \frac{x}{1!} + \frac{x^2}{2!} + \cdots\right)\right] \qquad (6A.20)$$

$$\simeq x \qquad (6A.21)$$

Hence it combines logarithmic behavior for large arguments with linear behavior for small arguments. The transition between linear and logarithmic behavior occurs at approximately $x = 1$ since $\sinh(1) = 1.18$.

6A.3   Choosing Between the Two Transformations

The base 10 log transformation

$$x = \frac{y}{|y|}\frac{\ln\left(|y| + \sqrt{y^2 + 4}\right) - \ln(2)}{\ln(10)} \qquad (6A.22)$$

and natural log transformation

$$x = \frac{y}{|y|}\ln\left(|y| + \sqrt{y^2 + 1}\right) \qquad (6A.23)$$

differ only in scaling constants. The base 10 log transformation (Eq. 6A.22) produces essentially base 10 logs for $|y| > 10$ (Table A.1 and Eq. 6A.13) and a linear relationship between x and y as $y \rightarrow 0$ (Eq. 6A.12). The base e transformation approaches equality $x \simeq y$ as $y \rightarrow 0$ in the linear range (Eq. 6A.21). In the logarithmic range the sinh transformation approaches

$$x = \ln(y) + \ln(2) + \frac{1}{4y^2} - \frac{3}{32y^4}\cdots \qquad y \rightarrow \infty \qquad (6A.24)$$

which is, of course, logarithmic but not as convenient a scaling as the base 10 case (Table A.1 and Eq. 6A.13).

So the choice of either Eqs. (6A.22) or (6A.23) depends on which end of the range is properly, or conveniently, scaled. Fig. 6.3 uses the arcsinh transformation. The scale is constructed by applying the transformation to 0, 1, 2, ... 10, 20, ... and plotting the results as tick marks.

7

Silica

7.1 INTRODUCTION

The production of ammonia, sulfide, and phosphate in sediments is the result of the mineralization of particulate organic matter by bacteria. The production of dissolved silica in sediments occurs via a different mechanism which is thought to be independent of bacterial processes. It occurs as the result of the dissolution of particulate biogenic silica (Hurd, 1973). The dissolution releases silica to the pore water.

Two classes of models have been proposed for the vertical distribution of silica in sediment pore waters. The first considers only dissolved silica and neglects the solid phase.[*] This approach is used initially for the simplified steady state model presented below. The more complete models consider both the solid phase and dissolved silica, and their interactions.[†] The final model presented below includes both these phases.

7.2 MODEL COMPONENTS

The schematic is presented in Fig. 7.1. The dissolution of particulate silica produces dissolved silica in the pore water of the sediment. However, silica has only a limited solubility in water $[Si]_{sat}$. It has been determined that the rate of biogenic silica dis-

---

[*]Anikouchine (1967), Berner (1974), Hurd (1973), Lerman (1975), Vanderborght et al. (1977).
[†]Boudreau (1990), Rabouille and Gaillard (1990), Schink and Guinasso Jr. (1980), Schink et al. (1975), Wong and Grosch (1978).

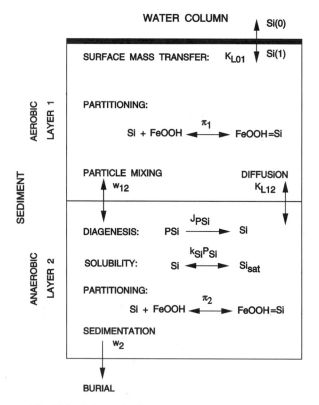

Fig. 7.1 Schematic diagram of the silica flux model.

solution is proportional to the silica solubility deficit $[\text{Si}]_{\text{sat}} - [\text{Si(aq)}]$ where $[\text{Si(aq)}]$ is the dissolved silica concentration.

To see that this is a reasonable formulation, consider the sequence of events as biogenic silica dissolves into water that is initially free of dissolved silica. At first, biogenic silica dissolves at its maximum rate, unimpeded by limited solubility. As the concentration of silica in pore water increases, however, the reverse reaction, the precipitation of particulate silica, begins to take place. This retards the overall rate of dissolution. As the pore water concentration continues to increase, the concentration eventually approaches the solubility limit of silica, and the reaction reaches a steady state where the rate of dissolution equals the rate of precipitation. The result is that there is no further increase of dissolved silica.

This formulation can be expressed as follows: Let S_{Si} be the rate of production of dissolved silica. The rate of biogenic silica dissolution is proportional to the silica solubility deficit: $[\text{Si}]_{\text{sat}} - [\text{Si(aq)}]$, and also the concentration of particulate biogenic silica P_{Si}. Thus

$$S_{\text{Si}} = k_{\text{Si}}\theta_{\text{Si}}^{(T-20)} P_{\text{Si}}\big([\text{Si}]_{\text{sat}} - [\text{Si(aq)}]\big) \qquad (7.1)$$

where

k_{Si} is the specific reaction rate for silica dissolution

θ_{Si} is the coefficient of temperature dependence

P_{Si} is the concentration of particulate biogenic silica

$[Si]_{sat}$ is the saturation concentration of silica in the pore water

$[Si(aq)]$ is the dissolved silica concentration.

For the initial steady state modeling analysis it is convenient to replace the product $k_{Si}P_{Si}$ with an overall first-order reaction rate k_{Si}. With this simplification Eq. (7.1) becomes

$$S_{Si} = k_{Si}\theta_{Si}^{(T-20)}\left([Si]_{sat} - [Si(aq)]\right)H_2$$

$$= k_{Si}\theta_{Si}^{(T-20)}[Si]_{sat}H_2 - k_{Si}\theta_{Si}^{(T-20)}[Si(aq)]H_2 \qquad (7.2)$$

This simplifies the analysis since the mass balance equation is now linear and an equation for P_{Si} is not required. This simplification, however, should be viewed only as an expedient. The basic principle guiding the development of these models is the principle of mass balance and, clearly, an adherence to this principle requires an explicit accounting of the source of silica. It is included in the numerical steady state calculations presented below, and in the time variable model discussed in Chapters 13 through 15.

7.3   SOLUTIONS

In order to use the general solutions obtained in Chapter 5, the two terms in this expression need to be related to their counterparts in the general solution, the layer 2 source term J_{T2}, and a layer 2 reaction rate k_2. Thus, comparing the terms in Eq. (7.2) to the definitions in Eqs. (5.1) yields the equivalences

$$J_{T2} = k_{Si}\theta_{Si}^{(T-20)}[Si]_{sat}H_2 \qquad (7.3)$$

$$k_2 = k_{Si}\theta_{Si}^{(T-20)}f_{d2} \qquad (7.4)$$

Note that the source term is the dissolution reaction and the sink is the precipitation reaction. The dissolved fraction f_{d2} is included to allow for the possibility that a fraction of the dissolved silica is sorbed to the particles in the sediment. In addition to the dissolution source, the source from the overlying water $s[Si(0)]$ must be included.

The mass balance equations for layers 1 and 2 are

$$H_1 \frac{d[\mathrm{Si}(1)]_T}{dt} = s\big([\mathrm{Si}(0)] - f_{d1}[\mathrm{Si}(1)]_T\big)$$

$$+ w_{12}\big(f_{p2}[\mathrm{Si}(2)]_T - f_{p1}[\mathrm{Si}(1)]_T\big)$$

$$+ K_{L12}\big(f_{d2}[\mathrm{Si}(2)]_T - f_{d1}[\mathrm{Si}(1)]_T\big)$$

$$- w_2[\mathrm{Si}(1)]_T \tag{7.5a}$$

$$H_2 \frac{d[\mathrm{Si}(2)]_T}{dt} = -k_2 H_2[\mathrm{Si}(2)]_T$$

$$- w_{12}\big(f_{p2}[\mathrm{Si}(2)]_T - f_{p1}[\mathrm{Si}(1)]_T\big)$$

$$- K_{L12}\big(f_{d2}[\mathrm{Si}(2)]_T - f_{d1}[\mathrm{Si}(1)]_T\big)$$

$$+ w_2\big([\mathrm{Si}(1)]_T - [\mathrm{Si}(2)]_T\big) + J_{T2} \tag{7.5b}$$

where

H_1 and H_2 are the depths of layers 1 and 2

$[\mathrm{Si}(0)]$ is the dissolved silica concentration in the overlying water

$[\mathrm{Si}(1)]_T$ and $[\mathrm{Si}(2)]_T$ are the total silica concentrations in layers 1 and 2

f_{d1} and f_{d2} are the dissolved fractions in layers 1 and 2

f_{p1} and f_{p2} are the particulate fractions in layers 1 and 2

s is the surface mass transfer coefficient

K_{L12} is the mass transfer coefficient between layers 1 and 2

w_{12} is the particle mixing velocity between layers 1 and 2

w_2 is the burial velocity

k_2 is the anaerobic layer decay rate, the dissolution sink (Eq. 7.4)

J_{T2} is the source from the dissolution of particulate silica (Eq. 7.3)

The solutions follow from the general equations given in Chapter 5. The layer 1 concentration of total dissolved silica (dissolved + sorbed) is Eq. (5.18)

$$[\mathrm{Si}(1)]_T = \frac{k_{\mathrm{Si}}\theta_{\mathrm{Si}}^{(T-20)}[\mathrm{Si}]_{\mathrm{sat}} H_2}{s f_{d1} + (\kappa_{\mathrm{Si},2} f_{d2} + w_2) r_{21}} \tag{7.6}$$

where (Eq. 7.4)

$$\kappa_{\mathrm{Si},2} = k_{\mathrm{Si}}\theta_{\mathrm{Si}}^{(T-20)} H_2 \tag{7.7}$$

the reaction velocity in layer 2. The net flux of silica to the overlying water is

$$J[\text{Si}] = s\left(f_{d1}[\text{Si}(1)]_T - [\text{Si}(0)]\right) \tag{7.8}$$

The result is

$$J[\text{Si}] = s f_{d1} \frac{k_{\text{Si}}\theta_{\text{Si}}^{(T-20)}[\text{Si}]_{\text{sat}}H_2}{s f_{d1} + (w_2 + \kappa_{\text{Si},2} f_{d2})r_{21}} - s[\text{Si}(0)] \tag{7.9}$$

where

$$r_{21} = \frac{w_2 + f_{d2}\kappa_{\text{Si},2} + w_{12}f_{p1} + K_{L12}f_{d1}}{w_{12}f_{p2} + K_{L12}f_{d2}} \tag{7.10}$$

7.3 A Simplified Solution

In addition to the dissolution source in layer 2, there is the source of dissolved silica that is transferred from the overlying water to layer 1. This can be included as though it were a layer 2 source for the sake of convenience. This approximation is discussed in Chapter 5 and used in the simplified phosphate flux model in section 6.4. Hence, Eq. (7.9) becomes

$$J[\text{Si}] = s f_{d1} \frac{k_{\text{Si}}\theta_{\text{Si}}^{(T-20)}[\text{Si}]_{\text{sat}}H_2 + s[\text{Si}(0)]}{s f_{d1} + (w_2 + \kappa_{\text{Si},2} f_{d2})r_{21}} - s[\text{Si}(0)] \tag{7.11a}$$

$$= s\left(\frac{k_{\text{Si}}\theta_{\text{Si}}^{(T-20)}[\text{Si}]_{\text{sat}}H_2 + s[\text{Si}(0)]}{s + \dfrac{f_{d2}r_{21}}{f_{d1}}k_{\text{Si}}\theta_{\text{Si}}^{(T-20)}H_2 + \dfrac{w_2 r_{21}}{f_{d1}}} - [\text{Si}(0)]\right) \tag{7.11b}$$

where Eq. (7.11b) follows from Eq. (7.11a) by dividing by f_{d1}. This equation can be further simplified by assuming that

$$\frac{f_{d2}}{f_{d1}}r_{21} \approx 1 \tag{7.12}$$

which corresponds to assuming that the silica partition coefficient is the same in both layers $f_{d1} = f_{d2}$ and that particle mixing is sufficiently intense so that $r_{21} \approx 1$. The result is

$$J[\text{Si}] = s\left(\frac{k_{\text{Si}}\theta_{\text{Si}}^{(T-20)}[\text{Si}]_{\text{sat}}H_2 + s[\text{Si}(0)]}{s + k_{\text{Si}}\theta_{\text{Si}}^{(T-20)}H_2 + \Omega_{\text{Si}}} - [\text{Si}(0)]\right) \tag{7.13}$$

where

$$\Omega_{\text{Si}} = \frac{w_2 r_{21}}{f_{d1}} \tag{7.14}$$

is the equivalent burial rate. These simplifications render the equation suitable for fitting to the silica flux data using nonlinear regression.

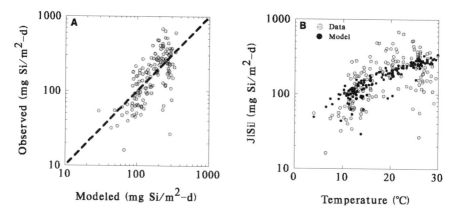

Fig. 7.2 (A) Comparison of observed and modeled silica fluxes. (B) Observed and modeled silica fluxes versus temperature of the overlying water.

7.3 B   Data Analysis

The parameters remaining to be estimated in Eq. (7.13) are the reaction rate parameters k_{Si}, θ_{Si} and the equivalent burial rate Ω_{Si}. Table 7.1 presents the results of a nonlinear regression fit to observed silica fluxes and compares them to values reported in the literature. The measured versus predicted fluxes are compared in Fig. 7.2A. Although there is substantial scatter, the comparison suggests that the simplified steady state model is capable of reproducing the general behavior of silica fluxes.

A common analysis procedure for fluxes is to correlate them to temperature variation. Fig. 7.2B presents the data and model results versus temperature. The approximately exponential variation is captured reasonably well by the model. This is due to the temperature dependency of the silica dissolution kinetics (Eq. 7.1). However, the predicted dependency is not exactly exponential. The relationship levels off at the higher temperatures. This is due to the appearance of the temperature correction term in both the numerator and denominator of the flux equation (7.13).

The model also predicts the magnitude of the silica flux. This is determined by the overall reaction rate for silica dissolution and the saturation concentration. The magnitude of the rate constant k_{Si} in turn depends on the quantity of particulate biogenic silica in the sediment and the specific rate constant k_{Si}. Thus the concentration of the particulate biogenic silica is required.

7.4   FINAL MODEL

The final model for silica flux includes a mass balance for particulate biogenic silica

$$H_2 \frac{dP_{Si}}{dt} = -S_{Si}H_2 - w_2 P_{Si} + J_{P_{Si}}$$ (7.15)

Table 7.1   Silica Model Parameters

Parameter	Symbol	Units	(a)	(b)	(c)	(d)	(e)
First order reaction rate	K_{Si}	d^{-1}	0.103	0.039	0.2	0.09	0.02 – 0.2
Temperature coefficient	θ_{Si}	–	1.059	1.059[f]	1.08	-	1.0836
Saturation concentration	$[Si]_{sat}$	mg Si/L	26.5[f]	26.5[f]	26.5	33.7	39.0
Equivalent burial velocity	Ω	m/d	0.0322	–	–	–	–
Half saturation constant	$K_{M,PSi}$	mg Si/L	–	19.8	100	–	–

[a]Nonlinear regression analysis using Eq. (7.13). [b]Nonlinear regression analysis of data from Conley and Schelske (1989), Conley et al. (1986). [c]Steady state model parameters. [d]Ullman and Aller (1989). [e]Lawson et al. (1978). [f]Assigned.

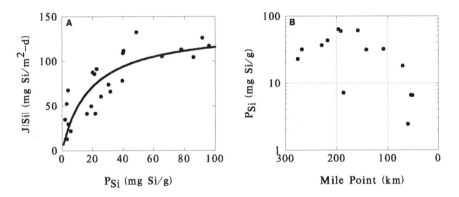

Fig. 7.3   (A) Silica flux versus particulate silica concentration. (B) Particulate silica concentration versus distance along the axis of Chesapeake Bay.

where P_{Si} is the concentration of particulate biogenic silica in the sediment and $J_{P_{Si}}$ is the depositional flux of particulate biogenic silica to the sediment. The loss terms are that due to dissolution (Eq. 7.1) and burial. This equation can be thought of as the analog of the diagenesis equations for particulate organic carbon, nitrogen, and phosphorus which are presented in Chapter 2 and discussed at length in Chapter 12. It specifies the rate at which particulate biogenic silica becomes dissolved silica.

The original formulation of the dissolution reaction was as a linear function of P_{Si} (Hurd, 1973). However, Conley and Schelske (1989) suggest that the rate of silica dissolution is not linear in particulate silica concentration but rather that the dependency saturates at higher concentrations. Data from Lake Michigan sediments, indicating that such a dependency is required, are shown in Fig. 7.3A. A Michaelis-Menton expression, which is fitted to the data, is also shown. Biogenic silica concentrations in Chesapeake Bay sediments range from less than 10 to 100 mg Si/g (Fig. 7.3B) which is similar to the sediments from Lake Michigan.

The expression which includes the Michaelis-Menton dependency of silica dissolution rate on particulate silica P_{Si} is

$$S_{Si} = k_{Si}\theta_{Si}^{(T-20)} \frac{P_{Si}}{P_{Si} + K_{M,P_{Si}}} ([Si]_{sat} - f_{d2}[Si(2)]) \qquad (7.16)$$

where the dissolution rate constant k_{Si} is now a first order constant with units of d^{-1}. The relationship between the specific rate constant k'_{Si} and the first order rate constant k_{Si} is

$$k'_{Si} = \frac{k_{Si}}{P_{Si} + K_{M,P_{Si}}} \qquad (7.17)$$

which follows from the definitions of the constants.

The partitioning of dissolved silica to iron oxyhydroxide is included using the same formulation as employed for phosphorus (Eqs. 6.19–6.20)

$$\pi_{Si,1} = \pi_{Si,2}(\Delta\pi_{Si,1}) \qquad\qquad [O_2(0)] > [O_2(0)]_{crit,Si} \qquad (7.18)$$

$$\pi_{Si,1} = \pi_{Si,2}(\Delta\pi_{Si,1})^{\beta_{li}} \qquad\qquad [O_2(0)] \leqslant [O_2(0)]_{crit,Si}$$

where

$$\beta_{Si} = \frac{[O_2(0)]}{[O_2(0)]_{crit,Si}} \qquad (7.19)$$

$\pi_{Si,2}$ is the partition coefficient of dissolved silica in the anaerobic layer, and $\Delta\pi_{Si,1}$ is the increase in the aerobic layer due to silica sorption to iron oxyhydroxide.

7.4 A Steady State Model Results

The silica steady state model is evaluated by comparing the variation of the silica flux with respect to the ammonia flux. This is the same technique that was applied to the analysis of phosphate fluxes (Section 6.5). The appropriate depositional flux of organic matter is chosen to reproduce a specific ammonia flux. The silica to carbon ratio of the organic matter is established using water column particulate data. Then the full model equations are solved at steady state to obtain the predicted silica flux.

In order to perform this computation, the exogenous variables are required as a function of ammonia flux, since they index the computation. The variation in overlying water silica concentration with respect to ammonia flux is shown in Fig. 7.4A,B. The other exogenous variables that are necessary for the calculation have been presented in the previous chapters.

As in the case of phosphate flux model, the partitioning of silica in the aerobic layer is larger than in the anaerobic layer. The reason that partitioning is included is that silica is known to sorb to iron oxyhydroxide (Sigg and Stumm, 1980). The

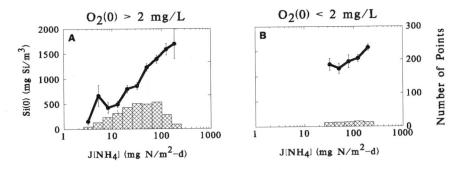

Fig. 7.4 Overlying water silica concentration versus ammonia flux.

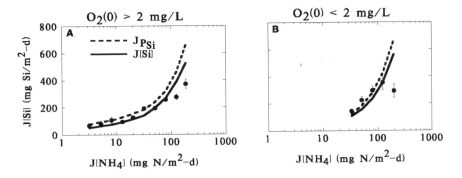

Fig. 7.5 Comparison of modeled (line) and observed (•) silica flux $J[Si]$ and depositional flux $J_{P_{Si}}$.

magnitude of the partition coefficients are determined using the time variable model as discussed subsequently.

The results are compared to observed fluxes in Fig. 7.5. Both the computed flux (solid line) and the estimated depositional flux (dashed line) are shown. The data and model for both the high and low dissolved oxygen subsets exhibit essentially the same behavior. Silica flux increases as ammonia flux increases. There appears to be no pronounced effect of overlying water dissolved oxygen on silica flux although the anaerobic fluxes are slightly larger than the corresponding aerobic fluxes. This can be seen by comparing the observed and computed fluxes to the depositional fluxes. For the computed fluxes, less silica is trapped and buried in the low dissolved oxygen concentration subset (Fig. 7.5B) than in the high dissolved oxygen concentration subset (Fig. 7.5A). This occurs for the same reason as for phosphate fluxes, except that the magnitude is smaller due to the smaller aerobic layer partition coefficient. The observations also appear to exhibit a slightly larger silica flux relative to the ammonia flux for the low dissolved oxygen subset although the effect is small.

The lack of a strong dissolved oxygen dependency is somewhat surprising because silica and phosphate sorb to iron oxyhydroxide to roughly the same extent

(Sigg and Stumm, 1980). A somewhat stronger dependency is exhibited by the time variable model, for the same reasons as the phosphate flux model, namely the effect of storage and release of sorbed silica.

7.5 CONCLUSIONS

The silica fluxes can be computed with reasonable accuracy by the simplified steady state model which relates the flux to temperature. The primary disadvantage is that the model does not consider particulate silica. Thus there is no tie to the depositional flux, as is required by mass balance considerations. Further the dissolution kinetics do not reflect the variation in particulate silica.

These deficiencies are corrected in the final model. The results of steady state computations, indexed by the ammonia flux, are in reasonable agreement with the observations. A small effect of overlying water dissolved oxygen is both computed and observed, which is somewhat surprising since the partitioning of silica to iron oxyhydroxide is almost as strong as for phosphate.

Part III

Oxygen

The consumption of oxygen by sediments – now called sediment oxygen demand or SOD – has been of concern from the beginnings of dissolved oxygen modeling in streams (Phelps, 1944, Streeter, 1935). The conventional approach was to treat SOD as an exogenous parameter, a rate that was either measured or estimated and supplied to the model. The first sediment flux model we constructed was part of the Lake Erie Eutrophication model (Di Toro, 1980b) in response to a question by the EPA project officer Nelson Thomas, an old SOD hand himself: "and when the phosphorus loading to the lake changes, and less algae settle to the sediment, what are you going to do to the SOD?" The result was the oxygen equivalents model for SOD (Chapter 8).

The sulfide oxidation model (Chapter 9) was developed during the Chesapeake Bay project, together with the nutrient models (Part II). The importance of sulfide in estuarine and marine sediments as an intermediate in the oxygen consumption reaction was apparent. The presence of sulfide as a solid phase, which transports at different rates than pore water, as well as pore water sulfide, clearly compromises the oxygen equivalence idea. The two-layer sediment flux model with sulfide as the intermediate storage reservoir was the result (Di Toro and Fitzpatrick, 1993). The Chesapeake Bay coupled water quality-sediment model (Cerco and Cole, 1993) has been very successful in simulating the long-term changes in water quality and eutrophication in Chesapeake Bay (Cerco, 1995b). Based on its successful performance, it has been used to predict the effects of nutrient loading reductions to the Bay (Cerco, 1995a).

The methane oxidation model (Chapter 10) was constructed as part of the Milwaukee River Comprehensive Study (SWRPC, 1987). Measurements of gas fluxes from the sediments of the Milwaukee River demonstrated the importance of methane as the endproduct of diagenesis. Clearly this finding is at variance with the idea that eventually all the endproducts of organic matter mineralization are oxidized using oxygen, which is the basis of the oxygen equivalent model, or that sulfides are the only important reduced intermediate. The loss of oxygen equivalents from the sediment via methane bubble formation is an important sink that must be accounted for.

The final refinement is the sulfide-methane oxidation model (Chapter 11) that produces the appropriate endproduct depending on the availability of sulfate. This is the model that is employed in the latest Chesapeake Bay coupled water column-sediment model. It incorporates both of the principal reduced intermediates that are eventually oxidized in the aerobic layer of the sediment.

8

Oxygen Equivalents

8.1 INTRODUCTION

The importance of sedimentary (or benthal) oxygen consumption in the oxygen balance of natural waters has been recognized since the beginning of the development of stream oxygen balance models – see Hatcher (1986) for an historical review, and Thomann and Mueller (1987) for the basics of water quality modeling. The procedure was simply to measure the sediment oxygen demand (SOD) as an areal flux of oxygen to the sediment and use that consumption rate as a sink in the mass balance models. If the contemplated control measures would not affect the SOD, this procedure is perfectly justifiable. However, most control alternatives affect the supply of organic particles to the sediment. The control of combined sewer overflows, for example, or the removal of nutrients from point and nonpoint sources directly or indirectly, reduces the supply of particulate organic matter (POM) to the sediment. Hence an important issue to be addressed in any of these studies is the effect this reduction has on the resulting SOD.

Attempts have been made to relate the SOD to the composition of the sediment itself – the expectation being that more organically rich sediments would have higher SODs. However only weak correlations are usually found, e.g. Rolley and Owens (1967). Nevertheless, some dissolved oxygen models include a formulation where SOD is a linear function of the rate of decaying sedimentary carbon – see Porcella et al. (1986) for a survey. In view of the conflicting observations of the relationship of SOD to sediment parameters, this assumption is somewhat tenuous.

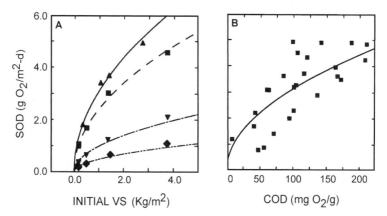

Fig. 8.1 (A) Sediment oxygen demand versus initial areal volatile solids. The time t indicates days after the start of the experiment. $t = 0$ (▲) Baity (1938). $t = 0$ (■), $t = 100$ (▼), $t = 200$ (◇) Fair et al. (1941). (B) SOD versus surface sediment COD (Gardiner et al., 1984).

8.2 PROPOSED MODELING FRAMEWORKS

A number of modeling approaches have been suggested for dealing with SOD. These are reviewed next.

8.2 A   Fair, Moore, and Thomas Experiments

The classical experiment by Fair et al. (1941) examined the variation of SOD in laboratory reactors as a function of sediment depth and time. The sediment employed was a mixture of sewage sludge, sand, and diatomaceous earth. They concluded that SOD varied as approximately the one-half power of the initial areal volatile solids concentration VS (kg/m²). Fig. 8.1A illustrates the fit of the data to a square root dependency for both the data from Fair et al. (1941) and from Baity (1938). The relationship is

$$SOD = a_1 \sqrt{VS} \tag{8.1}$$

where a_1 is a different constant for each time t measured from the start of the experiment.

The difficulty in applying this empirical result to natural sediments is that the dependent variable – areal volatile solids concentration in the experimental reactors – has no obvious field analog. The depth of sediment to be sampled in order to determine areal VS is undetermined. Therefore, although this work is widely known, it is not clear how the results (Eq. 8.1) are to be applied to field situations.

A similar relationship (Fig. 8.1B) has been observed in a set of field data from Green Bay (Gardiner et al., 1984). The SOD can be related to the square root of the surface sediment chemical oxygen demand COD.*

$$SOD = b_1 \sqrt{COD} \tag{8.2}$$

where b_1 is a constant. This relationship still implies a sampling depth that is difficult to specify. In both data sets a square root relationship is observed between SOD and measures of sediment organic matter concentration. This is quite different from the linear relationship of SOD to sediment organic carbon assumed in many dissolved oxygen models. We will return to this critical observation in Chapter 10, where the square root dependency is related to methane gas production and escape.

8.2 B   Models of the Electron Acceptors

A strategy that has been adopted for building models of SOD is to model only the concentration of the electron acceptors (O_2, NO_3^-, SO_4^{2-}, etc.) in the interstitial water of the sediment. The simplest model follows from assuming that the consumption of oxygen is zero-order with respect to oxygen concentration, that is, it is independent of the oxygen concentration so long as it is greater than zero, and that it is constant with respect to depth z below the sediment-water interface (Bouldin, 1968).

The derivation is instructive in the way that the boundary conditions are applied to determine the depth of the aerobic layer. The one-dimensional mass balance equation for the concentration of dissolved oxygen in the interstitial water of the sediment $[O_2(z)]$ is

$$-D_{O_2} \frac{d^2[O_2(z)]}{dz^2} = -R_0 \tag{8.3}$$

where D_{O_2} is the diffusion coefficient for dissolved oxygen, and R_0 is the zero-order consumption rate of oxygen. The convention for z is positive downward. The solution follows from two integrations of Eq. (8.3)

$$[O_2(z)] = A_0 + A_1 z + \frac{R_0 z^2}{2 D_{O_2}} \tag{8.4}$$

with A_1 and A_2 the arbitrary constants of integration. The three boundary conditions are

$$[O_2(z)]|_{z=0} = [O_2(0)] \tag{8.5}$$

which specifies the overlying water dissolved oxygen concentration $[O_2(0)]$;

*Chemical oxygen demand is measured by oxidation using potassium dichromate $K_2Cr_2O_7$ as the oxidant.

$$[O_2(H_1)] = 0 \tag{8.6}$$

which requires that the oxygen concentration be zero at the depth of the aerobic zone H_1; and

$$\left. \frac{d[O_2(z)]}{dz} \right|_{z=H_1} = 0 \tag{8.7}$$

which follows from the requirement that below H_1 the dissolved oxygen is zero. The final condition comes from the requirement that no oxygen is being consumed below H_1. Since oxygen cannot be consumed if it is not present, the flux of dissolved oxygen at $z = H_1$ – which would be a source to the sediment below $z = H_1$ – must be zero as well.

Eqs. (8.5) and (8.7) can be used to evaluate A_0 and A_1 so that Eq. (8.4) becomes

$$[O_2(z)] = [O_2(0)] + \frac{R_0 z}{D_{O_2}} \left(\frac{z}{2} - H_1 \right) \tag{8.8}$$

Eq. (8.6) can be used to obtain the depth of the aerobic layer

$$H_1 = \sqrt{\frac{2D_{O_2}[O_2(0)]}{R_0}} \tag{8.9}$$

The flux of oxygen to the sediment, which is the SOD, is obtained from the slope of the dissolved oxygen profile just below the sediment water interface at $z = 0$

$$SOD = -D_{O_2} \left. \frac{d[O_2(z)]}{dz} \right|_{z=0} = R_0 H_1 \tag{8.10}$$

The result is (Bouldin, 1968)

$$SOD = \sqrt{2D_{O_2} R_0 [O_2(0)]} \tag{8.11}$$

This equation makes several interesting predictions. It requires that SOD varies as the square root of the zero-order oxygen consumption rate R_0. If one supposes that the areal volatile solids concentration is proportional to the zero-order oxygen consumption rate, then this is just what Fair et al. (1941) found.

However, this model gives no guidance for specifying or measuring the zero-order oxygen consumption rate R_0. In fact at this level of development the model basically replaces a measurable flux, the SOD, with an unmeasurable quantity, R_0.

The model also predicts that SOD should vary as the square root of the overlying water dissolved oxygen. As shown in subsequent chapters, the observed variation of SOD with overlying water dissolved oxygen is more complicated than this model predicts. Nevertheless, this is an interesting formulation. In particular, it shows how the depth of the aerobic layer H_1 follows from the requirements of the oxygen mass balance.* We will see that this is a common feature of oxygen flux models.

Klapwijk and Snodgrass (1982) proposed a three-layer model where the layers are characterized by the electron acceptors used to react with the organic matter: oxygen (the aerobic layer), nitrate (the anoxic layer), and methane fermentation (the anaerobic layer). The SOD results from two reactions in the aerobic zone: the oxidation of carbonaceous reduced endproducts, and the nitrification of ammonia to nitrate. Since zero-order sources and sinks are used, the SOD model is equivalent to Eq. (8.11) with

$$R_0 = R_C + 4.57 R_N \qquad (8.12)$$

where R_C is the zero-order carbonaceous oxygen consumption rate, and R_N is the nitrification rate in the aerobic layer. The stoichiometric coefficient 4.57 g O_2/g NH_4-N, is the oxygen consumed by nitrification. In order to determine R_N, the maximum nitrification rate $R_{N,max}$ is specified as a model parameter. Then the ammonia production is estimated in the aerobic, anoxic, and anaerobic layers. If this exceeds $R_{N,max}$, then $R_N = R_{N,max}$, and the remaining ammonia production escapes as an ammonia flux. Otherwise, R_N is set equal to the total ammonia production and no ammonia escapes.

The rate parameters in the model are the aerobic carbonaceous oxidation rate R_C, the maximum nitrification rate $R_{N,max}$, and the anaerobic ammonia production rate. The use of this model requires either measurements or estimates of these three zero-order rates.

8.2 C   Oxygen and Temperature Dependency

The dependence of SOD on temperature and overlying water dissolved oxygen has been the subject of several laboratory and field investigations – see Porcella et al. (1986) for a summary. A mechanistic proposal (Walker and Snodgrass, 1986) for the relationship is

$$\text{SOD} = \mu(T) \frac{[O_2(0)]}{K_{O_2} + [O_2(0)]} + k_C(T)[O_2(0)] \qquad (8.13)$$

where $\mu(T)$ is the maximum biological SOD, K_{O_2} is the half saturation constant, and $k_C(T)$ is the chemical SOD at temperature T. The latter is measured by poisoning a duplicate core and measuring the remaining oxygen uptake. This is assumed to be due to nonliving processes and is called chemical SOD. The biological SOD is the total SOD minus the chemical SOD.

For an interesting use of this model in calculating the time course of oxygen uptake in a sediment incubation chamber see Hall et al. (1989).

8.2 D Diagenesis and Oxygen Equivalent Models

A model has been proposed (Di Toro, 1986, Di Toro and Matystik, 1980) that focuses on the production of sediment fluxes as a consequence of the flux of particulate organic carbon (POC) to the sediment and its decomposition. The idea is to model the transport of the reduced species (i.e., the electron donors) produced by the decomposition of organic carbon (e.g., $CH_4(aq)$, HS^-, Fe^{2+}, etc.) rather than the electron acceptors (O_2, NO_3^-, etc.). The diagenesis reaction can be thought of as producing oxygen equivalents of various reduced species. A mass balance equation is used to compute the flux of oxygen equivalents to the sediment-water interface. It is assumed that a fraction, $f_{O_2^*}$, is oxidized to produce SOD.

For situations where neither a gas phase nor any solid phases are formed, and to the approximation that the diffusion coefficients of the various reduced species are equal, the identity of the reduced species produced by diagenesis is immaterial as shown below. The SOD that results from the oxidation of this flux of oxygen equivalents is

$$SOD = J_{PCOD} f_{O_2^*} \frac{[O_2(0)]}{[O_2(0)] + \frac{5}{4}[NO_3(0)]} \tag{8.14}$$

where J_{PCOD} is the flux of reactive particulate COD to the sediment, and $[O_2(0)]$ and $[NO_3(0)]$ are the molar concentrations of dissolved oxygen and nitrate in the overlying water. Increasing the overlying water nitrate concentration is predicted to decrease the SOD since denitrification in the sediment consumes oxygen equivalents and reduces the flux to the sediment-water interface.

The principal focus of this model is to couple SOD explicitly to the input of particulate organic matter and the diagenesis reaction that makes it available for subsequent oxidation. The major weaknesses of the model are in not accounting for oxidation kinetics, other than including a parameter $f_{O_2^*}$ in Eq. (8.14) to account for possible incomplete oxidation, and in not explicitly considering ammonia oxidation.

8.2 E Discussion

All the above models, except for the diagenesis models of SOD, are based on modeling the transport and fate of the electron acceptors: O_2, NO_3^-, and SO_4^{2-}, and specifying the oxidation or consumption rates as model parameters. But it is not clear how to measure or model consumption rates or how to relate them to the flux of POM to the sediment. This is a severe weakness, since the causal chain starts with the input of POM to the sediment and results in sediment fluxes.

From a mass balance point of view, the consequence of POM diagenesis is the generation of reduced species: CH_4, H_2S, and NH_4^+. Hence the model should focus on the transport and oxidation of these reduced species. The distribution and fluxes of the electron acceptors, in particular the oxygen flux, result from the oxidation of these reduced species.

8.3 OXYGEN EQUIVALENTS

At first glance, developing an SOD model appears to be quite a complicated task since many aerobic and anaerobic reactions in sediments are involved – see Section 1.2 B. Biologically mediated reactions (e.g., methane production and consumption) as well as inorganic reactions (e.g., sulfide oxidation) all need to be considered. Layered models that distinguish zones of oxygen consumption and nitrate reduction[*] and a three-layer model including methane formation[†] have been proposed. The many other conceivable additional reactions, involving various electron donor and acceptor pairs[‡], are the most significant complicating feature, since their explicit inclusion seems unavoidable. We will also be forced to model certain of these electron acceptors explicitly, due to the inadequacies of the simple oxygen equivalent model presented below. Nevertheless, the oxygen equivalent idea is so attractive that a presentation is at least instructive.

A model of sediment oxygen demand can be constructed that ultimately dispenses with the apparent complexity by relating sediment oxygen demand to the flux of the oxygen equivalents of all reduced substances in the interstitial water without specific regard to their identity. In fact, since a simple aggregate measure of oxygen equivalents is available – chemical oxygen demand – a conclusion implicit in the model is that SOD can be determined by measuring the COD flux directly, or by measuring the interstitial water COD profile and measuring or estimating the sediment-water diffusion coefficient. Alternately, SOD can be calculated from a mass balance model of oxygen equivalents in the sediment itself. A calculation of the detailed redox chemistry of the sediment interstitial water is also possible and may be required for a detailed understanding of the situation. However, the COD flux methods may suffice in most cases.

The justification for any of these procedures is based on two assumptions:

1. that the redox chemistry is reasonably near thermodynamic equilibrium, which, as shown below, implies that COD flux and sediment oxygen demand are identical;

2. that the transport coefficients of the dissolved chemical species involved are independent of their identity, which dispenses with the requirement to compute the detailed redox chemistry.

If these assumptions are reasonable approximations, then sediment oxygen demand (the flux of dissolved oxygen from the overlying water to the sediment) is equal to the flux of dissolved oxygen equivalents, namely dissolved COD, from the sediment.

The purpose of the next section is to formulate a mass balance model of sediment-interstitial water interactions that is based on the above assumptions: thermodynamic

---

[*]Jahnke et al. (1982), Park and Jaffe (1996), Smith and Jaffe (1998), Soetaert et al. (1996), Vanderborght and Billen (1975). [†]Klapwijk and Snodgrass (1982). [‡]Park and Jaffe (1996).

equilibrium and species independent transport. Using this model, it can be demonstrated that SOD and dissolved COD flux are equivalent. The model is also useful for analyzing other interstitial water and solid phase chemical reactions. However, the principal purpose for presenting it below is that it provides the framework from which the SOD model follows.

8.3 A   Structure of the Analysis

The model presented below is based on two components: mass balance and chemical equilibrium. Mass balance models of substances that are conservative or, as in the case of radionuclides with have known decay rates, have a long history of application to sediments.[*] These models account for (1) diffusive mass fluxes arising from molecular and mechanical mixing of the interstitial waters, (2) advection due to sedimentation and compaction, and (3) the effects of the particulate organic matter decomposition (diagenesis) reactions.

Typically these models are applied to a single constituent of interest such as ammonia. For multiple constituents, a conceptual simplification is available if the reactions can be represented by the decay of organic matter of a fixed stoichiometry (Richards, 1965). The reason is that equivalents of each electron acceptor can be treated similarly and the quantities of substances used or produced are in stoichiometric ratios. Both of these approaches utilize the fundamental concept of mass balance expressed either as an algebraic stoichiometric relation or as a mass balance differential equation.

The second important assumption follows from observations that many important dissolved species in interstitial waters, which are involved in reversible reactions, are at approximate chemical equilibrium. For certain inorganic dissolved species and certain redox reactions, this has been tested by a number of investigators.[†] The evidence comes from evaluating the appropriate mass action equations. Mass balance is not usually a factor in these evaluations. While overall and complete thermodynamic equilibrium is never attained for all species in all settings, certain reactions occur so quickly that they are virtually in equilibrium over the time scale of sediment mass transport and diagenetic reactions. Thus, while not as universally applicable as the principle of chemical mass balance, chemical equilibrium is nonetheless a useful approximation in certain contexts since it dispenses with the need to specify detailed kinetic reaction sequences and rate coefficients.

8.3 B   Model Framework

The model of sediment behavior presented below is a synthesis of these ideas. The equations of mass balance and chemical equilibrium are combined into a single structure for the analysis of sediment interstitial water concentrations. Since bacterially

---

[*] Berner (1974, 1980a), Goldberg and Koide (1963), Lerman and Taniguchi (1972).
[†] See, for example, Garrells and Christ (1965), Kramer (1964), Thorstenson (1970).

mediated kinetics are in fact responsible for many of the redox reactions that affect interstitial water concentrations, it might seem at first glance that the assumption of chemical equilibrium is not applicable for these reactions. However, the thermo-dynamically predicted sequence of oxidation-reductions is commonly observed in nature as oxidation of organic material occurs (Stumm, 1966), so calculations based on chemical equilibrium can be a reasonable representation of the complex reac-tion kinetics that are actually taking place. The procedure is convenient because equilibrium calculations are independent of the reaction pathways and no detailed specification of the kinetics is necessary. Only the thermodynamic constants of the species of interest are required if detailed species concentration distributions are to be computed.

If reactions are known not to occur for kinetic reasons, even if they are ther-modynamically favored, they can be prevented from occurring in the equilibrium calculation as well. Thus a distinction is made among fast, slow, and prohibited reactions. Fast reactions are those that are assumed to be in (metastable) thermody-namic equilibrium over the time scale of analysis. Slow reactions are those for which the kinetics are important and must be specified. Prohibited reactions are those that are thought not to occur at all during the time scale of the analysis even if they are thermodynamically favored. The nitrogen system is the most important example for which thermodynamic equilibrium is not useful. If it happens that most reactions are of the fast type, then the approach presented below is a major simplification, since the number of parameters required to specify the behavior of the species of interest is substantially reduced, and a comprehensive analysis is possible.

The major complicating feature is that the chemical species involved in fast reac-tions are also subject to transport via diffusion and other mechanisms. Hence both mass transport and rapid reactions simultaneously affect concentrations. As various species are transported, the reactions adjust in complex ways to maintain equilibria among the species. The technique presented below for analyzing coupled mass trans-port equations with rapid reversible reaction kinetics is based on a transformation of the species mass balance equations into a set of smaller and simpler equations. These equations do not explicitly contain the sources and sinks due to the fast reversible ki-netic reactions that cause the analytical and computational difficulties. As a result they can be solved more directly together with the mass action equations for which there are available computational codes – e.g., DeLand (1967), Morel and Morgan (1972).

8.3 C   Conservation of Mass Equations

Consider the simplest setting, a one-dimensional steady state vertical analysis of N chemical species with names A_i. Let D_i be the diffusion coefficient of species A_i, and let w_i be the advective velocity of A_i, the velocity induced by the sedimentation of mass relative to a coordinate system fixed with respect to the sediment surface. Assume that these parameters are constant with depth. The use of more refined formulations that consider the vertical variation of porosity and the effect of com-paction (Imoboden, 1975) does not modify the argument presented below. Suppose that there exist N_r fast reversible chemical reactions involving the species A_i. The rate at which A_i is produced by reaction j is $\nu_{ji} R_j$ where R_j is the difference between the backward and forward reaction rates of the jth fast reversible reaction and ν_{ji} is the reaction stoichiometry, the quantity of A_i produced by the jth reaction. Let S_i be the net source of A_i due to other reactions that are occurring at slow rates and, therefore, not explicitly included in the fast reversible reaction set. Then the conservation of mass equations for the concentration of each species, $[A_i]$, are

$$-D_i \frac{d^2[A_i]}{dz^2} + w_i \frac{d[A_i]}{dz} = S_i + \sum_{j=1}^{N_r} \nu_{ji} R_j \qquad\qquad i = 1 \ldots N_r \qquad (8.15)$$

The difficulty with solving these coupled nonlinear equations numerically is that the forward and backward reaction rates for each fast reaction, R_j, correspond to very short time scales, whereas the slow reactions have long time scales. Numerical problems arise from the simultaneous presence of short and long time scales. Further, all that is usually known for these fast reactions is the ratio of the forward to backward reaction rate, namely the equilibrium constant. Finally, there are N_s coupled nonlinear equations to be solved simultaneously, which can be a substantial computational burden.

Various transformations have suggested to make Eq. (8.15) more tractable (Galant and Appleton, 1973, Shapiro, 1962). The crux of the idea is to eliminate the difficult terms and replace Eq. (8.15) with an alternate set of equations. The following fact (Di Toro, 1976) leads to a convenient choice for the transformation.

Let $B_k, k = 1, \ldots, N_C$ be the names of N_C components, and let a_{ij} be the quantity of component B_k in species A_i, that is, its stoichiometry in terms of the components. The components are the building blocks of the species. For example, the constituent atoms can be used as components. Then the formula matrix with elements a_{ik} is orthogonal to the reaction matrix with elements ν_{ji}

$$\sum_{i=1}^{N_s} a_{ik} \nu_{ji} = 0 \qquad\qquad k = 1 \ldots N_C, \ j = 1 \ldots N_r \qquad (8.16)$$

The physical fact embodied in these equations is that any reasonable reaction stoichiometry must conserve the component concentrations. This is apparent if the components that make up the species are thought of as the neutral atoms, together with electrons to provide the appropriate charges. Eq. 8.16 simply states that nonnuclear reactions must conserve atoms and charge. The specific choice of components is not important. Any consistent choice is a linear combination of neutral atoms and electrons and, therefore, must also be conserved.

This fact suggests that the conservation equations be transformed by multiplying Eq. (8.15) by the transpose of the formula matrix yielding

$$\sum_{i=1}^{N_s} \left(-D_i \frac{d^2}{dz^2} + w_i \frac{d}{dz} \right) a_{ik}[A_i] = \sum_{i=1}^{N_s} a_{ik} S_i + \sum_{j=1}^{N_r} R_j \sum_{i=1}^{N_s} a_{ik} \nu_{ji} \qquad (8.17)$$

But the orthogonality relation (Eq. 8.16) implies that the terms involving R_j are zero so that Eq. (8.17) becomes

$$\sum_{i=1}^{N_s} \left(-D_i \frac{d^2}{dz^2} + w_i \frac{d}{dz} \right) a_{ik}[A_i] = \sum_{i=1}^{N_s} a_{ik} S_i \qquad (8.18)$$

This is the fundamental simplification that makes the analysis tractable. The transformed equations are no longer functions of the fast reversible reactions. They are influenced only by the species mass transport coefficients and the slow reactions S_i.

It is clear from the derivation of the transformed equations that the simplifications introduced, namely steady state and spatially invariant transport coefficients, do not restrict the applicability of the transformation technique. The method is directly applicable to more general equations that consider temporally and spatially variable parameters.

8.3 D Species-Independent Transport

If D_i and w_i are not species dependent, so that $D_i = D$ and $w_i = w$ for all i, the summations $\sum_i a_{ik}[A_i]$ on the left-hand side of Eq. (8.18) become the component concentrations $[B_k]$ by the definition of a_{ik}

$$\sum_i a_{ik}[A_i] = [B_k] \qquad (8.19)$$

Therefore, the transformation yields conservation of mass equations for the N_c component concentrations

$$-D \frac{d^2[B_k]}{dz^2} + w \frac{d[B_k]}{dz} = \sum_{i=1}^{N_s} a_{ik} S_i \qquad k = 1 \ldots N_C \qquad (8.20)$$

These equations are independent of the fast reversible reactions. Therefore the fast reversible reactions can be assumed to be at equilibrium. Once the vertical distribution of the N_c components B_k are calculated from Eq. (8.20), the $N_s - N_c$ equilibrium mass action equations for A_i can be used in conjunction with the N_c stoichiometric algebraic component mass balance equations to solve for the concentrations of the N_s species, $[A_i]$, at any depth of interest, independent of any other depth, a considerably simplified computational task. Thus the components can be treated in exactly the same way as any other variable in mass transport calculations, without regard to the reversible reactions, so long as the mass transport coefficients are independent of the species.

For precipitation-dissolution reactions that may be of importance in sediments, this simplification amounts to assuming that the solid phase is subject to the same mass transport as the interstitial water, which is clearly not the case. For species-dependent mass transport Eqs. (8.17) must be addressed directly.

8.4 SEDIMENT OXYGEN DEMAND

The mass balance equations for the component concentrations (Eq. 8.20) represent a major conceptual simplification as well as a useful computational form. The assumption of species-independent transport yields a set of equations for the component concentrations that are independent of each other and the detailed species distribution. Therefore, one need only be concerned with the oxygen consuming component of the sediment chemistry. All the other chemical complexities, the many redox couples and the numerous species that are involved (Section 1.2 B), are of no consequence since they do not affect the mass balance equation for the components. It remains only to choose the relevant components for an SOD model.

Consider the following model formulation. Assume that the principal rate limiting kinetic reaction occurring in sediments is the bacterial conversion (diagenesis) of particulate sedimentary organic material, $C_aH_bO_cN(s)$, into available reactive form, $C_aH_bO_cN(aq)$, after which it reacts reversibly and rapidly with all other dissolved species in the interstitial water. For species-independent transport to be a reasonable assumption, the possibility of precipitation or dissolution of solid phase chemical species, such as $Fe(OH)_2(s)$, is specifically excluded, as is the possibility of gas phase formation (bubbles). The modifications required for consideration of solid phase interactions are discussed below. Schematically the reaction sequence is

$$C_aH_bO_cN(s) \rightarrow C_aH_bO_cN(aq) \rightleftharpoons A_1 \rightleftharpoons \cdots \rightleftharpoons A_{N_s} \qquad (8.21)$$

where $A_1 \cdots A_{N_s}$ are the rapidly reacting dissolved species being considered. The initial rate limiting reaction controls the rate of the entire reaction set, since it is assumed that all other reactions occur rapidly.

The key to a comprehensible result is the intelligent choice of the components that isolates oxygen in aerobic conditions. A number of possibilities are the neutral atoms

$$C_aH_bO_cN(s) \rightarrow a[C] + b[H] + c[O] + N \qquad (8.22)$$

or the more conventional choice for chemical equilibrium calculations

$$C_aH_bO_cN(s) \rightarrow a[CO_2] + \beta[H^+] + \beta[e^-] + \delta[H_2O] + [NH_3] \qquad (8.23)$$

However, the best choice is simply

$$C_aH_bO_cN(s) \rightarrow a[CO_2] + \beta[O_2] + \delta[H_2O] + [NH_3] \qquad (8.24)$$

so that particulate organic matter is thought of as being composed of the components: oxidized carbon, CO_2; the oxygen required to oxidize the carbon (the COD), O_2; and ammonia, NH_3. That is

$$C_aH_bO_cN(s) = (CO_2)_a(O_2)_\beta(H_2O)_\delta(NH_3) \qquad (8.25)$$

If species-independent transport is a reasonable assumption then each component distribution is independent of the others and can be considered separately. This is the principal simplification.

8.4 A Model Equations

Since oxygen consumption is our primary concern, consider the diagenetic decay of particulate COD via a first-order reaction for a purely advective sediment model which is the conventional formulation (Section 2.7 B). The mass balance equation for this situation is

$$w \frac{d[\text{PCOD}]}{dz} = -k_{\text{PCOD}}[\text{PCOD}] \tag{8.26}$$

so that the vertical distribution of particulate COD in the sediment is

$$[\text{PCOD}(z)] = \frac{J_{\text{PCOD}}}{w} \exp\left(-\frac{k_{\text{PCOD}}z}{w}\right) \tag{8.27}$$

where J_{PCOD} is the particulate COD flux from the overlying water to the sediment surface. The source of oxygen equivalents to the interstitial water is $-k_{\text{PCOD}}[\text{PCOD}(z)]$, the negative sign corresponding to the convention that positive COD is oxygen consumed so that positive COD produces negative oxygen equivalents. Thus the oxygen equivalent component equation is

$$-D \frac{d^2[\text{O}_2^*]}{dz^2} + w \frac{d[\text{O}_2^*]}{dz} = -k_{\text{PCOD}} \frac{J_{\text{PCOD}}}{w} \exp\left(-\frac{k_{\text{PCOD}}z}{w}\right) \tag{8.28}$$

where the superscript is used to distinguish O_2^*, the dissolved oxygen component, from $\text{O}_2(\text{aq})$, the dissolved oxygen species. The solution to Eq. (8.28) is (Chapter 2, Appendix A)

$$[\text{O}_2^*(z)] = [\text{O}_2^*(0)] - \frac{J_{\text{PCOD}}/w}{1 + k_{\text{PCOD}}D/w^2} \left\{ 1 - \exp\left(-\frac{k_{\text{PCOD}}z}{w}\right) \right\} \tag{8.29}$$

where $[\text{O}_2^*(0)]$ is the oxygen component concentration at $z = 0$.

The computation of the surface flux of oxygen equivalents follows from its definition

$$J[\text{O}_2^*] = -D \left. \frac{d[\text{O}_2^*]}{dz} \right|_{z=0} \tag{8.30}$$

and applying this to Eq. (8.29) yields

$$J[\text{O}_2^*] = \frac{\eta}{1 + \eta} J_{\text{PCOD}} \tag{8.31}$$

where

$$\eta = \frac{k_{\text{PCOD}}D}{w^2} \tag{8.32}$$

The fraction $\eta/(1 + \eta)$ is the portion of J_{PCOD} which is mineralized and returned to the sediment-water interface. The remainder is buried faster than it can decompose

Table 8.1 Decomposable Fraction and Sediment Reaction Rates

Component	Fraction Reacted	Reaction Rate yr^{-1}	η	Sediment Source
COD	$0.13 - 0.25^a$	$0.55 - 1.9$	–	River Muds[b]
COD	$0.13 - 0.48^a$	$2.4 - 3.6$	–	Sewage Solids[b]
Organic N	0.41	0.60^c	–	Long Island Sound[d]
		0.013^c	3.9	Foam Site
Organic C	0.67	0.0172	52.0	Lake Greifensee[d]

[a] Assuming 0.5 kg C/kg VS and 2.67 kg O_2/kg C. [b] Fair et al., 1941. [c] The larger value in the bioturbation zone. [d] Berner (1980a).

and diffuse to the sediment-water interface. Table 8.1 presents some examples of the fraction of organic matter that reacts diagenetically, the observed diagenetic reaction rates, and the resulting η computed from Eq. (8.32). The reacted fraction corresponds to the difference between the initial ($t = 0$ for batch reaction experiments) or surface ($z = 0$ for the analysis of sediment core data) and final organic matter concentrations for $t \to \infty$ or $z \gg w/k_{PCOD}$. The fact that unreacted sedimentary organic matter is found at depth indicates that only a fraction of PCOD reacts diagenetically. The reactivity of sedimentary organic matter is discussed in Chapter 12. Thus Eq. (8.31) should be modified to reflect this fact. Let f_ℓ be the reactive (or labile) fraction of the PCOD. Then Eq. (8.31) becomes

$$J[O_2^*] = \frac{\eta}{1 + \eta} f_\ell J_{PCOD} \qquad (8.33)$$

The importance of the correction $\eta/(1 + \eta)$ can be estimated from the examples listed in Table 8.1. For the lake examples $\eta/(1 + \eta) = 1$. For coastal and pelagic ocean sediments, Toth and Lerman (1977) have observed a correlation between k_{PON} and w for ammonia diagenesis $k_{PON}/w^2 = 0.01$ yr/cm². For neutral chemical species the molecular diffusivity is $D \simeq 2$ cm²/d (Table 2.1), $\eta \simeq 7.3$. Thus for coastal and oceanic sediments the burial correction may be marginally significant.

Fig. 8.2 illustrates the model structure with $f_\ell = 1$ for simplicity. Particulate COD decays as it advects downward due to sedimentation. It produces a negative concentration of dissolved oxygen equivalents in the interstitial water – negative since positive COD is, by convention, the oxygen consumed. Oxygen equivalents also advect downward, but in addition they are transported via diffusion. Since the oxygen equivalents concentration decreases with depth, the direction of the diffusive flux of oxygen equivalents is downward. This is the reverse of the situation where the concentration of a substance increases with depth in the interstitial water, such as ammonia, and its diffusive flux is upward. Hence oxygen equivalents are being transported downward from the water column to the sediment via diffusive exchange.

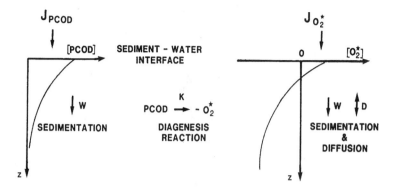

Fig. 8.2 Schematic diagram of the oxygen equivalents model.

8.5   OXYGEN EQUIVALENTS AND SOD

It remains only to argue that the flux of oxygen equivalents to the sediment is equal to the sediment oxygen demand. Consider the sediment-water interface with aerobic overlying water. At the surface there exists a region, however small, that is aerobic. For the choice of components CO_2^* and O_2^*, the thermodynamic equilibrium for a region where dissolved oxygen exists is that all carbon will exist as inorganic carbon, and that all oxygen equivalents are dissolved oxygen itself since no other aerobic species of oxygen equivalents exists for the C-H-O system. The reason for the choice of components CO_2^* and O_2^* is that the thermodynamic equilibrium solution can be obtained by inspection without any numerical calculations. It is simply that the dissolved oxygen concentration equals the concentration of the dissolved oxygen component $O_2(aq) = O_2^*$.

In general, the idea is to choose the components for all the species involved in redox reactions at the oxidation states which are known to be present in aerobic conditions. Thus, $SO_4^* = SO_4^{2-}$, $Fe(III)^* = Fe(III)^{3+}$ and so on, are chosen as components. Species in the sulfur system would be represented as

$$SO_4^{2-} = SO_4^*$$
$$HS^- = SO_4^* + H^+ - 2O_2^* \tag{8.34}$$

the carbon-hydrogen-oxygen system as

$$O_2(aq) = O_2^*$$
$$CO_2(aq) = CO_2^*$$
$$CH_4(aq) = CO_2^* + 2H_2O - 2O_2^* \tag{8.35}$$

Table 8.2 Comparison of Observed COD Flux and SOD

Anaerobic COD flux g O_2/m^2-d	Aerobic COD flux g O_2/m^2-d	Difference g O_2/m^2-d	Observed SOD g O_2/m^2-d	Reference
10.6–11.9	3.7–4.2	6.4-8.2	7.3	(a)
0.408	0.092	0.316	0.292	(b)
1.64±0.14	~0	1.64±0.14	1.68±0.56	(c)

[a] Fillos and Molof (1972). [b] Lauria and Goodman (1983). [c] Gelda et al. (1995)

the iron system as

$$Fe^{3+} = Fe(III)^*$$

$$Fe^{2+} = Fe(III)^* - H^+ + \frac{1}{2}H_2O - \frac{1}{4}O_2^* \tag{8.36}$$

and so on. Note that for this choice of components, the only species containing the component that is known to be present in significant concentrations in aerobic conditions is $O_2(aq)$ itself. Thus the molar concentration of O_2^* must be accounted for by $O_2(aq)$ so that $O_2(aq) = O_2^*$. It follows that their gradients are equal in aerobic conditions. Therefore SOD, which is the flux of dissolved oxygen at the sediment-water interface ($z = 0$), is

$$SOD = -D \left. \frac{d[O_2(aq)]}{dz} \right|_{z=0} \tag{8.37a}$$

$$= J[O_2(aq)] \tag{8.37b}$$

$$= J[O_2^*] \tag{8.37c}$$

$$= \frac{\eta}{1+\eta} f_\ell J_{PCOD} \tag{8.37d}$$

where the last equality Eq. (8.37d) comes from Eq. (8.33). Note that the argument requires that thermodynamic equilibrium is achieved where $O_2(aq) > 0$. This is equivalent to assuming that all reduced compounds are completely oxidized in the aerobic zone at or near the sediment-water interface.

8.5 A Extent of Oxidation and Experimental Confirmation

Complete oxidation may not actually occur as is suggested by experimental measurements summarized in Table 8.2. Aerobic dissolved COD fluxes have been measured together with anaerobic COD fluxes and SOD. This table illustrates two points. First, there is an aerobic COD flux so that only a portion of the total COD flux to the interface is oxidized. More important, however, is that the difference between the anaerobic COD flux and the aerobic COD flux equals the measured SOD.

These experiments directly confirm the model prediction that the reacted COD flux under aerobic conditions equals the SOD. Hence the use of the flux of oxygen equivalents as measured by COD for the prediction of SOD is shown by these experiments to be quantitatively justified. At present these data, and the model applications discussed subsequently, are the available support for the SOD model framework presented in this chapter.

A simple empirical correction to account for incomplete reaction of COD at the aerobic interface is to introduce a fraction, $f_{O_2^*}$, of the COD flux that actually reacts aerobically so that

$$\text{SOD} = f_{O_2^*} J[O_2^*] = f_{O_2^*} \frac{\eta}{1+\eta} f_\ell J_{\text{PCOD}} \tag{8.38}$$

From the data in Table 8.2, $f_{O_2^*} = 0.65 - 0.77$.

The specific geometric and transport model used for the sediment is not critical. For example, a well-mixed sediment layer of depth H may be more appropriate if the surface layer is mixed by biological or physical activity. Also the sequence of particulate COD oxidation need not be restricted to a single first-order reaction. For example if sedimenting algal and detrital carbon are the sources of particulate COD, then both labile (PCOD_ℓ) and refractory (PCOD_r) components can be considered. The reaction sequence is

$$\text{PCOD}_\ell \overset{k_\ell}{\to} f_r \text{PCOD}_r - (1 - f_r)O_2^* \tag{8.39}$$

$$\text{PCOD}_r \overset{k_r}{\to} -O_2^* \tag{8.40}$$

where f_r is the refractory PCOD component fraction of the algal carbon. The mass balance equations are analogous to Eq. (8.28), and at steady state the oxygen equivalents flux becomes (Di Toro and Matystik, 1980)

$$J[O_2^*] = \frac{J_{\text{PCOD}_r}}{(1 + wH/D)(1 + w/k_r H)}$$

$$+ \frac{J_{\text{PCOD}_\ell}}{(1 + wH/D)(1 + w/k_r H)} \left(1 - f_r + \frac{f_r}{1 + w/k_r H}\right) \tag{8.41}$$

where D is the effective interstitial water diffusion rate for the well-mixed layer. In fact, for this type of model the parameter D/H, a mass transfer coefficient, is the more fundamental parameter group (see Section 2.4).

The dimensionless group wH/D specifies the rate of burial of PCOD relative to the diffusion of O_2^*. The residence time/decay product of PCOD_r and PCOD_ℓ, respectively, in the active layer are $w/(k_r H)$ and $w/(k_\ell H)$. If these groups are all small relative to one, then

$$J[O_2^*] = J_{\text{PCOD}_r} + J_{\text{PCOD}_\ell} \tag{8.42}$$

and all the PCOD flux to the sediment returns as SOD. This single-layer model in its time variable form has been applied as a component in oxygen balance calculation for Lake Erie (Di Toro, 1980b).

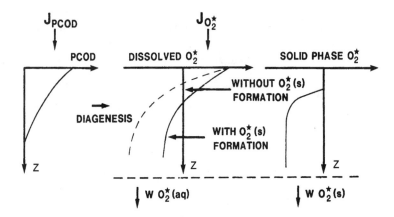

Fig. 8.3   Mass balance analysis including reduced solid phase formation.

8.5 B   Effect of Solid Phase Reactions

The validity of Eq. (8.38), which relates the SOD to the flux of particulate COD at the sediment-water interface, depends on the assumption of species-independent transport. It is this assumption that decouples the component equations from each other and allows the mass balance of O_2^* to be independent of the other components: CO_2^*, SO_4^*, $Fe(III)^*$ and so on. Species-independent transport is approximately true for dissolved species – ionic diffusion coefficients in sediments differ by approximately a factor of two to four (Table 2.1) – but it is surely not true for solid phase species. Thus a correction is required for the mass balance equations.

Instead of a detailed analysis of the species-dependent transport equations, consider the mass balance that results from the situation illustrated in Fig. 8.3. The flux of organic PCOD to the sediment is balanced by the flux of oxygen equivalents to the sediment and the loss via burial. To see this in terms of oxygen equivalents, consider Eq. (8.31) in the form

$$J[O_2^*] = \frac{\eta}{1+\eta} J_{\text{PCOD}} = J_{\text{PCOD}} - \frac{1}{1+\eta} J_{\text{PCOD}} \qquad (8.43)$$

Evaluating Eq. (8.29) for large depths ($k_{\text{PCOD}}z/w \gg 3$) indicates that

$$J_{\text{PCOD}} \frac{1}{1+\eta} = w\big([O_2(0)] - [O_2^*(\infty)]\big) \qquad (8.44)$$

where $[O_2^*(\infty)]$ is the oxygen equivalents concentration at large depth. Thus

$$J[O_2^*] = J_{\text{PCOD}} + w\big([O_2^*(\infty)] - [O_2^*(0)]\big) \qquad (8.45)$$

Note that Eq. (8.29) indicates that the oxygen equivalents concentration $[O_2^*(\infty)]$ would either be less than $[O_2^*(0)]$ if $J_{\text{PCOD}} > 0$, or $[O_2^*(\infty)]$ would be equal to

$[O_2^*(0)]$ if $J_{PCOD} = 0$. Hence the advective transport term is always negative or zero (i.e., $J[O_2^*] < J_{PCOD}$). This is physically reasonable since some dissolved COD is buried and does not exert an SOD.

When reduced solid phases form, such as $Fe(OH)_2(s)$, $FeS(s)$, the oxygen equivalents are in both dissolved $O_{2,aq}^*$ and solid $O_{2,s}^*$ phases. Both of these are buried via sedimentation so that

$$J[O_2^*] = J_{PCOD} + w\big([O_{2,aq}^*(\infty)] + [O_{2,s}^*(\infty)] - [O_2^*(0)]\big) \qquad (8.46)$$

The additional flux of solid phase oxygen equivalents $w[O_{2,s}^*(\infty)]$ reduces $J[O_2^*]$ since $[O_{2,s}^*(\infty)]$ is a negative number. From a practical point of view, the term $w[O_2^*(0)]$, the entrainment flux of overlying water, is usually negligible. For example $w < 10$ cm/yr, $[O_2^*(0)] < 10$ g O_2/m^3, so $w[O_2^*(0)] < 1$ g O_2/m^2-yr. Therefore, Eq. (8.46) becomes

$$J[O_2^*] = J_{PCOD} + w[O_2^*(\infty)] \qquad (8.47)$$

where $[O_2^*(\infty)] = [O_{2,aq}^*(\infty)] + [O_{2,s}^*(\infty)]$ is the bulk oxygen equivalents concentration of the sediment (aqueous and inorganic solid phase). Thus the proper correction is to decrease $J[O_2^*]$ to account for the loss due to burial of the reduced solid phases which form. With the exception of this correction, the formation of various reduced solid phases has no other effect on the steady state oxygen equivalents flux and, therefore, the SOD since ultimately the only source of oxygen equivalents is via the sediment-water interface.

8.5 C   Effect of Overlying Water Nitrate

Nitrate in the overlying water is subject to the same diffusive flux as is oxygen, so nitrate is transported to the anaerobic sediment layers where denitrification occurs. Its behavior is different than the other electron acceptors considered above. Nitrate reduction to nitrogen gas is not a reversible process. Therefore it cannot be included in the fast reaction set and must be considered explicitly. The rate of denitrification is rapid relative to the diagenesis rate (see Tables 4.2 and 12.2) so an explicit slow reactant approach with its attendant reaction rate is not necessary. Rather, it is assumed that all nitrate delivered to the anaerobic zone is rapidly reduced. Since the nitrogen gas produced at this location is a nonreversible endproduct, its formation is a permanent sink of dissolved COD and, therefore, a constant source of oxygen equivalents.

As a simple approximation, assume that all the denitrification occurs at the bottom of the aerobic zone H_1. The situation is illustrated in Fig. 8.4. The flux of nitrate to the sediment-water interface can also be approximated using the approximate expression (4.28)

$$J[NO_3] = -D\frac{d[NO_3]}{dz} = -\frac{D}{H_1}[NO_3(0)] \qquad (8.48)$$

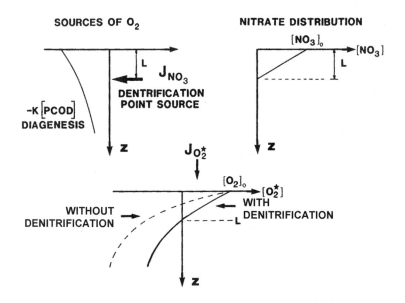

Fig. 8.4 Effect of overlying water nitrate concentration on SOD.

where $[NO_3(0)]$ is the overlying water nitrate concentration. The oxygen equivalents mass balance equation now has an additional term due to the source of oxygen equivalents provided by denitrification

$$NO_3 \rightarrow \frac{1}{2}N_2 - H^+ + \frac{1}{2}H_2O + \frac{5}{4}O_2 \tag{8.49}$$

Thus

$$-D\frac{d^2[O_2^*]}{dz^2} + w\frac{d[O_2^*]}{dz} = -k_{PCOD}\frac{J_{PCOD}}{w}\exp\left(-\frac{k_{PCOD}z}{w}\right)$$

$$+a_{O_2,NO_3}J[NO_3]\delta(z - H_1) \tag{8.50}$$

where $a_{O_2,NO_3}J[NO_3]\delta(z - H_1)$ is the delta function source of oxygen equivalents at location H_1. The stoichiometric coefficient $a_{O_2,NO_3} = 5/4$ from Eq. (8.49) and $J[NO_3]$ given by Eq. (8.48).

The use of the delta function (Kreyszig, 1972) assumes that the depth over which denitrification occurs is small and is representable by a point source of oxygen equivalents at $z = H_1$. The equivalent formulation is to break the problem into two regions: $0 \leqslant z \leqslant H_1$ and $z \geqslant H_1$. Then instead of the usual continuity condition for fluxes (Eq. 3.21c), the source is added at $z = H_1$

$$-D\frac{d[O_{2,1}^*(z)]}{dz}\bigg|_{z=H_1} + a_{O_2,NO_3}J[NO_3] = -D\frac{d[O_{2,2}^*(z)]}{dz}\bigg|_{z=H_1} \tag{8.51}$$

The solution is obtained using the methods in Appendix 3A. For $z < H_1$ the solution is

$$[O_2^*(z)] = [O_2^*(0)] - \frac{J_{PCOD}/w}{1 + k_{PCOD}D/w^2} \left\{ 1 - \exp\left(-\frac{k_{PCOD}z}{w}\right) \right\}$$

$$+ \frac{a_{O_2,NO_3}J[NO_3]}{w} \exp\left(-\frac{wH_1}{D}\right)\left(\exp\left(\frac{wz}{D}\right) - 1\right) \qquad (8.52)$$

The flux follows from its definition (Eq. 8.30)

$$J[O_2^*] = \frac{\eta}{1 + \eta}J_{PCOD} - a_{O_2,NO_3}J[NO_3] \qquad (8.53)$$

for $wH_1/D \ll 1$.

In order to evaluate the depth of the aerobic zone H_1, the condition that at $z = H_1$, $[O_2^*(H_1)] = 0$ is applied to Eq. (8.52) with $J[NO_3]$ given by Eq. (8.48). The result is

$$H_1 = \frac{1 + \eta}{\eta}\frac{D\left([O_2^*(0)] + \frac{5}{4}[NO_3(0)]\right)}{J_{PCOD}} \qquad (8.54)$$

for the case that $k_{PCOD}H_1/w \ll 1$ and $wH_1/D \ll 1$, corresponding to the case of a small aerobic layer depth. Using this expression, Eq. (8.54), in Eq. (8.48) and Eq. (8.53) yields

$$J[O_2^*] = J_{PCOD}\frac{\eta}{1 + \eta}\frac{[O_2^*(0)]}{[O_2^*(0)] + \frac{5}{4}[NO_3(0)]} \qquad (8.55)$$

Thus the presence of nitrate as an irreversible source of oxygen equivalents reduces the flux of O_2^* to the sediment by the ratio $[O_2^*(0)]/\left([O_2^*(0)] + \frac{5}{4}[NO_3(0)]\right)$. For example, if the overlying water oxygen and nitrate concentrations were $O_2(aq) = 10$ mg/L (0.313 mM) and $NO_3(0) = 1.0$ mg N/L (0.0714 mM) then the ratio = 0.778. If the nitrate concentration is 10 mg N/L then the ratio drops to 0.26. Hence, for large overlying water nitrate concentrations, is seems probable that the SOD would be reduced.

It is disturbing to note that two sets of experiments designed to test this effect (Andersen, 1978, Edwards and Rolley, 1965) both found no significant reduction of SOD as overlying water nitrate concentration increased (0-20 mg N/L) although increased nitrate flux to the sediment was observed. Since nitrate flux to the sediment was being observed and oxygen equivalents are required to convert the nitrate to nitrogen gas, it is likely that the oxygen equivalents were coming from some source which is not directly coupled to the immediate source of oxygen equivalents producing the SOD. This point will be examined in greater detail in Chapter 9. However at *steady state* it must be true that a nitrate flux to the sediment that is being denitrified must be consuming electrons and therefore reducing the need for oxygen as the electron acceptor.

It is interesting to note that the oxygen equivalents viewpoint predicts that the presence of sulfate in the overlying water has no direct effect on SOD, with the exception of the formation of reduced sulfur compounds that are lost by burial. The reason is that sulfate reduction produces HS^- and S^{2-} and so on, which are reversibly oxidizable. Hence their formation does not provide a sink of dissolved COD that escapes oxidation in the aerobic zone as does N_2. Rather, they function as reversible intermediates that transport dissolved COD upward and oxygen equivalents downward. Hence their absolute concentration is unimportant. If they are present only in small amounts, other species are available to perform the same function. This explains the observations that SOD values in fresh- and saltwater are the same order of magnitude, as we will see subsequently, whereas the overlying water sulfate concentrations differ by several orders of magnitude.

8.6 CONCLUSION

The oxygen equivalent model of SOD is an interesting first step in the development of a fully functional model. The basic idea is that dissolved COD, or equivalently, negative dissolved oxygen concentrations, can be used to model SOD. It is by no means apparent that this procedure is justifiable. The results presented above demonstrate that if species-independent transport is assumed – which effectively means no reduced solid or gaseous phases are involved – then the model is defensible. The problem is, and this is the subject of the next three chapters, that either a solid phase, iron sulfide, and/or a gaseous phase, methane bubbles, are ubiquitous features in sediments. Therefore, the oxygen equivalent model is of only limited practical importance. However, the techniques employed above will prove to be important for formulating and understanding more complex models for which chemical equilibrium and mass transport need to be considered together.

9

Sulfide

9.1 INTRODUCTION

This chapter presents a model of sulfide production and oxygen consumption. The equations for the general model developed in Chapter 5 are applied to the sulfide distribution in sediments. Previous models of sediment oxygen demand have been reviewed in Chapter 8. This model focuses on the formation and oxidation of sulfide as the principal endproduct of carbon diagenesis.

9.2 SULFIDE PRODUCTION

The sediment oxygen demand in marine waters is directly coupled to the production of sulfide as the endproduct of sulfate reduction.[*] The electrons liberated by carbon diagenesis are primarily accepted by sulfate which is reduced to sulfide. The reaction is

$$2CH_2O + SO_4^{2-} + H^+ \rightarrow 2CO_2 + HS^- + 2H_2O \qquad (9.1)$$

(see Section 1.2 B). The dissolved sulfide that is produced reacts with the iron in the sediment to form particulate iron sulfide (Morse et al., 1987). Therefore, the model must distinguish between the solid and dissolved sulfide phases.

---

[*]Howarth and Jorgensen (1984), Jorgensen (1977a, 1982), Jorgensen and Revsbech (1990), Jorgensen et al. (1983).

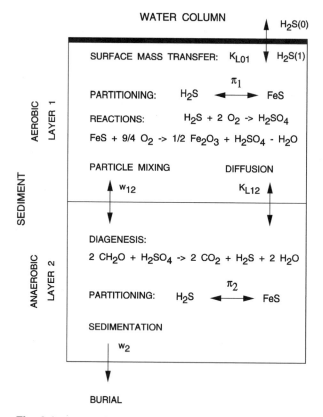

Fig. 9.1   Schematic diagram of the sulfide oxidation model.

The situation is illustrated in Fig. 9.1. Sulfide is produced in the anaerobic zone where a portion of it precipitates as iron monosulfide FeS(s). The remaining dissolved sulfide diffuses into the aerobic zone where it is oxidized to sulfate, consuming oxygen in the process. If the overlying water dissolved oxygen is low, then the dissolved sulfide is not completely oxidized and sulfide can diffuse to the overlying water. The particulate sulfide is also mixed into the aerobic zone where it can be oxidized to ferric oxyhydroxide $Fe_2O_3(s)$ consuming oxygen. Finally FeS(s) can be buried by sedimentation.

These are the only pathways for the reduced endproducts of carbon diagenesis considered in this model. Therefore, the model apportions the endproducts of carbon diagenesis to the oxidation of sulfide, diffusion of sulfide to the overlying water, and burial. The possibility of the formation of methane gas and its escape is not included. Actually only the escape of methane has been excluded from consideration. Carbon diagenesis can produce methane in the deeper part of the sediment. However, if it does not escape as methane bubbles, it diffuses into the zone of sulfate reduction.

Since no appreciable concentrations of dissolved methane are found in this zone[*], the methane must be oxidized, and sulfate, as the terminal electron acceptor, is reduced to sulfide. Therefore, the end result of carbon diagenesis is the production of sulfide as indicated in Eq. (9.1). We examine the extent and importance of methane formation and bubble generation in the next chapter.

9.3   SULFIDE OXIDATION

Both dissolved and particulate sulfide are oxidized in the aerobic layer. Therefore, dissolved and particulate sulfide reaction velocities are required. It has been found that the reaction rates for dissolved sulfide are linear in sulfide concentration and are either a linear or a fractional power of the oxygen concentration.[†] The catalytic effect of manganese and iron oxides has also been investigated.[#] A linear dependency is adopted.[‡]

For particulate sulfide oxidation, the reaction rate is also approximately linear in particulate sulfide concentration. The usual way of quantifying the concentration of reactive iron sulfide in sediment is as acid volatile sulfide (AVS). This is the solid phase sulfide that is extracted using cold hydrochloric acid (Morse et al., 1987). The normalized AVS concentration remaining as a function of time in aerobic sediment oxidation experiments (Di Toro et al., 1996a) is presented in Fig. 9.2. The exponential decrease in AVS (Figs. 9.2A and B) indicates that the reaction is first-order in AVS. The reaction rate for synthetic FeS is also first-order in AVS and first-order in dissolved oxygen concentration (Fig. 9.2B). Thus the kinetic expressions for both particulate and dissolved sulfide are similar. The aerobic reaction rate depth product is

$$k_1 H_1 = \left(k_{H_2S,dl} f_{dl} + k_{H_2S,pl} f_{pl}\right) \theta_{H_2S}^{(T-20)} \frac{[O_2(1)]}{K_{M,H_2S,O_2}} H_1 \qquad (9.2)$$

where $k_{H_2S,dl}$ and $k_{H_2S,pl}$ are the reaction rate constants for dissolved and particulate oxidation, respectively. A constant K_{M,H_2S,O_2} is used to scale the overlying water oxygen concentration. It is included for convenience only. At $[O_2(0)] = K_{M,H_2S,O_2}$ the sulfide oxidation reaction velocity is at its nominal value. The equivalent reaction velocities are

$$\kappa_{H_2S,dl} = \sqrt{D_1 k_{H_2S,dl}} \qquad (9.3)$$

$$\kappa_{H_2S,pl} = \sqrt{D_1 k_{H_2S,pl}} \qquad (9.4)$$

[*]Barnes and Goldberg (1976), Martens and Berner (1977), Reeburgh and Heggie (1974).
[†]Almegren and Hagstrom (1974), Millero (1991), Millero et al. (1987), Morse et al. (1987), O'Brien and Birkner (1977), Wilmot et al. (1988), Zhang and Millero (1991).
[#]Yao and Millero (1995).
[‡]Boudreau (1991), Cline and Richards (1969), Millero (1986).

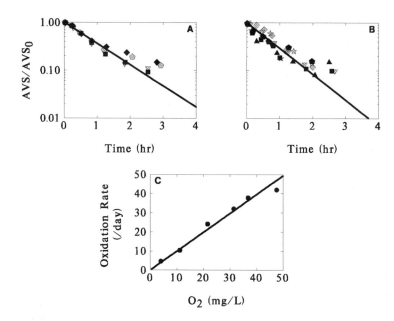

Fig. 9.2 Kinetics of FeS(s) oxidation for (A) Jamacia Bay and (B) Van Cortlandt Pond sediments (Di Toro et al., 1996a). (C) Effect of dissolved oxygen concentration on the oxidation rate (Nelson, 1978).

so that the fraction reacted in layer 1 $fr_{re,1}$, which will be denoted by fr_{ox}, is

$$fr_{ox} = \frac{\left(\kappa^2_{H_2S,d1}\, f_{d1} + \kappa^2_{H_2S,p1}\, f_{p1}\right)\theta^{(T-20)}_{H_2S}}{s}\frac{[O_2(1)]}{K_{M,H_2S,O_2}} \tag{9.5}$$

or

$$fr_{ox} = \frac{\left(\kappa^2_{H_2S,d1}\, f_{d1} + \kappa^2_{H_2S,p1}\, f_{p1}\right)\theta^{(T-20)}_{H_2S}}{s}\frac{[O_2(0)]}{2K_{M,H_2S,O_2}} \tag{9.6}$$

where the relationship: $O_2(1) = O_2(0)/2$ (Eq. 3.33) relates the aerobic layer oxygen concentration to the overlying water.

9.4 SOLUTIONS

The flux of sulfide J_{ox} oxidized in layer 1 is found using Eq. (5.22d)

$$J_{ox} = J_{re,1} = J_{T2}\frac{fr_{re,1}}{fr_{re,1} + fr_{aq} + fr_{br}} \tag{9.7}$$

Table 9.1 Parameters for Flux Apportionment Calculation

Parameter	Description	Value
m_1	Aerobic layer solids concentration (kg/L)	0.5
m_2	Anaerobic layer solids concentration (kg/L)	0.5
w_2	Sedimentation velocity (cm/yr)	0.25
w_{12}	Particle mixing velocity (m/d)	0.001
K_{L12}	Diffusive mass transfer coefficient (m/d)	0.01
$\kappa_{H_2S,dl}$	Dissolved H_2S oxidation reaction velocity (m/d)	0.20
$\kappa_{H_2S,pl}$	Particulate H_2S oxidation reaction velocity (m/d). Fig. 9.3A ,B, C, D	0.0, 0.0, 0.01, 0.40*
K_{M,H_2S,O_2}	Oxygen normalization constant (mg O_2/L)	4.0
$\pi_{H_2S,1}$	Aerobic layer partition coefficient for H_2S (L/kg). Fig. 9.3B	10,000
$\pi_{H_2S,2}$	Anaerobic layer partition coefficient for H_2S (L/kg). Fig. 9.3C, D	10,000, 100*

*Values refer to the indicated panels in Fig 9.3

The diffusion and burial fluxes are given by Eqs. (5.22a) and (5.22c)

$$J_{aq} = J_{T2}\frac{fr_{aq}}{fr_{re,1} + fr_{aq} + fr_{br}} \tag{9.8}$$

$$J_{br} = J_{T2}\frac{fr_{br}}{fr_{re,1} + fr_{aq} + fr_{br}} \tag{9.9}$$

9.4 A   Flux Apportionment

The distribution of carbon diagenesis among the various pathways is controlled by the magnitude of the sulfide partition coefficients and the oxidation reaction velocities. The partition coefficients determine the fraction of sulfide that is either in the dissolved or particulate fraction of the sediment.

An example which illustrates the importance of the extent of partitioning of sulfide into a solid phase is shown in Fig. 9.3A. The parameters used in the computation are listed in Table 9.1. These coefficients are justified subsequently. This is a cumulative plot of the proportion of carbon diagenesis that is either oxidized in the aerobic zone via sulfide oxidation J_{ox} that results in SOD, diffuses to the overlying water as a dissolved sulfide flux J_{aq}, or is buried J_{br}. For this example only dissolved sulfide is allowed to oxidize. For low partition coefficients, $\pi_1 = \pi_2 < 10^3$ L/kg, the burial is insignificant and only SOD and the diffusive flux are important. As the partition coefficient increases, the SOD and diffusive flux decrease and the burial flux increases. This is a consequence of the decrease in dissolved sulfide concentration so that less is available either for oxidation or for escape as an aqueous flux. Since there is no particulate oxidation, the only remaining possibility is loss by burial.

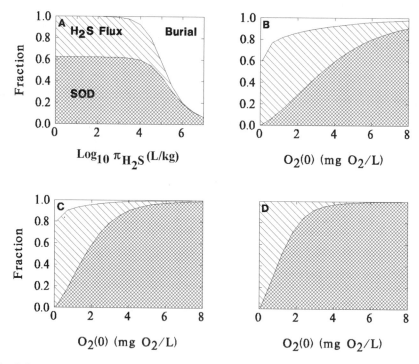

Fig. 9.3 (A) Effect of partition coefficients and (B) overlying water dissolved oxygen with increasing particulate sulfide oxidation rate (C-D) on the fraction of J_C that is oxidized as SOD J_{ox}, diffuses to the overlying water as an H_2S flux J_{aq}, or is buried J_{br}. See Table 9.1 for parameter values.

Fig. 9.3B illustrates the behavior of the fluxes at a fixed partition coefficient (10^4 L/kg) as a function of overlying water dissolved oxygen concentration $O_2(0)$. As dissolved oxygen decreases, the oxygen flux decreases, and both the diffusive flux and the burial flux increase. The reason is that as the oxygen concentration decreases the oxidation rate, and therefore the flux of oxygen to the sediment, decreases. As a consequence, the dissolved sulfide concentration increases. The result is that the aqueous flux of sulfide increases. The increased dissolved sulfide concentration also causes an increase in the particulate sulfide concentration – the ratio is the constant partition coefficient – which increases the loss of sulfide by burial.

The effect of increasing the oxidation rate of particulate sulfide is illustrated in Figs. 9.3C and D. A small particulate reaction velocity (Fig. 9.3C) increases the oxygen flux, but the effect of lowering overlying water dissolved oxygen is similar to that in the example above. Further increasing the particulate sulfide oxidation velocity (Fig. 9.3D) increases the oxidation flux until quite low dissolved oxygen, when the aqueous flux begins. For this case the burial flux is small enough to be negligible.

9.5   SEDIMENT OXYGEN DEMAND

Sediment oxygen demand is the common term for the flux of oxygen to the sediment. By convention, it is a positive number. The convention used in this book is that positive fluxes are from the sediment to the overlying water. Therefore, SOD $= -J[O_2]$. The flux of oxygen to the sediment is the result of the oxidation reactions in the aerobic layer. These reactions are presented below.

9.5 A   Sulfide Oxidation

The oxygen consumed by the oxidation of sulfide is one component of the total oxygen flux to the sediments. Carbon diagenesis J_C produces sulfide via the reaction given in Eq. (9.1). If oxygen equivalents, denoted by O_2^*, are adopted as the units for sulfide concentrations and fluxes, then the stoichiometric coefficient relating the flux of carbon diagenesis and sulfide production is

$$J_{T2} = a_{O_2,H_2S} J_C \tag{9.10}$$

where $a_{O_2,H_2S} = 2.67$ (mg O_2^*/mg C). The oxygen flux that results from the oxidation of sulfide is

$$J_{ox} = a_{O_2,H_2S} J_C \frac{fr_{ox}}{fr_{ox} + fr_{aq} + fr_{br}} \triangleq \text{CSOD} \tag{9.11}$$

Since the source of this oxygen flux is carbon diagenesis, it is called the carbonaceous sediment oxygen demand, CSOD. This is to distinguish it from the oxygen consumed by nitrification, which is discussed next.

9.5 B   Ammonia Oxidation and Denitrification

Oxygen is also consumed as a result of the oxidation of NH_4^+ to NO_3^-. The stoichiometry is $a_{O_2,NH_4} = 4.57$ g O_2 / g N. The quantity of ammonia that is nitrified is equal to the quantity of nitrate produced $S[NO_3]$. It can be calculated by evaluating the nitrification sink term in the mass balance equation (4.13)

$$S[NO_3] = k_{NH_4,1}[NH_4(1)] = \frac{\kappa_{NH_4,1}^2}{s} \theta_{NH_4}^{(T-20)}[NH_4(1)] \tag{9.12}$$

where $[NH_4(1)]$ is given by Eq. (3.40). The equivalent term for Michaelis-Menton kinetics is (Eq. 3.45)

$$S[NO_3] = \left(\frac{K_{M,NH_4}}{K_{M,NH_4} + [NH_4(1)]}\right)\left(\frac{[O_2(0)]}{2K_{O_2,NH_4} + [O_2(0)]}\right)$$
$$\frac{\kappa_{NH_4,1}^2}{s}\theta_{NH_4}^{(T-20)}[NH_4(1)] \tag{9.13}$$

The oxygen consumed by nitrification is called the nitrogenous sediment oxygen demand NSOD

$$\text{NSOD} = a_{O_2,NH_4} S[NO_3] \qquad (9.14)$$

The SOD of the sediment is the sum of the CSOD and NSOD

$$\text{SOD} = \text{CSOD} + \text{NSOD} \qquad (9.15)$$

9.5 C   Carbon Requirement for Denitrification

A final issue needs to be addressed. The denitrification of nitrate to nitrogen gas requires a carbon source. The reaction is (see Section 1.2 B)

$$\frac{10}{8}CH_2O + NO_3^- + H^+ \rightarrow \frac{10}{8}CO_2 + \frac{1}{2}N_2(g) + \frac{7}{4}H_2O \qquad (9.16)$$

and the carbon to nitrogen stoichiometric coefficient is $a_{C,N_2} = 10/8$ (mol C/mol N) = 1.071 (g C/g N) which has been confirmed experimentally (Copp and Dold, 1998). This requirement must be satisfied from carbon diagenesis, since it is the only source of reactive carbon. Hence the carbon diagenesis that reacts to form sulfide is that which remains

$$J_C - a_{C,N_2} J[N_2(g)] \qquad (9.17)$$

The rate of denitrification is equal to the flux of nitrogen gas $J[N_2(g)]$, which can be calculated by evaluating the denitrification sink term in the mass balance equations (4.5a–4.5b). The result, which is equivalent to the rate at which nitrogen gas is produced (Eq. 4.55), is

$$J[N_2(g)] = k_{NO_3,1} H_1[NO_3(1)] + k_{NO_3,2} H_2[NO_3(2)]$$

$$= \frac{\kappa_{NO_3,1}^2}{s}[NO_3(1)] + \kappa_{NO_3,2}[NO_3(2)] \qquad (9.18)$$

where $[NO_3(1)]$ and $[NO_3(2)]$ are given by Eqs. (4.6–4.7).

9.5 D   Final Equation

The equation for SOD is made up of the sum of the carbonaceous and nitrogenous components, with the former corrected for the denitrification sink of carbon diagenesis

$$\text{SOD} = a_{O_2,H_2S}\left(J_C - a_{C,N_2} J[N_2(g)]\right)\frac{fr_{ox}(s)}{fr_{ox}(s) + fr_{aq}(s) + fr_{br}(s)} + \text{NSOD} \qquad (9.19)$$

where NSOD is given by Eq. (9.14). This is the nonlinear equation that needs to be solved for SOD. The nonlinearity arises from the fact that the right hand side of the equation contains terms that are functions of $s = \text{SOD}/[O_2(0)]$ so that SOD appears on both sides of the equation. This equation can be solved numerically using standard root finding algorithms (Press et al., 1989).

9.6   DATA ANALYSIS

A commonly used technique to analyze sediment nutrient and oxygen fluxes is to examine the variation of one with respect to another (Nixon et al., 1976). This procedure is ideally suited for analyzing data using the models developed above because they predict ammonia flux as well as the oxygen flux within a comprehensive framework.

The idea is as follows: One of the flux measurements is used to estimate the diagenesis flux. For example, the ammonia flux $J[NH_4]$ is used to estimate J_N. Then carbon diagenesis J_C is estimated from nitrogen diagenesis using a suitable stoichiometric ratio. Once carbon diagenesis is known, the SOD can be computed (Eq. 9.19), and compared to the measured oxygen flux. Thus any set of laboratory or field measurements that includes simultaneous measurements of ammonia and oxygen fluxes can be compared to model predictions.

9.6 A   Methodology

The procedure uses the equation that relates ammonia flux to ammonia diagenesis Eq. (3.35)

$$J_N = \left(\frac{K_{M,NH_4}}{K_{M,NH_4} + [NH_4(1)]} \right) \left(\frac{[O_2(0)]}{2K_{O_2,NH_4} + [O_2(0)]} \right)$$

$$\left(\frac{\kappa_{NH_4,1}^2}{s} \theta_{NH_4}^{(T-20)} [NH_4(1)] \right)$$

$$+ s \left([NH_4(1)] - [NH_4(0)] \right) \tag{9.20}$$

where $[NH_4(1)]$ is computed from the ammonia flux (Eq. 3.7)

$$[NH_4(1)] = \frac{J[NH_4]}{s} + [NH_4(0)] \tag{9.21}$$

Using the stoichiometric ratio $a_{C,N}$ determined below, the carbon diagenesis flux J_C is determined. This is substituted into the SOD equation (9.19). The resulting nonlinear equation can be solved for SOD that corresponds to the starting ammonia flux.

In fact, this computation can be thought of as being indexed by the ammonia flux. That is, given an ammonia flux, the corresponding SOD is computed. In order to make the calculation, however, all the exogenous variables required in the computation of SOD are also required as a function of $J[NH_4]$. These relationships are established in the next section.

9.6 B   Exogenous Variables

Both the SOD and ammonia flux are strongly influenced by the overlying water oxygen concentration, $[O_2(0)]$. Lowering the oxygen concentration decreases the SOD

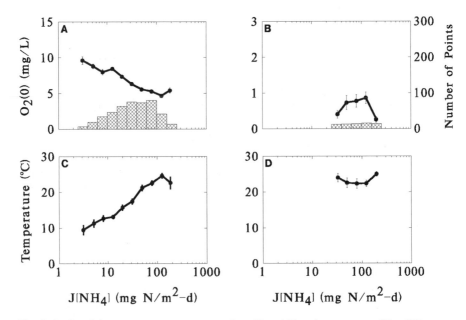

Fig. 9.4 Overlying water oxygen concentrations (A and B) and temperature (C and D) versus ammonia flux. (A and C) $[O_2(0)] > 2$ mg O_2/L, (B and D) $[O_2(0)] < 2$ mg O_2/L . Histogram presents the number of observations in each interval.

and increases the ammonia flux. Hence it is necessary to stratify the data with respect to $[O_2(0)]$. A division at $[O_2(0)] = 2$ mg O_2/L is chosen as a compromise between a suitably low dissolved oxygen concentration and the presence of a sufficient number of flux measurements in the Chesapeake Bay data set below that concentration.

Within the two subsets of observations, there may still be a systematic variation of the exogenous variables: overlying water concentrations: $[O_2(0)]$, $[NH_4(0)]$, $[NO_3(0)]$, and temperature T, with respect to ammonia flux. Fig. 9.4 presents the data versus $J[NH_4]$ for the variables with the most impact on the calculation: $[O_2(0)]$ and temperature. The data set is averaged over 0.2 $\log_{10}$ units of ammonia flux. The histograms are the number of data points in each interval.

For the subset $[O_2(0)] > 2$ mg/L, overlying water dissolved oxygen varies from 10 to 5 mg/L as the ammonia flux increases (Fig. 9.4A). This is a reflection of the increasing water temperature as shown in Fig. 9.4C. For the subset $[O_2(0)] < 2$ mg/L, overlying water dissolved oxygen is between 1 and almost 0 mg/L over the range of ammonia fluxes (Fig. 9.4B). The water temperature is high and almost constant (Fig. 9.4D), a reflection of the fact that anoxia occurs in the summer.

9.6 C Diagenesis Stoichiometry

The method being employed to calculate the SOD associated with an ammonia flux requires that the carbon diagenesis be estimated from the ammonia diagenesis. A

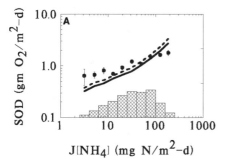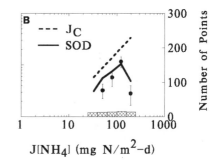

Fig. 9.5 Comparison of modeled (solid line) and observed (•) SOD. Carbon diagenesis J_C is also presented as a dashed line. (A) $[O_2(0)] > 2$ mg O_2/L. (B) $[O_2(0)] < 2$ mg O_2/L. Histogram presents the number of observations in each interval.

convenient approximation is that the ratio of carbon to nitrogen diagenesis fluxes follows Redfield stoichiometry: $a_{C,N} = J_C/J_N = 5.68$ (g C/g N). This approximation is shown in Chapter 12 to be applicable to Chesapeake Bay sediments. Although a stoichiometric ratio is not necessary when the sediment model is used as part of a coupled water column-sediment model, it provides a necessary relationship for the calculation of SOD from ammonia flux since it relates nitrogen diagenesis to carbon diagenesis.

9.6 D   SOD and Ammonia Fluxes

The calculations presented below are performed using the complete time variable model in steady state mode. The parameter values and formulations employed in the calculation are presented in Chapter 15, Tables 15.5 and 15.6. Since the complete model is forced by the depositional fluxes, the depositional flux of particulate nitrogen is back computed from the ammonia flux. The Redfield stoichiometric ratios are used to compute the other depositional fluxes. Rather than giving a brief and incomplete summary of the complete model formulation, it is more illuminating to examine the results of the calculation, using the insights and formulas obtained above.

The comparison of SOD and ammonia flux is shown in Fig. 9.5. The means and the standard errors of the data are shown. The histograms display the number of data points in each interval. The computed SOD and carbon diagenesis J_C which correspond to the ammonia flux are both illustrated. The relationship between SOD and $J[NH_4]$ for the two subsets is quite different and the model qualitatively reproduces the different behavior. The SOD for the low overlying water oxygen data set, $[O_2(0)] < 2$ mg/L (Fig. 9.5B), is smaller than the high dissolved oxygen data set (Fig. 9.5A) and the model successfully reproduces the trend. However, the model results are quantitatively less satisfactory. For the higher oxygen concentration (Fig. 9.5A) the data indicate a relationship that is less steep than the model predicts. And the model consistently overpredicts the observed SOD for the low dissolved oxygen data set (Fig. 9.5B).

9.7 COMMENTARY

The behavior of the SOD model can be understood as follows. The model apportions carbon diagenesis to the three terminal sinks: oxidation, sulfide flux to the overlying water, and burial. For the subset $[O_2(0)] > 2$ mg/L, the overlying water dissolved oxygen is actually greater than 5 mg/L (Fig. 9.4A), and essentially all the carbon diagenesis is oxidized (Fig. 9.3D). For $[O_2(0)] < 2$ mg/L, a substantial quantity is released as a sulfide flux (Fig. 9.3D), thereby lowering the SOD.

The model's inability to obtain the correct slope of the relationship between SOD and ammonia flux for the high dissolved oxygen data set points to a fundamental deficiency. Since the ammonia flux model appears to successfully reproduce the observations (Fig. 3.7), the problem must be with the SOD model. The SOD is overpredicted for the high ammonia fluxes. This could be remedied by choosing model parameters that apportion more of the diagenesis to burial, for example. But the underprediction of SOD for the small ammonia fluxes cannot be remedied in any way. All of the carbon diagenesis is being converted to SOD, and it is still not enough. There is no remedy within the context of a steady state model.

In fact, what we are seeing is a direct manifestation of a non-steady state effect. It is possible for the SOD to be larger than carbon diagenesis if it is due to the oxidation of previously stored particulate sulfide. In order for this to be the case, there must be periods where carbon diagenesis exceeds SOD. The excess production causes an increase in particulate sulfide that is stored in the sediment. This occurs during the time of rapid carbon diagenesis. As a consequence, larger SODs can be supported during periods of lower carbon diagenesis by the oxidation of the stored particulate sulfide.

Table 9.2 Average SOD

SOD (g O_2/m^2-d)	Observations (Standard Error)	Model (Stardard Error)
Without weighting	1.06 (0.126)	0.966 (0.233)
With weighting	1.14 (0.015)	1.01 (0.025)

For this explanation to be correct it must be true that the average SOD is correctly computed. The comparison is shown in Table 9.2 where the arithmetic average SODs are listed. The computation is performed both without weighting – each bin average in Fig. 9.5 is treated equally – and with weighting where the bin averages are weighted by the number of points in each bin average. The result is indeed that the average SODs are similar. Thus, it is the steady state assumption that is in error and this can only be remedied by employing a time variable calculation, which is presented in Chapter 14.

10

Methane

10.1 INTRODUCTION

In this chapter we focus on the production of methane and, in particular, methane gas from sediments. The oxidation of dissolved methane in the aerobic layer can consume oxygen, and this is analyzed as well. Finally the combination of methane and nitrogen gas production – the latter from denitrification – is used to estimate the composition of the gas escaping from sediments.

The model is formulated as a continuous one-dimensional diffusion model with three distinct layers (Fig. 10.1). That is, within each layer the concentration varies as a function of depth, rather than a single concentration. Oxidation of dissolved methane in the aerobic zone occurs as a first-order reaction. Transport within the interstitial water is by diffusion. The analysis begins with the fate of methane that is produced by the decay of POC. This model assumes that either there is no sulfate available or the rate of carbon diagenesis exceeds the rate of supply of sulfate to the sediments and that methane production is the dominant pathway of carbon diagenesis. For applications presented below, this appears to be a reasonable assumption. The situation where both sulfide and methane production occur is analyzed in Chapter 11.

10.2 STOICHIOMETRY AND OXYGEN EQUIVALENTS

The formation of methane is the result of a series of reactions that begin with the diagenesis of particulate organic carbon, represented by CH_2O, and end with the

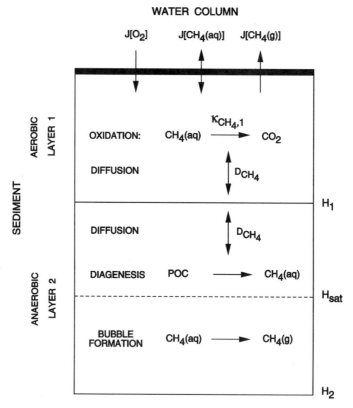

Fig. 10.1 Schematic diagram of the methane model. Diagenesis of POC produces CH_4 in the interstitial water. Oxidation, diffusion, and gas phase formation combine to produce the resulting fluxes of carbon species to the overlying water, and the resulting SOD. The depth of the aerobic layer H_1 is measured from $z = 0$, the sediment-water interface. The remaining depths are measured from H_1.

formation of methane. It is assumed that the reaction goes to completion and that no significant concentrations of intermediates such as volatile organic acids remain. The overall reaction in the anaerobic zone is

$$CH_2O \rightarrow \frac{1}{2}CO_2 + \frac{1}{2}CH_4 \qquad (10.1)$$

The oxygen consumption reaction in the aerobic zone is (see Section 1.2 B)

$$\frac{1}{2}CH_4 + O_2 \rightarrow \frac{1}{2}CO_2 + H_2O \qquad (10.2)$$

A convenient way to quantify (1) the amount of organic material that reacts, (2) the methane produced, and (3) the oxygen consumed is to use the oxygen equivalent of each reactant and endproduct. The definition of an oxygen equivalent, denoted by O_2^* to distinguish it from actual oxygen concentrations, is the oxygen required for complete oxidation. The stoichiometric coefficients for reactions (10.1) and (10.2) are all unity in this scheme.

By contrast, if carbon units are used, then 1 g CH_2O as carbon [= 2.67 g O_2^*] produces 1/2 g of CH_4 as carbon [= 1/2 (5.33 g O_2/g CH_4-C) = 2.67 g O_2^*] which consumes 2.67 g of O_2. Using oxygen equivalents simplifies the equations so we can express the concentrations in units that are appropriate for quantifying the impact on dissolved oxygen.

10.3   DISSOLVED METHANE MASS BALANCE

The analysis begins with a mass balance of methane in the interstitial water. In order to clarify the major features of the model, the oxidation of methane in the aerobic zone of the sediment is ignored for the moment. This will be remedied subsequently. Let S_{CH_4} be the rate of carbon diagenesis in oxygen equivalents that produces methane via Eq. (10.1) in the active layer of the sediment with depth H_2. The mass balance equation for the concentration of dissolved methane in the interstitial water $c(z)$ is

$$-D_{CH_4} \frac{d^2 c(z)}{dz^2} = S_{CH_4} \qquad\qquad 0 \leqslant z \leqslant H_2 \qquad (10.3)$$

where both the concentration of dissolved methane (g O_2^*/m^3) and the diagenesis source (g O_2^*/m^3-d) are in oxygen equivalent units. The solution is arrived at by two integrations

$$c(z) = A_{1C} + A_{2C} z - \frac{S_{CH_4} z^2}{2 D_{CH_4}} \qquad (10.4)$$

where A_{1C} and A_{2C} are the arbitrary constants of integration. The boundary conditions to be applied are

$$c(0) = 0 \qquad (10.5)$$

requiring the overlying water dissolved methane concentration to be zero – which is convenient and can be modified if $c(0)$ is significantly greater than zero. The boundary condition at the bottom of the active layer is

$$-D_{CH_4} \left. \frac{dc(z)}{dz} \right|_{z=H_2} = 0 \qquad (10.6)$$

which specifies that there is no flux of dissolved methane from below the active layer. The reason for this condition is that no methane is being produced by diagenesis below that depth. Applying these conditions to Eq. (10.4) yields

$$c(z) = \frac{S_{CH_4} z}{D_{CH_4}} \left(H_2 - \frac{z}{2} \right) \qquad (10.7)$$

The upward flux of dissolved methane at the sediment-water interface is

$$J[CH_4(aq)] = D_{CH_4} \left. \frac{dc(z)}{dz} \right|_{z=0} = S_{CH_4} H_2 \qquad (10.8)$$

which corresponds to all the methane production in the active sediment layer from $z = 0$ to $z = H_2$. If no other reactions were taking place, all the methane produced by diagenesis would be transported to the water column as a flux of dissolved methane. However, methane gas can form, and dissolved methane can be oxidized in the aerobic zone of the sediment. The first of these possibilities is considered next.

10.3 A   Gas Phase Formation

Methane is only slightly soluble in water. If its solubility, $[CH_4]_{sat} = 100$ mg O_2^*/L at $T = 20\,°C$ and one atmosphere pressure, is exceeded in the interstitial water, it forms a gas phase that escapes as bubbles.* Supersaturated conditions appear to be rare since the observed interstitial water methane concentrations do not exceed the solubility of methane at the in situ pressure.† Since the loss of methane as bubbles is an important sink, it is necessary to include it explicitly in the model structure.

The model, therefore, requires three layers (Fig. 10.1). Let H_{sat} be the depth at which the dissolved methane concentration reaches its solubility limit $[CH_4]_{sat}$. The appropriate concentration boundary condition in this case is

$$c(H_{sat}) = [CH_4]_{sat} \qquad (10.9)$$

Consider the flux condition that is required. The concentration of dissolved methane is constant and equal to $[CH_4]_{sat}$ below H_{sat}. Therefore, the concentration gradient just below $z = H_{sat}$ is zero. This requires that the flux of dissolved methane just above $z = H_{sat}$ also be zero. Hence

$$-D_{CH_4} \left. \frac{dc(z)}{dz} \right|_{z=H_{sat}} = 0 \qquad (10.10)$$

The solution to the mass balance equation (Eq. 10.4) with these boundary conditions (Eqs. 10.9–10.10) is

$$c(z) = [CH_4]_{sat} - \frac{S_{CH_4}}{2D_{CH_4}} (H_{sat} - z)^2 \qquad (10.11)$$

The final boundary condition (Eq. 10.5) can be used to determine the depth at which dissolved methane reaches its saturation concentration by setting $c(z) = 0$ at $z = 0$ in Eq. (10.11) and solving for H_{sat}

$$H_{sat} = \sqrt{\frac{2D_{CH_4}[CH_4]_{sat}}{S_{CH_4}}} \qquad (10.12)$$

---

*Kaplan (1974), Reeburgh (1969), Rudd and Taylor (1980). †Reeburgh (1969), Rudd and Taylor (1980).

The upward flux of dissolved methane at the sediment-water interface can be calculated from Eq. (10.11)

$$J[CH_4(aq)] = D_{CH_4} \left. \frac{dc}{dz} \right|_{z=0} = S_{CH_4} H_{sat} \qquad (10.13)$$

It corresponds to all the methane production in the unsaturated region of the sediment from $z = 0$ to $z = H_{sat}$. The remainder of the methane production must be escaping as a gas flux so

$$J[CH_4(g)] = S_{CH_4}(H_2 - H_{sat}) \qquad (10.14)$$

Note that for small rates of diagenetic methane production S_{CH_4} it is computationally possible for H_{sat} to be greater than H_2 – see Eq.(10.12) as $S_{CH_4} \to 0$. This results in a negative gas flux which is physically unrealistic. For these conditions the rate of escape of dissolved methane is large enough – and/or the diagenesis rate of production of methane is small enough – so that the interstitial water concentration of dissolved methane never exceeds $[CH_4]_{sat}$. Hence Eqs. (10.9) and (10.10) are no longer appropriate. Rather Eq. (10.7) describes the situation.

A more useful form of the flux equations arises if Eq. (10.12) is used for H_{sat} in Eq. (10.13)

$$J[CH_4(aq)] = \sqrt{2D_{CH_4}[CH_4]_{sat} S_{CH_4}} \qquad (10.15a)$$

and Eq. (10.14)

$$J[CH_4(g)] = S_{CH_4} H_2 - \sqrt{2D_{CH_4}[CH_4]_{sat} S_{CH_4}} \qquad (10.15b)$$

Note that these areal fluxes $J[CH_4(aq)]$ and $J[CH_4(g)]$ with units g O_2^*/m^2-d are expressed in terms of the volumetric rate of methane carbon diagenesis S_{CH_4} (g O_2^*/m^3-d). It is more appropriate to express the fluxes of methane out of the sediment in terms of the flux of methane production by carbon diagenesis in the sediment. Thus we define

$$J_{CH_4} = S_{CH_4} H_2 \qquad (10.16)$$

as the areal source of methane produced by diagenesis.

It is also convenient to define a diffusion mass transfer coefficient for dissolved methane

$$\kappa_D = \frac{D_{CH_4}}{H_2} \qquad (10.17)$$

where H_2 is the depth of the active sediment layer. Note the similarity of κ_D to K_{L12} used in the two-layer models (Eq. 2.56). Then the methane flux equations (10.15) become

$$J[CH_4(aq)] = \sqrt{2\kappa_D[CH_4]_{sat} J_{CH_4}} \qquad (10.18a)$$

$$J[CH_4(g)] = J_{CH_4} - \sqrt{2\kappa_D [CH_4]_{sat} J_{CH_4}} \qquad (10.18b)$$

These equations relate the flux of dissolved and gaseous methane to the rate of methane production by diagenesis, J_{CH_4}. Note that the gas flux is zero for

$$J_{CH_4} = 2\kappa_D [CH_4]_{sat} \qquad (10.19)$$

at which point $J[CH_4(aq)] = J_{CH_4}$. This completes the analysis of dissolved and gaseous methane fluxes, without regard to oxidation kinetics, which will be considered next.

10.3 B   Methane Oxidation Kinetics

The inclusion of oxidation kinetics in the aerobic zone results in model mass balance equations that are similar to the ammonia equations (3.20)

$$-D_{CH_4} \frac{d^2 c_1(z)}{dz^2} = -K_{CH_4,1} c_1(z) + S_{CH_4} \qquad 0 \leqslant z \leqslant H_1 \quad (10.20a)$$

$$-D_{CH_4} \frac{d^2 c_2(z)}{dz^2} = S_{CH_4} \qquad\qquad H_1 \leqslant z \leqslant H_2 \quad (10.20b)$$

The boundary conditions are

$$c_1(0) = 0 \qquad (10.21a)$$

corresponding to a zero concentration of dissolved methane in the overlying water. Continuity for concentrations and a flux balance at the aerobic-anaerobic boundary yields

$$c_1(H_1) = c_2(H_1) \qquad (10.21b)$$

$$-D_{CH_4} \left. \frac{dc_1(z)}{dz} \right|_{z=H_1} = -D_{CH_4} \left. \frac{dc_2(z)}{dz} \right|_{z=H_1} \qquad (10.21c)$$

Methane saturation at the depth of bubble formation requires that

$$c_2(H_{sat}) = [CH_4]_{sat} \qquad (10.21d)$$

This last condition (Eq. 10.21d) is the only difference between the ammonia (Eqs. 3.20 and 3.21) and methane mass balance equations. The corresponding ammonia bottom boundary condition (Eq. 3.21d) is a zero flux condition. The methane equations are solved in Appendix 10A. Both the exact and approximate solutions are given. The dissolved $J[CH_4(aq)]$ and gaseous $J[CH_4(g)]$ methane fluxes are

$$J[CH_4(aq)] = \sqrt{2\kappa_D [CH_4]_{sat} J_{CH_4}} \operatorname{sech}(\lambda_C H_1) \qquad (10.22)$$

$$J[CH_4(g)] = J_{CH_4} - \sqrt{2\kappa_D[CH_4]_{sat}J_{CH_4}} \tag{10.23}$$

where

$$\lambda_C = \sqrt{\frac{k_{CH_4,1}}{D_{CH_4}}} \tag{10.24}$$

$k_{CH_4,1}$ is the first-order rate constant for dissolved methane oxidation, and κ_D is the dissolved methane mass transport coefficient, defined in Eq. (10.17).

10.4 DISSOLVED OXYGEN MASS BALANCE

The oxidation of methane in the aerobic layer consumes dissolved oxygen. In this section various approaches to the problem are examined in order to formulate and solve, if possible, an equation for the SOD. In order to be complete, the contribution of ammonia oxidation will also be included. The initial approach is the most straightforward. Simply solve the continuous equation for dissolved oxygen with the appropriate sinks for methane and ammonia oxidation.

10.4 A   Continuous Mass Balance Equation

The results of the continuous one-dimensional models for methane (Section 10.3B) and ammonia (Section 3.3D) are available. These solutions can be used as sink terms in the dissolved oxygen mass balance equation

$$-D_{O_2}\frac{d^2[O_2(z)]}{dz^2} = -a_{O_2,NH_4}k_{NH_4,1}n_1(z) - k_{CH_4,1}c_1(z) \tag{10.25}$$

where a_{O_2,NH_4} is the stoichiometric ratio of oxygen consumed to ammonia oxidized. Since methane is computed in oxygen equivalents, no stoichiometric coefficient is necessary for the second term. It is interesting to compare this equation (10.25) with the first model for dissolved oxygen presented in Chapter 8 (Eq. 8.3)

$$-D_{O_2}\frac{d^2[O_2(z)]}{dz^2} = -R_0 \tag{10.26}$$

The difference is that the oxygen consumption term R_0, which was not related to any process, is replaced by sinks of known magnitudes: the consumption of oxygen by the oxidation of methane and ammonia.

The boundary conditions for Eq. (10.25) are

$$[O_2(z)]|_{z=0} = [O_2(0)] \tag{10.27a}$$

which specifies the overlying water dissolved oxygen concentration,

$$[O_2(z)]|_{z=H_1} = 0 \tag{10.27b}$$

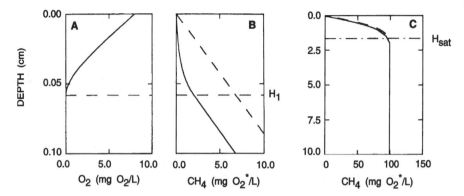

Fig. 10.2 Interstitial water concentrations profiles of dissolved oxygen and methane versus depth. (A) $O_2(z)$ from $z = 0$ to $z = 1$ mm. (B) $CH_4(aq)$ from $z = 0$ to $z = 1$ mm. (C) $CH_4(aq)$ from $z = 0$ to $z = 10$ cm. Dashed lines are for $\kappa_{CH_4,1} = 0$. Aerobic zone depth H_1 and depth of methane saturation H_{sat} are shown. Parameter values are given in Table 10.1.

which defines the depth of the aerobic layer, and

$$\frac{d[O_2(z)]}{dz}\bigg|_{z=H_1} = 0 \qquad (10.27c)$$

which follows from the requirement that below H_1 the flux of dissolved oxygen is zero since no oxygen is being consumed below $z = H_1$.

The solution is straightforward although somewhat involved, requiring two integrations of the right-hand side forcing functions (Appendix 10B) and a numerical method to determine H_1. Fig. 10.2 is an example of the interstitial water profiles of dissolved oxygen and methane computed using these equations. Figs. 10.2A and B present the solutions for the top millimeter of the sediment. Figs. 10.2C presents the methane solution for the full depth of the active sediment layer. The solutions for dissolved methane with $k_{CH_4,1} = 0$ are shown as dashed lines in Figs. 10.2B and C. Note that the solutions for $k_{CH_4,1} > 0$ and $k_{CH_4,1} = 0$ are quite different in the aerobic zone (B) but not in the deeper portions of the sediment (C). The reason is that the magnitude of the difference is not significant at the higher concentrations in the anaerobic layer – note the scale change between Figs. 10.2B and C. It is for this reason that Eq. (10.12), which was derived with $k_{CH_4,1} = 0$, can still be used for computing the depth of methane saturation H_{sat}.

The most noteworthy feature of the dissolved oxygen profile is the almost linear decrease of dissolved oxygen with respect to sediment depth in the aerobic layer. This approximation has been employed in the previous chapters and is employed in the next section to estimate the depth of the aerobic layer.

10.4 B   Maximum Oxygen Consumption

The full numerical solution is useful for assessing the accuracy of various approximations. But it does not provide any insight into the nature of the solution. In this section we examine an interesting approximation.

Suppose that the oxidation rate of dissolved methane in the aerobic zone was sufficiently large so that all the dissolved methane transported to the aerobic zone was oxidized. Then the flux of dissolved methane in oxygen equivalents $J[CH_4(aq)]$ would be equal to the maximum flux of oxygen to the sediment that would be required to oxidize this methane. Denote this flux $CSOD_{max}$ as the maximum – since we assume that all dissolved methane transported to the sediment water interface is oxidized – carbonaceous (due to organic carbon oxidation) SOD. Hence Eq. (10.18a) can be interpreted as

$$CSOD_{max} = \sqrt{2\kappa_D [CH_4]_{sat} J_{CH_4}} \tag{10.28}$$

This is an important result and the addition of oxidation kinetics of methane in the aerobic zone only modifies it somewhat. It relates the carbonaceous SOD to the square root of the diagenesis flux J_{CH_4}. But this is exactly the form of the relationship suggested by the data in Fig. 8.1A and B.

Eq. (10.28) also bears a superficial resemblance to the models based on the original oxygen diffusion model for SOD (Section 8.2 B, Eq. 8.11). In fact the derivation is also very similar with the depth of the aerobic layer following from the zero concentration boundary condition (Eq. 10.5). However, the square root dependency between SOD and diagenesis J_{CH_4} (or its surrogate R_0) arises for a completely different reason. In Eq. (8.11)

$$SOD = \sqrt{2D_{O_2}[O_2(0)]R_0} \tag{10.29}$$

the square root dependency is the result of a dissolved oxygen mass transfer limitation because the overlying water oxygen concentration $[O_2(0)]$ determines the magnitude of the SOD. A higher $[O_2(0)]$ results in a higher SOD. For the model based on methane production and oxidation, however, the cause of the square root dependency is the limited solubility of methane. Note that $[CH_4]_{sat}$ plays the same role in Eq. (10.28) as $[O_2(0)]$ does in Eq. (10.29).

For the cases where the methane gas flux is zero, Eq. (10.8) applies and

$$CSOD_{max} = J_{CH_4} \qquad\qquad J_{CH_4} < 2\kappa_D[CH_4]_{sat} \tag{10.30}$$

Eqs. (10.28–10.30) relate the methane carbon diagenesis flux, J_{CH_4}, to the maximum CSOD. Eq. (10.30) predicts a linear dependency of CSOD on diagenesis flux until the formation of a methane gas phase. For larger diagenesis fluxes, a square root dependence of CSOD on diagenesis flux is predicted (Eq. 10.28). As we will see below, essentially the same behavior occurs when methane oxidation kinetics are included.

10.5 SOD EQUATION

The consumption of dissolved oxygen from the overlying water, the sediment oxygen demand, is comprised of two components: the oxygen consumed by methane oxidation CSOD and ammonia oxidation NSOD

$$SOD = CSOD + NSOD \tag{10.31}$$

The carbonaceous sediment oxygen demand CSOD can be computed using a mass balance argument. The dissolved $J[CH_4(aq)]$ and gaseous $J[CH_4(g)]$ methane fluxes are (Eqs. 10.22–10.23)

$$J[CH_4(aq)] = \sqrt{2\kappa_D[CH_4]_{sat}J_{CH_4}}\,\text{sech}(\lambda_C H_1) \tag{10.32}$$

$$J[CH_4(g)] = J_{CH_4} - \sqrt{2\kappa_D[CH_4]_{sat}J_{CH_4}} \tag{10.33}$$

The source of methane is J_{CH_4}. A mass balance of methane requires that

$$J_{CH_4} = J[CH_4(g)] + J[CH_4(aq)] + CSOD \tag{10.34}$$

Methane produced by diagenesis either diffuses to the overlying water, escapes as a gas flux, or is oxidized. Solving for CSOD yields

$$CSOD = \sqrt{2\kappa_D[CH_4]_{sat}J_{CH_4}}\,[1 - \text{sech}(\lambda_C H_1)]$$
$$J_{CH_4} > 2\kappa_D[CH_4]_{sat} \tag{10.35a}$$

$$CSOD = J_{CH_4}[1 - \text{sech}(\lambda_C H_1)]$$
$$J_{CH_4} < 2\kappa_D[CH_4]_{sat} \tag{10.35b}$$

where if $J_{CH_4} < 2\kappa_D[CH_4]_{sat}$, the square root expression is replaced with J_{CH_4} (Eq. 10.35b).

In order to proceed further, the depth of the aerobic zone H_1 is required. It is related to the sediment oxygen demand by the equation

$$H_1 = D_{O_2}\frac{[O_2(0)]}{SOD} = \frac{D_{O_2}}{s} \tag{10.36}$$

where $s = SOD/[O_2(0)]$. This relationship is derived in Chapter 3 (Section 3.3 C, Eq. 3.15) and depends directly on the assumption of linearity of the dissolved oxygen concentration in the aerobic layer. Eq. (10.36) is substituted into Eq. (10.24) to evaluate the expression

$$\lambda_C H_1 = \left(\sqrt{\frac{k_{CH_4,1}}{D_{CH_4}}}\right)\left(\frac{D_{O_2}}{s}\right) = \frac{1}{s}\sqrt{k_{CH_4,1}\frac{D_{O_2}^2}{D_{CH_4}}} \tag{10.37}$$

The reaction velocity (see Section 3.3 C) for methane oxidation is defined as

$$\kappa_{CH_4,1} = \sqrt{k_{CH_4,1}\frac{D_{O_2}^2}{D_{CH_4}}} \approx \sqrt{k_{CH_4,1}D_1} \tag{10.38}$$

where $D_{O_2} \approx D_{CH_4} = D_1$ (see Table 2.1). Thus the argument of the sech function becomes

$$\lambda_C H_1 = \frac{\kappa_{CH_4,1}}{s} \tag{10.39}$$

In addition to the oxygen required for methane oxidation, the sediment oxygen demand due to ammonia oxidation is required. It can also be derived from a mass balance argument. Ammonia produced by diagenesis J_N either escapes to the overlying water as an ammonia flux $J[NH_4(aq)]$ or it is nitrified to nitrate, consuming oxygen in the process, which is the nitrogenous component (NSOD) of SOD. Hence

$$J_N = J[NH_4(aq)] + \frac{NSOD}{a_{O_2,NH_4 \to N_2}} \tag{10.40}$$

where $a_{O_2,NH_4 \to N_2}$ is the appropriate stoichiometric ratio for the oxygen consumed by ammonia oxidation and denitrification to nitrogen gas. Solving Eq. (10.40) for NSOD yields

$$NSOD = a_{O_2,NH_4 \to N_2}\left(J_N - J[NH_4(aq)]\right) \tag{10.41}$$

$$= a_{O_2,NH_4 \to N_2} J_N \left[1 - sech\left(\frac{\kappa_{NH_4,1}}{s}\right)\right] \tag{10.42}$$

where the solution from the continuous model for ammonia flux (Eq. 3.28) is used for consistency.

The equation for the SOD is the sum of CSOD (Eqs. 10.35) and NSOD (Eq. 10.42)

$$SOD = \sqrt{2\kappa_D[CH_4]_{sat}J_{CH_4}}\left[1 - sech\left(\frac{\kappa_{CH_4,1}}{s}\right)\right]$$

$$+ a_{O_2,NH_4 \to N_2} J_N \left[1 - sech\left(\frac{\kappa_{NH_4,1}}{s}\right)\right]$$

$$J_{CH_4} \geqslant 2\kappa_D[CH_4]_{sat} \tag{10.43a}$$

$$SOD = J_{CH_4}\left[1 - sech\left(\frac{\kappa_{CH_4,1}}{s}\right)\right] + a_{O_2,NH_4 \to N_2} J_N \left[1 - sech\left(\frac{\kappa_{NH_4,1}}{s}\right)\right]$$

$$J_{CH_4} < 2\kappa_D[CH_4]_{sat} \tag{10.43b}$$

where $a_{O_2,NH_4 \to N_2}$ is the stoichiometric ratio for the oxygen consumed by ammonia oxidation, which is discussed below. The two terms on the right-hand side of Eqs. (10.43) are the SOD exerted by organic carbon (methane) oxidation (CSOD), and the SOD exerted by ammonia oxidation (NSOD), respectively.

It is useful to relate the fluxes of ammonia J_N and methane J_{CH_4} produced by organic matter diagenesis so that there is only one source term in Eqs. (10.43). The ratio $a_{O_2^*,N} = J_{CH_4}/J_N$, which is the carbon (in oxygen equivalents) to nitrogen ratio, can be estimated from the Redfield ratio $a_{O_2^*,N} = 106$ mol C / 16 mol N = 5.68 g C/g N = 15.2 g O_2^*/g N (Section 1.2 A). Therefore

$$J_N = \frac{J_{CH_4}}{a_{O_2^*,N}} \tag{10.44}$$

is used in Eqs. (10.43).

10.5 A Nitrogen-Oxygen Stoichiometry

It remains to estimate the oxygen consumed from ammonia nitrification $a_{O_2,NH_4 \to N_2}$. The stoichiometry for the oxygen consumption due to ammonia oxidation is (Section 1.2 B, Eq. 1.25)

$$NH_4^+ + 2O_2 \to 2H^+ + NO_3^- + H_2O \tag{10.45}$$

This reaction implies a stoichiometry of $a_{O_2,NH_4 \to NO_3} = 2$ mol O_2/mol NH_4 = 4.57 g O_2/g NH_4-N. However, some of the nitrate that is produced is denitrified to nitrogen gas (Eq. 1.17)

$$\frac{1}{5}NO_3^- + \frac{6}{5}H^+ + e^- \to \frac{1}{10}N_2 + \frac{3}{5}H_2O \tag{10.46}$$

In order for this reaction to occur, an electron donor is required to supply the e^-. Presumably the methane that is being produced by diagenesis can be oxidized and donate the electrons required (Eq. 1.20)

$$\frac{1}{8}CH_4 + \frac{1}{4}H_2O \to \frac{1}{8}CO_2 + H^+ + e^- \tag{10.47}$$

Hence the stoichiometry of the denitrification reaction is

$$\frac{5}{8}CH_4 + H^+ + NO_3^- \to \frac{5}{8}CO_2 + \frac{1}{2}N_2 + \frac{7}{4}H_2O \tag{10.48}$$

with methane as the carbon source. It is not necessary that methane actually be the carbon source. Any other intermediate would suffice. However since denitrification consumes dissolved methane – or an intermediate – it reduces the quantity of dissolved methane that is available for direct oxidation via Eq. (10.2). Thus using $a_{O_2,NH_4 \to NO_3} = 4.57$ g O_2/g NH_4-N overestimates the oxygen consumption, if denitrification is taking place.

In order to proceed, therefore, it is necessary to quantify the amount of nitrate that is denitrified, so that the appropriate amount of methane can be deducted from the methane available for oxidation. For the complete model presented in Chapter 14, the amount of nitrate that is denitrified is computed using the nitrate flux model

(Chapter 4) and the appropriate amount of carbon diagenesis is consumed. With that procedure, the oxygen consumption due to nitrification is $a_{O_2,NH_4\rightarrow NO_3}$.

For the model formulated in this chapter, a simpler approach is adopted. We will assume that all the nitrate that is formed by nitrification is denitrified to nitrogen gas via Eq. (10.48). The resulting reduction in oxidizable methane can be accounted for by replacing the dissolved methane consumed in Eq. (10.48) by its oxygen equivalents, Eq. (10.2). That is, the NO_3^- in Eq. (10.45) is replaced using Eq. (10.48) and the CH_4 is replaced using Eq. (10.2). The result

$$NH_4^+ + OH^- + \frac{3}{4}O_2 \rightarrow \frac{1}{2}N_2 + \frac{10}{4}H_2O \qquad (10.49)$$

is simply the oxidation of ammonia to nitrogen gas directly with a stoichiometry of $a_{O_2,NH_4\rightarrow N_2} = 3/4$ mol O_2/mol N = 1.714 g O_2 consumed per g NH_4-N oxidized to nitrogen gas. This reaction can be thought of as the net oxidation of ammonia to nitrogen gas. The intermediate oxidation of ammonia to nitrate, and the subsequent denitrification of nitrate to nitrogen gas is bypassed. Ammonia is simply oxidized to nitrogen gas, and there is no stoichiometric requirement for methane. If some of the nitrate produced by nitrification is denitrified to ammonia instead of nitrogen gas, then there is no oxygen consumed and that reaction is not of concern.

The fact that ammonia is nitrified and then denitrified to nitrogen gas has an important consequence in terms of the nitrogenous SOD. The models discussed in Chapter 8 account for the 4.57 g O_2 consumed per g NH_4-N oxidized to nitrate (see Eq. 8.12). However since only the electron acceptors are considered in these models and not the consumption of oxygen demanding equivalents, e.g., methane as in Eq. (10.48), they compute a larger NSOD per unit of ammonia oxidized, but do not reduce the CSOD by the appropriate amount to account for the consumption of methane or other carbon intermediate that is required for denitrification.

10.5 B Solution Method

The equations that determine the SOD (Eqs. 10.43) have SOD on both sides of the equal sign, since $s = SOD/[O_2(0)]$. A simple iterative solution using back substitution has been found to be a satisfactory solution technique. A convenient starting value for SOD is

$$SOD_{max} = CSOD_{max} + \frac{a_{O_2,NH_4\rightarrow N_2}}{a_{O_2^*,N}}J_{CH_4} \qquad (10.50)$$

which corresponds to complete oxidation of dissolved methane and ammonia in the aerobic layer. Table 10.1 presents an example calculation. The parameters that are used are estimated from the Milwaukee River data set analyzed subsequently. The results of the iteration scheme are also included. The termination criterion is $|1-$ Ratio$| < 10^{-4}$ corresponding to three place accuracy.

The four parameters in the SOD model are the carbon diagenesis flux J_{CH_4}, the dissolved methane mass transfer coefficient κ_D, and the two reaction velocities, $\kappa_{CH_4,1}$ and $\kappa_{NH_4,1}$. Note that the depth of the active sediment layer H_2 is not an

Table 10.1   Example Computation

Parameters	Results
$a_{O_2,N} = 15.2$ g O_2^*/g N	SOD $= 1.709$ g O_2^*/m^2-d[a]
$a_{O_2,NH_4\to N_2} J_N = 1.714$ g O_2^*/g N	CSOD $= 0.8541$ g O_2^*/m^2-d
$[CH_4]_{sat} = 100$ mg O_2^*/L	NSOD $= 0.8553$ g O_2^*/m^2-d
$[O_2(0)] = 4.0$ mg O_2/L	$J[CH_4(aq)] = 0.8132$ g O_2^*/m^2-d
$\kappa_D = 0.139$ cm/d	$J[CH_4(g)] = 8.333$ g O_2^*/m^2-d
$\kappa_N = 0.897$ m/d	$J[NH_4] = 0.1589$ g N/m^2-d
$\kappa_C = 0.575$ m/d	$J[N_2(g)] = 0.4990$ g N/m^2-d
$J_{CH_4} = 10$ g O_2^*/m^2-d	$J_{GAS} = 3.316$ L/m^2-d

[a] $SOD = \sqrt{2\kappa_D [CH_4]_{sat} J_{CH_4}} \left[1 - \text{sech}\left(\frac{\kappa_{CH_4,1}}{s}\right)\right] + \frac{a_{O_2,NH_4\to N_2}}{a_{O_2,N}} J_{CH_4}\left[1 - \text{sech}\left(\frac{\kappa_{NH_4,1}}{s}\right)\right]$

$SOD = \sqrt{(2)(0.00139 \text{ m/d})(100 \text{ g/m}^3)(10 \text{ g/m}^2\text{-d})}\{1 - \text{sech}[(0.575 \text{ m/d})(4.0 \text{ g/m}^3)/(1.709 \text{ g/m}^2\text{-d})]\}$
$+(10 \text{ g/m}^2\text{-d})(1.714 \text{ g } O_2^*/\text{g N})/(15.2 \text{ g } O_2^*/\text{g N})\{1 - \text{sech}[(0.897 \text{ m/d})(4.0 \text{ g/m}^3)/(1.709 \text{ g/m}^2\text{-d})]\}$
$(1.709 \text{ g/m}^2\text{-d}) = 1.667 \text{ g/m}^2\text{-d}\{1 - 0.4877\} + 1.128 \text{ g/m}^2\text{-d}\{1 - 0.2415\}$

Iterations

i	SOD_i^a	SOD from Eq. (10.43a)	Ratio
1	2.794965	0.9871081	0.3531737
2	1.891036	1.555792	0.822719
3	1.723414	1.697036	0.9846941
4	1.710225	1.708646	0.9990769
5	1.709435	1.709343	0.999946

[a] $SOD_{[i+1]} = ([SOD_i] + [\text{SOD from Eq. (10.43a) evaluated using } SOD_i])/2$
$SOD_1 = CSOD_{max} + \frac{a_{O_2,NH_4\to N_2}}{a_{O_2,N}} J_{CH_4} = 1.667 + 1.128 = 2.795$ g/m^2-d

explicit model parameter. It is included in the definition of κ_D (Eq. 10.17), but it is not explicitly required since the diagenesis sources are expressed as fluxes. The methane saturation concentration $[CH_4]_{sat}$ (g O_2^*/m^3) is computed from

$$[CH_4]_{sat} = 100\left(1 + \frac{H_0}{10}\right)(1.024)^{(20-T)} \qquad (10.51)$$

where H_0 (meters) is the depth of the water column over the sediment that corrects for the in situ pressure, and T is the sediment temperature in °C. This equation is accurate to within 3% of the reported methane solubility – converted to oxygen equivalents – between 5 and 20 °C (Yamamoto et al., 1976).

10.5 C   Accuracy of the Approximation

The effect of the linear approximation (Eq. 10.36) on the accuracy of SOD computation is shown in Table 10.2.

Table 10.2 Comparison of Exact and Approximate Solutions for SOD

J_{CH_4} $(g\ O_2^*/m^2\text{-d})$	Approximate SOD $(g\ O_2/m^2\text{-d})$	Exact SOD $(g\ O_2/m^2\text{-d})$	Error (%)
	$\kappa_{CH_4,1} = 0.5$ m/d		
1.00	0.637	0.633	0.642
2.00	0.948	0.952	-0.439
5.00	1.550	1.625	-4.652
10.00	2.170	2.381	-8.866
20.00	2.968	3.401	-12.752
50.00	4.363	5.244	-16.795
100.00	5.731	7.073	-18.974
	$\kappa_{CH_4,1} = 2.0$ m/d		
1.00	0.639	0.636	0.475
2.00	0.970	0.965	0.460
5.00	1.717	1.720	-0.201
10.00	2.601	2.666	-2.440
20.00	3.759	4.017	-6.416
50.00	5.652	6.427	-12.060
100.00	7.354	8.697	-15.445

$\kappa_{NH_4,1} = 0.8$ m/d, $[O_2(0)] = 8.0$ g O_2/m^3, $\kappa_D = = 0.139$ cm/d, $[CH_4]_{sat} = 99$ g O_2^*/m^3, $H_2 = 10$ cm.

The exact SOD is obtained from the analytical solutions in Appendix 10B. A numerical procedure is used to solve for H_1 using the boundary condition equation: $[O_2(H_1)] = 0$. For all cases examined the error is never greater than 20% and for most of the cases the error is below 15%. Since the approximate equations are much more convenient and intelligible, they are preferable unless high accuracy solutions at large SODs are required. Both the exact and approximate equations conserve mass exactly so that this is not a criterion for choice.

10.5D Fluxes of the Other Components

The equations for the ammonia (Eqs. 3.28 and 10.44) and dissolved methane (Eq. 10.32) fluxes are

$$J[NH_4(aq)] = \frac{J_{CH_4}}{a_{O_2^*,N}}\text{sech}\left(\frac{\kappa_{NH_4,1}}{s}\right) \tag{10.52}$$

$$J[CH_4(aq)] = \sqrt{2\kappa_D[CH_4]_{sat}J_{CH_4}}\,\text{sech}\left(\frac{\kappa_{CH_4,1}}{s}\right) \tag{10.53}$$

Since nitrogen gas is being produced by denitrification, its magnitude can also be determined. At first glance, it appears that a mass balance analysis for N_2 is required that parallels the methane analysis. Nitrogen gas is produced in the anaerobic layer, and it diffuses to the aerobic layer and to the overlying water. If it exceeds its solubility in the anaerobic layer, it forms a gas phase.

However, the overlying water, and therefore the pore water, is initially almost completely saturated with respect to nitrogen gas – the atmosphere is 78% nitrogen gas – so very little nitrogen gas production is required to saturate the pore water. Hence the amount of nitrogen gas that diffuses to the overlying water is small, since the gradient – the concentration at atmospheric saturation versus completely saturated – is small. Instead, almost all the nitrogen gas produced by denitrification partitions into the gas phase. Therefore, to a good approximation, all the nitrogen gas produced by denitrification becomes a gas flux

$$J[N_2(g)] = \frac{J_{CH_4}}{a_{O_2^*,N}} \left[1 - \text{sech} \left(\frac{\kappa_{NH_4,1}}{s} \right) \right] \tag{10.54}$$

The total gas flux from the sediment can be computed by adding the methane gas flux (Eq. 10.33)

$$J[CH_4(g)] = J_{CH_4} - \sqrt{2\kappa_D [CH_4]_{\text{sat}} J_{CH_4}} \tag{10.55}$$

to the nitrogen gas flux. The volumetric gas flux (L/m^2-d) is

$$J_{Gas} = 22.4 \left(\frac{\frac{1}{2} J[CH_4(g)]}{32} + \frac{\frac{1}{2} J[N_2(g)]}{14} \right) \tag{10.56}$$

where one mole of gas = 22.4 liters, the molecular weights are for O_2 and N_2 as N are 32 and 14 respectively, and the coefficients of one half are from Eqs. (10.1) and (10.49). These fluxes are included in the example computation in Table 10.1.

10.6 DATA ANALYSIS

Three sets of data are analyzed with the model developed above. The first is a laboratory experiment designed to test the model prediction of an initially linear relationship between SOD and diagenesis (Eq. 10.43a), followed by a square root relationship (Eq. 10.43b). The Fair, Moore, and Thomas (1941) experiments that were discussed in Section 8.2 A are revisited in Sections 10.6A, B. Then the SOD and ammonia fluxes are examined as a function of total gas flux in Sections 10.6C, D. Finally field measurements from the Milwaukee River of SOD, total gas flux, and gas composition are examined in Section 10.6E.

10.6 A Sediment Dilution Experiment

The SOD model makes a specific prediction for the relationship between diagenesis flux J_{CH_4} and SOD. It should be linear for small diagenesis fluxes and have an approximately square root relationship for larger fluxes (Eq. 10.43). In order to perform

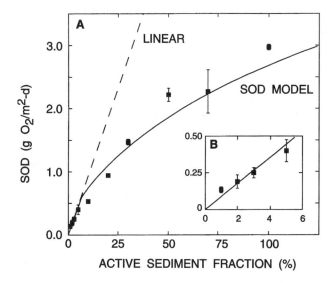

Fig. 10.3 Sediment dilution experiment. SOD versus percent of Milwaukee River sediment in the mixture. Data (mean ± standard deviation). SOD model (solid lines computed using Eqs. (10.43). Parameters are listed in Table 10.3. Axes and labels for the inset Fig. B are the same as (A). Dashed line in (A) is an extrapolation of the linear portion of the model result, shown in the inset figure (B).

a direct test, it is necessary to design an experiment in which the diagenesis flux can be varied systematically and the resulting SOD is observed.

The use of different depths of sediments – as was done by Fair et al. (1941) – was not chosen because another experimental variable, sediment depth, varies together with diagenesis flux. Instead a dilution procedure was employed (Di Toro et al., 1990b). A sediment sample from the Milwaukee River was diluted with clean sand to produce a sequence of sediments with known relative amounts of reactive organic matter. The diagenesis fluxes for these sediments are, therefore, in proportion to the fraction of active sediment they contain.

10.6 B   Results

The results of two replicated experiments are shown in Fig. 10.3. The initial linear behavior is shown in the inset. The dotted line is the extrapolated linear dependency. The model results are computed from the SOD model (Eqs. 10.43). Table 10.3 lists the parameters estimated from the nonlinear least square fit to these data (Marquardt, 1963).

The experiment was designed so that the diagenesis flux is proportional to the quantity of active sediment in each reactor. The nonlinear (approximately square root) relationship between SOD and diagenesis flux is clearly confirmed by the results (Fig. 10.3A) as is the linear relationship at low SODs (Fig. 10.3B). The fact that the lowest experimental SOD value is slightly larger than the linear relationship may be due to a small residual organic carbon in the sand that was not removed by the cleaning procedure. Since the active sediment concentration for this SOD is 1%, a small contamination could add a significant additional diagenesis flux.

10.6 C   Parameter Estimates Using Paired Fluxes

A commonly used technique to analyze sediment nutrient and oxygen fluxes is to examine the variation of one with respect to another (Nixon et al., 1976). This procedure is ideally suited for analyzing data using the SOD model developed above because it predicts other dissolved and gaseous fluxes as well: $J[CH_4(aq)]$, $J[NH_4(aq)]$, $J[CH_4(g)]$, and $J[N_2(g)]$. Thus any set of laboratory or field measurements that determine any two of these fluxes can be used to estimate the parameters of the model. The procedure is as follows.

The diagenesis flux J_{CH_4} can be expressed in terms of any of the measured fluxes. For example, the methane gas flux equation (Eq. 10.55) can be used to solve for J_{CH_4}

$$J_{CH_4} = \frac{\kappa_D[CH_4]_{sat}}{2}\left[1+\sqrt{1+\frac{2J[CH_4(g)]}{\kappa_D[CH_4]_{sat}}}\,\right]^2 \tag{10.57}$$

The total gas flux (Eq. 10.56) can also be used to solve for the diagenesis flux by substituting Eq. (10.54) for $J[N_2(g)]$. Either of these expressions can be substituted for J_{CH_4} in the SOD equation (Eqs. 10.43). The result is a relationship between SOD and either methane or total gas flux. The use of ammonia flux is also direct, since there is a relationship (Eq. 10.52) between $J[NH_4]$ and J_{CH_4}.

The three parameters to be estimated are reaction velocities, $\kappa_{CH_4,1}$ and $\kappa_{NH_4,1}$, and the mass transfer coefficient κ_D. The parameters $\kappa_{CH_4,1}$ and κ_D determine the CSOD and the methane fluxes. The reaction velocity $\kappa_{NH_4,1}$ determines the NSOD, ammonia, and $N_2(g)$ fluxes. A nonlinear least square procedure (Marquardt, 1963) is employed to estimate the parameters. In general, it is a three-parameter estimation problem, so a reasonable number of observations are required. The optimal data sets have both nitrogen and methane gas fluxes as well as SOD and ammonia fluxes. For the more restricted data sets, certain of the parameters are fixed at values obtained from the more complete sets.

10.6 D   SOD, Ammonia, and Gas Fluxes - Laboratory Data

The Fair et al. (1941) experiments relating SOD, ammonia, and total gas flux for various sediment depths can be analyzed directly. SOD and gas fluxes at $t = 20$ days, which corresponds to the initiation of the gas flux measurements, and at $t = 30$, 50, 75, 100 days after the start of the experiment are estimated by interpolating the reported data. The ammonia fluxes are geometric means of slopes estimated by differences in reported concentrations. Fig. 10.4 presents the SOD versus total gas flux data from Fair et al. (1941) and for the sediment dilution experiment described above. The lines are computed from Eqs. (10.43, 10.55–10.56) with the parameters

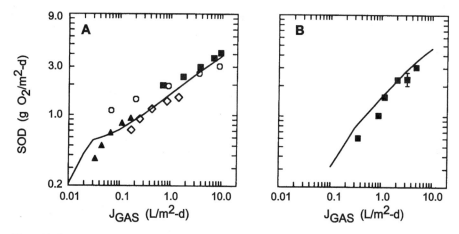

Fig. 10.4 SOD versus total gas flux.   (A) Fair et al. (1941) Reactor depth (cm) = 1.42($\blacktriangle$), 2.55($\blacksquare$), 4.75($\square$), 10.2($\diamond$). (B) Sediment dilution experiment. Lines are computed using Eqs. (10.43, 10.55, 10.56). Parameter values given in Table 10.3.

listed in Table 10.3.

Table 10.3   Kinetic and Transport Parameter Estimates Mean (Standard Error) at $20°C$

Variable	Fair et al., (1941)	Dilution Experiment	Milwaukee River
Temperature $°C$	22.5	20	15.5 (4.2)
$\kappa_{CH_4,1}$ (m/d)	0.839 (0.19)	>2.0	0.575 (0.156)
$\kappa_{NH_4,1}$ (m/d)	0.894 (0.097)	0.894*	0.897 (0.151)
κ_D (m/d)	0.268 (0.024)	0.268*	0.139 (0.047)
$\theta_{NH_4,1}$	–	–	1.079 (0.031)
J_{CH_4} (g O_2^* /m^2-d)	–	8.0	–

*Value is assumed

The Fair et al. (1941) data indicate, surprisingly, that the active layer depth H_2 plays no role in determining the relationship between SOD and gas flux. Even for the wide range of depths employed for the reactors (1.42 to 10.2 cm) and the wide range of observed gas fluxes over the duration of the experiment (0.03 to 10 L/m^2-d), the relationship between SOD and gas flux is remarkably consistent. No depth effect is evident in these data. A similar, although less direct, finding for the Milwaukee River field data is discussed below. These results prompted the definition of the mass transfer coefficient κ_D which appears to be constant, independent of the depth of the active diagenesis layer H_2.

The ammonia flux versus total gas flux are presented in Fig. 10.5 for the Fair et al. and the sediment dilution experiments. The lines are computed using Eqs. (10.52), and (10.55) to (10.56). The model results are not consistent with the Fair et al. data

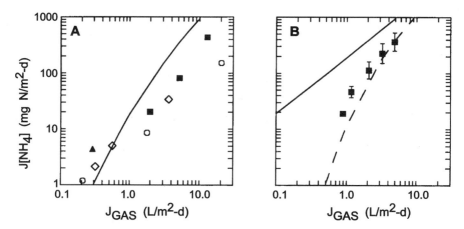

Fig. 10.5 Ammonia flux versus total gas flux. (A) Fair et al. (1941) data. Reactor depth (cm) = 1.42($\blacktriangle$), 2.55($\blacksquare$), 4.75($\square$), 10.2($\diamond$). (B) Sediment dilution experiment. Anaerobic (solid line) and aerobic (dashed line) flux. Lines are computed using Eqs. (10.43a, 10.55, 10.56). Parameter values given in Table 10.3.

but are in reasonable agreement for the sediment dilution experiment if the aerobic ammonia flux is considered. Unfortunately, the sediment dilution experiment was conducted in such a way that the ammonia flux cannot be uniquely associated with either aerobic or anaerobic conditions. During the observation period the oxygen in the reactor was depleted.

10.6 E   SOD and Gas Fluxes - Field Data

As part of an investigation of the effects of combined sewer overflows in the Milwaukee River (SWRPC, 1987) an extensive set of simultaneous field observations of SOD and gas fluxes was collected (Gruber et al., 1987). The relationship between SOD, gas composition, and total gas flux is shown in Fig. 10.6. The SOD-total gas flux relationship is similar to that observed in the Fair et al. and sediment dilution experiments shown in Fig. 10.4. A total gas flux of 1 L/m^2-d corresponds to an SOD of approximately 1 g O$_2$/m^2-d in the three data sets.

The SOD and gas composition data are used to estimate the reaction velocities and mass transfer coefficient. The individual in situ dissolved oxygen and temperature and water column depth at each station are used in the computation. The latter parameters are important, since the saturation concentration of methane is a function of both temperature and pressure (Eq. 10.51). It happens that it was possible to estimate the temperature dependency of $\kappa_{CH_4,1}$ and $\kappa_{NH_4,1}$ as well – assumed to be equal, since it was not possible to discriminate between the two temperature coefficients. The lines in Fig. 10.6 are the model results evaluated using the median (solid lines) and the median plus and minus the standard deviation (dashed lines) for dissolved oxygen, temperature, and methane saturation concentration given in Table 10.4.

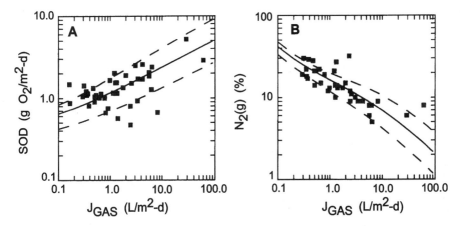

Fig. 10.6 Milwaukee River (A) SOD and (B) percent nitrogen gas versus total gas flux. Lines are computed using Eqs. (10.43, 10.55−10.56). Parameter values given in Table 10.3. Model results are evaluated using the median (solid) and the median plus and minus the standard deviation (dashed) of DO, temperature, and methane saturation concentration, given in Table 10.4.

Table 10.4 Environmental Parameters Mean (Standard Deviation)

Variable	Fair et al., (1941)	Dilution Experiment	Milwaukee River
Temperature $^{\circ}$C	22.5	20	15.5 (4.2)
Methane saturation $[CH_4]_{sat}$ (mg O_2^*/L)	93.1	100	179 (37.0)
Dissolved oxygen $[O_2(0)]$ (mg O_2/L)	6.0	6.0	5.0 (3.0)

The regular decrease in the percentage of nitrogen gas with increasing total gas flux (Fig. 10.6B) can be explained by examining Eqs. (10.54) and (10.55). In Eq. (10.54), the nitrogen gas flux is proportional to the diagenesis flux. However, the methane gas flux, Eq. (10.55), can be zero when $J_{CH_4} < 2\kappa_D[CH_4]_{sat}$. This corresponds to undersaturation of methane in the interstitial water. Hence for small diagenesis fluxes the gas flux is essentially all nitrogen gas. As diagenesis increases, the methane gas flux increases, but not proportionately due to the square root term in Eq. (10.55). As a result the nitrogen gas remains a significant proportion of the total gas flux. This accounts for the large percentage of nitrogen gas at low gas flux rates.

As the diagenesis flux, and also the total gas flux, continues to increase, the fraction of nitrogen gas in the gas phase can be estimated as follows. Suppose all the methane escaped without oxidation – which is approached as the gas flux becomes very large – and all the ammonia was oxidized to nitrogen gas. The reaction equations (10.1) and (10.49) indicate that ½ mole of N_2 is produced per mole of ammonia oxidized and ½ mole of CH_4 is produced per mole of CH_2O reacted. The fraction nitrogen gas $= \frac{1}{2}N_2 / (\frac{1}{2}CH_4 + \frac{1}{2}N_2) = J_N / (J[CH_4] + J_N)$ would approach a constant equal to the Redfield ratio of $16N/(106C+16N) \times 100\% = 13.1\%$. However, as the gas flux increases, the SOD also increases and the fraction of ammonia that oxidizes to nitrogen gas decreases as shown in Eq. (10.54). Hence the percentage of nitrogen gas can decrease still further. This accounts for the continual decline in nitrogen gas percentage (Fig. 10.6B) as gas flux increases.

10.7   RELATIONSHIP TO SULFIDE OXIDATION

The model for SOD presented in Chapter 9 is based on the oxidation of sulfides which are produced by the reduction of sulfate. The model constructed in this chapter assumes that methane is produced by the organic matter diagenesis. Therefore, for the data sets analyzed in this chapter, it is assumed that methane oxidation is the source of carbonaceous SOD and that the oxidation of ammonia results in nitrogen gas (i.e., complete denitrification). Methane that is produced in excess of the limit due to solubility escapes as methane gas. The composition of the escaping gas reflects the fraction of methane production that forms a gas phase and the production of nitrogen gas by denitrification.

These assumptions appear to be correct for the Fair et al. (1941) experiments and the results from the Milwaukee river. Both of these situations are characterized by rapid diagenesis. Therefore, it is likely that the sulfate supply was limited and carbon diagenesis proceeded to methane production.

It is not clear what the correct formulation should be in freshwater sediments. On the one hand, the sulfate concentration is much lower than in marine waters. On the other hand, freshwater sediments have concentrations of acid volatile sulfide that are similar to marine sediments (Di Toro et al., 1990a). It is clear that a model formulation is required that considers both methane and sulfide as intermediates that are oxidized.

With regard to the role of methane in affecting oxygen fluxes to the sediment, what is important is not whether methane forms but whether a methane *gas phase* forms and escapes to the overlying water. This constitutes a loss of electrons produced by carbon diagenesis. Therefore, they do not react with the electron acceptors in the sediment. This must be accounted for both in the sediment and in the overlying water mass balances. If methane gas does not form, then dissolved methane in the interstitial water would act as a reactive intermediate and would be oxidized with either sulfate or oxygen as the electron acceptor. A more complete model would combine both the formulations for sulfate reduction and methane production. This model is presented in the next chapter.

Appendix 10A: Positive and Negative Logarithmic Scale for Plotting

The dissolved methane mass balance equations are

$$-D_{CH_4}\frac{d^2c_1(z)}{dz^2} = -K_{CH_4,1}c_1(z) + S_{CH_4} \qquad 0 \leqslant z \leqslant H_1 \quad (10A.1a)$$

$$-D_{CH_4}\frac{d^2c_2(z)}{dz^2} = S_{CH_4} \qquad\qquad H_1 \leqslant z \leqslant H_2 \quad (10A.1b)$$

and the boundary conditions are

$$c_1(0) = 0 \qquad\qquad (10A.2)$$

$$c_1(H_1) = c_2(H_1) \qquad\qquad (10A.3)$$

$$-D_{CH_4}\left.\frac{dc_1(z)}{dz}\right|_{z=H_1} = -D_{CH_4}\left.\frac{dc_2(z)}{dz}\right|_{z=H_1} \qquad (10A.4)$$

$$c_2(H_{sat}) = [CH_4]_{sat} \qquad\qquad (10A.5)$$

The solutions for the differential equations are

$$c_1(z) = B_{1C}\exp(\lambda_C z) + B_{2C}\exp(-\lambda_C z) + \frac{S_{CH_4}}{k_{CH_4,1}} \qquad (10A.6)$$

$$c_2(z) = A_{1C} + A_{2C}z - \frac{S_{CH_4}z^2}{2D_{CH_4}} \qquad\qquad (10A.7)$$

Applying the boundary conditions sequentially yields

$$B_{1C} + B_{2C} + \frac{S_{CH_4}}{k_{CH_4,1}} = 0 \qquad\qquad (10A.8)$$

$$B_{1C}\exp(\lambda_C H_1) + B_{2C}\exp(-\lambda_C H_1) + \frac{S_{CH_4}}{k_{CH_4,1}} = A_{1C} + A_{2C}H_1 - \frac{S_{CH_4}H_1^2}{2D_{CH_4}} \qquad (10A.9)$$

$$\lambda_C B_{1C}\exp(\lambda_C H_1) - \lambda_C B_{2C}\exp(-\lambda_C H_1) = A_{2C} - \frac{S_{CH_4}H_1}{D_{CH_4}} \qquad (10A.10)$$

$$A_{1C} + A_{2C}H_{sat} - \frac{S_{CH_4}H_{sat}^2}{2D_{CH_4}} = [CH_4]_{sat} \qquad (10A.11)$$

This last condition complicates the solution to a remarkable extent. The most direct solution is obtained using Eq. (10A.8) for B_{2C}, then Eq. (10A.11) for A_{1C}, and Eq. (10A.10) for A_{2C}. Then B_{1C} comes from substituting these quantities into Eq. (10A.9)

$$B_{1C} = \frac{\begin{aligned}&\left\{[CH_4]_{sat} + S_{CH_4}(H_{sat} - H_1)^2/2D_{CH_4}\right.\\ &\left.+S_{CH_4}/k_{CH_4,1}[1 + \lambda_C(H_1 - H_{sat})]\exp(-\lambda_C H_1) - 1\right\}\end{aligned}}{2\sinh(\lambda_C H_1) + \lambda_C(H_{sat} - H_1)\cosh(\lambda_C H_1)} \tag{10A.12}$$

where the simplification

$$\frac{S_{CH_4}(H_{sat}^2 - H_1^2)}{2D_{CH_4}} + \frac{S_{CH_4}H_1(H_1 - H_{sat})}{D_{CH_4}} = \frac{S_{CH_4}(H_{sat} - H_1)^2}{2D_{CH_4}} \tag{10A.13}$$

has been used. The dissolved methane flux is

$$J[CH_4(aq)] = D_{CH_4}\left.\frac{dc_1(z)}{dz}\right|_{z=0} = D_{CH_4}\lambda_C(B_{1C} - B_{2C}) \tag{10A.14}$$

B_{1C} is given by Eq. (10A.12) and B_{2C} is obtained from Eq. (10A.8). The result is

$$J[CH_4(aq)] = D_{CH_4}\lambda_C\frac{\begin{aligned}&\left\{[CH_4]_{sat} + S_{CH_4}(H_{sat} - H_1)^2/2D_{CH_4}\right.\\ &+S_{CH_4}/k_{CH_4,1}[1 + \lambda_C(H_{sat} - H_1)]\sinh(\lambda_C H_1)\\ &\left.+ \exp(-\lambda_C H_1) - 1\right\}\end{aligned}}{\sinh(\lambda_C H_1) + \lambda_C(H_{sat} - H_1)\cosh(\lambda_C H_1)} \tag{10A.15}$$

10A.1 Simplification

A simplified solution results from replacing Eq. (10A.6) with

$$c_1(z) = B_{1C}\exp(\lambda_C z) + B_{2C}\exp(-\lambda_C z) \tag{10A.16}$$

where the term due to the source in the aerobic layer $S_{CH_4}/k_{CH_4,1}$ is neglected. Now the first boundary condition, Eq. (10A.8), is

$$B_{1C} + B_{2C} = 0 \tag{10A.17}$$

and

$$B_{1C} = \frac{[CH_4]_{sat} + S_{CH_4}(H_{sat} - H_1)^2/2D_{CH_4}}{2\sinh(\lambda_C H_1) + 2\lambda_C(H_{sat} - H_1)\cosh(\lambda_C H_1)}$$

$$\approx \frac{[CH_4]_{sat} + S_{CH_4}(H_{sat}^2)/2D_{CH_4}}{2\lambda_C H_{sat}\cosh(\lambda_C H_1)} \tag{10A.18}$$

The dissolved methane flux follows from the relationship

$$J[CH_4(aq)] = 2D_{CH_4}\lambda_C B_{1C} = \frac{D_{CH_4}}{H_{sat}}\frac{[CH_4]_{sat} + S_{CH_4}(H_{sat}^2)/2D_{CH_4}}{\cosh(\lambda_C H_1)} \tag{10A.19}$$

Substituting Eq. (10.12) for H_{sat} yields Eq. (10.22).

Appendix 10B: Solution of Dissolved Oxygen Mass Balance Equations

The dissolved oxygen mass balance equation is

$$-D_{O_2}\frac{d^2[O_2(z)]}{dz^2} = -a_{O_2^*,N}k_{NH_4,1}n_1(z) - k_{CH_4,1}c_1(z)$$

$$= -a_{O_2^*,N}k_{NH_4,1}\left(B_{1N}\exp(\lambda_N z) + B_{2N}\exp(-\lambda_N z) + \frac{S_N}{k_{NH_4,1}}\right)$$

$$-k_{CH_4,1}\left(B_{1C}\exp(\lambda_C z) + B_{2C}\exp(-\lambda_C z) + \frac{S_{CH_4}}{k_{CH_4,1}}\right)$$

$$(10B.1)$$

Two integrations yield

$$[O_2(z)] = F_1 + F_2 z + a_{O_2^*,N}\left(B_{1N}\exp(\lambda_N z) + B_{2N}\exp(-\lambda_N z) + \frac{S_N z^2}{2k_{NH_4,1}}\right)$$

$$+\left(B_{1C}\exp(\lambda_C z) + B_{2C}\exp(-\lambda_C z) + \frac{S_{CH_4}z^2}{2k_{CH_4,1}}\right) \qquad (10B.2)$$

The two arbitrary constants, F_1 and F_2 are evaluated using

$$[O_2(z)]|_{z=0} = [O_2(0)] \qquad (10B.3)$$

which specifies the overlying water dissolved oxygen concentration and

$$\frac{d[O_2(z)]}{dz}\bigg|_{z=H_1} = 0 \qquad (10B.4)$$

which follows from the requirement that below H_1 the flux of dissolved oxygen is zero because no oxygen is being consumed below $z = H_1$. Evaluating the first boundary condition yields

$$F_1 = [O_2(0)] - a_{O_2^*,N}[B_{1N} + B_{2N}] - [B_{1C} + B_{2C}] \qquad (10B.5)$$

and the second yields

$$F_2 = -a_{O_2^*,N}\lambda_N\left(B_{1N}\exp(\lambda_N H_1) - B_{2N}\exp(-\lambda_N H_1) + \frac{S_N H_1}{D_{NH_4}}\right)$$

$$-\lambda_C\left(B_{1C}\exp(\lambda_C H_1) - B_{2C}\exp(-\lambda_C H_1) + \frac{S_{CH_4} H_1}{D_{CH_4}}\right) \qquad (10B.6)$$

which completes the solution for the dissolved oxygen profile in the aerobic zone. The condition

$$[O_2(z)]|_{z=H_1} = 0 \qquad (10B.7)$$

is used to determine the depth of the aerobic layer H_1. This requires a numerical method.

11

Sulfide and Methane

11.1 INTRODUCTION

This chapter concludes the formulation of the oxygen consumption model. The electron acceptors in the anaerobic layer, sulfate and carbon dioxide, are both included. The formulation is based on the results of the previous two chapters. The basic problem is how to sequence the production of sulfide and methane. It is well known that sulfate is used before carbon dioxide as the electron acceptor. The evidence is the lack of dissolved methane in the zone of sulfate reduction. An example is shown in Fig. 1.13. As has been pointed out, this is the only plausible explanation for the observed profiles of sulfate and methane in the presence of mass transport.[*] Therefore, it is necessary to include sulfate explicitly in the model. If it is depleted, then methane formation occurs.

There is also the problem of layering. The actual vertical distribution of electron acceptors (oxidants) in sediments (Fig. 1.11) suggests a model formulation with explicit layers, delineated by the oxidation reaction taking place: aerobic oxidation, nitrate reduction, sulfate reduction, and methane formation.[†] The drawback with this formulation is that the concentrations of all the state variables in the model need to be computed in each layer, which increases the computational burden. This is a significant consideration when the sediment model is included as part of large eutrophication models with thousands of sediment-water interfaces. Each layer is the equivalent of adding a layer in the water column.

---

[*]Barnes and Goldberg (1976), Martens and Berner (1977), Reeburgh and Heggie (1974).
[†]Klapwijk and Snodgrass (1986), Smits and van der Molen (1993) .

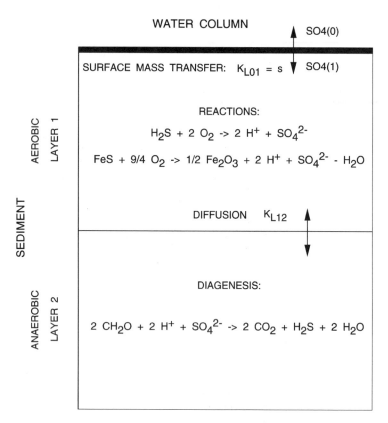

Fig. 11.1 Model of sulfate consumption.

The additional problem with an explicit layered model is how to compute the layer thicknesses at each time step. As shown below, this problem is unavoidable. However, it is much more difficult if the model explicitly includes state variables in each layer. Although one can imagine numerical approaches to the problem, it is prudent to ask whether *explicit* layers, with state variables in each layer, are necessary.

As an introduction to the problem, the next section examines the extent of sulfate consumption. The subsequent sections examine the layering issue.

11.2 SULFATE CONSUMPTION

Sulfate is depleted as a consequence of the reaction

$$2CH_2O + 2H^+ + SO_4^{2-} \rightarrow 2CO_2 + H_2S + 2H_2O \qquad (11.1)$$

For the two-layer model, the model framework for sulfate is presented in Fig. 11.1. The steady state mass balance equations for sulfate in this two-layer configuration are

$$0 = s([SO_4(0)] - [SO_4(1)]) + K_{L12}([SO_4(2)] - [SO_4(1)]) \quad (11.2a)$$

in the aerobic layer and

$$0 = -K_{L12}([SO_4(2)] - [SO_4(1)]) - a_{SO_4,C} J_C \quad (11.2b)$$

in the anaerobic layer. The terms in the aerobic layer mass balance Eq. (11.2a) are the exchange between the overlying water sulfate concentration $[SO_4(0)]$ and the aerobic layer concentration $[SO_4(1)]$, with mass transfer coefficient s, and the aerobic-anaerobic layer exchange with mass transfer coefficient K_{L12}. The terms in the anaerobic layer mass balance Eq. (11.2b) are the aerobic-anaerobic layer exchange and the consumption of sulfate as carbon is being oxidized and sulfate is being reduced (Eq. 11.1). The rate of sulfate consumption is $a_{SO_4,C} J_C$ where J_C is the carbon diagenesis flux and $a_{SO_4,C}$ is the sulfate to carbon stoichiometric coefficient. For the reaction listed above (Eq. 11.1) $a_{SO_4,C} = 1/2$ (mol S/mol C).

The solutions to these equations are

$$[SO_4(1)] = [SO_4(0)] - a_{SO_4,C} J_C \left(\frac{1}{s}\right) \quad (11.3a)$$

$$[SO_4(2)] = [SO_4(0)] - a_{SO_4,C} J_C \left(\frac{1}{s} + \frac{1}{K_{L12}}\right) \quad (11.3b)$$

Note the sum of the reciprocal mass transfer coefficients in Eq. (11.3b), for the mass transfers in series.

Sulfate is depleted when $[SO_4(2)] = 0$. A simple estimate for the overlying water sulfate concentration that produces this condition can be made using Eq. (11.3b) to compute $[SO_4(0)]$ for $[SO_4(2)] = 0$. The result is presented below in Table 11.1. An overlying water sulfate concentration of $[SO_4(0)] = 400$ mg SO_4/L is required for a carbon diagenesis of 1 g C/m^2-d. This is at the high range of freshwater streams (5–500 mg SO_4/L) (Davies and DeWiest, 1966). This computation suggests that methane formation is a common occurrence in freshwater sediments.

The above analysis assumes that no sulfate is produced by the oxidation of H_2S in the aerobic layer. The oxidation of sulfides produces various endproducts: elemental sulfur S^0, thiosulfate $S_2O_3^{2-}$, and polysulfides (Stumm and Morgan, 1996). However, if the oxidation proceeds to completion and the final product is sulfate

$$H_2S + 2O_2 \rightarrow SO_4^{2-} + 2H^+ \quad (11.4)$$

this would provide a source of sulfate to the aerobic layer. Assuming complete oxidation to sulfate in the aerobic layer, the mass balance equations become

$$0 = s([SO_4(0)] - [SO_4(1)]) + K_{L12}([SO_4(2)] - [SO_4(1)]) + a_{SO_4,C} J_C \quad (11.5a)$$

Table 11.1   Overlying water sulfate concentration required for $[SO_4(2)] = 0$

Parameter	Value	Units
J_C	1.0	gC/m^2-d
$a_{SO_4,C}$	$1/2$	mol/mol
s	0.1	m/d
D_d	10^{-3}	m^2/d
H_2	0.1	m
K_{L12}	0.01	m/d
$[SO_4(0)]$	400	mg SO$_4$/L

$$0 = -K_{L12}([SO_4(2)] - [SO_4(1)]) - a_{SO_4,C}J_C \qquad (11.5b)$$

The solutions are

$$[SO_4(1)] = [SO_4(0)] \qquad (11.6a)$$

$$[SO_4(2)] = [SO_4(0)] - a_{SO_4,C}J_C\left(\frac{1}{K_{L12}}\right) \qquad (11.6b)$$

The result is that the mass transfer resistance in the aerobic layer is removed (compare to Eqs. 11.3) and the transfer is controlled by the layer 1–2 mass transfer. However, the conclusions drawn from Table 11.1 are not materially changed since K_{L12} itself is small enough to limit the supply of sulfate. Hence, it is the supply of sulfate that limits the production of sulfide and causes the production of methane.

In addition we need to consider how to calculate the appropriate mass transfer coefficient for sulfate. In Table 11.1 it is assumed that the K_{L12} that is appropriate for ammonia is also appropriate for sulfate. This issue is directly related to the layering of the reactions in sediments.

11.3   LAYERS AND MASS TRANSFER RESISTANCES

The question is: What role do the layers play in the transport of solutes in a vertically segmented model of a sediment? The schematic in Fig. 11.2 diagrams the problem. A conservative solute of concentration c is transported from the overlying water to the sediment, which is idealized into three layers. There are zero-order sinks r_1, r_3 in layers 1 and 3. The mass balance equations for the concentrations in the three layers c_1, c_2, and c_3, respectively, are

Layer 1	$0 = (c_0 - c_1)s + (c_2 - c_1)K_{L12} - r_1$	(11.7a)
Layer 2	$0 = (c_3 - c_2)K_{L23} - (c_2 - c_1)K_{L12}$	(11.7b)
Layer 3	$0 = -(c_3 - c_2)K_{L23} - r_3$	(11.7c)

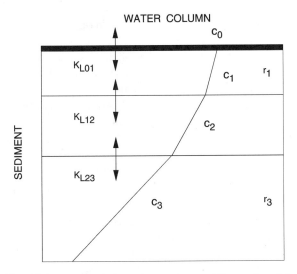

Fig. 11.2   Mass transfer resistance in a multilayered model.

where c_0 is the overlying water concentration and the substitution $s = K_{L01}$ for the surface mass transfer coefficient has been made. The solutions are derived in Fig. 11A.1 in Appendix 11A using MACSYMA. The result for layer 3 is

$$c_3 = c_0 - \frac{r_1}{s} - r_3 \left(\frac{1}{s} + \frac{1}{K_{L12}} + \frac{1}{K_{L23}} \right) \tag{11.8}$$

Noting the sum of reciprocals form of mass transfer resistances in series, we define an equivalent transfer coefficient

$$\frac{1}{K_L^*} = \frac{1}{K_{L12}} + \frac{1}{K_{L23}} \tag{11.9}$$

so

$$K_L^* = \left(\frac{1}{K_{L12}} + \frac{1}{K_{L23}} \right)^{-1} \tag{11.10}$$

which, when substituted into Eq. (11.8), yields

$$c_3 = c_0 - \frac{r_1}{s} - r_3 \left(\frac{1}{s} + \frac{1}{K_L^*} \right) \tag{11.11}$$

The important point is that Eq. (11.11) is the solution for an equivalent *two-layer* model

Layer 1 $0 = (c_0 - c_1) s + (c_3 - c_1) K_L^* - r_1$ (11.12a)

Layer 2 $0 = (c_3 - c_1) K_L^* - r_3$ (11.12b)

Thus the effect of intervening layers is simply to add (as reciprocals) additional mass transfer resistances (Eq. 11.8).

This analysis clarifies the consequences of the presence of intervening layers. They each contribute mass transfer resistances. Since mass transfer coefficients are diffusion coefficients divided by the layer depths, it is important that the layer depths be estimated in some way. However, it is clear from the above analysis that it is not necessary to *explicitly model* each layer. Only the depths are required.

11.3 A   Effect of Overlying Water Velocity

The analysis in the previous section clarifies the effect of layers through which solutes diffuse passively, without reacting. The layers contribute additional mass transfer resistance. This is the source of the influence of overlying water velocity on sediment fluxes in general and SOD in particular. The effect has been observed for oxygen fluxes (Boynton et al., 1981, Whittemore, 1986) and nitrate fluxes (Cerco, 1988). The data in Fig. 4.3 provide an example. The fluxes measured with rapid mixing are larger than those with slow mixing. If the mass transfer coefficient is known for the boundary layer at the sediment-water interface, then it can be added to the other mass transfer resistances (Mackenthun and Stefan, 1998, Nakamura and Stefan, 1994). For measuring devices, the stirring is chosen to be rapid enough so that the overlying water boundary layer does not influence the measurement.

11.4   MULTILAYER VERSUS TWO-LAYER MODELS

The implication of the analysis presented above is that a model that is computationally equivalent to a multilayer model can be formulated as shown in Fig. 11.3.

The multilayer approach (Fig. 11.3A) envisions a series of distinct layers that are explicitly included in the equation set. For the diagrammed situation, five segments would be required, one for each electron acceptor, and the fifth for the layer where methane gas is forming. But the conclusion reached in Section 11.3 suggests that only the interlayer mass transport coefficients are important. The equivalent mass balance equations can be formulated as a two-layer approximation (Fig. 11.3B) using the equivalent mass transfer coefficients (Eq. 11.10). Of course the layer depths are important, since they determine the mass transfer coefficients – see, for example, Eq. (11.14). The computational method for layer depths is examined in the next section.

One final point to consider. The analyses used to justify the two-layer approximation are all based on steady state equations. It is, therefore, not the case that two-layer and multilayer models are *exactly* the same. The analysis does suggest that for slowly varying situations, in which the time variable solutions approximately track a series of steady states, the models should behave similarly. The mathematical condition is that the time derivatives are small relative to the other terms in the equations.

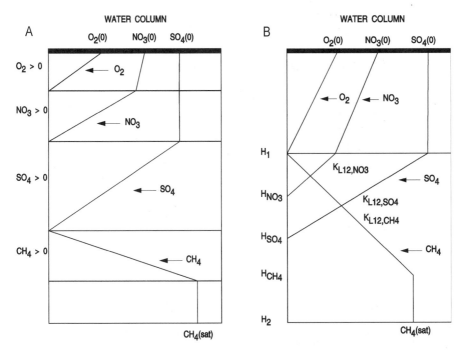

Fig. 11.3 (A) Multilayer model. (B) Equivalent two-layer model. The aerobic layer depth H_1 is measured from $z = 0$ at the sediment-water interface. The remaining depths are measured from H_1.

For pore water diffusion, an approximate estimate of the time scale τ of sediment model response is

$$\tau = \frac{H_2}{K_{L12}} \tag{11.13}$$

Eq. (11.13) is the residence time of pore water in the sediment of thickness H_2 and with mass transfer coefficient K_{L12}. It is the time it would take to replace all the pore water in the anaerobic layer H_2 (m) with a mass transfer coefficient K_{L12} (m/d). For $H_2 = 0.1$ m and $K_{L12} = 0.01$ m/d (Table 11.1), $\tau \simeq 10$ days. For seasonal simulations, this is rapid enough for the steady state approximations to be reasonable. More accurate estimates of times to steady state for the sediment model are presented in Chapter 16. For constituents with significant solid phase reservoirs, the time scales are much longer. But for the diffusion of electron acceptors, Eq. (11.13) is valid.

11.5 DEPTH OF SULFATE REDUCTION

The modeling approach described in the previous section requires a knowledge of the layer depths, in particular the depth of the layer where sulfate reduction is taking place. The derivation is based on the idealization presented in Fig. 11.4. Assuming

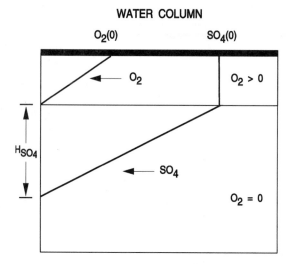

Fig. 11.4 Schematic of the vertical profiles of oxygen and sulfate.

place. The derivation is based on the idealization presented in Fig. 11.4. Assuming a linear concentration profile for simplicity, the flux of sulfate to the sediment is

$$J[SO_4] = D_{SO_4} \left. \frac{d[SO_4(z)]}{dz} \right|_{z=0} \tag{11.14a}$$

$$\simeq D_{SO_4} \frac{[SO_4(0)] - [SO_4(H_{SO_4})]}{H_{SO_4}} \tag{11.14b}$$

$$= D_{SO_4} \frac{[SO_4(0)]}{H_{SO_4}} \tag{11.14c}$$

where the fact that $[SO_4(H_{SO_4})] = 0$ – the concentration of sulfate at the bottom of the zone of sulfate reduction is zero – is used in Eq. (11.14b) to obtain Eq. (11.14c).

Eq. (11.14c) relates $J[SO_4]$ to H_{SO_4}. If another relationship could be found, then $J[SO_4]$ could be eliminated and H_{SO_4} would be known. Consider the following: Sulfate is consumed only in the sulfate reduction layer. In a two-layer model carbon diagenesis in the anaerobic layer occurs uniformly over H_2. The amount of carbon diagenesis that is being oxidized by sulfate is H_{SO_4}/H_2, since this is the fraction of the anaerobic layer where sulfate reduction is occurring. The sulfate flux to the sediment must equal this amount so that the source balances the sink. The equation that expresses this equality is

$$J[SO_4] = a_{SO_4,C} J_C \frac{H_{SO_4}}{H_2} \tag{11.15}$$

where $a_{SO_4,C}$ is the stoichiometric coefficient relating the sulfate consumed to carbon oxidized. Therefore, using Eq. (11.14c) for $J[SO_4]$ in Eq. (11.15) yields

$$D_{SO_4} \frac{[SO_4(0)]}{H_{SO_4}} = a_{SO_4,C} J_C \frac{H_{SO_4}}{H_2} \tag{11.16}$$

and solving for H_{SO_4} yields

$$H_{SO_4} = \sqrt{\frac{D_{SO_4}[SO_4(0)]H_2}{a_{SO_4,C} J_C}} \tag{11.17}$$

Of course, the depth must be limited by the depth of the active layer of the sediment

$$H_{SO_4} \leqslant H_2 \tag{11.18}$$

so either Eq. (11.17) or $H_{SO_4} = H_2$ is used.

Since the mass transfer coefficient for sulfate K_{L12,SO_4} is

$$K_{L12,SO_4} = \frac{D_{SO_4}}{H_{SO_4}} \tag{11.19}$$

Eq. (11.19) can be evaluated using Eq. (11.17) to give

$$K_{L12,SO_4} = \sqrt{a_{SO_4,C} J_C \frac{D_{SO_4}}{H_2[SO_4(0)]}} \tag{11.20}$$

which is the sulfate mass transfer coefficient.

Note that it varies with both the flux of carbon diagenesis J_C and the overlying water sulfate concentration $[SO_4(0)]$. Both small overlying water concentrations and large diagenesis fluxes enhance the mass transfer rate by making the zone of sulfate reduction smaller (Eq. 11.20). When $H_{SO_4} \leqslant H_2$ this is an important feedback mechanism that routes more carbon diagenesis to sulfate reduction and less to methane formation.

11.5 A Continuous Derivation

The depth of sulfate reduction can also be calculated without the assumption of a linear profile (Fig. 11.4). This derivation parallels the development of the SOD model by Bouldin (1968) presented in Section 8.2 B. It ignores the presence of the aerobic layer and proceeds directly from the top of the anaerobic layer. The one-dimensional mass balance equation for the concentration of SO_4 in the interstitial water of the sediment $[SO_4(z)]$ is

$$-D_{SO_4} \frac{d^2[SO_4(z)]}{dz^2} = -\frac{a_{SO_4,C} J_C}{H_2} \tag{11.21}$$

where D_{SO_4} is the diffusion coefficient, and $a_{SO_4,C} J_C / H_2$ is the zero-order consumption rate of sulfate. The convention for z is positive downward. The solution follows from two integrations

$$[SO_4(z)] = A_0 + A_1 z + \frac{a_{SO_4,C} J_C}{H_2} \frac{z^2}{2 D_{SO_4}} \tag{11.22}$$

with A_1 and A_2 the arbitrary constants of integration. The three boundary conditions are

$$[SO_4(z)]|_{z=0} = [SO_4(0)] \tag{11.23a}$$

which specifies the overlying water SO_4 concentration,

$$[SO_4(H_{SO_4})] = 0 \tag{11.23b}$$

which requires that the sulfate concentration is zero at the depth of the zone of sulfate reduction H_{SO_4}, and

$$\frac{d[SO_4(z)]}{dz}\bigg|_{z=H_{SO_4}} = 0 \tag{11.23c}$$

which follows from the requirement that no sulfate is being consumed below H_{SO_4}. Since sulfate cannot be consumed if it is not present, the flux of SO_4 at $z = H_{SO_4}$, which would be the source to the sediment below $z = H_{SO_4}$, must be zero. Eqs. (11.23a) and (11.23c) can be used to evaluate A_0 and A_1 so that Eq. (11.22) becomes

$$[SO_4(z)] = [SO_4(0)] + \frac{a_{SO_4,C} J_C}{H_2} \frac{z}{D_{SO_4}} \left(\frac{z}{2} - H_{SO_4} \right) \tag{11.24}$$

Eq. (11.23b) can be used in this equation to obtain the depth of the sulfate reduction layer $z = H_{SO_4}$ when $[SO_4(H_{SO_4})] = 0$. The result is

$$H_{SO_4} = \sqrt{\frac{2 D_{SO_4} [SO_4(0)] H_2}{a_{SO_4,C} J_C}} \tag{11.25}$$

This is larger by $\sqrt{2}$ than Eq. (11.17), which was obtained using the linear assumption.

Since this is presumably a more realistic estimate, we will use Eq. (11.25) as the estimate for H_{SO_4} and

$$K_{L12,SO_4} = \sqrt{a_{SO_4,C} J_C \frac{D_{SO_4}}{2 H_2 [SO_4(0)]}} \tag{11.26}$$

Table 11.2 Depth of Sulfate Reduction

Parameter	Value	Units
J_C	1	g C/m^2–d
$a_{SO_4,C}$	½	mol/mol
s	0.1	m/d
K_{L12}	0.01	m/d
$[SO_4(0)]$	50[a]	mg SO$_4$/L
H_{SO_4}	0.020[b]	m

[a] Median sulfate concentration in US rivers. [b] $H_{SO_4} < H_2 = 0.1$ m.

as the estimate for K_{L12,SO_4}. This new result is used to update the computation presented in Table 11.1, which evaluates the extent to which the sulfate supply to the sediment is limiting. The result is shown in Table 11.2. The overlying water concentration is set to 50 mg SO$_4$/L, the median river concentration. The resulting depth of sulfate reduction ($H_{SO_4} = 2$ cm) is considerably less than the total depth of the sediment. For this situation, we would expect production of methane, since there is a fraction of carbon diagenesis occurring between H_{SO_4} and H_2 for which no sulfate is available.

11.6 SULFATE AND METHANE MASS BALANCE EQUATIONS

The problem is to determine what fraction of carbon diagenesis is oxidized by sulfate and what fraction becomes methane, either dissolved methane or methane gas. The previous sections provided estimates of the depth of the sulfate reduction layer. In this section, a more complete mass balance analysis is presented.

The vertical distribution is illustrated in Fig. 11.5. The nitrate reduction layer is not explicitly included, since it is a thin layer under normal circumstances, and would provide little additional mass transfer resistance. The utilization of carbon diagenesis during nitrate reduction is taken into account in the nitrate flux model (Chapter 4). Carbon diagenesis J_C is either oxidized by sulfate J_{SO_4}, or becomes methane that is transferred to the aerobic layer $J_{CH_4(aq)}$, or becomes a methane gas flux $J_{CH_4(g)}$. If carbon diagenesis is assumed to be occurring uniformly in depth, then the apportionment of carbon diagenesis is proportional to the depths of these layers. The bar labeled J_C in Fig. 11.5 depicts the assumption being made. This condition can be used to determine the depths of each layer.

The mass balance equation for sulfate in the aerobic layer is

$$0 = s([SO_4(0)] - [SO_4(1)]) + \frac{D_{SO_4}}{H_{SO_4}}([SO_4(2)] - [SO_4(1)]) \qquad (11.27a)$$

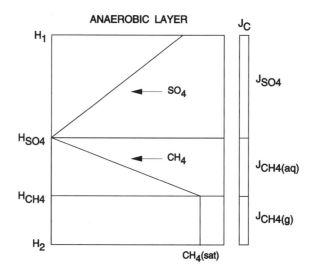

Fig. 11.5 Layers in the anaerobic zone: carbon diagenesis that is oxidized by sulfate J_{SO_4}, methane production that is transferred to the aerobic layer $J_{CH_4(aq)}$, and methane production that becomes gas flux $J_{CH_4(g)}$.

The terms represent the transfer to and from the overlying water, and the mass transfer between the aerobic and anaerobic layers, where the mass transfer coefficient is expressed in terms of the diffusion coefficient for sulfate D_{SO_4} and the depth of the sulfate reduction layer H_{SO_4}. The source of sulfate due to the oxidation to sulfide is ignored, since its inclusion dramatically complicates the solution of the equations and appears to not affect the solution very much – compare Eq. (11.3b) and Eq. (11.6b).

The anaerobic layer mass balance equation is

$$0 = -\frac{D_{SO_4}}{H_{SO_4}}([SO_4(2)] - [SO_4(1)]) - a_{SO_4,C} J_C \frac{H_{SO_4}}{H_2} \qquad (11.27b)$$

The quantity of sulfate consumed is proportional to the ratio of H_{SO_4} to H_2 as diagrammed in Fig. 11.5. A third equation is required to complete the equations for the three unknowns: $[SO_4(1)]$, $[SO_4(2)]$, and H_{SO_4}. If the linear profile for sulfate in Fig. 11.5 is realistic, then the average concentration in that layer is

$$[SO_4(2)] = \frac{[SO_4(H_1)] - [SO_4(H_{SO_4})]}{2} = \frac{[SO_4(H_1)]}{2} = \frac{[SO_4(1)]}{2} \qquad (11.27c)$$

This is the third equation.

The MACSYMA solution of Eqs. (11.27) is presented in Fig. 11B.1 of Appendix 11B. A most useful feature is illustrated: having MACSYMA render the solutions into FORTRAN, ready for use in numerical computations.

The resulting depth must satisfy the condition $H_{SO_4} \leqslant H_2$. For small diagenesis fluxes especially in marine sediments for which the overlying water sulfate concentration is large, the sulfate alone is sufficient to oxidize all the carbon diagenesis. For this case, $H_{SO_4} = H_2$, Eq. (11.27c) no longer applies, and the solutions are

$$[SO_4(1)] = [SO_4(0)] - \frac{a_{SO_4,C}J_C}{s} \tag{11.28a}$$

$$[SO_4(2)] = [SO_4(1)] - \frac{a_{SO_4,C}J_C}{D_{SO_4}/H_{SO_4}} \tag{11.28b}$$

11.6 A   Dissolved Methane

In the zone of methane formation, the remaining carbon diagenesis is converted into carbon dioxide and methane. The electron acceptor or oxidant for this case is carbon dioxide (Eq. 1.20)

$$\frac{1}{8}CO_2 + H^+ + e^- \rightarrow \frac{1}{8}CH_4 + \frac{1}{4}H_2O \tag{11.29}$$

because CO_2 is being reduced to CH_4. It accepts the electrons being produced by carbon diagenesis

$$\frac{1}{4}CH_2O + \frac{1}{4}H_2O \rightarrow \frac{1}{4}CO_2 + H^+ + e^- \tag{11.30}$$

The overall reaction is

$$CH_2O \rightarrow \frac{1}{2}CO_2 + \frac{1}{2}CH_4 \tag{11.31}$$

This reaction occurs if the depth of sulfate reduction is less than the depth of the anaerobic layer. In this case the carbon diagenesis flux that is available for methane formation is (Fig. 11.5)

$$J_{CH_4} = J_C\left(1 - \frac{H_{SO_4}}{H_2}\right) \tag{11.32}$$

Consider, first, the case of no methane gas formation (Fig. 11.6). Then the depth of methane production is the entire remaining depth

$$H_{CH_4} = H_2 - H_{SO_4} \tag{11.33}$$

and all of it contributes to the formation of dissolved methane. The mass balance equations for dissolved methane are

$$0 = s\left([CH_4(0)] - [CH_4(1)]\right) - \frac{\kappa^2_{CH_4,1}}{s}[CH_4(1)]$$

$$+ K_{L12,CH_4}\left([CH_4(2)] - [CH_4(1)]\right) \tag{11.34a}$$

$$0 = J_{CH_4} - K_{L12,CH_4}\left([CH_4(2)] - [CH_4(1)]\right) \tag{11.34b}$$

The solutions for the aerobic and anaerobic layers are

$$[CH_4(1)] = \frac{s^2[CH_4(0)] + s\,J_{CH_4}}{s^2 + \kappa_{CH_4,1}^2} \qquad (11.35a)$$

$$[CH_4(2)] = \frac{s^2 K_{L12,CH_4}[CH_4(0)] + \left(s^2 + s K_{L12} + \kappa_{CH_4,1}^2\right) J_{CH_4}}{K_{L12,CH_4}\left(s^2 + \kappa_{CH_4,1}^2\right)} \qquad (11.35b)$$

In this case the mass transfer coefficient for methane must reflect the two layers to be traversed in the anaerobic layer (Fig. 11.6). Since there are two layers, the mass

ANAEROBIC LAYER

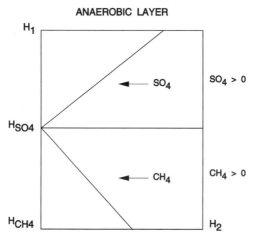

Fig. 11.6 Layers in the anaerobic zone: sulfate reduction and methane production. No gas phase forms for this case.

transfer coefficient for methane is (Eq. 11.10)

$$K_{L12,CH_4} = \left(\frac{1}{K_{L12}} + \frac{1}{K_{L23}}\right)^{-1} \qquad (11.36a)$$

$$= \left(\frac{H_{SO_4}}{D_{SO_4}} + \frac{H_{CH_4}}{D_{CH_4}}\right)^{-1} \qquad (11.36b)$$

where the relationship between mass transfer coefficient and layer depth (Eq. 11.19) is used in Eq. (11.36a) to obtain Eq. (11.36b).

Eqs. (11.35) are valid so long as the dissolved methane concentration is less than the saturated concentration $[CH_4(sat)]$

$$[CH_4(2)] \leqslant [CH_4(sat)] \qquad (11.37)$$

where $[CH_4(sat)]$ is determined by the solubility of methane and the pressure at the depth of the sediment (Eq. 10.51). If the computed methane concentration in the anaerobic layer exceeds the solubility of methane, then Eqs. (11.35) are no longer valid and methane gas production must be considered explicitly.

11.6 B Methane Gas Production

The situation is diagrammed in Fig. 11.5. As in the case of the sulfate reduction layer, the problem needs to be solved with an explicit H_{CH_4} in the equation set. Three unknowns are the aerobic layer methane concentration $[CH_4(1)]$, the methane gas flux $J[CH_4(g)]$, and the depth of dissolved methane production H_{CH_4}. The available equations begin with the layer 1 mass balance

$$0 = s\left([CH_4(0)] - [CH_4(1)]\right) - \frac{\kappa_{CH_4,1}^2}{s}[CH_4(1)]$$

$$+K_{L12,CH_4}\left([CH_4(sat)] - [CH_4(1)]\right) \tag{11.38a}$$

where

$$K_{L12,CH_4} = \left(\frac{H_{SO_4}}{D_{SO_4}} + \frac{H_{CH_4}}{D_{CH_4}}\right)^{-1} \tag{11.38b}$$

The production of methane gas can be computed using a total methane mass balance

$$J[CH_4(g)] = J_{CH_4} - s\left([CH_4(1)] - [CH_4(0)]\right) - \frac{\kappa_{CH_4,1}^2}{s}[CH_4(1)] \tag{11.38c}$$

This equation states that the total production of methane J_{CH_4} less the net loss to the overlying water and the loss to oxidation must be the gas flux. The third required equation follows from the proportionality assumption (Fig. 11.5)

$$\frac{H_{CH_4}}{H_2 - H_{SO_4}} = 1 - \frac{J[CH_4(g)]}{J_{CH_4}} \tag{11.38d}$$

The left-hand side of Eq. (11.38d) is the fraction of the total depth of methane production that is producing dissolved methane. The right-hand side is one minus the fraction of carbon diagenesis that is methane gas, which is the also the fraction producing dissolved methane.

The MACSYMA solution of Eqs. (11.38) is presented in Figs. 11C.1 and 11C.2, Appendix 11C. The result is a quadratic equation for $J[CH_4(g)]$. The depth H_{CH_4} follows from Eq. (11.38d) and $[CH_4(1)]$ from Eq. (11.38c). Although the solution is somewhat involved, it is neither computationally difficult nor expensive.

11.7 NUMERICAL EXAMPLES

The behavior of the coupled sulfate-methane model is examined in Fig. 11.7. Two values of the surface mass transfer coefficient are chosen: $s = 0.1$ and 100 m/d representing a larger and smaller aerobic layer thickness, respectively. Three overlying water sulfate concentrations are presented: $[SO_4(0)] = 10, 100, 1000$ mg SO_4/L. The carbon diagenesis in oxygen equivalent units is varied from $J_C = 0.1$ to 10

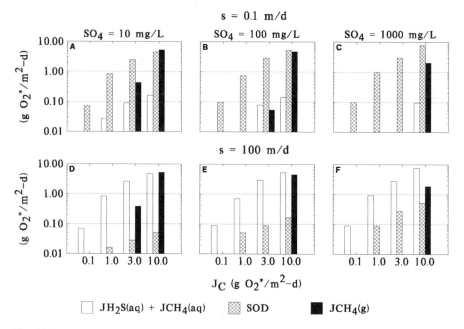

Fig. 11.7 Computed SOD, sulfide, and methane flux as a function of the surface mass transfer coefficient s (top and bottom rows), overlying water sulfate concentration [$SO_4(0)$] (the three columns), and carbon diagenesis J_C (abscissa).

g O_2^*/m^2-d. The quantities presented are the SOD, the sum of sulfide and methane oxidation in the aerobic layer, the dissolved flux, which is the sum of the dissolved methane and sulfide transferred to the overlying water, and the methane gas flux. The sum of these fluxes is the carbon diagenesis. All the fluxes are in oxygen equivalent units.

Increasing overlying water sulfate concentration (Figs. 11.7A to C and D to F) decreases the methane gas flux (solid bars) until at [$SO_4(0)$] = 1000 mg SO_4/L, which is approximately half the concentration in sea water, methane gas flux is predicted only at the highest carbon diagenesis. No gas flux is predicted for $J_C \leqslant 1.0$ g O_2^*/m^2-d.

The magnitude of the dissolved fluxes to the overlying water and the resulting SOD is controlled by the magnitude of s. For small s = SOD/[$O_2(0)$] (Figs. 11.7A, B, and C), the aerobic layer is thicker, the dissolved fluxes of sulfide and methane are oxidized, and SOD results (denoted by the cross-hatched bars). For large s (Figs. 11.7D, E, and F) the aerobic layer is thinner, and the residence time in the layer is insufficient to completely oxidize the fluxes of sulfide and dissolved methane to the aerobic layer. Rather they escape to the overlying water (the unfilled bars).

The next figure examines what fractions of the SOD and dissolved fluxes are due to sulfide and methane. The components of the SOD are displayed in Fig. 11.8(I). For aerobic overlying water and low sulfate concentration (A) the SOD is mainly due to methane oxidation (cross-hatching). Conversely, for high sulfate concentration (B), the SOD is due to sulfide oxidation (single hatching) until $J_C = 10$ g O_2^*/m^2-d when methane oxidation also is important. Approximately one-half of the methane produced is oxidized and one half escapes as a gas flux. For low overlying water dissolved oxygen ($s = 100$ m/d) and low sulfate concentrations (C) little SOD results, and what does is due to sulfide oxidation. Carbon diagenesis is converted almost entirely to methane flux. With higher sulfate concentration (D) more sulfide is produced and SOD due to sulfide oxidation increases.

The components of the dissolved flux are displayed in Fig. 11.8(II). For high overlying water oxygen concentrations (A–B), only dissolved methane fluxes are significant, and they are small. Both the methane and the sulfide produced are oxidized (Fig. 11.8(I) A–B). For low overlying water dissolved oxygen, the methane (C) and sulfide (D) produced escape as dissolved fluxes to the overlying water, rather than being oxidized (Fig. 11.8(I) C–D).

Thus the coupled sulfide-methane model smoothly transitions from the situations dominated by methane production, in low sulfate freshwaters, to sulfide domination in brackish and marine waters. It remains to be seen if the model conforms to observations. This is addressed in the next two sections.

11.8 UPPER POTOMAC ESTUARY

The Potomac Estuary has been the subject of many water quality investigations (see Thomann and Mueller (1987) for the history of modeling investigations). Not until recently have the modeling frameworks included an interactive sediment model. In this section, an analysis is presented of a data set that was collected specifically to aid in the calibration of the coupled sulfate-methane model (Boynton et al., 1995).

The sampling stations are illustrated in Fig. 11.9(I). The observed methane gas fluxes, overlying water sulfate concentrations, and SOD are shown in Fig. 11.9(II) for the indicated sampling stations. Note the inverse relationship between the methane fluxes (solid bars) and sulfate concentrations (crosshatched bars). No methane flux is measured at R-64, the station in Chesapeake Bay. The low SODs in July and August at R-64 reflect the low overlying water dissolved oxygen concentrations. Methane fluxes were always present at Gunston Cove. Gas fluxes were observed at Hedge Neck for May and August, and in August for Maryland Point.

The model is applied to these data in the usual way. The ammonia flux is used to compute ammonia diagenesis, which is then used to compute the flux of POM to the sediment. The basic idea is to use the steady state analytical solution relating ammonia flux to ammonia diagenesis (Eq. 3.19) to compute J_N. Then J_C is obtained using Redfield stoichiometry. Section 16.2 A presents the derivation of the equations. The model is then run to steady state.

The results for Gunston Cove, Maryland Point, and R-64 are shown in Fig. 11.10. Of course the model and data are essentially identical for ammonia, since the model depositional fluxes are derived from the ammonia fluxes. The comparison to SOD is reasonable, considering this is a steady state computation. The model essentially

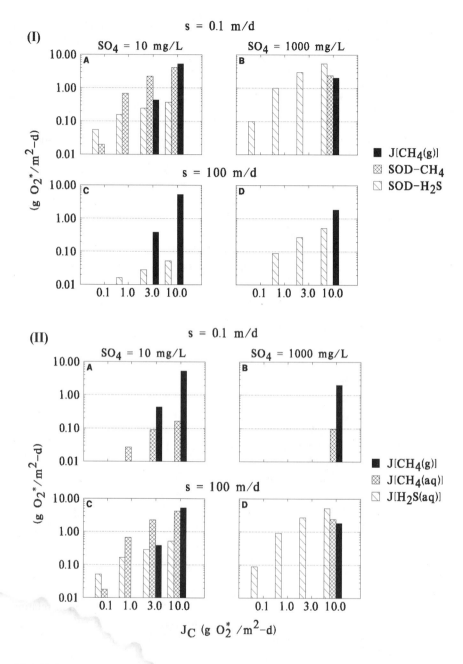

Fig. 11.8 (I) Components of the SOD: SOD due to sulfide oxidation (SOD-H$_2$S), methane oxidation (SOD-CH$_4$), and methane gas flux J[CH$_4$(g)]. (II) Dissolved fluxes to the overlying water: sulfide J[H$_2$S(aq)], methane J[CH$_4$(aq)], and the methane gas flux J[CH$_4$(g)]. As a function of the surface mass transfer coefficient s (top and bottom rows), overlying water sulfate concentration [SO$_4$(0)] (columns), and carbon diagenesis J_C (abscissa).

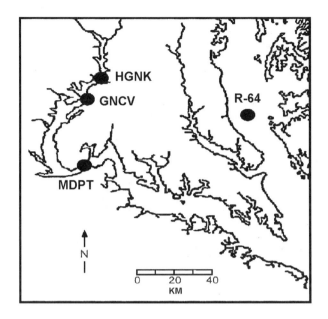

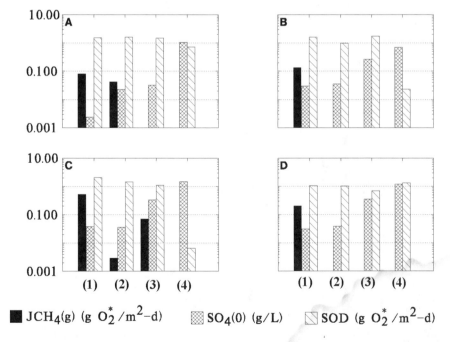

Fig. 11.9 Observations of methane gas flux, overlying water sulfate concentration, and SOD for the stations in the Potomac Estuary and Chesapeake Bay. (1) Gunston Cove (GNCV). (2) Hedge Neck (HGNK). (3) Maryland Point (MDPT). (4) R-64. (A) May. (B) July. (C) August. (D) October.

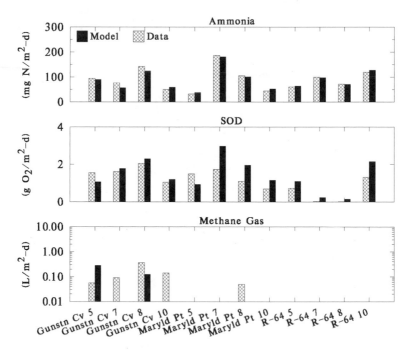

Fig. 11.10 Model and data comparison for Gunston Cove, Maryland Point, and R-64. J_C computed from $J[NH_4]$.

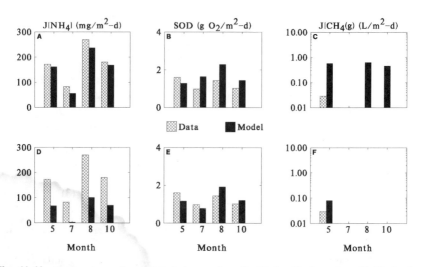

Fig. 11.11 Comparison of model and observations for Hedge Neck station. (A-C) J_C determined from $J[NH_4]$. (D-F) J_C reduced by one half.

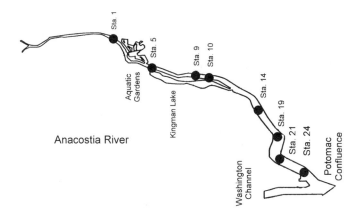

Fig. 11.12   Sampling stations in the Anacostia River.

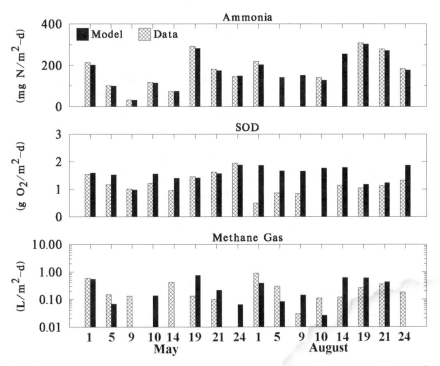

Fig. 11.13   Comparison of computed and observed ammonia, SOD, and methane gas flux for the eight Anacostia River sampling stations (Fig.11.12), May and August.

reproduces the observations: low methane fluxes in Gunston Cove, and essentially no methane fluxes at the other stations.

The results for Hedge Neck are shown in Fig. 11.11. Two computations are presented. The first (A–C) is carried out as in Fig. 11.10 with the ammonia flux determining the depositional flux. The model predicts large methane fluxes, and essentially none are observed. Note that the SOD is not reproduced well, with the model consistently higher than the data. The sensitivity of the model to this discrepancy is examined in (D–F) where the depositional flux is set to one half the value derived from the ammonia flux. The model is now computing SODs that are more consistent with the observations and the methane gas fluxes are also more faithfully reproduced.

The steady state model cannot reconcile the differences between the ammonia flux and the SOD in the Hedge Neck data. It is constrained by the stoichiometric relationships to produce high SODs when high ammonia fluxes are present. The discrepancies are likely due to time variable effects.

11.9 ANACOSTIA RIVER

As part of the District of Columbia's Combined Sewer Overflow (CSO) Abatement Program, the Metropolitan Washington Council of Governments has conducted studies that include the evaluation of water quality benefits resulting from CSO controls for the Anacostia River (COG , 1990). Combined sewer overflows discharge particulate organic material that can settle to the sediments of the river. The issue is: How would a change in the deposition of particulate organic material associated with CSO controls be related to changes in sediment oxygen demand, and therefore to dissolved oxygen concentrations in the overlying water?

A field study was conducted to measure SOD, ammonia, and methane fluxes (Sampou, 1990). The sampling stations are shown in Fig. 11.12. A total of eight stations were sampled in May and August 1990. The ammonia fluxes are used to estimate depositional fluxes and the results are shown in Fig. 11.13. Sulfate concentrations are quite low, so that only the methane portion of the model is active. The figure compares the longitudinal profiles for the stations listed in Fig 11.12, for the two surveys. The May SODs are matched quite nicely while the August SODs are lower than computed.

The methane fluxes in May are generally reproduced. They exhibit a pattern of declining fluxes from stations 1 to 14, then an increase at stations 19 to 21, and finally a decrease at station 24. For August the same pattern is present, and the model captures it quite nicely.

11.10 CONCLUSIONS

The coupled sulfide-methane model appears to be suitable for application to both fresh and marine waters. The overlying sulfate concentration determines whether sulfide or methane formation is the route that carbon diagenesis follows. The comparisons in this chapter, as well as in Chapter 10 for the methane model and Chapter 9 for sulfide oxidation – together with the results of subsequent Chapters 14 to 16 – indicate that this formulation can be used with some confidence to predict fluxes of methane, sulfide, and oxygen consumption.

Appendix 11A: MACSYMA Solution for the Three-Layer Equations

The MACSYMA solution for Eqs. (11.7) is presented in Fig. 11A.1. Lines c2–c4 define the equations and line c5 solves the resulting equation. Line c6 expands the solution to reveal its structure. Note for c_3 that r_3 is multiplied by the sum of reciprocals (see d6). The equivalent mass transfer coefficient is defined in line c7. In order to make the substitution, the most reliable technique is to solve the equation for one of the parameters and substitute for it. Thus, solving for k_{L12} produces the equation s3 in line c8. Evaluating c_3 which is s2[3], using ev(s2[3],s3)in line c9, simplifying using ratsimp, and expanding to obtain the sum of fractions yields the final solution.

Appendix 11B: MACSYMA Solution of the Sulfate Mass Balance Equations

The MACSYMA solution for Eqs. (11.27) is presented in Fig. 11B.1. Lines c3–c5 define the Eqs. (11.27). Lines c6–c7 eliminate so42 and so41. Line c8 solves the remaining equation for hso4 using solve(s2,hso4). The command rootscontract combines square roots in single expressions, which are then printed out. Line c11 solves for so41.

The commands that follow render the MACSYMA solutions into FORTRAN code that can be saved to a file for future use. The command FORTRAN( ) translates the argument into FORTRAN.

Appendix 11C: MACSYMA Solution of the Sulfide-Sulfate Mass Balance Equations

The MACSYMA solution for Eqs. (11.38) is presented in Figs. 11C.1 and 11C.2. Lines c3–c6 are the equations to be solved (Fig. 11C.1). Line c7 defines the set of equations eq. Lines c8–c10 progressively eliminate unknowns from the equation sets and define new sets, s0, s1, s2. Line c11 solves the remaining equation for jgas and factors it. Line c12 uses a for i:0 thru 2 do( ) loop to find the coefficients of the *i*th power of jgas in s2. ratexpand expands s2[1], the [1] selects the equations inside the [ ]. ratcoef(eq,x,i) finds the coefficient of the *i*th power of x in eq. The coefficients are listed in lines c13–c15. In line c16, rootscontract combines square roots in one expression, which is printed. Only one of the solutions s3[1] is listed

Mass Balance Equations for 3 layers

(c2)   eq1: 0=s*(c[0]-c[1])+k[l12]*(c[2]-c[1])-r[1]

(d2)
$$0 = \left(c_0 - c_1\right) s + \left(c_2 - c_1\right) k_{l12} - r_1$$

(c3)   eq2: 0=-k[l12]*(c[2]-c[1])+k[l23]*(c[3]-c[2])

(d3)
$$0 = \left(c_3 - c_2\right) k_{l23} - \left(c_2 - c_1\right) k_{l12}$$

(c4)   eq3: 0=-k[l23]*(c[3]-c[2])-r[3]

(d4)
$$0 = -\left(c_3 - c_2\right) k_{l23} - r_3$$

(c5)   s1:solve([eq1,eq2,eq3],[c[1],c[2],c[3]])[1]

(d5)
$$\left[c_1 = \frac{c_0 s - r_3 - r_1}{s}, c_2 = \frac{\left(c_0 k_{l12} - r_3\right) s + \left(-r_3 - r_1\right) k_{l12}}{k_{l12} s}, \right.$$
$$\left. c_3 = \frac{k_{l12}\left(c_0 k_{l23} - r_3\right) - r_3 k_{l23}\right) s + k_{l12}\left(-r_3 k_{l23} - r_1 k_{l23}\right)}{k_{l12} k_{l23} s} \right]$$

(c6)   s2:expand(s1)

(d6)
$$\left[c_1 = -\frac{r_3}{s} - \frac{r_1}{s} + c_0, c_2 = -\frac{r_3}{s} - \frac{r_1}{s} - \frac{r_3}{k_{l12}} + c_0, c_3 = -\frac{r_3}{s} - \frac{r_1}{s} - \frac{r_3}{k_{l23}} - \frac{r_3}{k_{l12}} + c_0 \right]$$

(c7)   sub1:k[leq]=1/(1/k[l12]+1/s+1/k[l23])

(d7)
$$k_{leq} = \frac{1}{\dfrac{1}{s} + \dfrac{1}{k_{l23}} + \dfrac{1}{k_{l12}}}$$

(c8)   s3:solve(sub1,k[l12])[1]

(d8)
$$k_{l12} = -\frac{k_{l23} k_{leq} s}{\left(k_{leq} - k_{l23}\right) s + k_{l23} k_{leq}}$$

(c9)   s4:expand(ratsimp(ev(s2[3],s3)))

(d9)
$$c_3 = -\frac{r_1}{s} - \frac{r_3}{k_{leq}} + c_0$$

Fig. 11A.1   MACSYMA solution for Eqs. (11.7).

Steady state equations

(c3)   eq1: 0=s*(so40-so41)+dd/hso4*(so42-so41)

(d3)
$$0 = \frac{dd\,(so42 - so41)}{hso4} + s\,(so40 - so41)$$

(c4)   eq2: 0=-dd/hso4*(so42-so41)-jcso4*hso4/h2

(d4)
$$0 = -\frac{dd\,(so42 - so41)}{hso4} - \frac{hso4\,jcso4}{h2}$$

(c5)   eq3: so42=so41/2

(d5)
$$so42 = \frac{so41}{2}$$

(c6)   s1:eliminate([eq1,eq2,eq3],[so42])$

(c7)   s2:eliminate(s1,[so41])$

(c8)   s3:rootscontract(solve(s2,hso4))$

(c9)   s3[1]

(d9)
$$hso4 = -\frac{\sqrt{4\,dd\,h2\,jcso4\,s^2\,so40 - 2\,dd\,h2\,jcso4\,s^2\,so41 + dd^2\,jcso4^2} + dd\,jcso4}{2\,jcso4\,s}$$

(c10)   s3[2]

(d10)
$$hso4 = \frac{\sqrt{4\,dd\,h2\,jcso4\,s^2\,so40 - 2\,dd\,h2\,jcso4\,s^2\,so41 + dd^2\,jcso4^2} - dd\,jcso4}{2\,jcso4\,s}$$

(c11)   s4:solve(s1[1],so41)[1]

(d11)
$$so41 = \frac{2\,hso4\,s\,so40 + dd\,so41}{2\,hso4\,s + 2\,dd}$$

(c13)   fortran(rootscontract(s3[1]))

```
HSO4 = -(SQRT(4*DD*H2*JCSO4*S**2*SO40-2*DD*H2*JCSO4*S**2*SO41+DD
1   **2*JCSO4**2)+DD*JCSO4)/(JCSO4*S)/2.0
```

(c14)   fortran(rootscontract(s3[2]))

```
HSO4 = (SQRT(4*DD*H2*JCSO4*S**2*SO40-2*DD*H2*JCSO4*S**2*SO41+DD*
1   *2*JCSO4**2)-DD*JCSO4)/(JCSO4*S)/2.0
```

(c15)   fortran(s4)

```
SO41 = (2*HSO4*S*SO40+DD*SO41)/(2*HSO4*S+2*DD)
```

(c16)   fortran(eq3)

```
SO42 = SO41/2.0
```

Fig. 11B.1   MACSYMA solution for Eqs. (11.27).

to conserve space (Fig. 11C.2). Line c18 solves the linear equations $[s0[1],s0[2]]$ for the unknowns $[ch41,ch42]$ which are printed out.

The commands that follow render the MACSYMA solutions into FORTRAN code that can be saved to a file for future use (Fig. 11C.2). The command FORTRAN( ) translates the argument into FORTRAN. The argument $ra0=ra[0]$ writes the literal $ra0$, since it is undefined, and the symbolic expression $ra[1]$.

Steady state equations

(c3)   eq0: k[l12ch4]=1/(h[so4]/d[d]+h[ch4]/d[d])

(d3)
$$k_{l12ch4} = \frac{1}{\dfrac{h_{so4}}{d_d} + \dfrac{h_{ch4}}{d_d}}$$

(c4)   eq1: 0=s*(ch4[0]-ch4[1])+k[l12ch4]*(ch4[sat]-ch4[1])-kch4[1]*ch4[1]

(d4)
$$0 = k_{l12ch4}\left(ch4_{sat} - ch4_1\right) + \left(ch4_0 - ch4_1\right)s - ch4_1\, kch4_1$$

(c5)   eq2: j[gas]=j[cch4]-kch4[1]*ch4[1]-s*(ch4[1]-ch4[0])

(d5)
$$j_{gas} = -\left(ch4_1 - ch4_0\right)s + j_{cch4} - ch4_1\, kch4_1$$

(c6)   eq3: h[ch4]=(h[2]-h[so4])*(1-j[gas]/j[cch4])

(d6)
$$h_{ch4} = \left(1 - \frac{j_{gas}}{j_{cch4}}\right)\left(h_2 - h_{so4}\right)$$

(c7)   eq:[eq0,eq1,eq2,eq3]$

(c8)   (s0:eliminate(eq,[k[l12ch4]]),s1:eliminate(s0,[ch4[1]]),s2:eliminate(s1,[h[ch4]]))$

(c9)   s3:facsum(solve(s2,j[gas]))$

Find the coefficients of the quadratic

(c10)   for i:0 thru 2 do(
         ra[i]:ratcoef(ratexpand(s2)[1],j[gas],i)
)$

(c11)   ra[0]

(d11)
$$\left(j_{cch4}\, d_d\, s + kch4_1\, j_{cch4}\, d_d\right)ch4_{sat} + \left(- ch4_0\, j_{cch4}\, d_d - h_2\, j_{cch4}^2\right)s - j_{cch4}^2\, d_d - kch4_1\, h_2\, j_{cch4}^2$$

(c12)   ra[1]

(d12)
$$\left(- j_{cch4}\, s - kch4_1\, j_{cch4}\right)h_{so4} + 2\, h_2\, j_{cch4}\, s + j_{cch4}\, d_d + 2\, kch4_1\, h_2\, j_{cch4}$$

(c13)   ra[2]

(d13)
$$\left(s + kch4_1\right)h_{so4} - h_2\, s - kch4_1\, h_2$$

(c14)   rootscontract(s3[1])

Fig. 11C.1   MACSYMA solution for Eqs. (11.38).

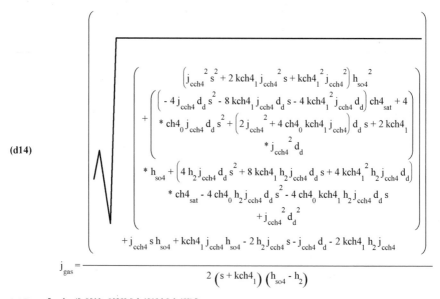

(d14)

$$j_{gas} = \frac{\left(\left(j_{cch4}^2 s^2 + 2\,kch4_1\,j_{cch4}^2\,s + kch4_1^2\,j_{cch4}^2 \right) h_{so4}^2 + \left(\begin{array}{c} \left(-4\,j_{cch4}\,d_d\,s^2 - 8\,kch4_1\,j_{cch4}\,d_d\,s - 4\,kch4_1^2\,j_{cch4}\,d_d \right) ch4_{sat} + 4 \\ *\,ch4_0\,j_{cch4}\,d_d\,s^2 + \left(2\,j_{cch4}^2 + 4\,ch4_0\,kch4_1\,j_{cch4} \right) d_d\,s + 2\,kch4_1 \\ *\,j_{cch4}^2\,d_d \end{array} \right) \right.}{2 \left(s + kch4_1 \right) \left(h_{so4} - h_2 \right)}$$

(c15)   s5:solve([s0[1],s0[2]],[ch4[1],h[ch4]])\$

(c16)   s5[1][1]

(d16)
$$ch4_1 = -\frac{-ch4_0\,s + j_{gas} - j_{cch4}}{s + kch4_1}$$

(c17)   s5[1][2]

(d17)
$$h_{ch4} = \frac{\left(j_{gas} - j_{cch4} \right) h_{so4} - h_2\,j_{gas} + h_2\,j_{cch4}}{j_{cch4}}$$

(c18)   fortran(ra0=ra[0])

```
RA0 = (J(CCH4)*D(D)*S+KCH4(1)*J(CCH4)*D(D))*CH4(SAT)+(-CH4(0)*J(
1  CCH4)*D(D)-H(2)*J(CCH4)**2)*S-J(CCH4)**2*D(D)-KCH4(1)*H(2)*J(
2  CCH4)**2
```

(c19)   fortran(ra1=ra[1])
```
RA1 = (-J(CCH4)*S-KCH4(1)*J(CCH4))*H(SO4)+2*H(2)*J(CCH4)*S+J(CCH
1  4)*D(D)+2*KCH4(1)*H(2)*J(CCH4)
```

(c20)   fortran(ra2=ra[2])
```
RA2 = (S+KCH4(1))*H(SO4)-H(2)*S-KCH4(1)*H(2)
```

Fig. 11C.2 MACSYMA solution for Eqs. (11.38).

Part IV

Time Variable Model Implementation

The models developed in the previous chapters are all steady state models. This simplifies the application considerably since constant conditions are assumed. Therefore, the fluxes can be related to either a single depositional flux or an estimate of a diagenesis flux. There are no time lags, no delayed effects, no varying storages, no hysteresis. However, as we have seen, time variable effects are significant in some cases – for example, the seasonal SOD variation – and critical in others – the transition from aerobic to anaerobic phosphate fluxes.

In principle, the modifications required to change a steady state into a time variable model are straightforward: Add a time derivative to each mass balance differential equation, account for time varying parameters due to temperature variation, and force the model with time varying inputs.

However, we also need to consider any important phenomena that have been neglected. The most important is the kinetics of diagenesis. For the steady state model applications, the diagenesis fluxes J_C or J_N are estimated, usually from the ammonia flux. For the time variable model, however, the input is the depositional flux of particulate organic matter J_{POM}. Hence the route by which J_{POM} becomes J_C, J_N, J_P, and J_{Si} – the mineralization kinetics of organic matter – must be elaborated. That is the subject of the first chapter of Part IV.

The next chapter is concerned with factors that influence the transport parameters, in particular, the influence of dissolved oxygen on bioturbation, and the varying thickness of the aerobic layer. As the layer thickens, mass from the anaerobic layer is entrained into the aerobic layer. The reverse occurs during aerobic layer thinning. The chapter ends with a presentation of the finite difference forms of the differential equations that are used for the numerical solutions in Part V.

12

Diagenesis

12.1 INTRODUCTION

The sediment flux modeling framework, diagrammed in Fig. 2.1, incorporates three processes. First, the sediment receives depositional fluxes of particulate organic carbon, nitrogen, phosphorus, and silica from the overlying water. Second, the mineralization of POM produces soluble intermediates which are quantified as diagenesis fluxes. Third, the intermediates react in the aerobic and anaerobic layers of the sediment and portions are returned to the overlying water as sediment fluxes.

The principal focus of the previous chapters has been on the last of these processes which leads to sediment fluxes. However, the computation of sediment fluxes requires that the magnitudes of the diagenesis fluxes be known. In the previous chapters, ammonia diagenesis is estimated from the ammonia flux corrected for the fraction that is nitrified. Carbon, phosphorus, and silica diagenesis fluxes are estimated using stoichiometric ratios.

In this chapter, diagenesis fluxes are explicitly computed using mass balance equations for the POM deposited to the sediment. A model for the diagenesis reaction is explicitly formulated. The source terms are the depositional fluxes of particulate C, N, and P to the sediment. The diagenesis fluxes result from the rate and extent of decay of particulate organic matter in the sediment. Since the mass balance equation and the kinetics of particulate silica mineralization have been formulated in Section 7.4, they are not considered in this chapter.

The integration of the mass balance equations for POC, PON, and POP provides the time variable diagenesis fluxes that are the inputs for the NH_4^+, NO_3^-, H_2S, and PO_4^{3-} mass balance equations. These equations are integrated to compute the sediment fluxes as functions of time. Because the model is no longer at steady state, the time variable model is capable of simulating the critical mechanism that modifies the temporal behavior of the fluxes, namely the storage and release of POM and diagenetically produced intermediates.

12.2   MASS BALANCE EQUATIONS

The mass balance equations for POC, PON, and POP include an expression for the diagenesis reaction that specifies the rate and extent of breakdown of particulate organic matter. The earliest model for this reaction employed a single first-order kinetic reaction rate (Berner, 1980a). However, it was found subsequently to be incomplete. Particulate organic matter initially mineralizes rather rapidly, but then the reaction slows down. This has been successfully modeled by assigning a fraction of the POM to various reactivity classes (Westrich and Berner, 1984). These are termed "G classes" after the symbols used to identify POM in each class.

Each class represents a portion of the organic material that reacts at a specific rate. The reaction rate for each successive class is approximately an order of magnitude smaller than the previous class. For the sediment flux model application three G classes are chosen representing three scales of reactivity: G_1, rapidly reactive (20 day half-life); G_2, more slowly reactive (1 year half-life); and G_3, which for this model, is taken to be nonreactive.

The varying reactivity of the G classes controls the time scale over which changes in depositional fluxes will be reflected in changes in diagenesis fluxes. If the reactive POM fractions were reacting rapidly, then the diagenesis flux would equal the reactive fraction of the depositional flux since there would be no time lag introduced by mineralization.

The mass balance equations for particulate organic carbon, nitrogen, and phosphorus are similar. Consider POC, and let POC_i be the concentration of POC in the ith diagenesis class ($i = 1, 2,$ or 3). The mass balance equation for POC_i in the anaerobic layer is

$$H_2 \frac{dPOC_i}{dt} = -k_{POC,i}\theta_{POC,i}^{(T-20)} POC_i H_2 - w_2 POC_i + f_{POC,i} J_{POC} \qquad (12.1)$$

where

POC_i is the concentration of POC in reactivity class i in layer 2

$k_{POC,i}$ is the first-order reaction rate coefficient

$\theta_{POC,i}$ is the temperature coefficient w_2 sedimentation velocity

J_{POC} is the depositional flux of POC from the overlying water to the sediment

$f_{POC,i}$ is the fraction of J_{POC} that is in the ith G class

Table 12.1 Diagenesis Kinetic Parameters

$k_{POC,1}$	reaction rate constant for $G_{POC,1}$	0.035	d^{-1}
$\theta_{POC,1}$	temperature coefficient for $G_{POC,1}$	1.100	–
$f_{POC,1}$	fraction in G_1	0.65	–
$k_{POC,2}$	reaction rate constant for $G_{POC,2}$	0.0018	d^{-1}
$\theta_{POC,2}$	temperature coefficient for $G_{POC,2}$	1.150	–
$f_{POC,2}$	fraction in G_2	0.20	
$k_{POC,3}$	reaction rate constant for $G_{POC,3}$	0.0	d^{-1}
$\theta_{POC,3}$	temperature coefficient for $G_{POC,3}$	–	–
$f_{POC,3}$	fraction in G_3	0.15	

The aerobic layer is not included because of its small depth relative to the anaerobic layer: $H_1 \approx 0.1$ cm relative to $H_2 \approx 10$ cm. Even if aerobic diagenesis were occurring at a more rapid rate, say ten times faster, the contribution would still be small ($\frac{1}{10}$) relative to that in the anaerobic layer.

The kinetic coefficients employed in the model are given in Table 12.1. The reaction rates and temperature coefficients for POC_1 and POC_2 are representative of values reported in the literature (see Tables 12.2 and 12.3). The G fractions are derived from the calibration as discussed below.

Once the mass balance equations for POC_1 and POC_2 are solved, the carbon diagenesis flux J_C is computed from the rate of mineralization of the two reactive G classes

$$J_C = \sum_{i=1}^{2} k_{POC,i}\theta_{POC,i}^{(T-20)} POC_i H_2 \tag{12.2}$$

The mass balance equations for particulate nitrogen and phosphorus are completely analogous

$$H_2\frac{dPON_i}{dt} = -k_{PON,i}\theta_{PON,i}^{(T-20)} PON_i H_2 - w_2 PON_i + f_{PON,i} J_{PON} \tag{12.3}$$

$$H_2\frac{dPOP_i}{dt} = -k_{POP,i}\theta_{POP,i}^{(T-20)} POP_i H_2 - w_2 POP_i + f_{POP,i} J_{POP} \tag{12.4}$$

as are the equations for the diagenesis fluxes

$$J_N = \sum_{i=1}^{2} k_{PON,i}\theta_{PON,i}^{(T-20)} PON_i H_2 \tag{12.5}$$

$$J_P = \sum_{i=1}^{2} k_{POP,i}\theta_{POP,i}^{(T-20)} POP_i H_2 \tag{12.6}$$

The reaction rates and temperature coefficients for particulate organic nitrogen and phosphorus are assumed to be identical to those listed above for particulate organic carbon. The appropriate G fractions are discussed below.

Table 12.2   Three G Model Reaction Rates K and Fractional Composition $f^{(a)}$

Aerobic Algae Decay – ($T = 20$–$22°C$)					
Medium	f_1 (%)	k_1 (d^{-1})	f_2 (%)	k_2 (d^{-1})	Reference
Marine	–	0.038	–	0.0088	(b)
Fresh	50	0.019–0.06	–	–	(c)
	–	0.030–0.07	–	0.0049	(d)
	–	0.041	–	–	(e)
Marine	50	0.066	16	0.0038	(a)
Average	50	0.046	16	0.0058	
Half-life (d)		15		120	
Anaerobic Sediment Decay – Sulfate Reduction					
Medium	f_1 (%)	k_1 (d^{-1})	f_2 (%)	k_2 (d^{-1})	Reference
Marine	–	0.024	–	0.0023	(a)
	–	0.020	–	0.0030	(a)
	–	–	–	0.0016	(f)
	–	–	–	0.0012	(g)
	–	–	–	0.0026	(h)
	–	0.027	–	0.0026	(i)
Average		0.024		0.002	
Half-life (d)		29		345	

[a]Westrich and Berner (1984). [b]Grill and Richards (1964). [c]Jewell and McCarty (1971). [d]Otsuki and Hanya (1972). [e]Fallon and Brock (1979). [f]Berner (1980b). [g]Turekian et al. (1980). [h]Billen (1982). [i]Roden and Tuttle (1995).

12.3   DIAGENESIS STOICHIOMETRY

As pointed out by Berner (1977), the ratio of the changes in depth of two constituents in pore water can be used to deduce the stoichiometry of the decaying organic matter. This can be seen by examining the solution of a one-dimensional model for organic matter decay and endproduct accumulation. Consider two reactive G classes. Each class $G_i(z)$ is assumed to be decaying following first-order kinetics. The vertical transport term represents the burial by sedimentation. For simplicity, no other particle transport terms are considered. The mass balance equation for POC is

$$w_2 \frac{d\text{POC}_i}{dz} = -k_{\text{POC},i} \text{POC}_i \tag{12.7}$$

Table 12.3 Temperature Dependence*

Reaction	ΔH kcal/mol	θ	References
Ammonia flux	22	1.138	(a)
	19	1.118	(b)
	24	1.151	(c)
Ammonia diagenesis	23	1.144	(a)
	19	1.118	(d)
	23	1.144	(e)
Sulfate reduction	21	1.131	(e)
	19	1.118	(e)
Sulfate reduction	17	1.104	(f)
	24	1.150	(g)
	16	1.098	(h)
	20	1.133	(n)
	22	1.137	(i)
	25.1	1.158	(j)
	17.9–20.1	1.111–1.125	(k)
	75–84	1.111–1.125	(d)
	18.6–26.3	1.115–1.166	(l)
Sulfate reduction	13.5	1.082	(m)
Depth < 10 cm	(8.6–16.3)	(1.052–1.100)	
Sulfate reduction	21.7	1.136	(m)
Depth > 10 cm	(16.0–31.5)	(1.098–1.203)	
G_1	16.3	1.10	This work
G_2	23.9	1.15	This work

*Klump and Martens (1983), Westrich (1983). [a] Klump (1980), Klump and Martens (1981). [b] Aller and Benninger (1981). [c] Nixon et al. (1976). [d] Aller and Yingst (1980). [e] Klump and Martens (1989). [f] Wheatland (1954). [g] Kaplan and Rittenberg (1964). [h] Nedwell and Floodgate (1972). [i] Vosjan (1974). [j] Goldhaber et al. (1977). [k] Jorgensen (1977b). [l] Abodollahi and Nedwell (1979). [m] Westrich and Berner (1988). [n] Roden and Tuttle (1995).

with the analogous equation for PON

$$w_2 \frac{d\text{PON}_i}{dz} = -k_{\text{PON},i} \text{PON}_i \tag{12.8}$$

The solutions are

$$\text{POC}_i(z) = \text{POC}_i(0) \exp\left(-\frac{k_{\text{POC},i} z}{w_2}\right) \tag{12.9}$$

and

$$\text{PON}_i(z) = \text{PON}_i(0) \exp\left(-\frac{k_{\text{PON},i} z}{w_2}\right) \tag{12.10}$$

where $POC_i(0) = J_{POC,i}/w_2$ and $PON_i(0) = J_{PON,i}/w_2$, the concentrations at $z = 0$. The depositional fluxes are $J_{POC,i}$ and $J_{PON,i}$.

The decay of PON produces ammonia and the decay of POC consumes sulfate. Thus the relationship between ammonia generation and sulfate depletion should be a measure of the nitrogen to carbon ratio of the decaying organic matter. The mass balance equation for pore water ammonia is

$$-D_{NH_4}\frac{d^2[NH_4(z)]}{dz^2} + w_2\frac{d[NH_4(z)]}{dz} = \sum_{i=1}^{2} k_{PON,i}PON_i \qquad (12.11)$$

and for pore water sulfate is

$$-D_{SO_4}\frac{d^2[SO_4(z)]}{dz^2} + w_2\frac{d[SO_4(z)]}{dz} = -a_{SO_4,C}\sum_{i=1}^{2} k_{POC,i}POC_i \qquad (12.12)$$

where D_{NH_4} and D_{SO_4} are the pore water diffusion coefficients for ammonia and sulfate respectively, and $a_{SO_4,C}$ is the stoichiometric ratio of SO_4 reduced to POC oxidized. Note the assumption that there are no other sources or sinks of ammonia or sulfate. The solutions of these equations are

$$[NH_4(z)] = [NH_4(0)] + \frac{PON_1(0)}{1 + D_{NH_4}k_{PON,1}/w_2^2}\left\{1 - \exp\left(-\frac{k_{PON,1}z}{w_2}\right)\right\}$$

$$+\frac{PON_2(0)}{1 + D_{NH_4}k_{PON,2}/w_2^2}\left\{1 - \exp\left(-\frac{k_{PON,2}z}{w_2}\right)\right\} \qquad (12.13)$$

and

$$[SO_4(z)] = [SO_4(0)]$$

$$-a_{SO_4,C}\frac{POC_1(0)}{1 + D_{SO_4}k_{POC,1}/w_2^2}\left\{1 - \exp\left(-\frac{k_{POC,1}z}{w_2}\right)\right\}$$

$$+\frac{POC_2(0)}{1 + D_{SO_4}k_{POC,2}/w_2^2}\left\{1 - \exp\left(-\frac{k_{POC,2}z}{w_2}\right)\right\} \qquad (12.14)$$

The key to evaluating the stoichiometry of the decaying POM is to find the ratio of sulfate to ammonia change: $d[SO_4(z)]/d[NH_4(z)]$. This can be found by dividing $d[SO_4(z)]/dz$ by $d[NH_4(z)]/dz$. The result is

$$\frac{d[SO_4(z)]}{d[NH_4(z)]} = -a_{SO_4,C} \times$$

$$\frac{\dfrac{POC_1(0)k_{POC,1}\exp\left(-k_{POC,1}z/w_2\right)}{w_2^2 + D_{SO_4}k_{POC,1}} + \dfrac{POC_2(0)k_{POC,2}\exp\left(-k_{POC,2}z/w_2\right)}{w_2^2 + D_{SO_4}k_{POC,2}}}{\dfrac{PON_1(0)k_{PON,1}\exp\left(-k_{PON,1}z/w_2\right)}{w_2^2 + D_{NH_4}k_{PON,1}} + \dfrac{PON_2(0)k_{PON,2}\exp\left(-k_{PON,2}z/w2\right)}{w_2^2 + D_{NH_4}k_{PON,2}}}$$

$$(12.15)$$

As it stands, this ratio does not provide a useful result. However, for most situations, the following simplification is available: $w_2^2 \ll Dk$ for each of the denominator terms. This follows from order of magnitude estimates of the various parameters presented in Table 12.4. A comparison of the entries: w_2^2, $D_{SO_4}k_{POC,1}$, and

Table 12.4 Order of Magnitude of the Diagenesis Parameters

Parameter	Value	Units	Parameter	Value	Units
w_2	10^{-5}	m/d	$k_{POC,2}$	10^{-3}	/d
D_{SO_4}	10^{-4}	m²/d	w_2^2	10^{-10}	(m/d)²
D_{NH_4}	10^{-4}	m²/d	$D_{SO_4}k_{POC,1}$	10^{-6}	(m/d)²
$K_{POC,1}$	10^{-2}	/d	$D_{SO_4}k_{POC,2}$	10^{-7}	(m/d)²

$D_{SO_4}k_{POC,2}$, demonstrates that the approximation $w_2^2 \ll Dk$ is valid for both G_1 and G_2. The sedimentation velocity used in the analysis corresponds to $w_2 = 0.4$ cm/yr, and an order of magnitude increase would not change the conclusion. Thus, Eq. (12.15) becomes

$$\frac{d[SO_4(z)]}{d[NH_4(z)]} = -a_{SO_4,C}\frac{D_{NH_4}}{D_{SO_4}}$$

$$\times \frac{POC_1(0)\exp\left(-k_{POC,1}z/w_2\right) + POC_2(0)\exp\left(-k_{POC,2}z/w_2\right)}{PON_1(0)\exp\left(-k_{PON,1}z/w_2\right) + PON_2(0)\exp\left(-k_{PON,2}z/w_2\right)} \quad (12.16)$$

Denote the ratio of carbon to nitrogen in POM by

$$a_{C,N} = \frac{POC_1(0)}{PON_1(0)} = \frac{POC_2(0)}{PON_2(0)} \quad (12.17)$$

where we assume for simplicity that G_1 and G_2 organic matter have the same C to N stoichiometry, Eq. (12.16) becomes

$$\frac{d[SO_4(z)]}{d[NH_4(z)]} = -a_{SO_4,C}\frac{D_{NH_4}}{D_{SO_4}}a_{C,N}$$

$$\times \frac{PON_1(0)\exp\left(-k_{POC,1}z/w_2\right) + PON_2(0)\exp\left(-k_{POC,2}z/w_2\right)}{PON_1(0)\exp\left(-k_{PON,1}z/w_2\right) + PON_2(0)\exp\left(-k_{PON,2}z/w_2\right)} \quad (12.18)$$

If the decay rates are the same for carbon and nitrogen, then the term in the second line equals one and

$$a_{C,N} = -a_{C,SO_4}\frac{D_{SO_4}}{D_{NH_4}}\frac{d[SO_4(z)]}{d[NH_4(z)]} \quad (12.19)$$

where $a_{C,SO_4} = a_{SO_4,C}^{-1}$.

Thus the carbon to nitrogen ratio $a_{C,N}$ can be found from an analysis of the ratio of vertical changes in sulfate and ammonia (Eq. 12.19). Similarly, $a_{C,P}$ can be found using the phosphate and sulfate pore water profiles. Since the data set to be analyzed also includes alkalinity, the relationship between sulfate consumption and alkalinity generation can also be investigated. Note that this analysis is appropriate only to the extent that ammonia, sulfate, phosphorus, and alkalinity in pore water are being influenced only by diagenesis.

12.3 A   Data Analysis

The Bricker et al. (1977) data set used subsequently consists of pore water observations at approximately 1 to 5 cm depth intervals for a number of locations in Chesapeake Bay (see Part VII). The data are processed as follows: The changes in sulfate, ammonia, phosphate, and alkalinity over a depth interval z_1 to z_2 are computed from formulas of the form

$$\Delta[SO_4(\bar{z}_{12})] = [SO_4(z_2)] - [SO_4(z_1)] \tag{12.20}$$

where $\bar{z}_{12} = 1/2(z_1 + z_2)$, the average depth. They are then multiplied by the ratio of the diffusion coefficients (Table 2.1). In addition, the sulfate to carbon stoichiometry, a_{C,SO_4}, is required. The stoichiometric equation that describes sulfate reduction (Eq. 11.1) indicates that 2 moles of CH_2O react with 1 mole of sulfate, so $a_{C,SO_4} = 2$ mol C/mol SO_4. Hence Eq. (12.19) and the analogous equations for $a_{C,P}$ and $a_{SO_4,Alk}$ become

$$a_{C,N} = -2\frac{D_{SO_4}}{D_{NH_4}}\frac{\Delta[SO_4(\bar{z}_{12})]}{\Delta[NH_4(\bar{z}_{12})]} \tag{12.21}$$

$$a_{C,P} = -2\frac{D_{SO_4}}{D_{PO_4}}\frac{\Delta[SO_4(\bar{z}_{12})]}{\Delta[PO_4(\bar{z}_{12})]} \tag{12.22}$$

$$a_{SO_4,Alk} = -\frac{D_{SO_4}}{D_{HCO_3}}\frac{\Delta[SO_4(\bar{z}_{12})]}{\Delta[Alk(\bar{z}_{12})]} \tag{12.23}$$

for concentrations in molar units. The results are estimates of the ratio of the carbon to nitrogen and phosphorus, and the ratio of sulfate consumed to alkalinity produced by diagenesis, in the depth interval $z_1 - z_2$.

The full Bricker et al. (1977) pore water data set is used. The exceptions are the measurements affected by Hurricane Agnes – the steady state assumption is questionable – and any pore water interval for which the sulfate concentration at z_1 is < 5 mM. The latter restriction is to insure that sulfate is the primary electron acceptor for the change that occurs in the interval z_1 to z_2, which is assumed in the analysis.

The stoichiometric coefficients that result from this analysis are plotted versus $\bar{z}_{12}$ in Fig. 12.1. The number of data points are nearly 100 in the 0 to 1 cm interval to approximately 10 at the lower depths. The decrease in number of points is due to the decrease of sulfate below 5 mM at the lower depths of the sediment.

The computed ratios are compared to Redfield stoichiometry

$$C_{106}H_{263}O_{110}N_{16}P_1 = (CH_2O)_{106}(NH_3)_{16}(H_3PO_4) \qquad (12.24)$$

that is, $a_{C,N}$ = 6.62 mol C/mol N = 5.68 g C/g N, and $a_{C,P}$ = 106 mol C/mol P = 41 g C/g P. The alkalinity stoichiometry can be deduced from the redox reaction for sulfate reduction (Section 1.2 B)

$$2CH_2O + SO_4^{2-} + 2H^+ \rightarrow 2CO_2 + H_2S + 2H_2O \qquad (12.25)$$

Two moles of carbon reacts with one mole of sulfate, and with two moles of H^+, liberating two moles of alkalinity in the process. The sources and sinks of alkalinity in sediments is discussed more completely in Chapter 17. For this analysis, alkalinity is assumed to have no other sources or sinks.

As shown in Fig. 12.1, the observed C/N ratios (A) and the alkalinity stoichiometry (D) are reasonably close to Redfield stoichiometry. The carbon to phosphorus stoichiometry (B) and as a consequence the nitrogen to phosphorus stoichiometry (C) is enriched in P relative to the Redfield ratio. This result is due to the assumption used in this analysis, namely that phosphate is behaving analogously to sulfate and ammonia in pore water. In particular, it is assumed that phosphate escapes from the sediment by passive diffusion to the overlying water. However, since phosphate is trapped in sediments (Chapter 6) it builds up to a larger concentration than does ammonia. This is reflected in the larger phosphate concentrations and an apparent deviation from Redfield stoichiometry.

12.4 DIAGENESIS KINETICS

The rate at which organic material mineralizes can also be determined by measuring the rate at which reactants are consumed and endproducts accumulate in a closed reaction vessel. The situation is first analyzed theoretically, and then the results are applied to a set of data from Chesapeake Bay sediments.

12.4 A Theory

Consider an experiment in which a sample of sediment is retrieved, an anaerobic incubation is started at $t = 0$, and the production of ammonia is monitored. The initial concentrations of reactive PON fractions are $PON_1(0)$ and $PON_2(0)$. G_3 is assumed not to be reacting appreciably during the experiment. The decay of each PON fraction follows first-order kinetics

$$\frac{dPON_i}{dt} = -k_{PON,i}PON_i \qquad (12.26)$$

so that

$$PON_i(t) = PON_i(0) \exp\left(-k_{PON,i}t\right) \qquad (12.27)$$

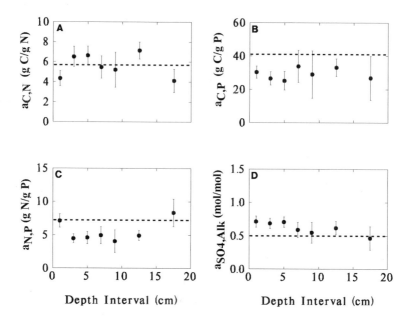

Fig. 12.1 Stoichiometric coefficients (A) $a_{C,N}$ (B) $a_{C,P}$ (C) $a_{N,P}$ (D) $a_{SO_4,Alk}$ of reactive POM deduced using pore water profiles from Chesapeake Bay. Mean $\pm$ standard error of the mean. Redfield ratio is indicated by the dashed line.

The mass balance equation for ammonia is

$$\frac{d[NH_4]}{dt} = k_{PON,1}PON_1 + k_{PON,2}PON_2 \qquad (12.28)$$

so that

$$[NH_4(t)] = [NH_4(0)] + PON_1(0)\left(1 - \exp(-k_{PON,1}t)\right)$$
$$+PON_2(0)\left(1 - \exp(-k_{PON,2}t)\right) \qquad (12.29)$$

Ammonia increases as both G_1 and G_2 mineralize at their individual rates.

It is interesting to examine which G component contributes the majority of the ammonia released. It depends on their relative amounts in the sediment. This can be approximated by solving the PON mass balance Eq. (12.3) at steady state

$$PON_i = \frac{f_{PON,i}J_{PON}}{k_{PON,i}H_2 + w_2} \qquad (12.30)$$

where a constant $T = 20^\circ C$ is assumed. Substituting this result in Eq. (12.29) and noting that $k_{PON,i} H_2 \gg w_2$ yields

$$[NH_4(t)] = [NH_4(0)] + \frac{J_{PON}}{H_2}$$

$$\left\{ f_{PON,1} \frac{\left(1 - \exp(-k_{PON,1}t)\right)}{k_{PON,1}} + f_{PON,2} \frac{\left(1 - \exp(-k_{PON,2}t)\right)}{k_{PON,2}} \right\} \qquad (12.31)$$

where $f_{PON,i}$ is the fraction of PON in component G_i.

Two points of interest emerge. Initially t is small and

$$\frac{1 - \exp\left(-k_{PON,1}t\right)}{k_{PON,1}} \approx \frac{1 - \left(1 - k_{PON,1}t + \cdots\right)}{k_{PON,1}} \approx t \qquad (12.32)$$

so that

$$[NH_4(t)] \approx [NH_4(0)] + \frac{J_{PON}}{H_2} \left(f_{PON,1}t + f_{PON,2}t\right) \qquad (12.33)$$

Thus, initially each G component contributes to the ammonia increase in proportion to its PON fraction $f_{PON,i}$ in the sediment.

This is also what is occurring when the sediments are continuously receiving depositional fluxes of PON and generating ammonia fluxes. To see this, consider the formula for ammonia diagenesis, Eq. (12.5)

$$J_N = k_{PON,1} H_2 PON_1 + k_{PON,2} H_2 PON_2 \qquad (12.34)$$

and substituting the steady state concentrations Eq. (12.30) yields

$$J_N = k_{PON,1} H_2 \frac{f_{PON,1} J_{PON}}{k_{PON,1} H_2 + w_2} + k_{PON,2} H_2 \frac{f_{PON,2} J_{PON}}{k_{PON,2} H_2 + w_2}$$

$$\approx f_{PON,1} J_{PON} + f_{PON,2} J_{PON} \qquad (12.35)$$

so that each component contributes in proportion to its fractional composition in J_{PON}.

By contrast, as $t \to \infty$, Eq. (12.31) becomes

$$[NH_4(\infty)] = [NH_4(0)] + \frac{J_{PON}}{H_2} \left\{ \frac{f_{PON,1}}{k_{PON,1}} + \frac{f_{PON,2}}{k_{PON,2}} \right\} \qquad (12.36)$$

Now the fractional contribution includes the inverse of the reaction rates of the components. Since $k_{PON,2} \ll k_{PON,1}$, the G_2 component dominates the contribution. The reason is that G_2 is the most plentiful reactive component in the sediment, and it all eventually reacts to produce ammonia.

The difference between the two extremes can be understood as follows: Initially the results are analogous to the field situation where the depositional flux continuously supplies PON to the sediment. For this case the G_1 fraction is largest and

Table 12.5   Kinetic Parameters and Sediment Components

Parameter	Units	G_1	G_2	G_3
$f_{PON,i}$	–	0.65	0.25	0.10
$k_{PON,i}$	d^{-1}	0.035	0.0018	0.0
$G_i(0)$	mg N/g	0.019	0.136	1.46
$G_i(0)$	%	1.2	8.4	90.4
Percentage contributions of the G components*				
J_N	%	72	28	0
$NH_4(\infty)$	%	12	88	0

*J_N (Eq. 12.35), $NH_4(\infty)$ (Eq. 12.36).

it accounts for the largest fraction. In a kinetics experiment, however, the depositional flux is not present and as time passes the fractions contribute in proportion to their concentrations in the sediment at the time of collection. Since G_1 decays more rapidly than G_2, there is more of the latter in the sediment at the time of collection. Hence, in the kinetic experiment, the G_2 source dominates.

12.4 B   Numerical Example

A numerical computation can clarify the situation. For a depositional flux of $J_{PON} = 50$ mg N/m^2-d, a sedimentation velocity of $w_2 = 0.25$ cm/yr, and $m_2 = 0.5$ kg/L, the sediment composition is given in Table 12.5, where $G_1(0)$ is computed using Eq. (12.30)

$$G_1(0) = \frac{J_{PON}\, f_{PON,1}}{k_{PON,1}\, H_2 m_2} = \frac{(50 \text{ mg N/m}^2\text{-d})\,(0.65)}{(0.035 \text{ d}^{-1})\,(0.1 \text{ m})\,(0.5 \text{ kg/L})} \tag{12.37}$$

The sediment is approximately 90% G_3, 10% G_2, and 1% G_1. A similar percentage (3–4%) of metabolizable carbon was estimated for the 0–10 cm depth interval from mesohaline Chesapeake Bay sediments (Roden and Tuttle, 1995). The fractional contributions to diagenesis flux and total ammonia release are given in Table 12.5. The difference between the two cases is entirely due to the lack of a depositional flux in the kinetics experiment. This does not diminish the utility of kinetic experiments; it just clarifies the analysis to which the data should be subjected.

12.4 C   Application to Chesapeake Bay Sediments

A set of sediment mineralization experiments using sediments from lower Chesapeake Bay has been performed by Burdige (1989). This section presents an analysis of the results of these experiments. For more details see Burdige (1991).

Sediments from five stations were retrieved. Three depth intervals were chosen to represent various ages of sediment organic material: 0 to 2 cm, 5 to 7 cm, and 12 to 15 cm. Samples from each depth interval were composited. A slurry was made

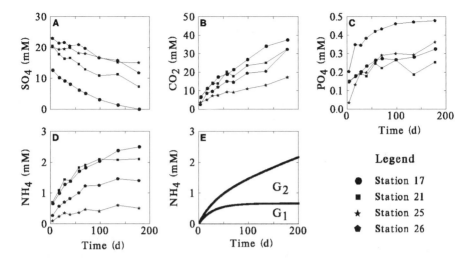

Fig. 12.2 Temporal variation of dissolved constituents in a sediment incubation experiment. (A) SO_4, (B) CO_2, (C) PO_4, (D) NH_4. Data from Burdige (1989). (E) Ammonia generated from G_1 and $G_1 + G_2$ decay.

with additional seawater and a series of 50 mL vessels were filled and incubated at 25°C. At various times during a 180-day incubation, a vessel was centrifuged and the concentrations of NH_4, CO_2, SO_4, and PO_4 in the filtrate were measured. Data from the surface layer incubations are presented in Fig. 12.2A–D for four stations. The decrease in sulfate and the increases in the other constituents are the result of POM mineralization.

Reaction Rates   With two reactive G components, it is expected that the concentrations would increase or decrease following equations of the form

$$[c(t)] = [c(0)] \pm \frac{J_{POM}}{H_2}$$

$$\left\{ f_{POM,1} \frac{\left(1 - \exp(-k_{POM,1}t)\right)}{k_{POM,1}} + f_{POM,2} \frac{\left(1 - \exp(-k_{POM,2}t)\right)}{k_{POM,2}} \right\}$$

$$(12.38)$$

where POM represents POC, PON, POP as appropriate and the minus sign applies to sulfate consumption. Fig. 12.2E presents a cumulative plot of the two terms representing the contributions of G_1 and G_2 (Eq. 12.31). The parameters used to compute this curve are listed in Table 12.5. Note the similarity in shape of the cumulative ammonia curve (E) to the data (D): an initial rapid increase, followed by a slower increase in concentration.

In principle, equation (12.38) should be fit to the data from each station and depth interval to determine the relevant parameters: $c(0)$, J_{POM}, $f_{POM,1}$, $k_{POM,1}$, $f_{POM,2}$, and $k_{POM,2}$. Unfortunately, there are an insufficient number of data points to reliably estimate this many parameters. As a consequence a simplified equation is fit to the data

$$c(t) = c(0) \pm m_2 G(0)\big(1 - \exp(-k_{POM,1}t)\big) \qquad (12.39)$$

where m_2 is the solids concentration in the slurry and $G(0)$ is the reactive organic matter on a dry weight basis. Fig. 12.3 presents the results as probability plots. The mineralization rates (I) basically span the range from G_1 to G_2 reactivity. The rates decrease from the 0–2 cm interval (A) to the 5–7 cm interval (B) and are quite low – below the G_2 mineralization rate – in the 12–15 cm interval (C). The low reactivity in the 12–15 cm depth interval supports the use of $H_2 = 10$ cm for the depth of the active layer in the applications that follow. Also the nitrogen and phosphorus mineralization rates are systematically larger than the carbon mineralization rates (D). It has been previously observed that nitrogen mineralization seems to occur relatively more rapidly than carbon mineralization (e.g., see Berner (1980a)).

In addition to the reaction rate, an estimate of the fraction of the sediment that can be mineralized $G(0)$ is made (Fig. 12.3 II). This can be compared to the expected fraction of $G_1 + G_2$ in a sediment sample, which is estimated in the previous section to be 10% – the remaining 90% is the G_3 component (Table 12.5). The comparison to the results of the experiment is made in Fig. 12.3 (II). The reactive fraction declines with depth, and it is quite small in the 12–15 cm interval (C). The composited data for the 0–2 and 5–7 cm intervals (depths < 10 cm) is also shown (II D). The median reactive fraction is on the order of 10% which confirms the model results.

Thus, although kinetic experiments of this sort cannot be used to determine the reaction rates and reactive fractions to be used in a multi-G diagenesis model, they can be used to confirm that the choices made for these parameters are not drastically contradicted by the experimental information.

Stoichiometry   One additional analysis is possible using these data. Since the various endproducts of diagenesis are measured simultaneously, it is possible to examine the stoichiometry of the decaying organic matter. This is similar to the analysis of pore water profiles presented in Section 12.3.

Since any pair of variables can be chosen, consider the relationship between sulfate reduced and carbon dioxide produced. The concentrations are given by

$$[CO_2(t)] = [CO_2(0)] + \frac{J_{POC}}{H_2}$$

$$\left\{ f_{POC,1}\frac{\big(1 - \exp(-k_{POC,1}t)\big)}{k_{POC,1}} + f_{POC,2}\frac{\big(1 - \exp(-k_{POC,2}t)\big)}{k_{POC,2}} \right\} \qquad (12.40)$$

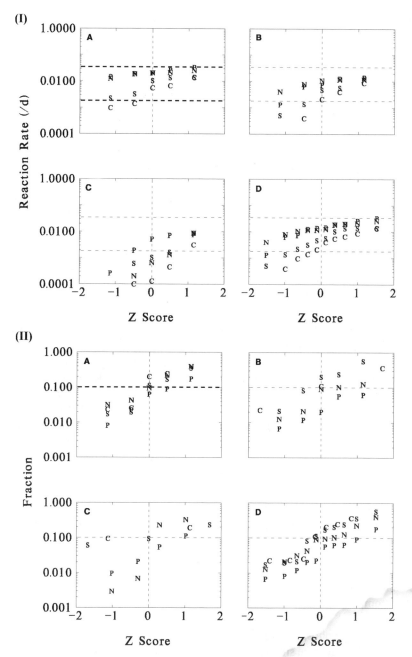

Fig. 12.3 Probability plots of the reaction rates (I) and fraction $G_1(0)/(G_1(0)+G_2(0))$ (II) for carbon (C), sulfate (S), and phosphorus (P). Organic matter from the depth intervals (A) 0–2 cm, (B) 5–7 cm, (C) 12–15 cm, (D) < 10 cm. Z score corresponds to 0, $\pm 1\sigma$, $\pm 2\sigma$, for a standard normal deviate. The horizontal dashed lines correspond to the reaction rates (I) and G_1 fractions (II) in Table 12.5.

and

$$[SO_4(t)] = [SO_4(0)] - a_{SO_4,C} \frac{J_{POC}}{H_2}$$

$$\left\{ f_{POC,1} \frac{(1 - \exp(-k_{POC,1}t))}{k_{POC,1}} + f_{POC,2} \frac{(1 - \exp(-k_{POC,2}t))}{k_{POC,2}} \right\} \qquad (12.41)$$

Since the bracketed terms are equal, Eq. (12.40) can be used to substitute for the bracketed term into Eq. (12.41)

$$[SO_4(t)] = [SO_4(0)] - a_{SO_4,C} \frac{J_{POC}}{H_2} \frac{[CO_2(t)] - [CO_2(0)]}{J_{POC}/H_2} \qquad (12.42)$$

so that

$$[SO_4(t)] = -a_{SO_4,C}[CO_2(t)] + [SO_4(0)] + a_{SO_4,C}[CO_2(0)] \qquad (12.43)$$

This suggests that a plot of $SO_4(t)$ versus $CO_2(t)$ can be used to determine the POM stoichiometry $-a_{SO_4,C}$. In order for one straight line to apply, each station is analyzed individually and the y-intercept, $[SO_4(0)] + a_{SO_4,C}[CO_2(0)]$, is found from a linear regression. This concentration is subtracted from $SO_4(t)$ so that the initial concentration is zero for each station. Fig. 12.4A presents the data from the 0–2 cm and 5–7 cm intervals. They are compared to a straight line with slope = $1/2$. Note that all the data conform to this relationship. These data justify the use of Redfield stoichiometry (Eq. 12.25) to represent the stoichiometry of sulfate reduction.

The comparisons of NH_4 and PO_4 to CO_2 produced are also presented in Fig. 12.4B and C. The analysis procedure is the same. The lines correspond to Redfield stoichiometry. The ammonia concentrations are adjusted for ammonia sorption (Burdige, 1989) since the concentration representing the total ammonia production is required. The nitrogen to carbon stoichiometry (Fig. 12.4B) is approximately Redfield as indicated by the approximate conformity with the straight lines. However, there is a clear bias: The shallow depths are comparatively nitrogen rich and the deeper depth interval are comparatively nitrogen poor. As pointed out above, this is a common observation – the nitrogen component mineralizes more rapidly than the carbon component. A more refined diagenesis model would account for this behavior explicitly.

The carbon to phosphorus stoichiometry (Fig. 12.4C) also appears to be roughly Redfield, although some significant departures are observed. However, these data are more difficult to interpret since the extent of phosphorus partitioning to solid phases during the experiment is unknown. It is not known how to correct the concentrations to reflect this and other chemical phenomena. Hence it is likely that the lack of a uniform stoichiometry is a reflection of chemical reactions that are occurring. Therefore, on balance, Redfield stoichiometry seems an acceptable approximation.

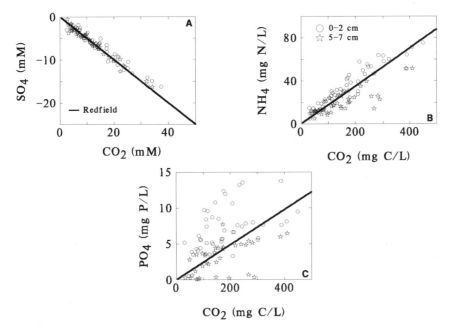

Fig. 12.4 Stoichiometry of the reactive organic matter deduced from the change in solute concentrations. (A) Carbon-sulfate. (B) Carbon-ammonia. (C) Carbon-phosphorus. The lines represent Redfield stoichiometry.

12.5 DEPOSITIONAL FLUX

The following sections present the results of the calibration of the diagenesis portion of the sediment model to the Chesapeake Bay data set. The diagenesis model receives the depositional fluxes of particulate organic carbon, nitrogen, and phosphorus, and computes the quantity of the solutes that are liberated by mineralization. These diagenesis fluxes are the source terms for the flux model equations. The diagenesis model also computes the concentrations of G_1 through G_3 carbon, nitrogen, and phosphorus, which can be compared to appropriate observations.

The object of the calibration is to estimate the magnitudes of the depositional fluxes. They are assumed to vary from station to station and from year to year. But they are assumed to be constant within the year. This rather unrealistic choice is made for two reasons. First, there is no obvious way to include a seasonal variation. Second, despite a constant depositional flux, there is a strong seasonal modulation that is imposed on the resulting diagenesis fluxes due to the temperature dependency of the diagenesis rate constants. Thus the effect of assuming a constant versus a time varying depositional flux is greatly diminished. As shown below, the seasonal variation of diagenesis is reasonably well reproduced using a constant yearly average depositional flux.

Table 12.6 Yearly Average Particulate Organic Nitrogen Depositional Fluxes J_{PON} (mg N/m^2-d)

Station	1985	1986	1987	1988
Point No Point	66.6	61.3	34.1	50.0
R-64	114.2	110.0	50.0	110.0
R-78	71.7	52.2	30.0	40.0
Still Pond	57.0	80.0	47.4	30.0
St. Leo	64.0	47.1	72.3	57.9
Buena Vista	97.5	120.0	90.0	90.0
Ragged Point	75.0	125.0	40.0	30.0
Maryland Point	82.5	81.0	77.9	60.0

The PON depositional flux J_{PON} is estimated by fitting the ammonia diagenesis flux. Estimates of ammonia diagenesis can be made from the observations of ammonia flux and the other necessary variables as follows (Eq. 3.35)

$$J_N = J[NH_4] + \left(\frac{K_{M,NH_4} \theta_{K_{M,NH_4}}^{(T-20)}}{K_{M,NH_4} \theta_{K_{M,NH_4}}^{(T-20)} + [NH_4(1)]} \right) \left(\frac{[O_2(0)]}{2K_{O_2,NH_4} + [O_2(0)]} \right)$$

$$\frac{\kappa_{NH_4,1}^2 \theta_{NH_4}^{(T-20)}}{s} [NH_4(1)] \tag{12.44}$$

Ammonia diagenesis is the sum of the ammonia flux and the quantity of ammonia that is nitrified to nitrate. The aerobic layer ammonia concentration is estimated using (Eq. 3.47)

$$[NH_4(1)] = \frac{J[NH_4]}{s} + [NH_4(0)] \tag{12.45}$$

The kinetic coefficients are listed in Tables 3.1 and 3.2. These equations can be applied pointwise to each ammonia flux observation. The result is a time series of estimates of ammonia diagenesis that can serve as the calibration data for estimating the depositional fluxes.

The ammonia diagenesis flux is computed from the diagenesis model by integrating the mass balance Eq. (12.3) and using Eq. (12.5) for J_N. The model results are compared to the estimates in Fig. 12.5. The depositional fluxes of PON are listed in Table 12.6. The estimated depositional flux at R-64 compares quite well with that estimated from sediment trap data (Roden et al., 1995) and Redfield stoichiometry ($J_{POC} = 784$ mg C/m^2-d; $J_{PON} = 138$ mg N/m^2-d). The seasonal variation of ammonia diagenesis appears to be reasonably well reproduced with low rates during the cold periods of the year and maximal rates during midyear. There appears to be no systematic problem that can be attributed to the use of yearly average depositional fluxes.

The depositional fluxes of carbon, phosphorus, and silica are established using constant stoichiometric ratios. If $J_{PON}(i, j)$ is the depositional flux of PON for sta-

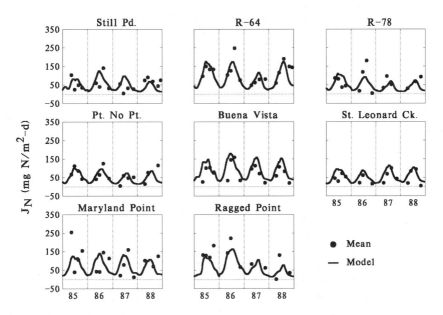

Fig. 12.5 Computed and observed temporal variation of ammonia diagenesis J_N. Stations are identified in Fig. 14.1 and listed in Table 14.1. The abscissa denotes the year.

tion i during year j, then the carbon depositional flux is given by

$$J_{POC}(i, j) = a_{C,N} J_{PON}(i, j) \qquad (12.46)$$

and equivalently for phosphorus and silica by

$$J_{POP}(i, j) = a_{C,P}^{-1} J_{POC}(i, j) \qquad (12.47)$$

$$J_{PSi}(i, j) = a_{C,Si}^{-1} J_{POC}(i, j) \qquad (12.48)$$

The depositional fluxes are apportioned into the three G classes using the stoichiometric coefficients which are listed in Table 12.7. For carbon and phosphorus, these are the Redfield stoichiometries. The carbon to silica ratio is based on a limited amount of overlying water particulate biogenic silica data. The fractions of the depositional fluxes in the G_1, G_2, and G_3 classes are listed in Table 12.7. Note that there is slightly less nitrogen in G_3 relative to carbon and phosphorus. This nitrogen-poor G_3 is necessary to reproduce the nitrogen-poor stoichiometry of the sediment organic matter as shown below.

12.6 SEDIMENT COMPOSITION

Table 12.7 Depositional Flux — Stoichiometry and G Class Fractions

$a_{C,N}$	5.68	g C/g N	
$a_{C,P}$	41.0	g C/g P	
$a_{C,Si}$	2.0	g C/g Si	
$f_{POM,i}$	G_1	G_2	G_3
Carbon	0.65	0.20	0.15
Nitrogen	0.65	0.25	0.10
Phosphorus	0.65	0.20	0.15

The most important calibration of the diagenesis model is to compare the resulting diagenesis fluxes to estimates derived from observations. This was done in the previous section. The diagenesis fluxes are critical, since they are the inputs to the mass balance equations that determine the sediment fluxes. It is essential that they be correctly specified if realistic fluxes are to be computed. This is the reason that ammonia diagenesis is used to establish the magnitude of the nitrogen depositional fluxes. There are, however, additional data which can be used for calibration, namely the sediment composition.

It is important to note that the gross sediment composition is almost entirely due to G_3 POM. The reason is that the reactive fractions have decayed to produce the diagenesis flux (Table 12.5). The comparisons are presented in Fig. 12.6 for the four main stem Chesapeake Bay stations. Remarkably, the agreement with the observed PON is almost perfect (A). However, the POC data (B) show an enrichment at the upstream stations. This may be due to an additional source of POM from the Susquehanna River which is terrigenous and relatively poor in PON relative to POC.

The phosphorus measurements are total phosphorus TP, the sum of particulate organic phosphorus POP and particulate inorganic phosphorus PIP. Both the computed TP and PIP are shown (Fig. 12.6). The agreement for TP is quite reasonable. At the upper bay station (Still Pond), a substantial fraction of TP is PIP, as shown. However, at the further downstream stations, the majority is POP. The causes of the variation in the forms of phosphorus are discussed in Section 14.6 where the results of the phosphate flux model are presented.

12.7 SEDIMENT ALGAL CARBON

The primary source of POC to the sediments of Chesapeake Bay is algal POC. Hence, the sediment should have a corresponding concentration of chlorophyll a (Chl_a). The utility of sediment Chl_a has been demonstrated using data obtained from Long Island Sound (Sun et al., 1991). The decay kinetics of Chl_a in sediments has been found to be relatively independent of temperature with a first-order decay constant of approximately 0.03 d^{-1}. It is fortuitous that this is also the average rate

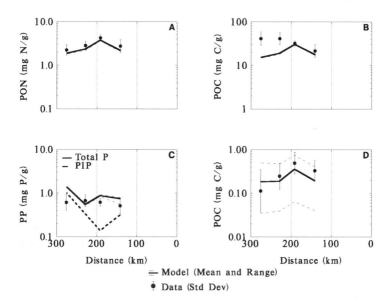

Fig. 12.6 Comparison of computed and observed sediment (A) particulate organic nitrogen [PON(2)], (B) particulate organic carbon [POC(2)], (C) total [POP(2)] and inorganic [PIP(2)] phosphorus, and (D) particulate algal carbon.

constant for the mineralization of G_1 carbon. Hence the concentration of sediment Chl$_a$ should be a direct measure of the concentration of G_1 carbon in the sediment.

The measurements that are available for Chesapeake Bay are for the top one half to one centimeter of the sediment (Part VII). The model computations are for the average concentrations in the active sediment layer of 10 cm depth. Hence it is necessary to convert the surface chlorophyll *a* measurements to depth averages. The data presented by Sun et al. (1991) for Long Island Sound can be used to compute the ratio of surface (0–1 cm) chlorophyll, Chl$_s$, to depth averaged (0–10 cm) chlorophyll, Chl$_{av}$. The results are presented in Fig. 12.7. The cosine fit to Chl$_{av}$/Chl$_s$ (Fig. 12.7B) is used to convert the Chesapeake Bay surface chlorophyll data to the 10 cm depth averages.

The resulting average particulate algal chlorophyll concentration is converted to carbon using a carbon to chlorophyll ratio of 60 mg C/mg Chl$_a$ (Table 1.1), which has been found to be representative of Chesapeake Bay plankton. The results are compared to the computed G_1 carbon concentration in Fig. 12.6D. Note that the G_1 concentration is approximately two orders of magnitude less than the total POC (Fig. 12.6), in accord with the analysis presented in Table 12.5. The magnitudes and spatial distribution are well reproduced by the model. This result provides additional support that the depositional fluxes are reasonable and that the diagenesis model appropriately describes POM mineralization.

The seasonal variation of total and algal POC are examined in Fig. 12.8. The data

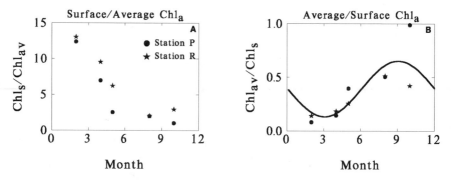

Fig. 12.7   (A) Ratio of surface to 10 cm depth average chlorophyll a Chl_{av}/Chl_s. (B) Sinusoidal fit to the annual variation of the ratio.

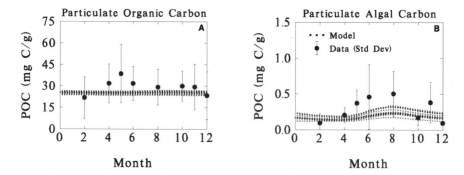

Fig. 12.8   Temporal distribution of computed and observed (A) particulate organic carbon $\sum_i [POC_{G_i}(t)]$ and (B) particulate algal carbon $[POC_{G_1}(t)]$.

are from four main Chesapeake Bay stations that have been averaged by month in order to detect seasonal variations. Total POC (Fig. 12.8A) exhibits no discernible seasonable variation and the model computes no variation. This is not unexpected, since there is no mechanism in the model by which G_3 carbon can vary. The depositional flux and the sedimentation velocity are both constants.

The comparison of the seasonal variation of algal POC and $[POC_1(t)]$ (Fig. 12.8B) reveals a systematic difference. Computed G_1 carbon peaks in the spring while the temperature is still low and then declines as the loss by diagenesis exceeds the rate of supply. Algal POC appears to exhibit a different pattern with the maximum occurring in summer. This may be a reflection of the differing temperature dependencies between G_1 carbon and algal POC. The former has a large temperature coefficient: $\theta_{POC,1} = 1.10$, while the latter is reported to be practically temperature independent (Sun et al., 1991).

Nevertheless, the extent of the agreement as exhibited in the spatial distribution (Fig. 12.6D) demonstrates the utility of sediment chlorophyll as a measure of G_1 carbon. More detailed vertical profiles can also be used to quantify the rate of particle mixing.* If the ratio of surface to depth averaged chlorophyll is large, then little particle mixing is occurring. However, if the ratio approaches unity, then the mixing is intense. Such data would be ideally suited for quantification of the particle mixing velocity.

12.8 CONCLUSIONS

The diagenesis of POM deposited to the sediment is formulated as a three G component reaction. The depositional flux is assumed to be a constant within each year, at each station. The yearly average depositional fluxes are chosen to reproduce the average ammonia diagenesis that is estimated from the observed ammonia fluxes. The comparison to the annual cycle of ammonia diagenesis indicates that the use of constant within year depositional fluxes produces acceptable results.

The stoichiometry of the deposited POM can be deduced in a number of ways. The changes in pore water concentrations reflect the composition of organic matter that has mineralized. This is also the case for the anaerobic incubation experiments. The results indicate that the assumption of Redfield stoichiometry is an adequate approximation.

The validity of the diagenesis model is examined in a number of ways. The most important is the comparison to ammonia diagenesis. However, the composition of the sediment POM is also important. The comparison is reasonable for carbon, nitrogen, and phosphorus. This basically validates the G_3 components, since they dominate the gross sediment composition. The anaerobic mineralization experiments can be used to estimate the quantity of G_2 in the sediment, since it dominates the reactive portion. The G_2 reaction rates and sediment fraction also compare favorably with the model results. The final validation is the comparison of G_1 carbon to the quantity of algal carbon in the sediment, which is estimated from the sediment chlorophyll. Again the comparison is satisfactory.

These comparisons indicate that the parameters used for the diagenesis model are realistic. They duplicate the available observations with reasonable fidelity. Therefore, the diagenesis model is used to drive the sediment flux model so that time variable fluxes can be computed.

---

*Gerino et al. (1998), Stephens et al. (1997), Sun et al. (1991, 1994).

13

Mass Transport and Numerical Methods

13.1 INTRODUCTION

The flux models presented in the previous chapters are all steady state solutions of the mass balance equations. The time variable diagenesis model is presented in Chapter 12. In this chapter, the additional information that is necessary for constructing the time variable model will be presented. The simplified steady state solutions used a lumped parameter Ω that included the mass transfer K_{L12} and particle mixing w_{12} coefficients between the aerobic and anaerobic layers. However, these parameters are explicitly required for the time variable calculations. The anaerobic layer depth H_2 is also essential. The time variation of the thickness of the aerobic layer H_1 requires that the mass balance equations be modified to account for the entrainment that occurs. Finally, the numerical methods for specifying the overlying water concentrations, the initial conditions for the sediment state variables, and the finite difference algorithm, are presented.

13.2 TRANSPORT PARAMETERS

The particulate and dissolved phase mixing coefficients between the two layers determine the rate at which solutes stored in the anaerobic layer are transferred to the aerobic layer and potentially to the overlying water. They influence the time variable model results more than they do the steady state solutions. Therefore, a more detailed description of these processes is required.

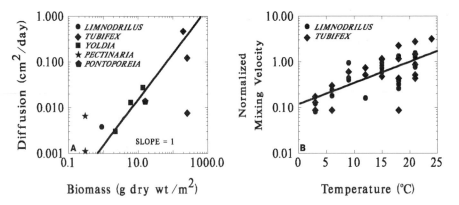

Fig. 13.1 (A) Diffusion coefficient for particle mixing versus benthic biomass. Data from Matisoff (1982). (B) Mixing velocity normalized to the mixing velocity at 20°C versus temperature. Data from McCall and Tevesz (1982).

13.2 A   Particulate Phase Mixing

The rate of mixing of sediment particles by macrobenthos (bioturbation) is quantified in this model by estimating the apparent particle diffusion coefficient D_p. It has been found that the variation appears to be proportional to the biomass of the benthos, as shown in Fig. 13.1A. In addition, it has been found that benthic biomass is correlated to the carbon input to the sediment (Maughan, 1986, Robbins et al., 1989). In order to make the sediment model self-consistent – that is, to use only internally computed variables in the parameterizations – it seems reasonable to assume that benthic biomass is proportional to the labile carbon in the sediment that is calculated by the model as POC_1. Seasonal variation of the magnitude of particle mixing has been observed (Balzer, 1996). This assumption is also made in an interesting theoretical model that predicts a constant $H_2 = 9.7$ cm (Boudreau, 1998) (Section 13.2 D). While this may not be as satisfactory as modeling the benthic biomass directly, it appears to be a reasonable first step.

The temperature dependency of particle mixing has also been investigated (Fig. 13.1B). The data sets are all normalized with respect to the mixing velocity at $T = 20$°C. The straight line corresponds to an Arrhenius temperature dependency with $\theta_{D_p} = 1.117$. Similar temperature dependency ($\theta_{D_p} = 1.107$–1.109) has been found for marine sediments (Gerino et al., 1998). Hence the particle mixing velocity w_{12} can be expressed as

$$w_{12}^* = \frac{D_p \theta_{D_p}^{(T-20)}}{H_2} \frac{POC_1}{POC_{1,R}} \tag{13.1}$$

where $POC_{1,R}$ is the reference G_1 concentration at which $w_{12}^* = D_p/H_2$ at 20°C. The superscript * is used to distinguish this formulation from the final expression for w_{12} that is developed below.

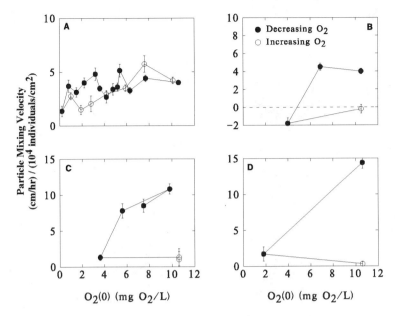

Fig. 13.2   Particle mixing velocity versus overlying water dissolved oxygen concentration. Dissolved oxygen first decreased and then increased. Rates of decreasing and increasing varied in A−D. Data from Robbins et al. (1984).

A series of experiments by Robbins et al. (1984) examined the relationship between particle mixing due to benthic organisms and the overlying water dissolved oxygen concentration. The results of four experiments are shown in Fig. 13.2, a plot of particle mixing velocity versus dissolved oxygen. The solid symbols denote the data during the initial declining phase of the dissolved oxygen, and the open symbols denote the data during the subsequent increase. There is a general dependency of mixing rate on dissolved oxygen, with the lower rates occurring at the lower dissolved oxygen concentrations. This dependency will be modeled using a Michaelis-Menton expression. Note, however, that there is a hysteresis in the results. The particle mixing rate does not return to the same magnitude when the dissolved oxygen is increased following the decrease. This behavior is addressed in the next section.

The particle mixing mass transfer coefficient that includes the temperature dependence, the benthic biomass dependence, and the Michaelis-Menton oxygen dependency is

$$w_{12}^* = \frac{D_p \theta_{D_p}^{(T-20)}}{H_2} \frac{POC_1}{POC_{1,R}} \frac{[O_2(0)]}{K_{M,D_p} + [O_2(0)]} \tag{13.2}$$

with units [L/T]. The parameter values are listed in Table 13.1. The particle diffusion coefficient is established via calibration as described subsequently in Chapter 14. The particle mixing half-saturation constant K_{M,D_p} appears to be representative

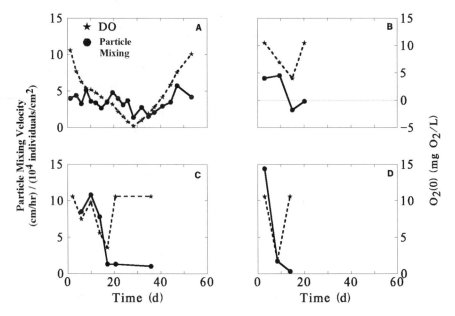

Fig. 13.3 Mixing velocity and dissolved oxygen versus time. Dissolved oxygen first decreased and then increased. Rates of decreasing and increasing varied in A−D. Data from Robbins et al. (1984).

of the data in Fig. 13.2 although it is by no means a precise representation. One difficulty is discussed next.

13.2 B Benthic Stress

In addition to the reduction in particle mixing velocity due to the instantaneous oxygen concentration, it has been found necessary to include a more lasting effect. In particular, if anoxia occurs, then the benthic faunal population is reduced or eliminated and cannot recover. This cannot be modeled using a functional relationship between particle mixing rate and dissolved oxygen. Fig. 13.3 presents the time history of the particle mixing rate and dissolved oxygen for the results analyzed in Fig. 13.2. As the dissolved oxygen declines, the particle mixing rate also declines. But as the dissolved oxygen subsequently increases, the particle mixing rate either increases more slowly, experiments A and B, or not at all, experiments C and D.

Table 13.1 Parameter values for particle mixing

D_p	Diffusion Coefficient for Particle Mixing	1.2×10^{-4}	m^2/d
θ_{D_p}	Temperature coefficient for D_p	1.117	−
$G_{POC,R}$	Reference concentration for $G_{POC,1}$	0.1	mg C/g
K_{M,D_p}	Particle mixing O_2 half-saturation constant	4.0	mg/L

The same type of behavior is exhibited by the benthic populations in Chesapeake Bay. Fig. 13.4 presents the bottom water dissolved oxygen and the mean abundance of benthic organisms at a station in the deep trough. Benthic abundance increases

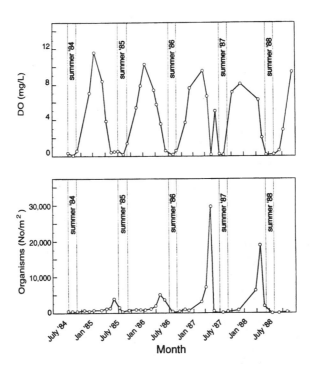

Fig. 13.4   Benthic organism density and bottom water dissolved oxygen for a deep water station in Chesapeake Bay. Data from Versar (1990).

as the summer progresses. However, the occurrence of anoxia reduces the population dramatically. After overturn, the dissolved oxygen increases, but the population does not recover. Since the particle mixing rate is proportional to the population abundance (Fig. 13.2) it presumably also does not increase in response to the increased dissolved oxygen. The same information is presented for a station near the deep trough in Fig. 13.5. The dissolved oxygen decline is not as pronounced at this station, and the population does exhibit some recovery, but not to the levels before the dissolved oxygen decline.

Benthic Stress Model   A simple model of this phenomenon can be based on the idea of modeling the stress that low dissolved oxygen imposes on the population. The model is analogous to the formulation employed in modeling the toxic effect of chemicals on organisms (Mancini, 1983). A first-order differential equation is employed that accumulates stress S when overlying water dissolved oxygen is below the particle mixing half-saturation constant for oxygen K_{M,D_p}. Stress accumulates as the oxygen concentration decreases, and is dissipated at a first-order rate with rate

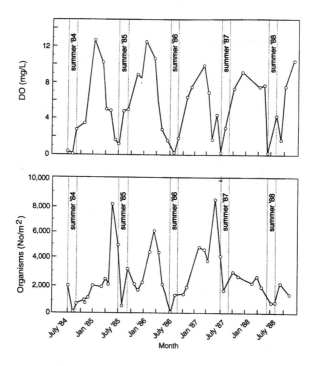

Fig. 13.5   Benthic organism density and bottom water dissolved oxygen for a station near the deep trough in Chesapeake Bay. Data from Versar (1990).

constant k_S when conditions improve. Thus

$$\frac{dS}{dt} = -k_S S + \frac{K_{M,D_p}}{K_{M,D_p} + [O_2(0)]} \qquad (13.3)$$

where

S is the accumulated benthic stress (T)

k_S is the first-order decay coefficient for accumulated stress (T^{-1})

K_{M,D_p} is the particle mixing half-saturation constant for oxygen (mg O$_2$/L)

The behavior of this formulation can be understood by evaluating the steady state stresses

$$0 = -k_S S + \frac{K_{M,D_p}}{K_{M,D_p} + [O_2(0)]} \qquad (13.4)$$

at the two oxygen extremes. For oxygen concentrations that are large relative to K_{M,D_p}, $k_S S \to 0$. However, as $[O_2(0)]$ approaches zero at the onset of anoxia then $k_S S \to 1$ (Eq. 13.4) and the term $(1 - k_S S)$ approaches zero. This suggests that

$(1 - k_S S)$ is the proper variable to quantify the degree of benthic stress. The expression is unitless and requires no additional parameter – for example, a half-saturation constant for benthic stress.

The final formulation for the particle mixing velocity, which includes the benthic stress, is

$$w_{12} = w_{12}^* \min_{\text{each year}} (1 - k_S S) \qquad (13.5)$$

where w_{12}^* is defined above (Eq. 13.2). The stress $(1 - k_S S)$ is initially close to unity, since $k_S S \to 0$ for $[O_2(0)] \gg K_{M,D_p}$. As dissolved oxygen decreases, $k_S S \to 1$, $(1 - k_S S) \to 0$ and stress increases. After the period of low oxygen has passed, the stress is continued at its minimum value through the end of the year. That is the meaning of the "min" in Eq. (13.5). This modification is made in order to conform to the observation that once the benthic population has been suppressed by low oxygen, it does not recover until the next year (Fig. 13.4).

13.2 C   Dissolved Phase Mixing

Dissolved phase mixing between layers 1 and 2 is via passive molecular diffusion which is enhanced by the mixing activities of the benthic organisms (bioirrigation). This is modeled by increasing the diffusion coefficient relative to the molecular diffusion coefficient. The mass transfer coefficient can be expressed in terms of the diffusion coefficient via (Eq. 4.48)

$$K_{L12} = \frac{D_d \theta_{D_d}^{(T-20)}}{H_2/2} \qquad (13.6)$$

where a temperature dependency has been added. Since it has been demonstrated that the pore water ammonia concentrations are primarily determined by K_{L12} (Eq. 4.43–4.47), the dissolved phase mixing can be calibrated directly (see Section 4.7). The result is $D_d = 5.0$ (cm^2/d) and $\theta_{D_p} = 1.08$. The resulting diffusion coefficient is roughly twice to three times the molecular diffusivity (Table 2.1), an indication of the importance of benthic enhancement. This degree of increase has been measured in laboratory microcosms (Matisoff and Wang, 1998). The temperature coefficient is chosen to be typical of biological reactions.

13.2 D   Active Layer Depth

The depth of the active layer H_2 does not appear as a parameter in the steady state solutions for the general sediment model (Eq. 5.18–5.19). It is implicitly included in the layer 2 reactions rates, via the definition of the reaction velocity $\kappa_2 = k_2 H_2$. However, since κ_2 is the parameter estimated from the data and used in the equations, the value of H_2 does not appear. Hence its value has no direct effect on the steady state fluxes.

However, H_2 directly influences the time variable behavior of the model. This occurs because it multiplies the time derivative of the layer 2 particulate organic matter mass balance equation (Eq. 12.1)

$$H_2 \frac{d\text{POC}_i}{dt} = -k_{\text{POC},i} \theta_{\text{POC},i}^{(T-20)} \text{POC}_i H_2 - w_2 \text{POC}_i + f_{\text{POC},i} J_{\text{POC}} \qquad (13.7)$$

and the layer 2 solute mass balance equation (Eq. 5.1b)

$$H_2 \frac{dC_{T2}}{dt} = -k_2 H_2 C_{T2} - w_{12}(f_{p2} C_{T2} - f_{p1} C_{T1})$$

$$-K_{L12}(f_{d2} C_{T2} - f_{d1} C_{T1}) + w_2(C_{T1} - C_{T2}) + J_{T2} \qquad (13.8)$$

Its importance can be understood as follows: At any instant in time, the magnitude of the product $H_2 dC_{T2}/dt$ is fixed by the magnitude of the terms on the right-hand side of the layer 2 mass balance equation. Hence, H_2 and dC_{T2}/dt are inversely related. Consider the case where H_2 is small. Then, since the product has a fixed magnitude, dC_{T2}/dt must be large. Therefore, C_{T2} changes rapidly and the model responds quickly to changes. Conversely, a large H_2 produces a smaller dC_{T2}/dt and changes occur more slowly.

The physical reason for the importance of H_2 is that it controls the volume of the anaerobic layer reservoir. The quantity of solute stored in the layer determines the time it takes for changes in inputs to be reflected in changes in stored solute. Changes in stored solute are eventually reflected in changes in fluxes. Thus the magnitude of H_2 controls the long-term response time of the sediment. This is just a restatement of the more mathematical reasoning presented in the previous paragraph. The computational consequences are examined in subsequent chapters. In particular, Chapter 16 examines the transient response of the model.

The mechanisms that determine the active layer depth are those that influence the depth to which sediment solids are mixed. These mixing mechanisms establish a homogeneous layer within which the diagenesis and other reactions take place. The principal agents of deep sediment mixing are the large benthic organisms (Figs. 1.14–1.15). Hence H_2 is chosen to represent the depth of organism mixing. Active layer depths of 5 to 15 cm have been reported for estuaries (Aller, 1982). An interesting theoretical model based on the assumption that the particle mixing diffusion coefficient is proportional to the available food $D_p \sim POC_{\text{reactive}}$ (Eq. 13.1) yields $H_2 = 9.7$ cm (Boudreau, 1998). An accompanying statistical analysis of reported values from widely varying locations – from shallow to water depths > 5000 m – yielded a mean $\pm$ standard deviation of 9.8 ± 4.5 cm. Thus, a value of 10 cm seems appropriate. Particles below this depth cannot be recycled into the active layer of the sediment. They are assumed to be permanently buried and lost from the system.

13.3 SEDIMENT SOLIDS

The most important feature of the sediment that is directly related to the solid fraction of the sediment is the rate at which solids accumulate in the sediment. Of secondary importance is the concentration of solids in the sediment which is related to the volume fraction of solids in a sediment, namely the porosity (Section 1.1 A).

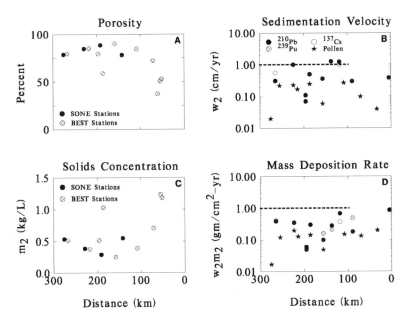

Fig. 13.6 Chesapeake Bay longitudinal plots of porosity ϕ, sedimentation velocity w_2, sediment solids concentrations m_2, and mass deposition rate $w_2 m_2$.

13.3 A Solids Sedimentation and Burial

The deposition of solids from the water column to the sediment causes an increase in the depth of the sediment relative to a fixed datum – say the depth of bedrock. Consider a layer of fixed depth measured from the sediment-water interface, for example, the active layer. As sediment solids are deposited, the new solids increase the overall depth of sediment. From the point of view of the active layer, which is a constant depth from the sediment-water interface, this layer moves upward as the sediment depth increases. The velocity at which the layer moves is termed the sedimentation velocity. The mathematical derivation is given in Section 2.3 B.

This phenomenon is important because it causes a loss of mass from the active layer. The sediment layer is moving vertically as the sediment depth increases, and the vertical motion causes sediment solids to be lost from the bottom of the active layer.

The sedimentation velocity for Chesapeake Bay has been measured using a number of methods. The results are displayed in Fig. 13.6B. There is considerable variability in the estimates. This is not unexpected since the rate at which solids are deposited can depend on site specific features. For the stand-alone calibration presented in Chapter 14, an average value of 0.25 cm/yr is selected.

13.3 B Solids Concentrations

The partitioning model (Eqs. 5.2) that is used to determine the fraction of the solutes that are in the particulate or dissolved form requires the concentration of sorbing solids. Figs. 13.6A and C present observed 0 to 10 cm average sediment porosity and solids data for the main stem of the Chesapeake Bay. A solids concentration of $m_2 = 0.5$ kg/L seems representative of the upper bay stations. Since m_1 and m_2 appear as products with the partition coefficients that are determined by calibration, it is reasonable to use sediment solids concentrations that are representative of the upper bay stations.

The product of the sedimentation velocity and the solids concentration is the mass flux of solids to the sediment. The results are shown in Fig. 13.6D. It is interesting to note that the solids' mass flux is more nearly constant along the axis of the bay. The reason is that lower sedimentation velocities are associated with higher solids concentrations.

13.4 EFFECT OF VARYING LAYER THICKNESS

The time variation introduces an additional feature besides the need to consider the derivatives in the mass balance equations (Eqs. 5.1). It is related to the variation in the thickness of the aerobic and anaerobic layers, H_1 and H_2, and the mass balance consequences.

13.4 A Entrainment Flux

The time variable form of the mass balance equations can be derived by considering the rate of change of the mass of the constituent being considered in each layer. For a concentration c_1 and a depth H_1 the product $H_1 c_1$ is the mass/unit area (e.g., mg /m^2) in layer 1. The mass balance equations equate the rate of change of the mass to the sources and sinks

$$\frac{d\,(H_1 c_1)}{dt} = S_1 \tag{13.9a}$$

$$\frac{d\,(H_2 c_2)}{dt} = S_2 \tag{13.9b}$$

where S_1 and S_2 represent all the source-sink terms. First, consider what would happen if the layer thicknesses change and there are no sources or sinks. The situation is illustrated in Fig. 13.7. Eqs. (13.9) become

$$\frac{d\,(H_1 c_1)}{dt} = 0 \tag{13.10a}$$

$$\frac{d\,(H_2 c_2)}{dt} = 0 \tag{13.10b}$$

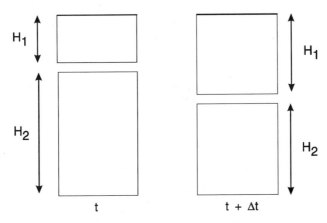

Fig. 13.7 Volume changes from t to $t + \Delta t$.

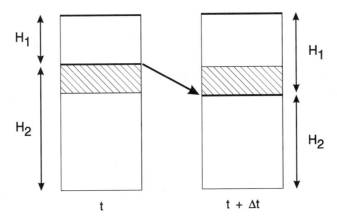

Fig. 13.8 Entrainment of layer 2 mass (shaded region) into layer 1 during the time step Δt as H_1 increases.

and the solutions are

$$H_1(t)\, c_1(t) = H_1(0)\, c_1(0) \tag{13.11a}$$

$$H_2(t)\, c_2(t) = H_2(0)\, c_2(0) \tag{13.11b}$$

That is, the mass in each layer would remain constant over the time interval Δt. Therefore, the volumes would not be connected in any way. As the volumes expand or contract, the concentrations would decrease or increase, respectively, to keep the mass constant.

However, this is not the correct result if the depth of one layer is changing at the expense of the other. In the sediment, the motion of the H_1–H_2 boundary causes an entrainment of mass into and out of the layers. This is illustrated in Fig. 13.8. Consider the sediment column at time t. Suppose that during the time Δt the H_1–H_2 boundary moves downward as illustrated. For this case H_1 is increasing and

$dH_1/dt > 0$. Then the shaded layer of sediment, which was in H_2 at time t, moves into H_1 at time $t + \Delta t$. This amounts to a loss of mass from H_2 and a gain for H_1. Thus the mass balance equations become

$$\frac{d\,(H_1 c_1)}{dt} = +c_2 \frac{dH_1}{dt} \tag{13.12a}$$

$$\frac{d\,(H_2 c_2)}{dt} = -c_2 \frac{dH_1}{dt} \tag{13.12b}$$

For the case that the boundary moves upward and $dH_1/dt < 0$ the source and sink terms are reversed

$$\frac{d\,(H_1 c_1)}{dt} = -c_1 \left| \frac{dH_1}{dt} \right| = +c_1 \frac{dH_1}{dt} \tag{13.13a}$$

$$\frac{d\,(H_2 c_2)}{dt} = +c_1 \left| \frac{dH_1}{dt} \right| = -c_1 \frac{dH_1}{dt} \tag{13.13b}$$

Mass is lost from the top layer and gained by the bottom layer. The absolute value signs are employed to emphasize that the amount of mass entrained is determined by the magnitude of dH_1/dt and not its sign.

It is possible to write Eqs. (13.13) as one set of equations by defining the functions

$$\dot{H}_1^+ = \begin{cases} \dfrac{dH_1}{dt} & \text{if } \dfrac{dH_1}{dt} \geqslant 0 \\[2ex] 0 & \text{if } \dfrac{dH_1}{dt} < 0 \end{cases} \tag{13.14a}$$

$$\dot{H}_1^- = \begin{cases} -\dfrac{dH_1}{dt} & \text{if } \dfrac{dH_1}{dt} \leqslant 0 \\[2ex] 0 & \text{if } \dfrac{dH_1}{dt} > 0 \end{cases} \tag{13.14b}$$

where the dot denotes the time derivative. A straightforward way of computing this function is to use the magnitude of the derivative so that positive and negative terms subtract as appropriate to give zero*

$$\dot{H}_1^+ = +\frac{1}{2} \left(\frac{dH_1}{dt} + \left| \frac{dH_1}{dt} \right| \right) = +\frac{1}{2} \left(\dot{H}_1 + \left| \dot{H}_1 \right| \right) \tag{13.15a}$$

$$\dot{H}_1^- = -\frac{1}{2} \left(\frac{dH_1}{dt} - \left| \frac{dH_1}{dt} \right| \right) = -\frac{1}{2} \left(\dot{H}_1 - \left| \dot{H}_1 \right| \right) \tag{13.15b}$$

---

*A straightforward computational implementation of Eqs. (13.14) involves the use of if( ) statements. For compilers that parallelize computations it is imperative that if( ) statements be avoided. The abs( ) function is faster and accomplishes the same thing.

Using these functions, the mass balance equations which include entrainment can be written as

$$\frac{d\,(H_1 c_1)}{dt} = +c_2 \dot{H}_1^+ - c_1 \dot{H}_1^-$$ (13.16a)

$$\frac{d\,(H_2 c_2)}{dt} = -c_2 \dot{H}_1^+ + c_1 \dot{H}_1^-$$ (13.16b)

and employing the chain rule for the derivatives yields

$$H_1 \frac{dc_1}{dt} + c_1 \frac{dH_1}{dt} = +c_2 \dot{H}_1^+ - c_1 \dot{H}_1^-$$ (13.17a)

$$H_2 \frac{dc_2}{dt} + c_2 \frac{dH_2}{dt} = -c_2 \dot{H}_1^+ + c_1 \dot{H}_1^-$$ (13.17b)

and finally

$$H_1 \frac{dc_1}{dt} = +c_2 \dot{H}_1^+ - c_1 \left(\dot{H}_1 + \dot{H}_1^- \right)$$ (13.18a)

$$H_2 \frac{dc_2}{dt} = -c_2 \left(\dot{H}_2 + \dot{H}_1^+ \right) + c_1 \dot{H}_1^-$$ (13.18b)

where $\dot{H}_1 = dH_1/dt$ and $\dot{H}_2 = dH_2/dt$.

13.4 B Layer Thickness

The time derivative of the aerobic layer thickness $\dot{H}_1$ is computed from the equation specifying H_1 (Eq. 3.15)

$$H_1 = \frac{D_1\,[O_2\,(0)]}{SOD}$$ (13.19)

Taking the derivative of this equation yields

$$\dot{H}_1 = \frac{dH_1}{dt} = \frac{SOD\dfrac{d\,(D_1\,[O_2\,(0)])}{dt} - (D_1\,[O_2\,(0)])\dfrac{dSOD}{dt}}{SOD^2}$$ (13.20)

$$= \frac{1}{SOD} \left(\frac{d\,(D_1\,[O_2\,(0)])}{dt} - \frac{dSOD}{dt} H_1 \right)$$ (13.21)

The variation in H_2 comes from the assumption of a constant active layer depth H_T

$$H_T = H_1 + H_2$$ (13.22)

Differentiating with respect to t yields

$$0 = \frac{dH_1}{dt} + \frac{dH_2}{dt} \tag{13.23}$$

or

$$\frac{dH_2}{dt} = -\frac{dH_1}{dt} \tag{13.24}$$

which simply states that as H_1 increases H_2 decreases, and vice versa.

13.5 NUMERICAL CONSIDERATIONS

The time variable solutions of the sediment mass balance equations (5.1a) and (5.1b) are computed using numerical integration methods. These require that the exogenous variables in the equations (e.g., the overlying water concentration and temperature) be available as smooth functions of time. In addition, the initial conditions – the values of the concentrations at the start of the integration – are required. Finally, a finite difference scheme for the differential equations is required. These three topics are discussed next.

13.5 A Boundary Conditions

In order to calibrate the sediment model in stand-alone mode, it is necessary to specify the overlying water concentrations and temperature as a function of time at each station for the period of calibration. Many choices are possible, such as interpolation formulas – see Press et al. (1989) for a discussion. We have chosen to use a four-term Fourier series. Fig. 13.9 presents an example of the result: the overlying water dissolved oxygen data and the Fourier series fit to the Chesapeake Bay stations. The approximately sinusoidal variation motivated the choice of a Fourier series as the interpolating function. The graphical displays for the other variables (Figs. 13A.1 to 13A.5) are presented in Appendix 13A, together with a more detailed discussion of the computational method employed to fit the data.

13.5 B Sediment Initial Conditions

The time variable solutions of the sediment model equations require initial conditions to start the computations. These are the concentrations at $t = 0$ of all the state variables: The particulate concentrations $G_{POC,i}(0)$, $G_{PON,i}(0)$, $G_{POP,i}(0)$, and $P_{Si}(0)$, and the total concentrations for ammonia, nitrate, sulfide, phosphate, and silica, $C_{T1}(0)$ and $C_{T2}(0)$. Strictly speaking these initial conditions should reflect the past history of the depositional fluxes and overlying water conditions for the particular application being analyzed. Since this is usually impractical owing to lack of data for these earlier years, it is necessary to adopt some other strategy to obtain initial conditions.

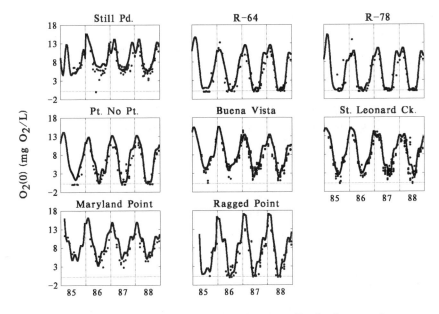

Fig. 13.9 Fourier series fit of the overlying water dissolved oxygen data.

Two possibilities are available. The first is to assign these initial conditions. However, there appears to be no unique way to establish these concentrations. Therefore, the results can be quite arbitrary and subject to a large uncertainty.

The alternate is to equilibrate the model using the first year for which inputs and overlying water data are available. The procedure is as follows. The model equations can be solved using an arbitrary set of initial conditions. In order to speed up the convergence, the first set of initial conditions is chosen to be the steady state solution for the first year's average conditions. This ensures reasonable initial conditions. The model is then integrated for one year. The final concentrations at the end of the first year are then used as the initial conditions and the equations are solved again for the first year. This procedure is repeated until the final conditions at the end of the year are equal, within a tolerance, to the initial conditions. At this point the model is at periodic steady state. The solution represents the situation that would be reached if the conditions for the initial year of the simulation had repeatedly occurred and the sediment had equilibrated to these conditions.

The utility of this method is that the initial conditions result from a well-specified requirement – that of periodic steady state – rather than a more arbitrary procedure. When changes are made in the kinetic parameters to improve the calibration, the initial conditions are recalculated with the new model coefficients. This removes the initial conditions from the parameters that require calibration. They are always set at the concentrations that produce a periodic steady state for the first year's inputs.

13.5 C   Finite Difference Equations

Choosing finite difference schemes for numerically solving differential equations is a matter of balancing accuracy and convenience of solution (Hamming, 1962). The initial choice determines the nature of the resulting equations to be solved.

Explicit and Implicit Schemes   The choice is to use an implicit, rather than an explicit, integration scheme. The difference is best understood using a simple differential equation as an example

$$\frac{dC(t)}{dt} = -kC(t) \tag{13.25}$$

An explicit finite difference approximation is

$$\frac{C(t + \Delta t) - C(t)}{\Delta t} = -kC(t) \tag{13.26a}$$

whereas an implicit approximation is

$$\frac{C(t + \Delta t) - C(t)}{\Delta t} = -kC(t + \Delta t) \tag{13.26b}$$

The time step Δt is chosen to be suitably small so the difference is a reasonable approximation to the derivative.

Note the difference in the right-hand sides of Eqs. (13.26). The explicit scheme (Eq. 13.26a) evaluates the concentration $C(t)$ at time t which is explicitly known. For example, for the first step in the integration $C(0)$ is the initial condition. The concentrations at times $t = \Delta t, 2\Delta t, \ldots$ are found by successively solving Eq. (13.26a) for $C(t + \Delta t)$

$$C(t + \Delta t) = C(t) - \Delta t k C(t) = C(t)(1 - k\Delta t) \tag{13.27a}$$

in terms of $C(t)$.

For the implicit scheme (Eq. 13.26b) the concentration is evaluated at time $t + \Delta t$, which is not known. It is implicit in Eq. (13.26b), that is, it can be found from the solution of Eq. (13.26b)

$$C(t + \Delta t) = \frac{C(t)}{1 + k\Delta t} \tag{13.27b}$$

So why bother with an implicit scheme? The answer can be seen by comparing Eqs. (13.27) for large k. The explicit scheme (Eq. 13.27a) will produce a negative concentration, which renders the solution useless. The implicit scheme (Eq. 13.27b) produces a small positive concentration that may not be accurate – and if fact is almost certainly not accurate – but is not wildly wrong, i.e., negative. When integrating sets of equations with a mixture of fast and slow reactions, implicit schemes are the method of choice. Otherwise, impracticably small Δt's are required. Also, small inaccurate concentrations can usually be tolerated, but negative concentrations stop the solution.

So why not use implicit finite difference methods as a matter of course? Each time step requires the solution of usually a set of equations which can be computationally burdensome, and for nonlinear equations, very difficult.

Application to the Sediment Flux Model   The presence of the small aerobic layer, and the surface mass transfer coefficient $s = \mathrm{SOD}/[O_2(0)]$, which can be a large number if $[O_2(0)]$ is small, can produce situations for which explicit schemes would be inappropriate. Fortunately the choice is easy in this case. This is due to the similarity of the implicit finite difference equations to the steady state equations for which a simple solution algorithm is available. Given the concentrations at t, the finite difference equations are solved for the unknown concentrations at $t + \Delta t$. Since layer 1 is quite thin, $H_1 \approx 1$ mm $= 10^{-3}$ m, and the surface mass transfer coefficient is of order $s \approx 0.1$ (m/d), the residence time in the layer is: $H_1/s = 10^{-2}$ days. Hence, it can be assumed to be at steady state without any loss of accuracy (Eq. 5.1a)

$$0 = H_1 \frac{dC_{T1}^{t+\Delta t}}{dt} = s\left(f_{d1}C_{T1}^{t+\Delta t} - C_{d0}^{t+\Delta t}\right) + w_{12}\left(f_{p2}C_{T2}^{t+\Delta t} - f_{p1}C_{T1}^{t+\Delta t}\right)$$

$$+K_{L12}\left(f_{d2}C_{T2}^{t+\Delta t} - f_{d1}C_{T1}^{t+\Delta t}\right) - w_2 C_{T1}^{t+\Delta t} - \frac{\kappa_1^2}{s}C_{T1}^{t+\Delta t} + J_{T1}^{t+\Delta t}$$

$$+C_{T2}^{t+\Delta t}\dot{H}_1^+ - C_{T1}^{t+\Delta t}\left(\dot{H}_1 + \dot{H}_1^-\right) \tag{13.28}$$

The times at which the concentrations are evaluated are denoted by superscripts t and $t + \Delta t$. The layer 2 mass balance finite difference equation (Eq. 5.1b) which is implicit in time is

$$H_2 \frac{C_{T2}^{t+\Delta t} - C_{T2}^t}{\Delta t} = -w_{12}\left(f_{p2}C_{T2}^{t+\Delta t} - f_{p1}C_{T1}^{t+\Delta t}\right)$$

$$-K_{L12}\left(f_{d2}C_{T2}^{t+\Delta t} - f_{d1}C_{T1}^{(t+\Delta t)}\right)$$

$$-\kappa_2 C_{T2}^{t+\Delta t} + w_2 C_{T1}^{(t+\Delta t)} - w_2 C_{T2}^{t+\Delta t} + J_{T2}^{t+\Delta t}$$

$$+C_{T1}^{t+\Delta t}\dot{H}_1^- - C_{T2}^{t+\Delta t}\left(\dot{H}_2 + \dot{H}_1^+\right) \tag{13.29}$$

which can be put into a form that is similar to the steady state equations

$$0 = -w_{12} \left(f_{p2} C_{T2}^{t+\Delta t} - f_{p1} C_{T1}^{t+\Delta t} \right) - K_{L12} \left(f_{d2} C_{T2}^{t+\Delta t} - f_{d1} C_{T1}^{t+\Delta t} \right)$$

$$-\kappa_2 C_{T2}^{t+\Delta t} + w_2 C_{T1}^{t+\Delta t} - w_2 C_{T2}^{t+\Delta t} - \frac{H_2 C_{T2}^{t+\Delta t}}{\Delta t} + J_{T2}^{t+\Delta t} + \frac{H_2 C_{T2}^{t}}{\Delta t}$$

$$+ C_{T1}^{t+\Delta t} \dot{H}_1^- - C_{T2}^{t+\Delta t} \left(\dot{H}_2 + \dot{H}_1^+ \right) \tag{13.30}$$

The terms corresponding to the derivatives $H_2 C_{T2}^{t+\Delta t}/\Delta t$ and $H_2 C_{T2}^t/\Delta t$, and the layer variations $C_{T1}^{t+\Delta t} \dot{H}_1^- - C_{T2}^{t+\Delta t} \left(\dot{H}_2 + \dot{H}_1^+ \right)$ simply add to the layer 2 removal rate and the forcing function, respectively. Hence the solution algorithm for these equations is the same as the steady state model. $C_{T1}^{t+\Delta t}$ and $C_{T2}^{t+\Delta t}$ are the two unknowns in the two equations (Eq. 13.28 and 13.30) which are solved at every time step.

For the sake of symmetry the diagenesis equations are also solved in implicit form

$$H_2 \frac{\text{POC}_i^{t+\Delta t} - \text{POC}_i^t}{\Delta t} = -k_{\text{POC},i} \theta_{\text{POC},i}^{(T-20)} \text{POC}_i^{t+\Delta t} H_2 - w_2 \text{POC}_i^{t+\Delta t} + J_{\text{POC},i} \tag{13.31}$$

so that

$$\text{POC}_i^{t+\Delta t} = \left(\text{POC}_i^t + \frac{\Delta t}{H_2} J_{\text{POC},i} \right) \left(1 + \frac{w_2 \Delta t}{H_2} + k_{\text{POC},i} \theta_{\text{POC},i}^{(T-20)} \Delta t \right)^{-1} \tag{13.32}$$

Similarly the particulate biogenic silica equation becomes

$$P_{\text{Si}}^{t+\Delta t} = \left(P_{\text{Si}}^t + \frac{\Delta t}{H_2} J_{\text{PSi}} \right)$$

$$\left(1 + \frac{w_2 \Delta t}{H_2} + k_{\text{Si}} \theta_{\text{Si}}^{(T-20)} \Delta t \frac{[\text{Si}]_{\text{sat}} - f_{d2}[\text{Si}(2)]^t}{P_{\text{Si}}^t + K_{\text{M,PSi}}} \right)^{-1} \tag{13.33}$$

where P_{Si} in the Michaelis-Menton term has been kept at time level t to simplify the solution. This is normal practice in using implicit finite difference methods: Keep the troublesome nonlinear terms at time level t to avoid solving nonlinear equations at each time step. Of course, an accuracy price is paid, but the simplification more than compensates.

Solution Technique   The solutions of the layer 1 and layer 2 mass balance equations still require an iterative method, since the surface mass transfer coefficient $s = \text{SOD}/[O_2(0)]$ is a function of the SOD which, in turn, is a function of the ammonia and sulfide mass balance equations. A simple back-substitution method can be used to solve the equations at each time step. The procedure is

1. Start with an initial estimate of SOD. For example: $SOD = a_{O_2,C} J_C$, or the previous time step SOD.

2. Solve layer 1 and 2 equations for ammonia, nitrate, and sulfide.

3. Compute the SOD that results: SOD = NSOD + CSOD.

4. Refine the estimate of SOD. A root finding method is used to make the new estimate (Press et al., 1989).

5. Go to (2) if no convergence.

6. Compute the phosphate and silica fluxes.

This method has been found to be quite reliable. Since it is implicit, it can be used to compute the steady state solution very easily by setting Δt to a large number. And, by comparison to an explicit scheme, it adds only a small amount of additional computation. A computer program employing this method is presented in Appendix B.

Appendix 13A: Fourier Series and the Boundary Conditions

We have chosen to use a four-term Fourier series (Kreyszig, 1972) to interpolate the overlying water concentrations. The equation is

$$C_{d0}(t) = a_0 + \sum_{k=1}^{4} a_k \sin\left(\frac{2\pi kt}{T}\right) + b_k \cos\left(\frac{2\pi kt}{T}\right) \qquad (13A.1)$$

where

$C_{d0}(t)$ is the function to be approximated

a_0 is the constant term in the function

a_k and b_k are the coefficients of the periodic terms

T is the period of the sine and cosine functions. $T = 1$ yr in this case.

Four terms have been found to be sufficient. Too few terms and the function is not flexible enough, too many terms and extraneous fluctuations are added. The data for each year are fit separately. Since the data are not sampled at regular intervals, the usual formulas for the Fourier coefficients (Hamming, 1962) are not applicable. Rather, a straightforward multiple linear regression can be used to estimate the nine coefficients: $a_0, \ldots, a_4; b_1, \ldots, b_4$. Graphical displays (Figs. 13A.1 to 13A.5) for variables other than dissolved oxygen (Fig. 13.9) demonstrate the utility of this procedure.

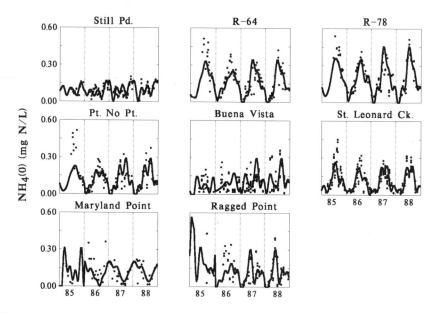

Fig. 13A.1   Fourier series fit to Chesapeake Bay overlying water ammonia concentrations.

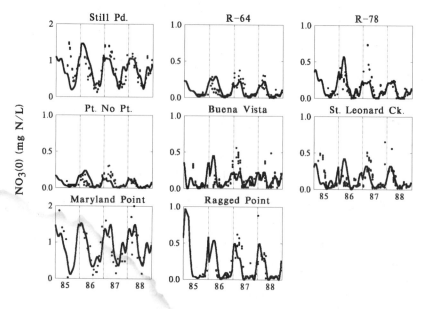

Fig. 13A.2   Fourier series fit to Chesapeake Bay overlying water nitrate concentrations.

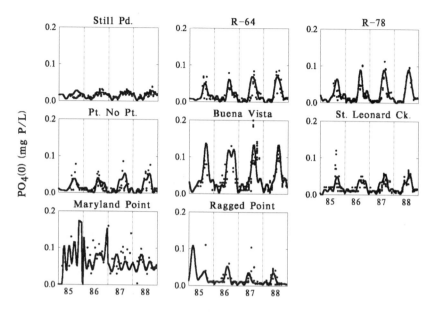

Fig. 13A.3   Fourier series fit to Chesapeake Bay overlying water phosphate concentrations.

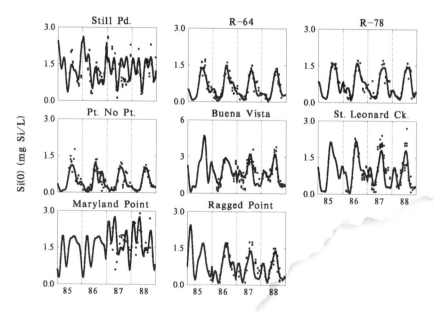

Fig. 13A.4   Fourier series fit to Chesapeake Bay overlying water silica concentrations.

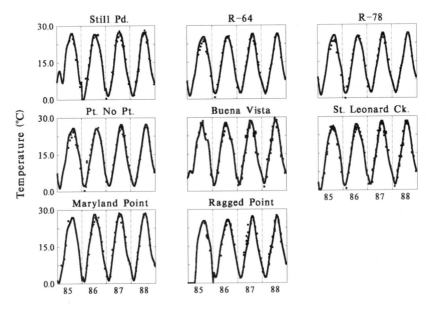

Fig. 13A.5   Fourier series fit to Chesapeake Bay overlying water temperature concentrations.

Part V

Model Calibration and Applications

The following three chapters apply the sediment flux model to a number of large, and not so large, data sets. The diagenesis kinetics and mass transport parameters are established in Part IV. Certain of the nutrient and oxygen parameters are determined in Parts II and III. The purpose of the next chapter is to finish the calibration and make a detailed comparison of the time variable model results to the Chesapeake Bay flux and sediment data. These data sets provide the primary calibration data for the final parameter estimations. The performance of the model is analyzed in detail, both visually and using various statistical comparisons.

Chapter 15 compares the model results to a large and very interesting data set from a nutrient addition experiment conducted at the Mesocosm Experimental Research Laboratory (MERL) of the University of Rhode Island, and to smaller data sets from Long Island Sound and Lake Champlain. The idea is to determine whether the model parameters can be applied either without any alterations – the best of all possible worlds – or with only selective parameters needing adjustments. The latter turns out to be the case. A complete tabulation is presented of the model parameters used in these applications, and two others where a coupled water column-sediment eutrophication model has been applied: Massachusetts Bay and Jamaica Bay. The model is remarkably robust. The only parameters that need tuning appear to be the phosphate partition coefficients, a not unexpected result considering the various chemical variables that affect phosphate partitioning.

The final chapter revisits the steady state model and compares the steady state predictions to all the data sets. A number of sensitivity calculations are presented that clarify the model's behavior. Finally the time to steady state is examined using approximate analytical solutions and numerical results. The question is: How rapidly does the model respond to changes in depositional flux?

14

Chesapeake Bay

14.1 INTRODUCTION

This chapter presents the results of the calibration of the time variable sediment flux model to the Chesapeake Bay data set. The primary calibration data are the observed sediment fluxes. However, the model also computes the organic and inorganic particulate and dissolved concentrations in the anaerobic layer. These are compared to observed pore water and particulate phase measurements. The stations are shown in Fig. 14.1 and listed in Table 14.1. There are four stations in the main bay: Still Pond is in the oligohaline portion of the bay and experiences no anoxia. R-78, R-64, and Point No Point are further down the bay and become anoxic each year (Fig. 13.9). There are two stations in the Patuxent River: Buena Vista and St. Leonard Creek, and two stations in the Potomac River: Maryland Point and Ragged Point. Both of the lower estuary stations (SL and RP) experience summer anoxia (Fig. 13.9).

The calibration of the time variable model is constrained by the interrelationships between the fluxes. These arise from the stoichiometric dependencies of the deposi-

Table 14.1 Chesapeake Bay Stations and Abbreviations

Location	Station	Symbol	Location	Station	Symbol
Main Bay	Still Pond	SP	Patuxent	Buena Vista	BV
	R-78	R78		St. Leonard Creek	SL
	R-64	R64	Potomac	Maryland Point	MP
	Point No Point	PP		Ragged Point	RP

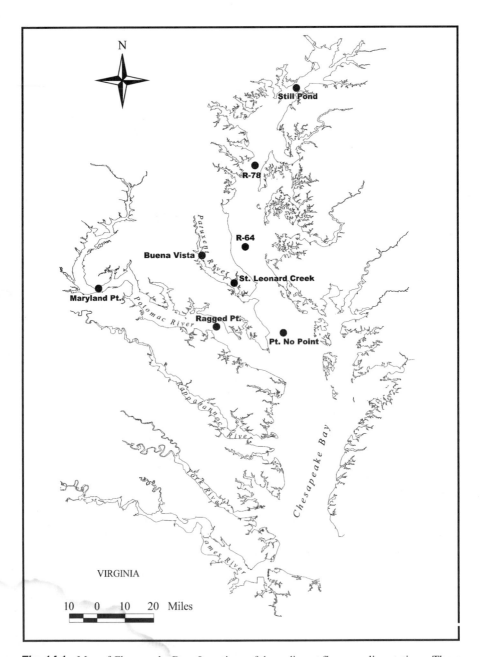

Fig. 14.1 Map of Chesapeake Bay: Locations of the sediment flux sampling stations. There are four Chesapeake Bay main stem stations: Still Pond, R-78, R-64, and Point No Point; two Patuxent Estuary stations: Buena Vista and St. Leonard Creek; and two Potomac Estuary stations: Maryland Point and Ragged Point.

tional fluxes, which supply carbon, nitrogen, phosphorus and silica in fixed proportions. In addition, the mechanisms that determine the fluxes are interdependent. The depth of the aerobic layer, which regulates the extent of all oxidation reactions, is a function of the SOD and the overlying water dissolved oxygen. The fluxes of ammonia, nitrate, sulfide, oxygen, phosphate, and silica are computed using the product of the difference in overlying water and computed aerobic layer concentrations, and the surface mass transfer coefficient s. But s itself is computed using the ratio of computed oxygen flux and the observed overlying water dissolved oxygen concentration. Hence, the model fluxes are interrelated due to their implicit dependency on s.

The calibration involves choosing model parameters that best reproduce the observations. Some of the model parameters have been established using the steady state version of the model (Parts II, III, and IV). Others are the result of fitting the time variable model fluxes to the observations as discussed below.

The model computation is performed as follows. For each station the model is equilibrated to the 1985 inputs as discussed in Section 13.5 B. Then using the equilibrated 1985 initial conditions, and the depositional fluxes listed in Table 12.6, the model equations are integrated for the four-year period: from 1985 to 1988. The resulting fluxes and concentrations are displayed in the figures that follow.

14.2 AMMONIA

In order to illuminate the inner workings of the model, plots of various concentrations and fluxes of nitrogen for station R-78 are presented in Fig. 14.2. The depositional flux of PON provides the source to the diagenesis model (Fig. 14.2D). It is assumed to be a constant for each year. The flux is apportioned to the three G classes with the appropriate stoichiometry (Table 12.7), which react at the appropriate reaction rates (Table 12.5). The concentrations for the three particulate organic nitrogen G classes that result are shown in Fig. 14.2A. Note that almost all of the PON in the sediment is the nonreactive G_3 component consistent with the previous analysis (Chapter 12). Since the decay rates for the two reactive classes are temperature dependent, the annual temperature variation (Fig. 13A.5) produces an annual variation with maxima in the spring and minima in the fall. This occurs because the mineralization reactions are slow in the early part of the year when the temperatures are low, and unreacted PON builds up. Then, during the high temperature periods, mineralization exceeds production and the PON concentration decreases.

The diagenesis flux that results is shown in Fig. 14.2B, a plot of the components of J_N due to G_1 and G_2, denoted by J_{N,G_1} and J_{N,G_2}, and J_N itself. The majority of the flux is produced by G_1, followed by G_2. The fractions are in proportion to $f_{PON,1}$ and $f_{PON,2}$, the fractions of J_{PON} that are in the two G classes (Table 12.7). The reason is that essentially all of G_1 and G_2 react away in H_2 so that the fraction buried is negligible. Hence, by mass balance each component must be converted to J_N. G_3, on the other hand, does not react. It just passes through H_2 and is buried.

The ammonia concentrations in the overlying water, aerobic, and anaerobic layers are shown in Fig. 14.2C. In addition the surface mass transfer coefficient s is

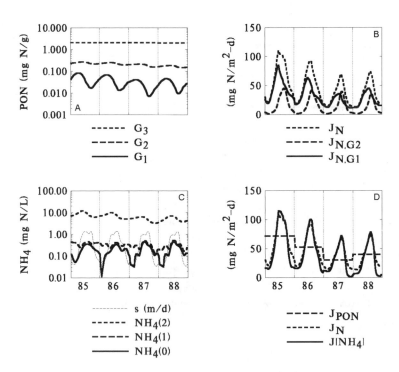

Fig. 14.2 Temporal distributions. (A) G_1, G_2, and G_3 PON concentrations. (B) Ammonia diagenesis J_N and the contributions J_{N,G_1} and J_{N,G_2} from G_1 and G_2, respectively. (C) Overlying water [$NH_4(0)$], aerobic layer [$NH_4(1)$], and anaerobic layer [$NH_4(2)$] ammonia concentrations and surface mass transfer coefficient s. (D) PON depositional flux J_{PON}, ammonia diagenesis J_N, and ammonia flux $J[NH_4]$. Station R-78. The abscissa is time in years.

plotted. It is important to realize that the magnitude of the aerobic and anaerobic layer ammonia concentrations do not determine the magnitude of the ammonia flux to the overlying water. Rather, they are determined by the magnitude of the mixing coefficients. This is clear from the steady state solutions, Eqs. (3.5) and (3.6). The magnitude of the ammonia flux is determined by the rate of production by diagenesis and the fraction that is nitrified. Therefore, it is misleading to interpret the layer concentrations as causing the flux. Note that even though the ammonia flux peaks in the summer, the gradient between the overlying water and aerobic layer concentrations is smallest. This is due to the large surface mass transfer coefficient that reduces the aerobic layer concentration. Conversely, the gradient is maximum in the winter, which corresponds to the smallest ammonia fluxes, but also to the smallest s.

A comparison of the three fluxes representing the input, J_{PON}, the result of mineralization, J_N, and the output, $J[NH_4]$, is shown in Fig. 14.2D. The depositional flux, J_{PON}, is assumed to be constant within the year, as shown. The diagenesis flux varies seasonally due to the temperature dependence of the reaction rates. The ammonia flux is also shown. During the cold periods, the ammonia flux is substantially below the diagenesis flux. The difference is being nitrified or is causing an increase in the anaerobic layer ammonia concentration. During the summer, the peak ammonia flux actually slightly exceeds the diagenesis flux. The extra ammonia is being supplied from storage in the anaerobic layer.

14.2 A   Data Comparisons

The ammonia flux data are compared to observations in Fig. 14.3. The first four

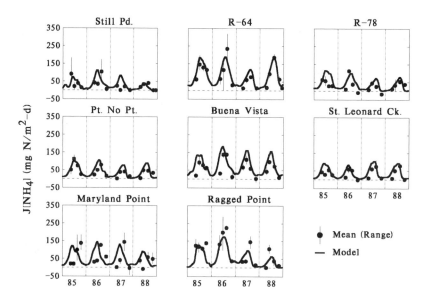

Fig. 14.3   Observed and computed temporal distribution of ammonia fluxes. The horizontal dashed line represents zero flux. The abscissa is time in years.

stations are the Chesapeake Bay main stem stations, followed by the two Patuxent estuary stations and the two Potomac estuary stations (see Fig. 14.1). The mean of the triplicated measurements is shown, together with the range denoted by the vertical line. If no line is shown, the range is smaller than the symbol. In general, the model reproduces the very small fluxes during the cold periods and the peaks in the summer. The temporal variation is due primarily to the variation in ammonia diagenesis (Fig. 14.2B), and to a lesser extent by the variation in the fraction of ammonia diagenesis that is nitrified.

The sediment nitrogen concentrations are examined in Fig. 14.4. The longitudinal profiles of particulate organic nitrogen PON and pore water ammonia concentrations for the main stem of the Chesapeake Bay are compared to the computations for the four main bay stations (A and C). The data are for the top 10 cm, corresponding to the anaerobic layer of the model. The model computations for the four years are averaged and the mean and range are presented.

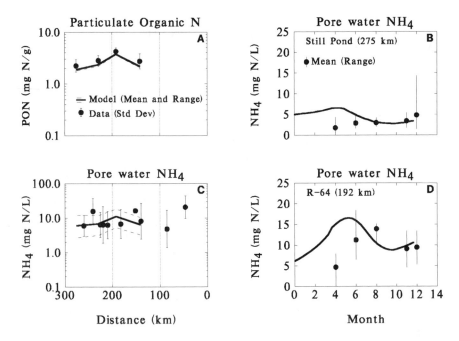

Fig. 14.4 Particulate and pore water nitrogen concentrations: Observed and computed. PON (A) and pore water ammonia (C) versus distance from the mouth of the bay. Temporal plots of pore water ammonia for Still Pond (B) and R-64 (D).

The spatial variation of increasing and then decreasing PON concentrations is reproduced by the model computations (A). This is a direct result of the variation in the depositional fluxes for the four stations. In addition, the magnitude of the PON concentrations computed by the model depends on the fraction of the depositional flux that is the G_3 component $f_{PON,3}$ and the sedimentation velocity w_2. The agreement suggests that these parameters are consistent with the observations.

The spatial variation of anaerobic layer ammonia concentration is compared to the pore water data from the Bricker et al. (1977) data set in Fig. 14.4C. As pointed out in Chapter 3, the anaerobic layer ammonia concentration is used to estimate the layer 1–2 diffusive mass transfer coefficient K_{L12}. The pore water data exhibits more variability than the computations. However, these measurements are from many stations from the mid 1970s, not just the four Chesapeake Bay stations for which flux data are available. Therefore, the comparison should be viewed more as an order of magnitude check that the diffusive exchange coefficient is reasonable. Some pore water data for the Chesapeake Bay stations are available for 1988. These are compared to the model computations in Fig. 14.4B and D. The model correctly reproduces the smaller concentrations at Still Pond (B), corresponding to a smaller depositional flux, than at R-64 (D). The temporal variation seems to be reasonably well reproduced, although there appears to be a phase difference for R-64.

Integrated Comparisons In addition to time series comparisons of the observed and modeled fluxes, there are many other possible comparisons. Four ways are presented in Fig. 14.5. Fig. 14.5A is a pointwise comparison. The different

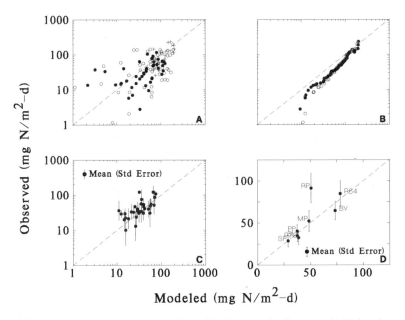

Fig. 14.5 Comparison of observed and modeled ammonia fluxes. (A) Pointwise comparison: observed versus modeled, main stem (•) and tributary (o) and stations (+) with $[O_2(0)] < 2.0$ mg O_2/L. (B) Quantile comparisons, main stem (•) and tributary (o). (C) Yearly averages: means ± standard errors of the means. (D) Station averages: means ± standard errors of the means.

symbols represent main stem and tributary stations with overlying water dissolved oxygen greater than 2 mg/L, and all samples less than 2 mg/L. There is considerable scatter in the comparison, especially at the lower ammonia fluxes. This appears to be mostly a matter of mismatches in timing between the data and the model. A more rigorous statistical analysis of the goodness of fit is presented below.

Fig. 14.5B is a comparison of the probability distributions of the data and model values. It compares the ordered set of observations to the ordered set of model predictions. The plot is constructed as follows. The model values are ordered from lowest to highest. The data are also ordered from lowest to highest. Then the ordered model values and data are plotted against each other. Thus, the lowest computed model flux is plotted against the lowest observed flux. Then the next in order are plotted against each other, and so on until the largest values are plotted. This type of plot is called a quantile plot (Wilkinson, 1990) since it compares the quantiles (the ordered values) of two samples.

The main stem data are analyzed separately from the tributaries and both data sets are plotted in Fig. 14.5B. The modeled fluxes are slightly larger, in general, than the observed fluxes. However, the range of values is well represented. This comparison indicates that, considered as whole without regard to station or time, the distribution of the main stem and tributary ammonia fluxes is reproduced by the model. Since this is quite a weak form of calibration, it is reassuring that this comparison is reasonable.

The bottom plots compare averages: yearly averages (C) and station averages (D). The model averages are computed using the model output at every 10 days. The data averages are either 4 points for the yearly averages, or 16 points for the station averages. The symbols are the means ± standard errors of the means. It is interesting to note that the yearly average comparisons seems to indicate that the observed fluxes are slightly larger than the modeled fluxes whereas the quantile plot indicates the opposite. The difference is that the yearly average model fluxes are computed using the full year computation, whereas the quantiles compare only pointwise observations and model output. The station averages (D) are identified by a two letter code identified in Table 14.1. The model correctly reproduces the station averages with one exception, Ragged Point, for which the standard error of the mean is quite large.

Parametric Variations   Ammonia fluxes vary in response to changes in other variables. The patterns of the fluxes versus temperature, overlying water dissolved oxygen, and surface mass transfer coefficients are presented in Fig. 14.6, which compares the observed (top row) and modeled (bottom row) distributions. The tempera-

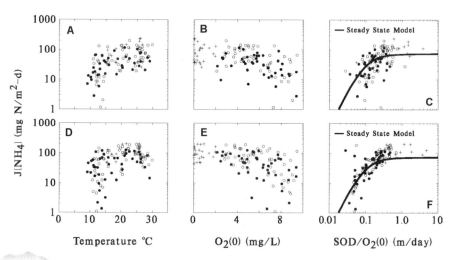

Fig. 14.6 Comparison of observed (top row) and modeled (bottom row) ammonia fluxes vs the indicated parameters, where $s = SOD/[O_2(0)]$. Main bay (•), tributaries (○), stations for which $[O_2(0)] < 2.0$ mg/L (+).

ture dependence (A and D) is expected, since the mass transport coefficients and reaction rates are all temperature dependent. Note, however, that there is considerable spread, particularly at low temperatures, and the model reproduces that behavior.

The middle column (B and E) presents the observed and modeled fluxes versus overlying water dissolved oxygen concentration. The distributions appear to be reasonably similar. Both the model and the data display a rough inverse relationship between $J[NH_4]$ and $[O_2(0)]$ but the scatter is quite large.

The right column (C and F) compares the ammonia fluxes to the surface mass transfer coefficient. The simplest steady state model (Eq. 3.36) is included as well. The time variable model results show a strong relationship to s as might be expected whereas there is more scatter in the observations. It is interesting that the time variable model results conform reasonably closely ($\pm3\times$) to the steady state model, despite the significantly more complex kinetics and the varying diagenesis fluxes.

Hysteresis Effects The ammonia flux-temperature relationship is examined more closely in Fig. 14.7. It has been pointed out (Boynton et al., 1990, Cowan and

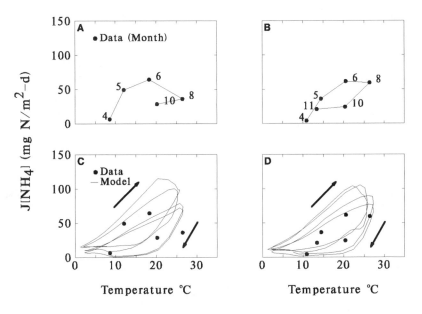

Fig. 14.7 Ammonia flux vs temperature over an annual cycle for stations R-78 (A and C) and Point No Point (B and D). Data only with months indicated (A and B). Model and data comparisons without months indicated (C and D). Arrows define the progression in time. Lines are model results for the four years simulated.

Boynton, 1996) that ammonia fluxes are not a single function of temperature but rather display hysteresis. The observed average monthly fluxes for two main stem stations are plotted versus temperature (Figs. 14.7A and B). The ammonia fluxes are generally higher in the spring months than in the fall months at the same temperature. The lines in the bottom plots (C and D) are the ammonia fluxes for the four years of model calculations. The hysteresis effect is qualitatively reproduced by the ammonia flux model. The cause is the seasonal variation of the G_1 component of PON (Fig. 14.2A). The spring concentrations are much higher than the fall concentrations corresponding to the same temperature. Therefore, ammonia diagenesis will exhibit some hysteresis, and consequently, so will ammonia flux.

Components of the Flux The flux components that make up the mass balance equations are presented in Fig. 14.8. The depositional flux J_{PON}, the loss of

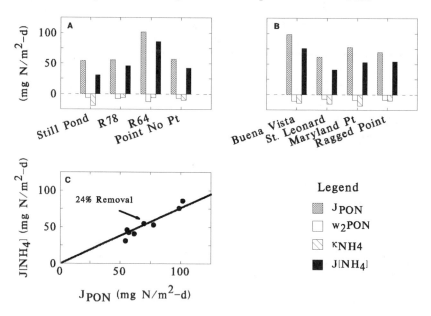

Fig. 14.8 Bar plots of the flux components for the main stem (A) and tributary (B) stations. Ammonia flux versus depositional flux (C). The line represents a 24% loss to nitrification and burial.

PON by sedimentation w_2PON, the loss via nitrification labeled as κ_{NH_4}, and the ammonia flux $J[NH_4]$ are shown for each main stem (Fig. 14.8A) and tributary (Fig. 14.8B) stations. Note that the burial flux is roughly comparable to the loss via nitrification. The influence of overlying water anoxia is also apparent. The stations that experience anoxia in the summer (R-64, R-78, PP, and RP) exhibit significantly less loss by nitrification. The overall loss of deposited nitrogen can be quantified by comparing the ammonia flux to the depositional flux (Fig. 14.8C). Approximately 24% of the depositional flux J_{PON} is lost either as PON buried or via nitrification. As shown below, very little of the nitrified ammonia is returned to the overlying water. Therefore, this component is also essentially a permanent sink of nitrogen.

14.3 NITRATE

The observed and computed time series of nitrate fluxes are shown in Fig. 14.9. Still Pond, the station nearest the head of the bay, exhibits a strong seasonal distribution of nitrate fluxes to the sediment. This is due to the large overlying water nitrate concentrations at this station ranging from 0.5 to 1.5 mg N/L (Fig. 13A.2). The other main bay stations are characterized by almost zero nitrate fluxes throughout the year, which the model reproduces. The overlying water nitrate concentrations are typically

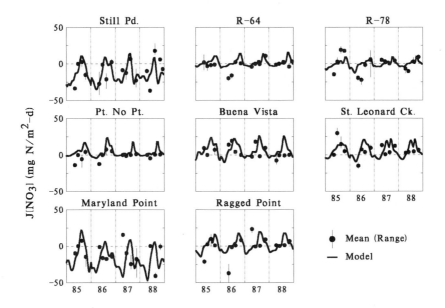

Fig. 14.9 Computed and observed nitrate fluxes.

less than 0.25 mg N/L and these stations have significant periods of anoxia in the summer. There appear to be infrequent positive or negative spikes which the model is unable to capture.

The Patuxent river stations (BV and SL, see Table 14.1) in Fig. 14.9, are computed to have nitrate fluxes that are slightly positive and the data seem to reflect that behavior. The Potomac river stations (MP and RP) are quite different. The upstream station at Maryland Point is predicted to have substantial fluxes to the sediment, due to a high overlying water nitrate concentration (0.5 to 1.5 mg N/L). The Ragged Point station is predicted to have zero flux during the period of anoxia and slightly positive fluxes in the fall. The time series of observations appear to reflect this behavior.

Integrated Comparisons The pointwise comparison (Fig. 14.10A) indicates that the model has almost no ability to predict a particular nitrate flux at a specific time and station. However the quantile comparison (Fig. 14.10B) is satisfactory, indicating that the model reproduces the observed distribution of fluxes. This suggests that the global behavior of the model is correct, but that the pointwise predictions are very noisy.

The following observations may help to explain this result: The nitrate flux is determined by the difference of two processes: the flux of overlying water nitrate into the sediment, and the flux of nitrate produced by ammonia nitrification out of the sediment. Errors in either of the fluxes are magnified because the net flux is the difference between these fluxes. Hence any individual flux prediction has a relatively large error associated with it.

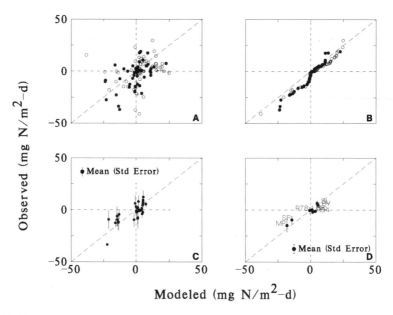

Fig. 14.10 Comparison of observed and modeled nitrate fluxes. (A) Pointwise comparison: observed versus modeled, main stem (•) and tributary (○) and stations (+) with $[O_2(0)] < 2.0$ mg O_2/L. (B) Quantile comparisons, main stem (•) and tributary (○). (C) Yearly averages: means ± standard errors of the means. (D) Station averages: means ± standard errors of the means.

The comparison of the yearly averages (Fig. 14.10C) suggests that the model can roughly reproduce the observations. However, the comparison of the station averages (Fig. 14.10D) indicates that the model can indeed capture the salient features that distinguish stations. The two stations with the largest fluxes to the sediment are distinguished from the stations with essentially zero fluxes, and from the two with slightly positive fluxes. At this level of averaging, the model is quite successful.

Parametric Variations The relationship between nitrate flux and temperature and overlying water dissolved oxygen and nitrate concentrations are examined in Fig. 14.11. Neither the model nor the data show any strong systematic pattern with respect to temperature or dissolved oxygen. The relationship with overlying water nitrate concentration is more apparent. The pattern of positive nitrate fluxes associated with small overlying water nitrate concentrations and negative fluxes associated with large overlying water nitrate concentrations is apparent in the modeled fluxes (F) and less strongly evident in the observed fluxes (C). The model also predicts that for low overlying water dissolved oxygen concentrations (the + symbol) the nitrate flux is essentially zero and almost all of the observations conform. The reason for the zero fluxes is the low overlying water nitrate concentrations at these stations and also the low overlying water dissolved oxygen concentrations that reduces ammonia nitrification and, therefore, the production of nitrate in the sediment.

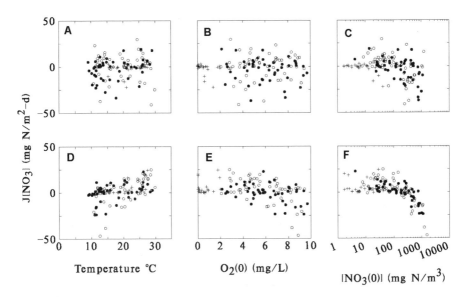

Fig. 14.11   Observed (top row) and computed (bottom row) nitrate flux vs temperature (A and D), overlying water dissolved oxygen (B and E), and nitrate concentrations (C and F). Symbol legend: • Main bay, ○ tributaries, + stations for which $[O_2(0)] < 2.0$ mg/L.

Components of the Flux   The flux components are presented in Fig. 14.12. They are: the source of nitrate due to nitrification $S[NO_3]$, the source due to surface mass transfer from the overlying water $s[NO_3(0)]$, the loss due to denitrification in both the aerobic and anaerobic layers denoted by κ_{NO_3}, the sink to surface mass transfer to the overlying water $s[NO_3(1)]$, and the net nitrate flux $J[NO_3] = s([NO_3(1)] - [NO_3(0)])$. From these results, it is possible to understand what controls the nitrate flux. The two stations with significant nitrate fluxes to the sediment (Still Pond and Maryland Point) have large inputs from the overlying water $s[NO_3(0)]$. The stations with essentially zero fluxes (R-78, R-64, Ragged Point) have an intermediate overlying water source. The remaining stations with the positive fluxes to the overlying water have small overlying water sources.

14.4   SULFIDE

The time series of observed and computed sulfide fluxes are presented in Fig. 14.13. The overlying water dissolved oxygen is also plotted for reference. Only two observations are available for main stem stations (R-64 and Point No Point) and the model computes fluxes of comparable magnitudes. The sulfide fluxes occur when the overlying water dissolved oxygen is sufficiently low to limit the oxidation of sulfide in the aerobic layer. The result is that sulfide is transferred to the overlying water by surface mass transfer.

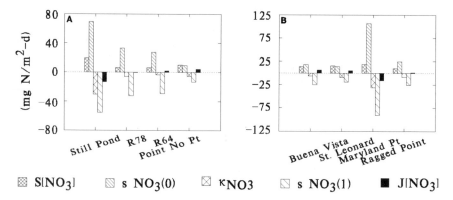

Fig. 14.12   Bar plots of the nitrate flux components for the main stem (A) and tributary (B) stations.

Fig. 14.14 compares the sediment data for organic carbon and particulate sulfide to the model computations. The particulate organic (A) and algal (C) carbon results have been discussed in Chapter 12. The comparison of the sulfide data (B) highlights the fact that the sulfur cycle in the model is not complete. The model computations are substantially in excess of the observations for acid volatile sulfide (AVS) which is a measure of iron monosulfide FeS. The model forms FeS using a partitioning equilibria. FeS is considered to be reactive and can be oxidized. This is the only reaction considered in the model. However, iron monosulfide can also react with elemental sulfur to form iron pyrite FeS_2, which is much less reactive (Morse et al., 1987). The result would be a buildup of FeS_2 in the sediment. The chromium reducible sulfide (CRS) plotted in Fig. 14.14B is a measure of both FeS and FeS_2 (Giblin and Wieder, 1992). Including the reaction for the formation of pyrite would lower the concentration of FeS computed by the model and bring it into closer agreement with the AVS observations, and at the same time allow a buildup of FeS_2, to match the observed total inorganic sulfide in the sediment.

Fig. 14.15 examines the seasonal variation of solid phase sulfide (Fig. 14.15A,B) and pore water sulfide concentrations (Fig. 14.15C,D). The model predicts almost no seasonal variation, whereas the pore water data (Fig. 14.15C) appear to indicate a seasonal variation. However, the model does capture the difference in pore water concentrations at Still Pond (Fig 14.15D) and R-64 (Fig. 14.15C).

It is apparent that a price has been paid for simplifying the sulfide cycle and using linear partitioning to determine the particulate and dissolved species. The model calculates solid phase sulfide concentrations that are between the observed FeS and FeS_2 concentrations. The fact that the FeS pool is too large prevents it from responding to the seasonal variations of the sources and sinks. As a consequence, pore water sulfide concentrations cannot vary either. Finally, while constant linear partitioning is convenient, it cannot reproduce the variation to be expected in partitioning due to, for example, the variation in iron content of the sediment. Considering these deficiencies, it is somewhat surprising that the sulfide model is at all representative.

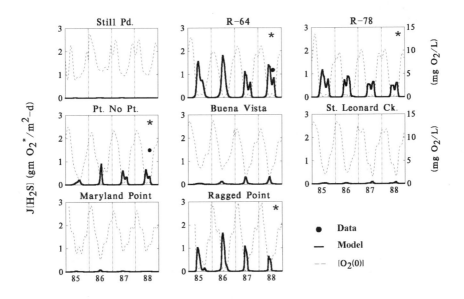

Fig. 14.13 Computed and the two observed sulfide fluxes. Lightweight dashed line is the overlying water dissolved oxygen concentration. The * denotes a station with periods of anoxia.

Components of the Flux   The flux components are presented in Fig. 14.16. They are: the depositional flux of POC in oxygen equivalents J_{POC}, the loss of POC via sedimentation $w_2[POC]$, the loss via consumption by denitrification $S[NO_3]$, the loss via oxidation of sulfide denoted by κ_{H_2S}, the loss of particulate sulfide via sedimentation $w_2[PS]$, and the sulfide flux to the overlying water $J[H_2S]$. The significant removal component is the loss of POC by burial. The burial of inorganic sulfide is small, as is the denitrification consumption. This is confirmed by the comparison by the relationship between the depositional flux J_{POC}, and the amount that is either oxidized CSOD, or escapes to the overlying water, $J[H_2S]$ (Fig. 14.16C). The results indicates that 18% of the depositional flux is not recycled as either carbonaceous SOD (via the oxidation of sulfide) or as a sulfide flux to the overlying water. This is slightly in excess of the 15% of POC that is G_3 carbon that is inert and is completely removed by burial. The remaining 3% is lost by burial of particulate sulfide. This compares quite well with an estimate of 14–21% based on comparing the depositional flux estimated using sediment trap data and the estimated burial flux (Roden et al., 1995).

14.5 OXYGEN

The time series of observed and computed oxygen fluxes are shown in Fig. 14.17. There is a different pattern of oxygen fluxes from the stations that are aerobic through-

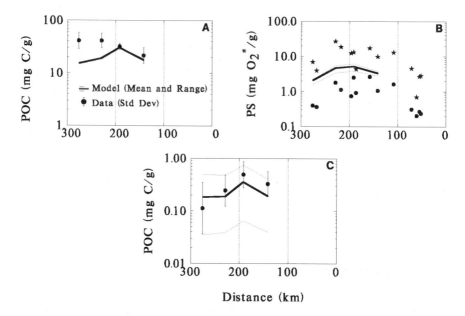

Fig. 14.14 (A) Particulate organic carbon, (B) Particulate sulfide: acid volatile sulfide AVS (•) chromium reducible sulfide CRS (★), (C) Particulate algal carbon versus distance from the bay mouth.

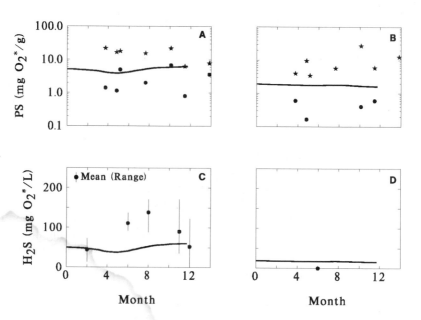

Fig. 14.15 Temporal variation of particulate PS and pore water H_2S sulfide concentrations in oxygen equivalents. Acid volatile sulfide (•) and chromium reducible sulfide CRS (★). Station R-64 (A) and (C), Still Pond (B) and (D).

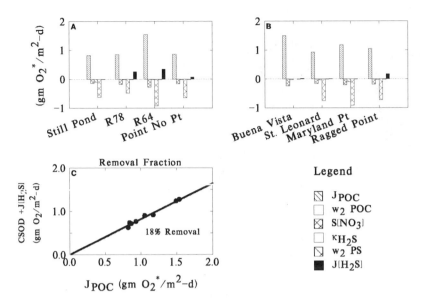

Fig. 14.16 Flux components for the sulfide flux: (A) Main bay and (B) tributary stations. The bars are, left to right for each station, J_{POC}, $w_2[POC]$, $S[NO_3]$, κ_{H_2S}, $w_2[PS]$, and $J[H_2S]$. (C) Depositional carbon flux J_{POC} vs total carbon diagenesis derived oxygen equivalent flux CSOD $+J[H_2S]$.

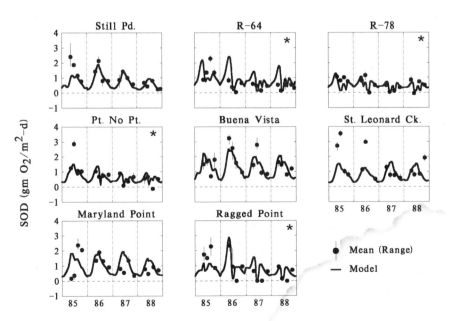

Fig. 14.17 Observed and computed oxygen fluxes to the sediment (SOD). Stations with periodic anoxia (*).

out the year and those that experience hypoxia or anoxia, which are identified with an asterisk (*). The aerobic stations, Still Pond in the main bay, and all but Ragged Point in the tributaries, exhibit a seasonal distribution that is similar to the ammonia fluxes. However, the stations that experience anoxia, the remaining main stem stations and Ragged Point, lack a strong seasonal cycle. The model reproduces this contrasting behavior reasonably well. The mechanisms involved are the lack of overlying water oxygen which prevents oxidation, directly reduces particle mixing, and produces benthic stress.

Fig. 14.18 displays the particle mixing velocity with w_{12} (Eq. 13.5), and without w_{12}^* (Eq. 13.2), the effect of benthic stress. For the anoxic stations (*), the particle mixing is strongly inhibited—compare the light and dark shaded curves. As a con-

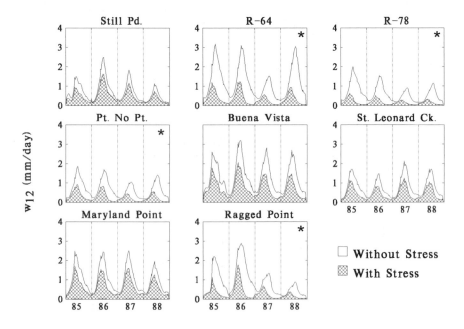

Fig. 14.18 Particle mixing velocity with and without benthic stress. Stations with periodic anoxia (*).

sequence, the particulate sulfide is not mixed into the aerobic layer where oxidation can occur. Hence, the summer peak of SOD does not occur. Rather the SOD is spread out over the year. This is a significant time variable effect that the steady state models are unable to capture (Section 9.6 D).

Integrated Comparisons Fig. 14.19 presents the pointwise, quantile, and average comparisons. The results are similar to the ammonia and nitrate fluxes. The pointwise comparison (Fig. 14.19A) is scattered, whereas the quantile distributions (Fig. 14.19B) are comparable. The yearly average comparison (Fig. 14.19C) is less scattered than the pointwise comparison. The station average comparison (Fig.

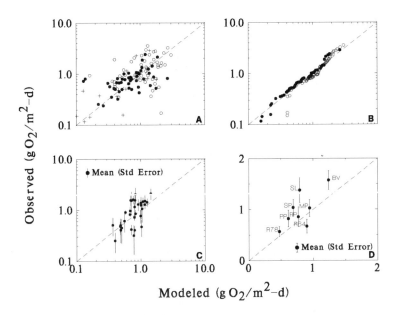

Fig. 14.19 Comparison of observed and modeled SOD. (A) Pointwise comparison: observed versus modeled, main stem (•) and tributary (○) and stations (+) with $[O_2(0)] < 2.0$ mg O_2/L. (B) Quantile comparisons, main stem (•) and tributary (○). (C) Yearly averages: means ± standard errors of the mean. (D) Station averages: means ± standard errors of the mean.

14.19D) indicates that the observations are slightly larger than the model results. This may be due to the unexplained spikes of SOD (see Fig. 14.17) that increase the average observed SOD.

Parametric Variations   Fig. 14.20 presents the relationship between SOD and temperature, overlying water dissolved oxygen and ammonia flux. Neither the data nor the model show any strong temperature dependence. There is a consistent dependency of SOD on overlying water dissolved oxygen for the lower concentrations in both the observations and the model results (middle column). It is reasonable to expect that SOD will decrease as the overlying water oxygen decreases. In the limit as $O_2(0)$ approaches zero, the SOD must also approach zero since there is no oxygen to consume.

The relationship between SOD and ammonia flux for aerobic cases is more consistent in the model results (F) than in the observations (C). However, both the model and the observations indicate that low oxygen concentration favors a lowered SOD and an increased ammonia flux (see Fig. 14.6B).

Components of the Flux   The components of sediment oxygen demand are shown in Fig. 14.21. The nitrogenous component NSOD is a small fraction of the carbonaceous (i.e., sulfide oxidation) component CSOD. The two make up the direct

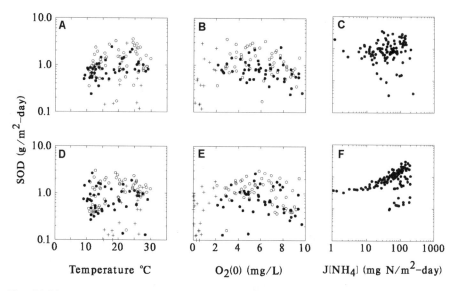

Fig. 14.20 Comparison of observed (top row) and computed (bottom row) SOD versus temperature (A and D), overlying water dissolved oxygen (Band E), and ammonia flux (C and F). Symbol legend: • Main bay, ○ tributaries, + stations for which $[O_2(0)] < 2.0$ mg/L.

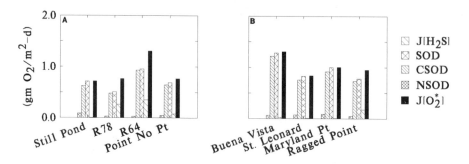

Fig. 14.21 Flux components for the SOD. (A) Main bay and (B) tributary stations.

oxygen uptake, the SOD, of the sediment. The sulfide flux $J[H_2S]$ is also shown, which for the anoxic stations can be a significant component. The sum of NSOD, CSOD, and $J[H_2S]$, is the total oxygen equivalent sediment flux $J[O_2^*]$.

14.6 PHOSPHATE

The time series of phosphate fluxes are shown in Fig. 14.22. The dramatic effect of hypoxic and anoxic conditions is apparent. Phosphate fluxes are small during aerobic conditions. However, anoxia produces dramatic increases, approaching 50 to 100 mg

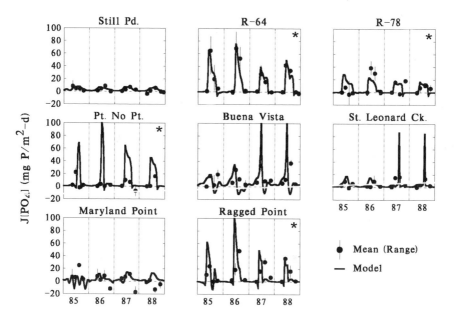

Fig. 14.22   Observed and computed phosphate flux. Stations with periodic anoxia (*).

P/m^2-d. This is nearly one half of the ammonia fluxes at that time (Fig. 14.3). Since the ratio of ammonia to phosphate production by diagenesis is 7.23 g N/g P, the excess phosphate is being released from the phosphate stored in the sediment during the aerobic periods. This is the mechanism that produces the large anoxic fluxes. By contrast, the steady state model cannot produce fluxes that exceed the diagenetic production of phosphate (Chapter 6).

The comparison to sediment phosphorus concentrations is shown in Fig. 14.23. The total particulate phosphorus TP (A), is made up of particulate organic POP and inorganic PIP phosphorus. Inorganic phosphorus comprises a large fraction of the total phosphorus at the upstream stations, but it is less further downstream. The data for PIP (B) confirm this observation. The model captures this behavior, in particular, the decline of sediment inorganic phosphorus from Still Pond to R-78 and the rest of the main stem stations. The reason is that Still Pond is an aerobic station and the phosphate flux is quite small. As a consequence, the stored phosphorus increases relative to the rest of the main stem stations which all experience anoxic periods and high phosphate fluxes.

The calculated pore water phosphate concentrations (C) are proportional to the PIP concentrations, since they are related in the model by a linear partition coefficient. A comparison of the data for PIP and pore water PO_4 indicate that this is not the case. The furthest upstream pore water concentration is lowest, whereas the PIP concentration is highest. This can be seen more directly in Fig. 14.24 which presents the seasonal distribution of solid phase (A, B) and pore water (C, D) phosphate for Still Pond and R-64 during 1988. The partition coefficient can be chosen to repre-

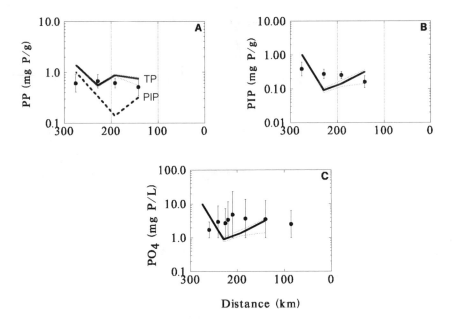

Fig. 14.23 Longitudinal profiles of (A) observed and computed total particulate phosphorus (TP), (B) particulate inorganic phosphorus (PIP), and (C) pore water phosphorus. Particulate inorganic phosphorus is also shown in (A) for comparison purposes. Solid line ± light line: model mean and range. Data: (•) mean ± standard deviation.

sent the situation at R-64 (A, C) but not simultaneously at Still Pond (B, D). This suggests that the partition coefficient is largest at the upstream station and decreases in the downstream direction.

These results indicate that the use of partition coefficients with empirical oxygen dependencies can produce realistic flux models. However, reproducing the details of the pore water and solid phase composition of the sediment requires a more sophisticated chemical calculation.

Integrated Comparisons Fig. 14.25 presents the pointwise, quantile, average comparisons. The scales employed are for an arcsinh transformation of the data (Chapter 6). The pointwise comparison (A) is much like those seen previously, very little coherence between observed and predicted fluxes. A number of cases occur where the model predicts a negative flux and the observation is positive (top left quadrant). This occurs just after turnover when the overlying water oxygen increases. The model recreates the aerobic layer immediately, with its high partition coefficient. The resulting low aerobic layer phosphate concentration causes a flux to the sediment. A more realistic formulation would involve a model of the iron cycle. The formation of iron oxyhydroxide would take place more slowly, and the aerobic layer partition coefficient would increase more slowly.

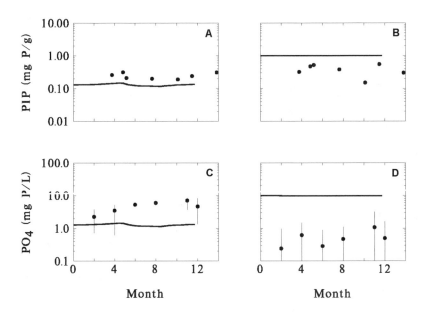

Fig. 14.24 Temporal distribution of computed (line) and observed (•) solid phase (A, B) and pore water inorganic phosphorus (C, D) concentrations for stations R-64 (A, C) and Still Pond (B, D).

The quantile comparison (B) shows a bias toward higher model fluxes in the main stem. However, the yearly averages (C), which are based on the yearly average model flux rather than the pointwise modeled fluxes, indicate a bias toward higher observed fluxes. The station averages (D) are in reasonable agreement with the observed averages, with the exception of Point No Point. It is interesting to note that for this station, the observations straddle the predicted large anoxic fluxes (see Fig. 14.22). Therefore, the observed station average is smaller than the model yearly average.

Parametric Variations   Fig. 14.26 presents the relationship of phosphate flux to overlying water dissolved oxygen, the surface mass transfer coefficient, and ammonia flux. Both the model and the observations feature large (>10 mg P/m^2-d) fluxes for low overlying water dissolved oxygen. No relationship is apparent between phosphate flux and SOD/[$O_2(0)$]. There is, however, a relationship to ammonia flux with generally increasing phosphate fluxes, with the highest fluxes associated with the periods of low dissolved oxygen. The difficulty with the negative fluxes can also be seen. They occur at intermediate ammonia fluxes, whereas they are modeled to occur at the highest ammonia fluxes.

Components of the Flux   The flux components are displayed in Fig. 14.27A, B. The depositional flux J_{POP}, burial of organic w_2POP, and inorganic w_2PIP phosphorus, and the phosphate flux $J[PO_4]$ are included. For the aerobic stations, burial

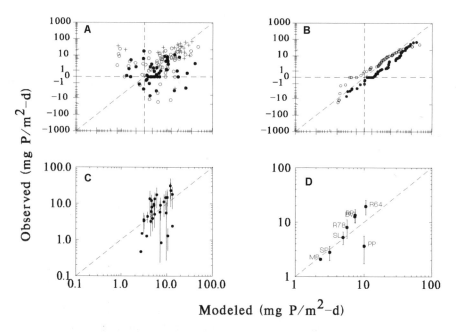

Fig. 14.25 Comparison of observed and modeled phosphate fluxes. (A) Pointwise comparison: observed versus modeled, main stem (•) and tributary (○) and stations (+) with $[O_2(0)] < 2.0$ mg O_2/L. (B) Quantile comparisons, main stem (•) and tributary (○). (C) Yearly averages: means ± standard errors of the means. (D) Station averages: means ± standard errors of the means. (A) and (B) axes use the arcsinh transformation (Appendix 6A).

of PIP is more significant than burial of POP as a sink of phosphorus. The reason is that phosphate retention in the sediments is larger for these stations since no large phosphate fluxes occur.

The relationship between the depositional source and the resulting flux during the four years of simulation is quite variable, reflecting the varying efficiency of phosphorus trapping (Fig. 14.27C). For Point No Point (PP), the flux from the sediment exceeded the flux to the sediment. This occurs at the expense of the stored phosphate. This can be seen in Fig. 14.28 which presents the time history of POP and PIP. The reason for the release of stored phosphorus is that the station had significantly longer periods of low dissolved oxygen in the latter years (Fig. 13.9). Since the model is equilibrated to the 1985 conditions, the state of the sediment, and in particular, the stored phosphate, reflects the fluxes for that year. As the period of anoxia increased, the flux to the overlying water increased, and the stored phosphorus decreased in response.

This figure also highlights the difference in phosphorus composition for the aerobic and anaerobic stations. Aerobic stations have PIP concentrations that are significantly larger than POP concentrations. The reverse is true for the anoxic (*) stations.

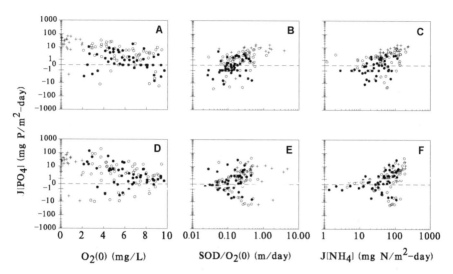

Fig. 14.26 Comparison of observed (top row) and computed (bottom row) phosphate flux versus overlying water dissolved oxygen (A and C), SOD/$O_2(0)$ (B and D), and ammonia flux (C and F). Symbol legend: • Main bay, ○ tributaries, + stations for which $[O_2(0)] < 2.0$ mg/L.

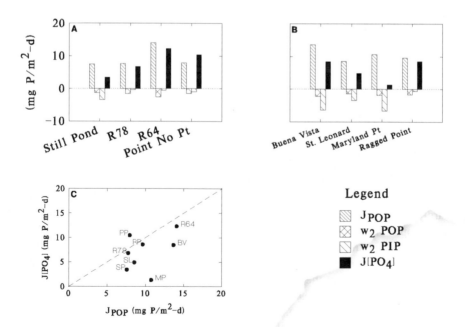

Fig. 14.27 Components of the phosphate flux for (A) Main Bay and (B) tributary stations. (C) Station average flux versus depositional flux J_{POP}.

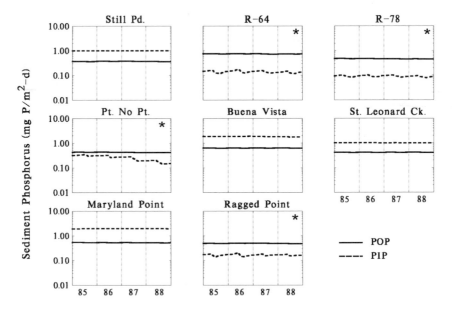

Fig. 14.28 Temporal variation of particulate organic (POP) and inorganic (PIP) sediment phosphorus fluxes. Stations with periodic anoxia (*).

14.7 SILICA

The time series of silica fluxes are shown in Fig. 14.29. The seasonal cycle, which is present at all the stations, arises from the temperature dependency of the dissolution reaction. The depositional flux of particulate silica is constant in this calibration, so that only temperature variation produces the seasonal variability. Silica partitioning in the aerobic sediment layer causes enhanced fluxes during periods of anoxia as can be seen in the anoxic stations (*).

Fluctuations are also caused by variations in overlying water silica concentrations as shown in Fig. 14.30, which compares the overlying water concentration $[Si(0)]$ with the dissolved aerobic layer concentration $f_{d1}[Si(1)]$. When the overlying water concentration approaches or exceeds the aerobic layer concentration, the flux is sharply reduced, since it is proportional to the difference in concentrations. The sharp drops at Buena Vista are caused by this effect.

The sediment silica data are presented in Fig. 14.31. The longitudinal distribution of biogenic particulate silica from a survey in the fall of 1988 is compared to the model calculation at the same time (A). The observed silica is slightly greater than the model computations. The contribution of sorbed silica to the total silica concentration is small as shown.

The longitudinal distribution of pore water silica (C) indicates that computed pore water silica is lower than observations. The silica saturation concentration, $[Si]_{sat} = 40$ mg Si/L, is shown as a dashed line. There appears to be a slight increasing trend

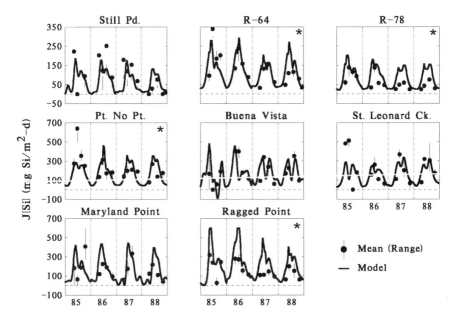

Fig. 14.29 Observed and computed silica fluxes. Stations with periodic anoxia (*).

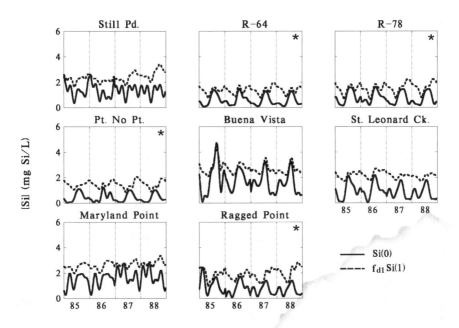

Fig. 14.30 Temporal plots of overlying water Si(0) and aerobic layer pore water f_{d1}Si(1) silica concentrations. Stations with periodic anoxia (*).

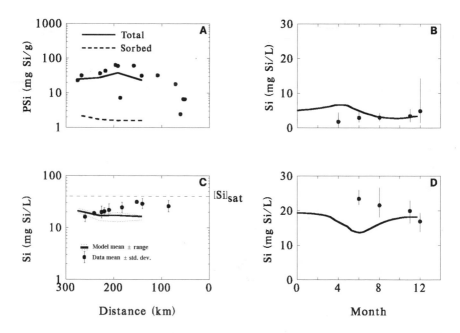

Fig. 14.31 Particulate biogenic (A) and pore water silica (C) vs. distance from the bay mouth. Temporal plots of pore water silica for stations Still Pond (B) and R-64 (D).

in pore water concentration toward the mouth of the bay that is not reproduced by the model. It may be that the saturation concentration is increasing in the downstream direction. The seasonal distribution of pore water silica is shown in Fig. 14.31B, D. The model correctly computes larger concentrations at R-64 (D) relative to Still Pond (B) although the seasonal variation does not appear to be correct.

Integrated Comparisons   The pointwise and quantile comparisons (Fig. 14.32) are much like the previous results: a substantial amount of scatter for the pointwise comparison (A) and a slight bias of the model fluxes exceeding the observations as indicated from the quantile plots (B). The yearly average results (C) form a cluster with not much variation. The station average comparisons (D) indicate that the model both over- and underestimates the observations, but, on balance seems to reproduce the general trend.

Parametric Variations   The relationship of silica flux to temperature, overlying water dissolved oxygen, and ammonia flux are shown in Fig. 14.33. The temperature dependence in the observed fluxes, while scattered, is somewhat more pronounced than in the model results. The opposite is true for the dependency to $O_2(0)$. The data show a weak enhancement at low dissolved oxygen whereas the model exhibits a somewhat stronger relationship.

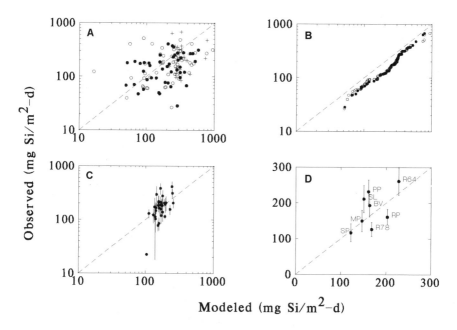

Fig. 14.32 Comparison of observed and modeled silica fluxes. (A) Pointwise comparison: observed versus modeled. main stem (•) and tributary (○) and stations (+) with $[O_2(0)] < 2.0$ mg O_2/L. (B) Quantile comparisons, main stem (•) and tributary (○). (C) Yearly averages: means ± standard errors of the means. (D) Station averages: means ± standard errors of the means.

The silica flux is compared to the ammonia flux in the last column. The relationship is evident in both the data and model computations, although the model relationship is stronger due to the relationship between depositional fluxes of nitrogen and silica. The plateau in the model fluxes at 80 mg Si/m²-d is due to the additional detrital silica flux, J_{DetrSi}, which is assumed to exist at all stations, in addition to the depositional flux J_{PSi} that is stoichiometrically related to POM fluxes.

Components of the Flux The flux components are shown in Fig. 14.34A, B. The components are: the sources due to biogenic J_{PSi} and detrital J_{DetrSi} silica deposition, the burial of particulate biogenic w_2[PSi] and sorbed w_2[Si(2)] silica, and the resulting silica flux J[Si]. Burial of particulate biogenic silica is the only significant sink, since the concentration of sorbed silica is considerably smaller (Fig. 14.31). A comparison of the total silica input $J_{TSi} = J_{PSi} + J_{DetrSi}$ to that which is recycled J[Si] is shown in Fig. 14.34C. The removal fraction is quite variable and does not appear to be strongly related to the total silica input. This is because there is a limitation to the quantity of silica that can be recycled, which is determined by the solubility of silica. Hence, the silica fluxes are less variable than the total fluxes to the sediment.

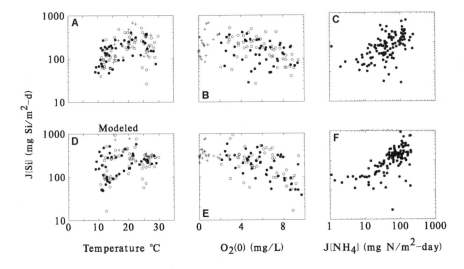

Fig. 14.33 Observed (top row) and modeled (bottom row) silica flux vs temperature (A and D), overlying water dissolved oxygen concentration (B and E), and ammonia flux (C and F). Symbol legend: ● Main bay, ○ tributaries, + stations for which $[O_2(0)] < 2.0$ mg/L.

14.8 STATION COMPOSITE PLOTS

The sediment flux time series for ammonia, oxygen, phosphate, and silica are grouped by stations in Figs. 14.35–14.36. Since each of the fluxes are driven by the same depositional flux, modified by the appropriate stoichiometric ratios, the relationships between the various fluxes are determined by the overlying water concentrations and the kinetics.

For Still Pond (Fig. 14.35), the fluxes have a seasonal variation which are all in phase. They are not disrupted by overlying water hypoxia or anoxia. By contrast, the relationships among the fluxes at station R-64 are distinct. The ammonia and silica fluxes show a seasonal variation related to temperature. However the oxygen and phosphate fluxes are different. The oxygen flux is almost constant through the latter part of each year. The very large phosphate fluxes relative to the ammonia flux are the result of the storage of phosphorus during aerobic periods and its release during anoxia.

There is a difficulty with the calibration to the SOD data at R-64. The model cannot reconcile the observations of high ammonia, phosphate, and silica fluxes that occur during the first part of each year, and the lack of variation in the oxygen flux during the same time period. The fact that the depositional fluxes of nitrogen, silica, phosphate, and carbon are all in constant stoichiometric ratio requires that the model predicts a substantial oxygen flux occurs as the sulfide that is produced in the early part of each year is oxidized during the first half of the year. The onset of anoxia and the persistence of benthic stress suppresses the oxygen flux for the latter half of

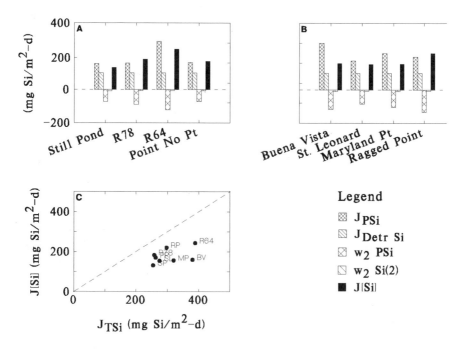

Fig. 14.34 Components of the silica flux for (A) Main bay and (B) tributary stations (C) station average silica flux $J[\text{Si}]$ versus total silica input flux to the sediment J_{TSi}.

the year. This inability to account for an observed anomaly points to an area that warrants further investigation.

Stations R-78 and Point No Point (Fig. 14.35) also exhibit this difference between the seasonal variation of the ammonia and silica flux, and the oxygen flux. However, since the depositional fluxes are smaller at these stations, as indicated by the smaller ammonia fluxes, the spring increase in SOD is not as dramatic and does not contradict the observations.

The fluxes at Buena Vista and St. Leonard on the Patuxent estuary are presented in Fig. 14.36. By and large the model is a reasonable representation of the data. However, the 1985 data for St. Leonard Creek illustrate an inconsistency which the model cannot reconcile. The large observed fluxes of oxygen and silica suggest a large depositional flux. However, the ammonia and phosphate fluxes suggest a smaller flux. These discrepancies cannot be reconciled within a framework that is restricted to constant stoichiometric ratios for the particulate organic matter that settles into the sediment.

For the Maryland Point and Ragged Point stations on the Potomac estuary (Fig. 14.36) the magnitudes of the fluxes are in reasonable agreement with the large diagenesis flux suggested by the ammonia fluxes. The exception is the Ragged Point silica flux which is computed to be larger than the observations.

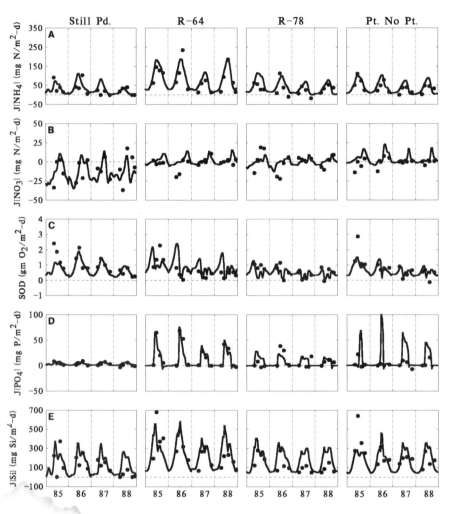

Fig. 14.35   Temporal variation of calculated (line) and observed (•) sediment fluxes grouped by station. Main bay stations.

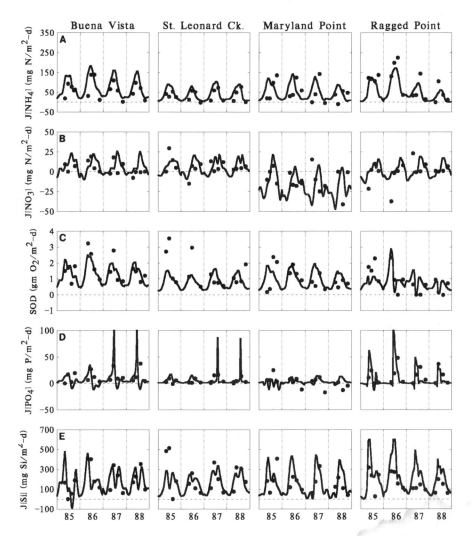

Fig. 14.36 Temporal variation of calculated (line) and observed (•) sediment fluxes grouped by stations. Tributary stations.

It is possible that these discrepancies are related to the assumption of a constant stoichiometric relationship between the depositional fluxes. This simplification is unavoidable if the depositional fluxes are to be estimated from ammonia diagenesis. The alternate choice – estimating the depositional fluxes independently using the observed fluxes to the overlying water – introduces too many degrees of freedom in the stand-alone calibration, thereby weakening it severely. When the sediment model is coupled to an overlying water eutrophication model, the constant stoichiometric assumption is not made. Rather the depositional fluxes result from the water column processes that produce particulate organic matter, and these need not be in Redfield stoichiometric ratios.

14.9 CONCLUSIONS

The stand-alone calibration of the sediment flux model highlights both the strengths and weaknesses of the model. The relationships between the concentrations of solutes in the solid phase, pore water, and the sediment fluxes are rationalized within the framework of a mass balance analysis. The seasonal patterns are reproduced with reasonable fidelity for the oxic stations. The influence of anoxia on phosphate and oxygen fluxes – enhancing the former and suppressing the latter – is captured as well. The phosphate flux model employs a parameterization of the aerobic layer phosphate partitioning that depends on the overlying water dissolved oxygen. The suppression of the oxygen flux that persists after the anoxic period relies on the formulation of benthic stress. Although these formulations are empirical, they appear to produce reasonable simulations.

The model is not able to reproduce the pointwise distribution of the fluxes. Plots of observed versus modeled fluxes display significant scatter. This appears to be related to a lack of precise timing between computed and observed fluxes. A visual inspection of the time series plots supports this observation. By contrast, the quantile plots demonstrate that the model reproduces the overall distribution of fluxes in the main stem and the tributaries if station location and timing are not considered. The comparison of predicted and observed yearly means and station means reveals that as the degree of averaging increases, the model is usually better able to predict the observations.

This is examined quantitatively in Fig. 14.37, a plot of the squares of the correlation coefficients between observed and modeled fluxes for pointwise, yearly averages, and station averages. The square of the correlation coefficient R^2 is the fraction of the observed variance that is removed by the model predictions. If R^2 is small, very little variability is removed by the model and therefore it has little predictive power. If, however, R^2 approaches one, then the model is capturing all the variability. Both Pearson (A and B) – the usual correlation coefficient – and Spearman's rank correlation coefficient (C) (Wilkinson, 1990) are computed for the arithmetic and the arcsinh transformed variables. Since the Spearman's rank correlations are identical for the nontransformed and transformed data – this is because the arcsinh transfor-

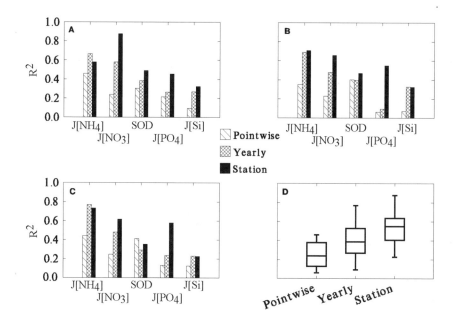

Fig. 14.37 Correlation coefficients comparing pointwise, yearly, and station averages. (A) Pearson correlation coefficient. (B) Pearson correlation coefficient, arcsinh transformed variables. (C) Spearman rank correlation coefficient. (D) Box plots of all correlation coefficients.

mation is monotonic and the rank orders are preserved – only the nontransformed results are presented.

In general, R^2 improves as the averaging increases, although sometimes the station averages decrease slightly or remain the same. Since the station averages comprise only 8 points, these correlation coefficients are quite uncertain.

The box plot (D) that combines the results of the three computational methods for R^2, summarizes the probability distributions of R^2. The median (25th percentile, 75th percentile) R^2 increases from 0.24 (0.13, 0.40) to 0.39 (0.27, 0.58) to 0.55 (0.35, 0.66) as the averaging interval increases.

The overall impression of the calibration is that the fine scale variations cannot be captured, but that the overall quantitative relationships between the fluxes, together with the solid phase and pore water concentrations, are successfully rationalized. The seasonal behavior and the relative variations are reproduced.

15

MERL, Long Island Sound, and Lake Champlain

15.1 INTRODUCTION

The purpose of this chapter is to present applications of the sediment flux model to other data sets. Two points of interest: whether the model can reproduce the fluxes measured in other settings, and how general is the parameterization. That is, can the model be expected to apply to situations for which it is not calibrated?

We examine, first, a large and complete data set that was collected as part of a eutrophication experiment conducted at the Mesocosm Experimental Research Laboratory (MERL) of the University of Rhode Island, School of Oceanography. Then we examine two field collected data sets: one from Long Island Sound, and a second from Lake Champlain, a freshwater lake.

15.2 MERL

The Mesocosm Experimental Research Laboratory was established to conduct experiments in large (meso) experimental enclosures that could be manipulated systematically in various ways. The idea was to provide an experimental component to estuarine research that complements the more traditional, observationally based, approaches. These large, outdoor mesocosms have been shown to reproduce the major features of the nutrient cycles in Narrangansett Bay (Pilson et al., 1980). For the nutrient addition experiment, a wide range of loadings of inorganic nutrients were added over a three-year study period (Nixon et al., 1986, Oviatt et al., 1986). Extensive measurements were made in the water column and the sediments. These

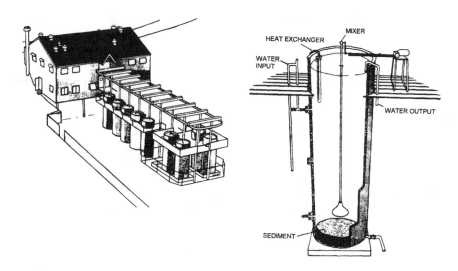

Fig. 15.1   The MERL facility at the Graduate School of Oceanography, University of Rhode Island.

data are available as data reports (Frithsen et al., 1985a,b,c), a practice for which the investigators are to be applauded. The facility is diagrammed in Fig. 15.1.

The MERL mesocosms are large outdoor tanks approximately 1.8 m in diameter with an interior water depth of 5.0 m. They are located on the shoreline of Narragansett Bay on the campus of the University of Rhode Island's Graduate School of Oceanography. Bay water is pumped through the tanks. The hydraulic detention time is approximately 30 days. The tanks have mixers to reproduce the vertical mixing regime in Narragansett Bay and, in particular, the lack of stratification. Sediments are collected using a large box core that maintains the vertical orientation of the sample. The top 40 cm are placed into containers in the bottom of the tanks (Nixon et al., 1986). Sediment fluxes are measured using a benthic chamber that completely encloses the bottom sediment. It is installed by divers to make the measurements and then removed. Thus the measured sediment fluxes represent the entire contribution of approximately 2.5 m^2 of sediment surface area. By contrast, field-based measurements that employ core tubes sample a much smaller area, which increases variability due to small-scale sediment variations.

15.2 A   Nutrient Addition Experiment

The MERL mesocosms have been used for many experiments.[*] The data analyzed below comes from the Nutrient Addition Experiment.[†] Its purpose was to examine

---

[*]Nowicki and Oviatt (1990), Oviatt et al. (1993, 1995, 1984, 1987), Sampou and Oviatt (1991)
[†]Nixon et al. (1986), Oviatt et al. (1986), Pilson et al. (1980)

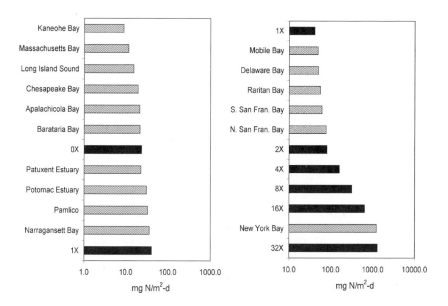

Fig. 15.2 Comparison of nitrogen areal loadings to selected estuaries and the MERL experiment (Nixon et al., 1986). Note the scale change.

the consequences of nutrient enrichment to coastal estuaries. The duration was approximately 2 ¼ years over three calendar years. The nutrient dosing was increased in a geometric series, 1X, 2X, 4X, 8X, 16X, and 32X, in addition to three control tanks, denoted as 0X. The nutrients added were inorganic nitrogen, phosphorus, and silica in a molar ratio of 12.8 N : 1.0 P : 0.91 Si to match the stoichiometry of sewage entering the bay (Nixon et al., 1986). Areal loading rates of total nitrogen to the tanks varied from 23 mg N/m^2-d for the controls, 63 mg N/m^2-d for 1X, 103 mg N/m^2-d for 2X and so on geometrically to 1308 mg N/m^2-d for the 32X (Kelly et al., 1985). These loading rates are compared to estimates of areal loading rates from other estuaries in Fig. 15.2. The control to 2X loading rates span the majority of the estuaries listed. The higher loading rates are characteristic of heavily loaded estuaries such as the waters surrounding New York City. The range of loadings actually examined was in excess of 50-fold. This was a very important feature of the experimental design. Alternate experimental designs that mimic agricultural plot experiments – for example, treatments with and without added nutrients – can be replicated and analyzed using analysis of variance methods. Such a framework would detect changes but would be entirely unsuitable for the analysis applied below, or any other analysis that seeks to understand the causes of the effects due to increased nutrient loadings.

As a result of the range in nutrient inputs, mean annual water column dissolved inorganic nitrogen (DIN) concentrations increased from 56.0 to 4200 μg N/L, mean annual chlorophyll *a* ranged from 4.0 to 70 μg/L and total system carbon production ranged from 0.55 to 2.2 g C/m^2-d (Nixon et al., 1986). The average concentrations in the overlying water are shown in Fig. 15.3. The phosphate concentrations (D) are

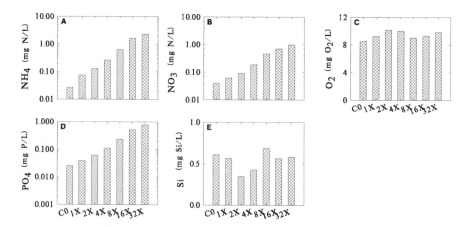

Fig. 15.3 Average overlying water concentrations for the nutrient addition experiment. (A) ammonia, (B) nitrate, (C) oxygen, (D) phosphate, (E) silica. The x-axis denotes the treatment starting with the control.

varying approximately 50-fold across the treatments, which is consistent with the increases in loading. By contrast, the ammonia concentrations (A) are varying by more than 100-fold. The reason is that some of the ammonia is being nitrified, which accounts for the increasing nitrate in the overlying water (B). Silica concentrations (E) are high enough in the bay water so that the additions from the loadings are not significant.

Sediment processes were also examined during the experiment. Sediment oxygen demand, nutrient fluxes, pore water, and solid phase concentrations were also measured. In addition, manganese flux and sediment compositional data were collected (Hunt and Kelly, 1988). The manganese data are analyzed in Chapters 18 and 19.

15.2 B   Application of the Sediment Flux Model

The sediment model is applied in stand-alone mode. That is, it is not dynamically coupled to the overlying water, as it would be if it were being used as part of a larger coupled water quality-sediment model. In fact a coupled modeling analysis has been completed (Di Toro et al., 2001, Lowe and Di Toro, 2001). For the stand-alone application, the overlying water concentrations are specified externally. In addition, the average annual depositional flux of particulate organic nitrogen to the sediment J_{PON} is specified and Redfield stoichiometry is assumed in specifying the carbon, phosphorus, and silica fluxes.

Depositional Flux   For application to the MERL data set, the magnitudes of J_{PON} are chosen to fit the measured ammonia fluxes. The kinetic and transport parameter values are set to the values obtained from the calibration to the Chesapeake

Bay data set (Chapter 14). The required depositional fluxes are shown in Fig. 15.4. Fluxes vary from less than 50 to 130 mg N/m^2-d, less than a 3-fold variation.

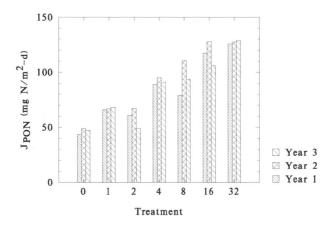

Fig. 15.4 Depositional flux of particulate organic nitrogen.

In order to run the model, it is necessary to specify the initial concentrations of all the state variables. Since these concentrations were not measured – indeed it would be difficult if not impossible to measure the concentrations in the aerobic layer – they must be arrived at in another way. The objective is to produce the initial conditions in the sediments at the start of the experiment. The method chosen is to equilibrate the model by cycling the model until a periodic steady state is achieved. The same forcing functions are used for each annual cycle. Since the objective is to simulate the state of the sediments at the start of the experiment, the obvious choice is the inputs for the first year from the control tanks. That is, the overlying water concentrations observed in a control tank and the depositional flux that reproduces the control tank ammonia flux are used to initialize all the tanks. This is equivalent to assuming that the state of the sediments in the year when they were collected is similar to that which is produced by a cyclical repetition of the conditions that were observed in the control tank for the first year of the experiment. Since the controls are a reasonable replication of Narragansett Bay, this seems to be a reasonable assumption.

This is essentially the same procedure employed for the Chesapeake Bay sediments – using a specific year and cycling the model to initialize the sediment. For that case, the procedure was adopted as an expedient. For the MERL experiments, it is a reasonable way to represent the state of the sediments taken from the bay before the start of the experiment.

It is also a convenient and parsimonious way of proceeding. The initial conditions are not externally specified by some arbitrary means. Therefore, they are not an additional set of calibration parameters that need to be determined by fitting to the observations. This greatly reduces the degrees of freedom in the model and renders the calibration less of a curve-fitting exercise. Also the initial conditions are consistent with all the parameters used in the model. Thus initially all the sediments have the same state variable concentrations as the control. The differences in subsequent behavior occur solely due to the increased nutrient loading applied during the experiment.

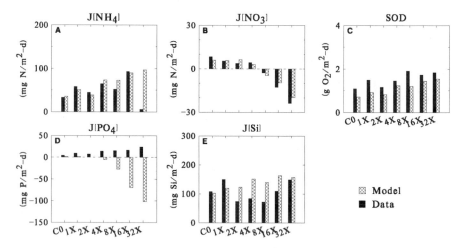

Fig. 15.5 Comparison of sediment flux model calculations (hatched bars) and observations ■ using Chesapeake Bay parameter set. Modeled fluxes and observations averaged over the period of the experiment.

Initial Application   The initial application of the model uses the Chesapeake Bay parameter set *exactly*. A summary of the resulting fluxes, and the comparable data, are shown in Fig. 15.5. These are the average fluxes for the period of the experiment. The model reproduces the ammonia fluxes (A), as it should since the PON flux is calibrated from these data. The exception is 32X for which significant negative fluxes were observed during the winter. The model does not capture this feature and we have no good explanation for the observation.

The nitrate fluxes are reproduced quite nicely. The progression from positive (out of the sediment) to negative (into the sediment) is caused by the increasing nitrate concentrations in the overlying water (Fig. 15.3B). The oxygen fluxes (C) are also reproduced quite nicely, and the silica fluxes, which do not vary very much across the treatments, are also reasonably well reproduced.

It is the phosphorus fluxes (D) that are completely wrong. While the data show an increase with increased loading, the model starts low and reverses direction – from out of the sediment to into the sediment – as loading increases. This is the result of the overlying water phosphorus concentrations increasing (Fig. 15.3D), the same behavior as the nitrate fluxes. However, the data do not support this behavior.

The comparison of sediment and pore water concentrations is also instructive (Fig. 15.6). Sediment PON concentrations are reproduced nicely, whereas the sediment

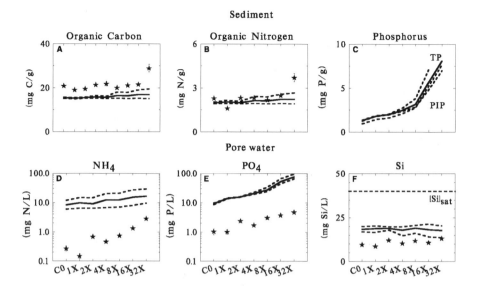

Fig. 15.6 Sediment solid phase (top row) and pore water (bottom row) comparisons of model computations (lines) and observations (★). The dashed lines are the range in concentrations computed by the model over the period of simulation.

POC is predicted to be slightly lower than observed. A slight change in the G_3 POC to PON ratio is indicated. The pore water concentrations, however, are not reproduced at all. Both ammonia and phosphorus are both incorrect by one to two orders of magnitude, and the silica concentration is also overestimated. What is interesting about this result is that the sediment PON and the ammonia fluxes (Fig. 15.5A) are both reproduced quite nicely.

Pore Water Diffusion The analysis of the steady state model for ammonia flux (Chapter 3) indicates that the flux of ammonia is determined primarily by the magnitude of the diagenesis source J_N and the overlying water concentration $[NH_4(0)]$ and not at all by the mass transfer coefficient from the anaerobic to the aerobic layer K_{L12} (Eq. 3.9). However, the anaerobic layer pore water concentration is inversely proportional to K_{L12} (Eq. 3.6). Thus the reason the model is not reproducing at least the ammonia concentrations is due to an incorrect K_{L12}.

The consequences of increasing K_{L12} by a factor of 5 are shown in Fig. 15.7 for the pore water concentrations. The fluxes, which are not shown, are not changed at all. The results are indistinguishable from Fig. 15.5. The phosphorus fluxes are still incorrect. However, the pore water concentrations are dramatically improved. Both the ammonia and silica concentrations are now correctly calculated, an increasing concentration for ammonia and a constant concentration for silica. However the phosphate concentration is still incorrect by an order of magnitude.

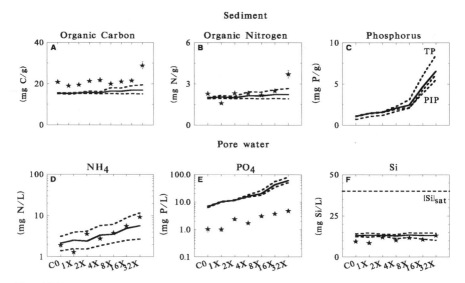

Fig. 15.7 Comparison of model computations (lines) and observations (★). Pore water diffusion coefficient adjusted. The dashed lines are the range in concentrations computed by the model over the period of simulation.

Phosphate Partition Coefficient This incorrect result for phosphorus fluxes suggests that the problem is with the partition coefficients for phosphate. If the partition coefficients are too large, then the pore water concentration would be too low, and the fluxes would tend to be increasingly from the overlying water to the sediment as the overlying water concentrations increase. Fig. 15.8 present the results if the partition coefficients in both the aerobic and anaerobic layers are decreased as listed under MERL mesocosms in Table 15.1. The phosphate fluxes are now reproduced quite nicely (Fig. 15.8I, D) as are the pore water concentrations (II, E). There is also an order of magnitude change in the sediment POP and PIP concentrations, from 1–10 mg P/g, (Fig. 15.7C) to 0.5–1.0 mg P/g (Fig. 15.8II, C), a more reasonable range of values.

Table 15.1 Phosphate Sorption Parameters used in Fig. 15.8

Parameter	MERL Mesocosms	Chesapeake Bay[a]	
$\Delta \pi_{PO_4,1}$	20.0	100.0	-
$\pi_{PO_4,2}$	20.0	300.0	L/kg
$\pi_{PO_4,1} = \Delta \pi_{PO_4,1} \times \pi_{PO_4,2}$	4.0×10^2	3.0×10^4	L/kg

[a] Used as the initial values.

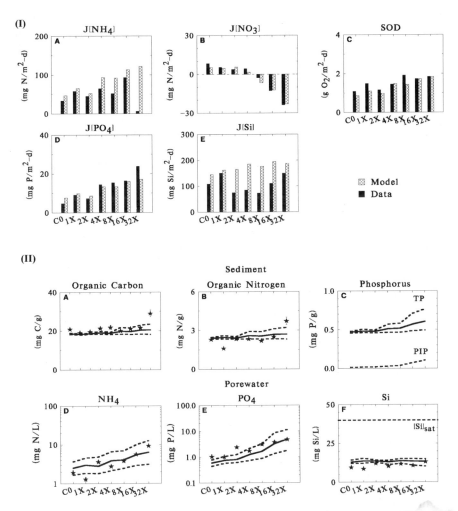

Fig. 15.8 Model results with adjusted phosphate partition coefficient. (I) Comparison of observed and modeled fluxes. (II) Comparison of model computations (lines) and observations. The dashed lines are the range in concentrations computed by the model over the period of simulation.

Seasonal Variation   The previous figures presented the average fluxes and concentrations. The model must also reproduce the seasonal distribution of fluxes to the extent that it can. The comparison in Fig. 15.9 indicates that, in general, they are nicely reproduced with some exceptions. The negative ammonia fluxes in the winter – a consistent feature of the winter data from the 4X treatment onward – are not reproduced. The seasonal distributions of pore water concentrations are shown in Fig. 15.10. The data are too infrequent to determine whether the seasonal variations are properly captured. The depth of the aerobic layer (D) is also presented. The layer thickness decreases with increased loading due to the increasing SOD.

15.2 C   Analysis of the Results

The analysis of the predicted versus observed fluxes is presented in Fig. 15.11. The pointwise comparison for ammonia (A1) is similar to the Chesapeake Bay result (Fig. 14.5) although the range of ammonia fluxes is larger. Nitrate fluxes are again not predictable on a pointwise basis – compare (A2) and Fig. 14.10. The SOD are surprisingly good (A3) with very good agreement over nearly two orders of magnitude. The Chesapeake Bay results (Fig. 14.19) are also not bad, but are more scattered. Predicted versus observed phosphate fluxes are somewhat scattered with occasional incorrect sign predictions – upper left and lower right quadrant in (A4). This is also a feature for the Chesapeake Bay results (Fig. 14.25). The results for silica fluxes are also somewhat scattered – compare (A5) and Fig. 14.32.

The quantile plots (B) – cross plots of the ordered data and computed fluxes – indicate that the model results, taken as a whole, have essentially the same probability distribution as the data. They would be exactly the same if the points followed the 1 to 1 line exactly. The plots comparing yearly averages (C) and treatment averages (D) examine the ability of the model to reproduce average fluxes. As in the case with the Chesapeake Bay results, the averages correspond more closely than the pointwise results.

The variations in fluxes as a function of temperature variations are examined in Fig. 15.12. Both the observations and modeled fluxes are shown. The idea is to see if the model reproduces similar patterns, which for these fluxes, it does. This result suggests that temperature is an important variable influencing sediment fluxes – a well known and well documented result (Nixon and Pilson, 1983) – and that the model's parameterization for temperature dependencies appears to be adequate to the task of predicting temperature's influence.

The other critical process that affects the magnitude of the fluxes is diagenesis. Although there is no direct measurement of diagenetic production rate, the ammonia flux can serve as a surrogate. Since ammonia is nitrified, it is not a perfect analog. Nevertheless, it was used in the development and calibration of the steady state versions of the model (Chapter 3 to 10). The variations of SOD, phosphate, and silica fluxes with ammonia flux are presented in Fig. 15.13. The model demonstrates a strong relationship between ammonia flux and SOD, phosphate and, to a lesser extent, silica. This is due to the underlying assumption of Redfield stoichiometry that determines the ratios of nitrogen, phosphorus, and silica that are released per mole

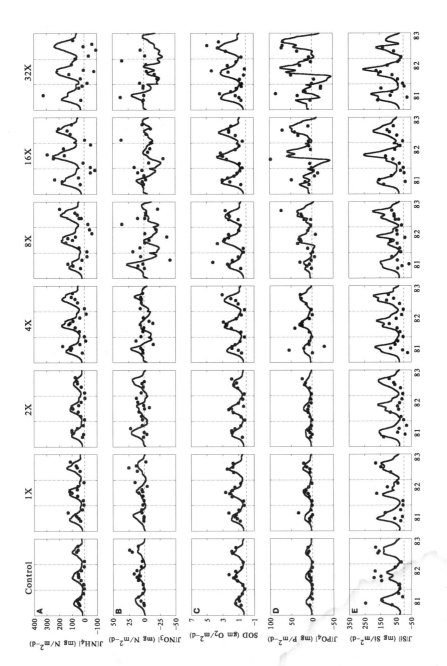

Fig. 15.9 Temporal variation of sediment fluxes. Comparison of model computations (lines) and observed data (•).

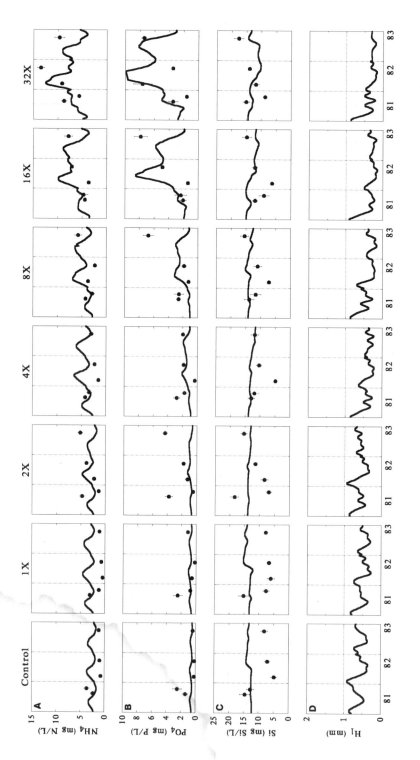

Fig. 15.10   Temporal distribution of pore water concentrations. Comparison of model computations (lines) and observed data (●).

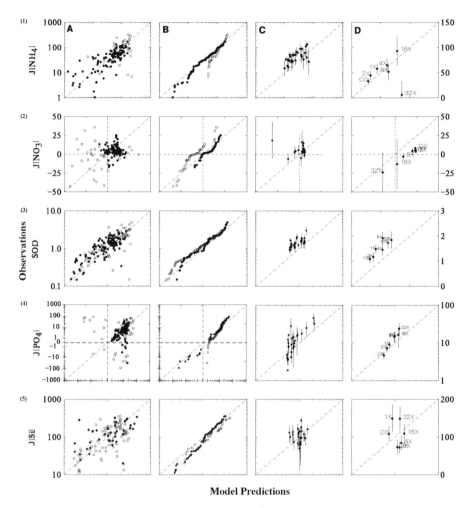

Fig. 15.11 Comparison of observed and modeled fluxes. (A) Pointwise comparison: 0X-4X (●), 8X-32X (○). (B) Quantiles.0X-4X (●), 8X-32X (○). (C) Yearly averages. (D) Treatment averages. Units as in Fig. 15.9. The scales on the right-hand side apply to the treatment average plots (D) only.

of carbon mineralized (Table 1.1). These stoichiometric relationships are shown in Fig. 15.13 on both the model and data plots. Since neither ammonia nor SOD is exactly proportional to nitrogen and carbon diagenesis, the relationship cannot be exact. Nevertheless, there is a clear relationship that is stronger for SOD and phosphate, and weaker for silica. Since the fraction of silica deposition that is recycled is variable, because of the solubility limitation, on the one hand, and an additional source that is unrelated to the flux of particulate organic matter, this discrepancy is to be expected.

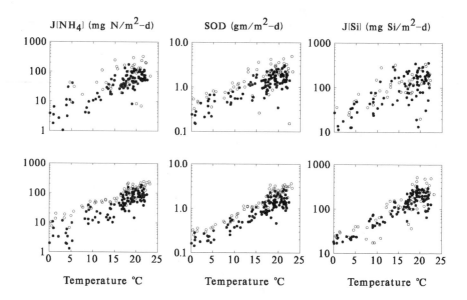

Fig. 15.12   Variation of ammonia, SOD, and silica fluxes with temperature. Observed (top row) and modeled (bottom row). 0X-4X (●), 8X-32X (○).

The differences between the fluxes of PON, POC, and POP that settle to the sediment – for silica the total silica source is the sum of the particulate depositional flux and the endogenous silica source $J_{TSi} = J_{PSi} + J_{DetrSi}$ – and the computed fluxes that result are shown in Fig. 15.14. These are average results over the period of simulation. The lines that are included are from the Chesapeake Bay sediment model. The MERL model produces essentially the same results for ammonia and SOD. For phosphorus, an increasing fraction is retained at the higher treatments, which is a reflection of the effect of increasing overlying water concentration (Fig. 15.3). As the overlying water concentration increases, the diffusive gradient becomes smaller, and the flux to the overlying water is reduced. For silica, the silica flux is essentially constant, independent of the magnitude of the depositional source. The flux is not limited by the source but rather by silica solubility and partitioning.

15.2 D   Conclusions

The application of the sediment flux model to the MERL data set with the Chesapeake Bay calibration parameters has been remarkably successful. Only three parameters needed adjustment: the pore water diffusion coefficient (actually the aerobic-anaerobic mass transfer coefficient K_{L12}) and the phosphate partition coefficients in the aerobic and anaerobic layers $\Delta\pi_{PO_4,1}$, $\pi_{PO_4,2}$. The pore water diffusion coeffi-

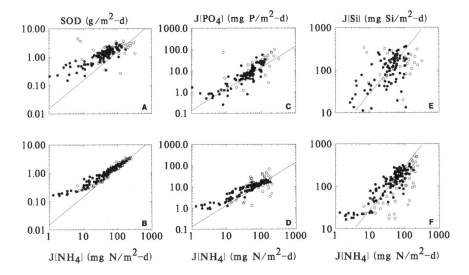

Fig. 15.13 Variation of SOD, phosphate and silica fluxes with ammonia flux. Observed data (top row) and modeled fluxes (bottom row) are presented. 0X-4X (•), 8X-32X (o). The lines are the Redfield stoichiometric relationships (Table 1.1).

cient was increased fivefold. This presumably reflects the additional mixing caused by the pumping action of the plunger that mixes the overlying water.*

The changes necessary in the phosphate partition coefficient are the consequence of not explicitly considering the chemical factors that affect phosphorus partitioning. In particular, the higher pHs that are characteristic of the more heavily loaded treatments would cause a decrease in phosphate (HPO_4^{2-}) sorption due, presumably, to the increased competition with OH^-. This is the price paid for using partition coefficients that are independent of the pore water chemistry. It is a clear example of the trade-off between simplicity and generality. For the manganese flux models developed in Chapters 18 and 19, the pH dependency turns out to be equally important, and it is explicitly incorporated.

15.3 LONG ISLAND SOUND

Collecting large and comprehensive sets of sediment flux measurements is a difficult and expensive undertaking. This is what makes the Chesapeake Bay program data unique. The MERL data set is also unusual in its completeness: due, in part, to the accessibility of the mesocosms and the intelligence of the experimental design. In

---

*This explanation was suggested by Professon Scott Nixon. The MERL investigators were concerned about this artifact but did not know the magnitude of the effect.

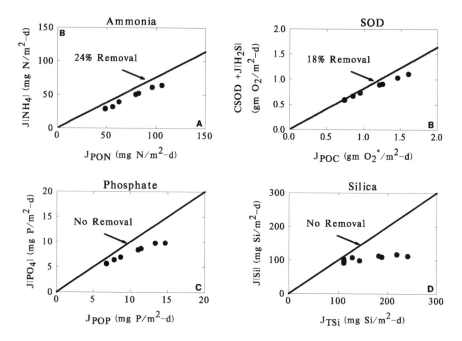

Fig. 15.14   Removal fractions computed by the model. Fraction of flux of PON, POC, POP, and TSi returned to the water column. The removal fractions are from the Chesapeake Bay sediment flux model (Figs. 14.8 and 14.16).

this and the next section, we present an analysis of two data sets that are more representative of the quantity of data that one is likely to encounter in practice: more limited in frequency and time, and without many ancillary measurements. The purpose is to examine the applicability of the sediment flux model to these new data sets.

The Long Island Sound data set was collected as part of a National Estuary Study. Its focus was an investigation of the summer hypoxia that develops in the Western Sound. It is the result of excessive phytoplankton growth in the spring, the resulting deposition of POM to the sediment, the development of stratification in the summer, and the consumption of hypolimnetic DO by water column respiration and SOD. In order to quantify the causes of the hypoxia, and to evaluate the efficacy of remedial actions, a eutrophication model was built that incorporated the sediment flux model (HydroQual, 1996). A set of oxygen and nutrient flux measurements were collected to support the sediment model application (Mackin et al. (1991) , see Aller (1994) for more details).

The data collection program was designed in order to reconcile two requirements: a fixed sampling budget, and the desire for both detailed spatial and temporal coverage. This leads to a sampling design in which a few master stations are sampled more frequently in time, and more stations sampled less frequently. To these are

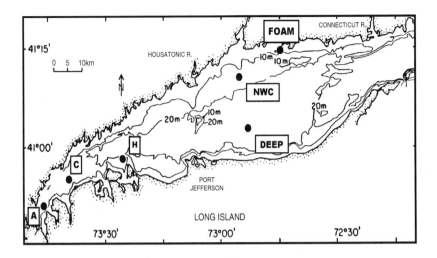

Fig. 15.15 Long Island Sound sediment sampling station locations.

usually added stations where sampling has been done in previous studies for the sake of extending the data record.

The master stations for the Long Island study are presented in Fig. 15.15. The stations in the Western Sound are in the region of the seasonal hypoxia. Those in the Central Sound are locations where extensive measurements have been made previously (Aller, 1980a,b). The application of the sediment model is made in stand-alone mode. The overlying water concentrations are fit using Fourier series. The overlying water dissolved oxygen concentrations are shown in Fig. 15.16. Similar curves are generated for the rest of the overlying water concentrations. The fitted curves are repeated for the two years due to lack of data.

The depositional flux of organic matter is estimated from the ammonia fluxes, using the simplified steady state model. The ammonia flux is corrected for the loss due to nitrification. The equations are presented in the next chapter (Section 16.2 A). At steady state, this is the nitrogen diagenesis J_N corresponding to the ammonia flux. Since each ammonia flux measurement produces an estimate of J_N, the average is used to estimate the annual average depositional flux of POM for that station. The J_N is produced only by the mineralization of the G_1 and G_2 organic matter fractions $f_{PON,1} + f_{PON,2}$, but not the G_3 fraction $f_{PON,3}$ which is deposited but not reacted. Thus the flux of organic matter is increased accordingly

$$J_{PON} = J_N \frac{f_{PON,1} + f_{PON,2} + f_{PON,3}}{f_{PON,1} + f_{PON,2}} = J_N \frac{1}{f_{PON,1} + f_{PON,2}} \qquad (15.1)$$

The sedimentation fluxes are listed in Table 15.2. The results are shown in Fig. 15.17. The parameters are from the Chesapeake Bay calibration, with the phosphate partition coefficient in the aerobic layer lowered to $\Delta\pi_{PO_4,1} = 100$ L/kg. The fit ap-

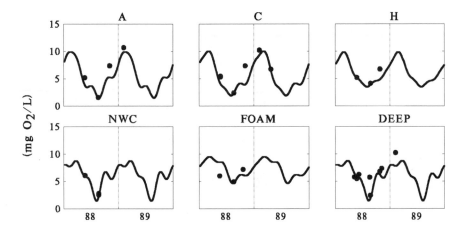

Fig. 15.16 Overlying water dissolved oxygen (•) with Fourier series fit (line) for the Long Island Sound stations identified in the map (Fig. 15.15).

Table 15.2 Estimated Diagenesis and Depostional Fluxes for Long Island Sound Stations

Station	J_N (mg N/m²-d)[a]	J_{PON} (mg N/m²-d)[b]
A	117.0	130.0
C	49.3	54.8
H	21.4	23.8
NWC	45.6	50.7
FOAM	68.2	75.8
DEEP	42.0	46.7

[a] Eqs. 16.5. [b] Eq. 15.1.

pears to be reasonable, considering the few data points. The trends between stations are captured nicely although there is the occasional aberrant observation, e.g., a high ammonia flux coupled with a low SOD for the summer at station A.

In addition to the flux data, it would be useful to compare pore water and solid phase results. Some observations for other Western Long Island Sound locations are available. Pore water data are compared in Fig. 15.18 (top). For station A and the data station at MP 23, the concentrations are similar. Low nitrate concentrations, and in increasing order, phosphate, ammonia, silica, and total sulfide (in oxygen equivalents) are comparable. For the locations further into Long Island Sound, the observed concentrations are significantly lower. Since these locations are not paired to the modeled locations the differences are not unexpected.

Solid phase data are compared in Fig. 15.18 (bottom). A novel feature of these observations is the biogenic silica measurements. These compare quite favorably with the model computations for station A and the MP 23 data as do the other constituents: particulate organic carbon, nitrogen, total phosphorus and particulate inor-

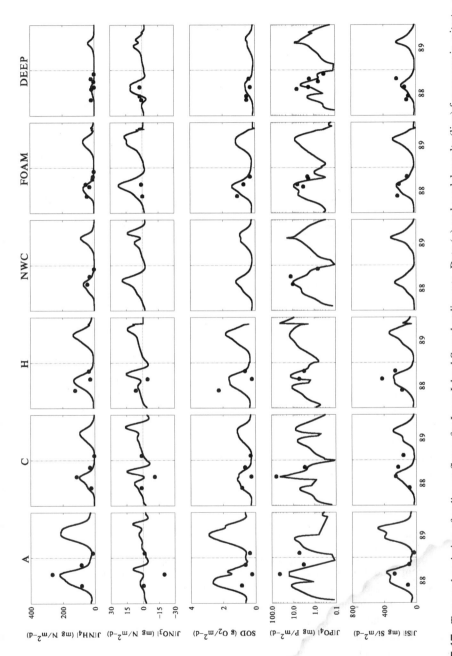

Fig. 15.17 Temporal variation of sediment fluxes for Long Island Sound sediments. Data (•) and model results (line) for ammonia, nitrate, oxygen, phosphate, and silica fluxes. The abscissa is time in years.

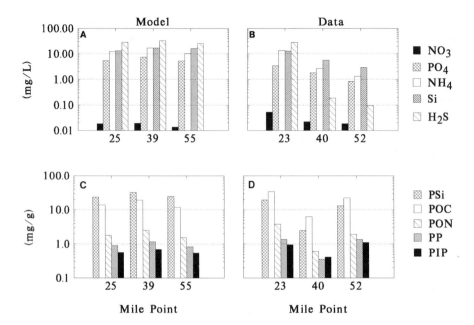

Fig. 15.18 Pore water (A and B) and sediment (C and D) concentrations. Model (A and C) and data (B and D) for three stations. Model stations are A, C, and H (Fig. 15.15). Pore water concentration units: mg N/L, mg P/L, mg N/L, mg Si/L, and mg O_2^*/L. Sediment concentration units: mg Si/g, mg C/g, mg N/g, and mg P/g.

ganic phosphorus. As with the pore water data, the observations decline at station 40. Curiously, the concentrations increase at station 52, in contrast to the pore water data. The model is incapable of this behavior unless K_{L12} changes, possibly due to an increase in the biomass of irrigating biota.

The Long Island data set is interesting because it contains measurements of the depth of the aerobic layer H_1. These were made using an oxygen microelectrode. The comparison is given in Fig. 15.19. The depth of the aerobic layer is computed from (Eq. 3.15)

$$H_1 = D_{O_2} \frac{[O_2(0)]}{\text{SOD}} = \frac{D_{O_2}}{s} \qquad (15.2)$$

The sediment model does not explicitly use the aerobic layer diffusion coefficient D_{O_2}. Instead it is incorporated into the definition of the reaction velocity (Eq. 3.17). As a consequence, the aerobic layer depth is not explicitly computed. However if D_{O_2} is specified then Eq. (15.2) can be used. Fig. 15.19 presents the results using the molecular diffusion coefficient for oxygen (Table 2.1) with a temperature dependence of $\theta = 1.08$. The overlying water dissolved oxygen concentration is included for comparison purposes. The seasonal variation is reasonably well represented although the large H_1 values at stations NWC and H in the winter are not reproduced.

Both the variation in SOD and $[O_2(0)]$ – compare the difference between stations A and DEEP – contribute to the seasonal variation in H_1.

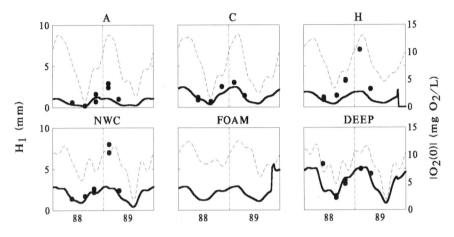

Fig. 15.19 Temporal variation of aerobic layer depth H_1. Model (solid line) and data (•) comparison. Overlying water dissolved oxygen (dashed line) for comparison.

15.4   LAKE CHAMPLAIN

Lake Champlain is a large freshwater body of water bordered by New York State to the west, Vermont to the east, and the Province of Quebec, Canada to the north. With a length of 170 km and a water surface area of 1,130 km², it is one of the larger lakes in North America. The average depth of Lake Champlain is 22.8 m with a maximum depth of 122 m. Lake Champlain is a morphologically complex body of water with numerous embayments (Fig. 15.20I). Due to its size, morphology, and loading sources, a variety of water quality conditions are found within its shores. Details of the lake's limnology, environment, and water uses have been previously reviewed by Meyer and Gruendling (1979).

As part of a study of the eutrophication of the lake, a series of sediment flux measurements were made using the same techniques employed in Chesapeake Bay (Cornwell and Owens, 1998a). Measurements were made in July and October of 1994 and April 1995 at the stations shown in Fig. 15.20I. The overlying water concentrations and the Fourier series approximations are shown in Fig. 15.20II. For the periods with no observations, synthetic data were employed. Either the early spring data were assumed to be applicable in the winter, or atmospheric saturation of dissolved oxygen and the normal temperature seasonal cycle were assumed. Also the April '95 data were assumed to occur in April '94 for convenience of application.

Since this is a stand-alone application of the model, the deposition of organic matter needs to be specified. A depositional flux for each station could be estimated. However, in the interests of simplicity we examine the results using the same depo-

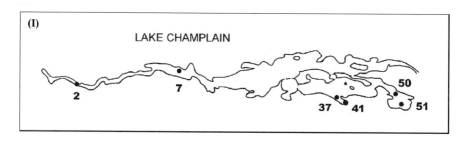

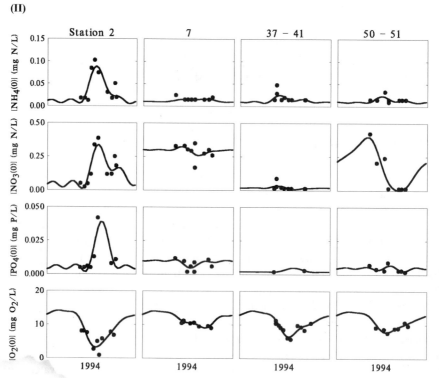

Fig. 15.20 Lake Champlain (I) Station locations. (II) Overlying water concentrations (•) and Fourier series approximations (lines).

sitional flux for each station: $J_{PON} = 25$ mg N/m^2-d. The model parameters are from the Chesapeake Bay calibration except for the phosphate partition coefficients, which are obtained by calibrating to these data specifically, as discussed below.

15.4 A   Fluxes, Pore Water, and Sediment Concentrations

The fluxes are shown in Fig. 15.21. The two flux measurements shown at each time reflect different overlying water conditions in the laboratory: at the in situ dissolved oxygen concentration and at atmospheric saturation. The changes in fluxes are not always in the expected direction. Perhaps the sediments did not equilibrate to the higher dissolved oxygen properly. Therefore, the model analysis is not definitive, but still instructive nonetheless.

The differences in model results are due solely to the variations in overlying water conditions (Fig. 15.20II). It is instructive and, perhaps somewhat surprising, to see how much variation results. The differing dissolved oxygen distribution is the principal cause, with the nitrate concentration influencing the nitrate fluxes as well.

The pore water and solid phase concentrations are shown in Fig. 15.22. The phosphorus results are satisfactory. The particulate organic phosphorus POP is reproduced nicely. Both the pore water and particulate inorganic phosphorus are also reproduced by calibrating the phosphate partition coefficients. However the pore water ammonia and nitrate concentrations are overestimated by the model. The pore water nitrate concentrations can be reduced by increasing the denitrification rates in the sediment: $\kappa_{NO_3,1}$ and $\kappa_{NO_3,2}$. Fig. 15.23 presents the results of doubling both these rates. The nitrate flux is now computed to be mostly into the sediment, rather than mostly out of the sediment (Fig. 15.21). It is difficult to decide definitively if this is more representative of the data. However, there is a dramatic improvement in the fit to the pore water nitrate concentration. A factor of two change in denitrification reaction velocities produced almost an order of magnitude change in concentration. The pore water ammonia concentration, however, is still higher than observed. It is possible to remedy this problem by increasing K_{L12} the mass transfer coefficient between layer 1 and 2 and adjusting other parameters as well. However, this model is designed to model fluxes first, and then pore water concentrations in an average way. The coarseness of the vertical segmentation (two layers) precludes a more refined analysis.

Table 15.3   Phosphate Partition Coefficients [a]

	Units	Calibration	MERL Mesocosms	Chesapeake Bay
$\Delta \pi_{PO_4,1}$	–	20.0	20.0	100.0
$\pi_{PO_4,2}$	(L/kg)	1000.0	20.0	300.0
$\Delta \pi_{PO_4,1} \times \pi_{PO_4,2}$	(L/kg)	2.0×10^4	4.0×10^2	3.0×10^4

[a] Used in Fig. 15.24

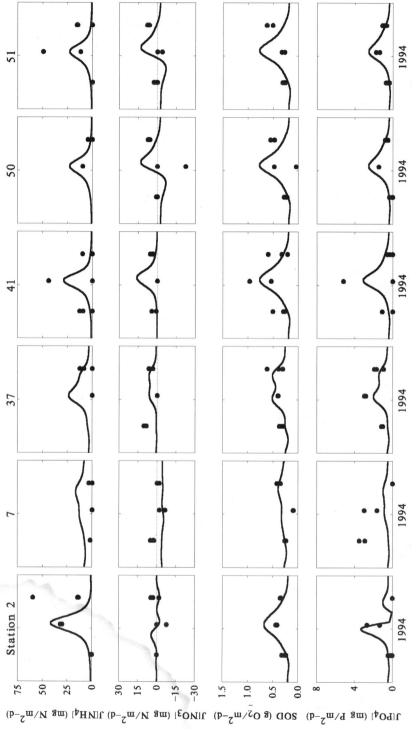

Fig. 15.21 Temporal variation of Lake Champlain sediment fluxes. Abscissa is time in years. Data (•) and model (lines) for ammonia, nitrate, SOD, and phosphate fluxes.

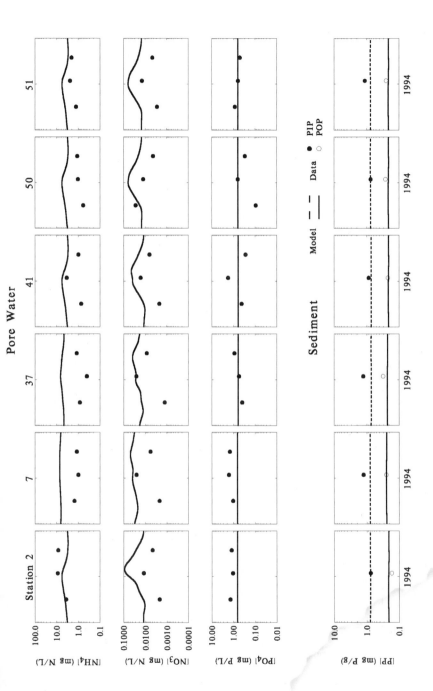

Fig. 15.22 Temporal variation of Lake Champlain pore water and sediment concentrations. Abscissa is time in years. Data (●) and model (lines) for ammonia, nitrate, and phosphate pore water concentrations, and sediment particulate inorganic phosphorus (PIP) and particulate organic phosphorus (POP).

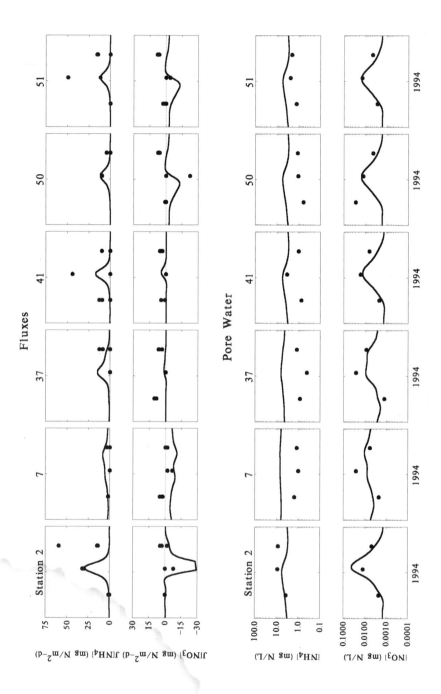

Fig. 15.23 Effect of a two-fold increase in $\kappa_{NO_3,1}$ and $\kappa_{NO_3,2}$ on ammonia and nitrate fluxes and pore water concentrations. Compare to Figs. 15.21 and 15.22.

Sensitivity to Phosphate Partitioning Of more interest is the effect of parameter changes on the modeled phosphorus fluxes. Fig. 15.24 presents the Station 2 results for three sets of phosphate partition coefficients $\Delta \pi_{PO_4,1}$ and $\pi_{PO_4,2}$: the calibrated values that result from fitting the Lake Champlain (A), and the MERL (B) and Chesapeake Bay (C) parameters (Table 15.3). The fluxes (1), pore water (2), and particulate phosphate (3) concentrations in the anaerobic layer are shown. The MERL partition coefficients (B) result in a higher flux, an almost identical pore water concentration [$PO_4(2)$], and a much lower particulate inorganic phosphate concentration PIP. The lower PIP is a consequence of the higher flux to the overlying water, and consequently less phosphate storage in the sediment. The Chesapeake Bay partition coefficients (C) are at the opposite end of the scale. They produce a lower flux, a higher pore water concentration, and the correct PIP result. It is interesting to note that the magnitudes of the layer 1 aerobic partition coefficients $\pi_{PO_4,1} = \Delta \pi_{PO_4,1} \times \pi_{PO_4,2}$ for the three cases (Table 15.3) suggest that it is the aerobic layer partition coefficient that controls the extent of phosphorus trapping and, therefore, the concentration of PIP in the sediment. Cases (A) and (C) have the larger $\pi_{PO_4,1}$'s and the correct PIP. Note also the variation in $\Delta \pi_{PO_4,2}$ between (A) and (C) – a factor of 4 – has almost exactly a four fold effect on layer 2 pore water phosphate concentration, but a much smaller effect on the flux.

The impression one gains from these results is that the pore water concentrations are a more sensitive function of the parameters than are the fluxes. This is indeed fortunate since it is highly desirable for a model to be transportable, without many – or in the best case any – parameter changes.

15.5 SUMMARY OF PARAMETER VALUES USED IN ALL APPLICATIONS

The parameters used in the applications discussed above are presented in Tables 15.5 and 15.6. Parameters from two other applications are also included. These are from eutrophication models for Massachusetts Bay (HydroQual, 1995) and Jamaica Bay, NY (HydroQual, 1996). The sediment data for these studies were collected by Giblin et al. (1997) and Cornwell and Owens (1998b) respectively.

The values which differ from the consensus are highlighted. The source of some of these differences is due to changes that were made in order to improve the model's performance. In many cases, the changes did not materially change the results. However, the parameters were not reset to their original values. These vestigial changes are not, therefore, necessarily indicative of significant differences. They simply are remnants from failed attempts at improving the calibration.

There are a few insignificant differences in the recycle fractions (Table 15.5). A change in the G_3 fraction affects the POM concentration of the sediment but does

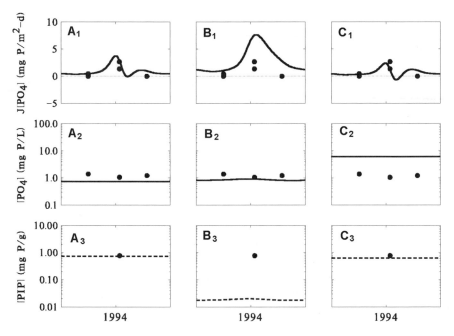

Fig. 15.24 Sensitivity of the station 2 phosphate flux (1), anaerobic layer pore water phosphate (2), and particulate inorganic phosphorus (3) concentrations to variations in phosphate partition coefficients (Table 15.3). Lake Champlain calibration (A), MERL (B), Chesapeake Bay (C).

little to affect the fluxes, since a change in the small nonreactive fraction makes little difference. There are no differences in the diagenesis kinetic constants.

The sedimentation velocities and solids concentrations are, in principle, measured values so any differences reflect the different settings. The differences in the mixing coefficients are restricted to the coupled models, which may be vestigial, and the increased pore water mixing in the MERL mesocosms due, presumably, to the action of the plunger. The benthic stress parameters are the same. These are important only during periods of severe anoxia, which is characteristic only of the Chesapeake Bay data set. Thus the fact that they have not been changed is no reflection on their universality.

The kinetic parameters are listed in Table 15.6. The nitrification parameters are essentially constant. For the freshwater sections of the coupled Chesapeake Bay model, it was found to be necessary to increase the rates (Cerco and Cole, 1995) somewhat, as indicated. If the denitrification rates used for Lake Champlain are increased, the pore water nitrate concentrations are reproduced (Fig. 15.23). Higher nitrification rates for freshwater sediments, when compared to estuarine sediments have been reported by Seitzinger et al. (1991) and confirmed by direct experiments (Rysgaard et al., 1999). Also, slightly higher denitrification rates appear to be appropriate for Jamaica Bay. No significant differences have been found for the sulfide and silica parameters.

For phosphorus, however, the variation is extreme (Table 15.4). The aerobic layer

Table 15.4 Phosphate Partition Coefficients (L/kg)

Variable	Chesa-peake Bay	Lake Cham-plain	Long Island Sound	Jamaica Bay	MERL	Massa-chusetts Bay
$\pi_{PO_4,1}$	3×10^4	2×10^4	1×10^4	1×10^4	4×10^2	4×10^2
$\pi_{PO_4,2}$	1×10^2	1×10^3	1×10^2	1×10^2	2×10^2	2×10^2

partition coefficients $\pi_{PO_4,1}$ vary from 3×10^4 for Chesapeake Bay sediments, to 1–2×10^4 for Lake Champlain, Long Island Sound and Jamaica Bay, to the low 4×10^2 for the MERL nutrient addition experiment. By contrast, the anaerobic layer partition coefficient $\pi_{PO_4,2}$ varies from 10^2 to 10^3, an order of magnitude. The variables that affect phosphate partitioning include pH, sediment composition, and solution ionic strength. The low $\pi_{PO_4,1}$ and $\pi_{PO_4,2}$ for the MERL experiment are almost certainly due to the high pH produced by the large algal biomass. The high $\pi_{PO_4,2}$ for Lake Champlain is probably due to the low solution ionic strength of freshwater when compared to seawater. However, these speculations would need to be confirmed by including a chemical model that actually computes the extent of phosphate sorption into the sediment flux model.

Table 15.5 Parameter Summary — Diagenesis, Sediment Properties, Mixing

Variable	Chesapeake Bay	MERL	Long Island Sound	Lake Champlain	Massachusetts Bay	Jamaica Bay
			Recycle fractions			
$f_{POC,N,P,1}$	0.65	0.65	0.65	0.65	0.65	0.65
$f_{POC,2}$	0.20	0.20	0.20	0.20	**0.25**	0.20
$f_{PON,2}$	0.25	0.25	0.25	0.25	0.25	0.25
$f_{POP,2}$	0.20	0.20	0.20	0.20	0.20	**0.25**
$f_{POC,3}$	0.15	0.15	0.15	0.15	**0.10**	0.15
$f_{PON,3}$	0.10	0.10	0.10	0.10	0.10	0.10
$f_{POP,3}$	0.15	0.15	0.15	0.15	0.15	**0.05**
			Diagenesis			
$k_{POC,N,P,1}$ (1/d)	0.035	0.035	0.035	0.035	0.035	0.035
$\theta_{POC,N,P,1}$	1.10	1.10	1.10	1.10	1.10	1.10
$k_{POC,N,P,2}$ (1/d)	0.0018	0.0018	0.0018	0.0018	0.0018	0.0018
$\theta_{POC,N,P,2}$	1.15	1.15	1.15	1.15	1.15	1.15
k_{Si} (1/d)	0.50	0.50	0.50	0.50	0.50	0.50
θ_{Si}	1.10	1.10	1.10	1.10	1.10	1.10
a_{CN} (g C/g N)	5.68	5.68	5.68	5.68	5.68	5.68
a_{CP} (g C/g P)	41.0	41.0	41.0	41.0	41.0	41.0
a_{CSi} (g C/g Si)	2.0	2.0	2.0	2.0	2.0	2.0
			Solids			
w_2 (cm/yr)	0.25	0.25	0.25	**0.20**	0.25	**0.25–0.75**
m_1 (kg/L)	0.50	0.50	0.50	0.50	0.50	0.2–1.2
m_2 (kg/L)	0.50	0.50	0.50	0.50	0.50	0.2–1.2
			Mixing			
D_d (cm^2/d)	5.0	**25.0**	5.0	5.0	**12.5**	**6.0–50.0**
θ_{D_d}	1.08	1.08	1.08	1.08	**1.15**	**1.1**
D_p (cm^2/d)	0.6	0.6	0.6	0.6	0.6	0.6
θ_{D_p}	1.117	1.117	1.117	1.117	**1.15**	**1.08**
H_2 (cm)	10.0	10.0	10.0	10.0	10.0	10.0
			Benthic stress			
k_S (1/d)	0.03	0.03	0.03	0.03	0.03	0.03
K_{M,D_p} (mg O$_2$/L)	4.0	4.0	4.0	4.0	4.0	4.0

Table 15.6 Parameter Summary – Kinetics

Variable	Chesa-peake Bay[#]	MERL	Long Island Sound	Lake Cham-plain	Massa-chusetts Bay	Jamaica Bay
Ammonia						
κ_{NH_4} (m/d)	0.131*	0.131	0.131	0.131	0.131	0.131
θ_{NH_4}	1.123	1.123	1.123	1.123	1.123	1.123
π_{NH_4} (L/kg)	1.0	1.0	1.0	1.0	1.0	1.0
K_{M,NH_4} (mg N/ m^3)	728.0	728.0	728.0	728.0	728.0	728.0
$\theta_{K_{M,NH_4}}$	1.125	1.125	1.125	1.125	1.125	1.125
K_{M,NH_4,O_2} (mg O$_2$/L)	0.37	0.37	0.37	0.37	**0.74**	0.37
Nitrate						
$\kappa_{NO_3,1}$ (m/d)	0.10[†]	0.10	0.10	**0.1–0.2**	0.10	**0.125**
$\kappa_{NO_3,2}$ (m/d)	0.25	0.25	0.25	**0.25–0.5**	0.25	0.25
θ_{NO_3}	1.08	1.08	1.08	1.08	1.08	1.08
Sulfide						
$\kappa_{H_2S,1}$ (m/d)	0.20	0.20	0.20	0.20	0.20	0.20
$\kappa_{H_2S,2}$ (m/d)	0.40	0.40	0.40	0.40	0.40	0.40
θ_{H_2S}	1.08	1.08	1.08	1.08	1.08	1.08
$\pi_{H_2S,1}$ (L/kg)	100	100	100	100	100	100
$\pi_{H_2S,2}$ (L/kg)	100	100	100	100	100	100
K_{M,H_2S,O_2} (mg O$_2$/L)	4.0	4.0	4.0	4.0	4.0	4.0
Silica						
k_{Si} (1/d)	0.50	0.50	0.50	0.50	0.50	**0.75**
θ_{Si}	1.10	1.10	1.10	1.10	1.10	1.10
[Si]$_{sat}$ (mg Si /L)	40.0	40.0	40.0	40.0	40.0	40.0
$K_{M,PSi}$ (g Si/L)	50.0	50.0	50.0	50.0	50.0	50.0
$\pi_{Si,1}$ (L/kg)	10.0	10.0	10.0	10.0	10.0	10.0
$\pi_{Si,2}$ (L/kg)	100	100	100	100	100	100
J_{DetrSi}	100	100	100	100	100	**90**
[O$_2$]$_{crit,Si}$	2.0	2.0	2.0	2.0	2.0	2.0
Phosphate						
$\Delta\pi_{PO_4,1}$	**300**[‡]	**20.0**	**100**	**20.0**	**20.0**	**100**
$\pi_{PO_4,2}$ (L/kg)	**100**	**20.0**	**100**	**1000**	**20.0**	**100**
[O$_2$]$_{crit,PO_4}$ (mg O$_2$/L)	2.0	2.0	2.0	2.0	2.0	2.0

[#]Saltwater–Freshwater values used in the Chesapeake Bay coupled model calibration (Cerco and Cole, 1994): *0.14–0.20 (m/d), [†]0.125–0.30 (m/d), [‡]300–3000.

16
Steady State and Time Variable Behavior

16.1 INTRODUCTION

The model calibrations presented in Chapters 14 and 15 demonstrate that the time variable sediment flux model is capable of reproducing the observations with reasonable fidelity. In this chapter we return to the steady state model for two reasons: to summarize the data sets and perform some sensitivity analyses. The model equations used for the steady state model are summarized in Appendix B. These are intended to further illuminate the model behavior and highlight some weaknesses.

We conclude with an analysis of the time to reach steady state. This is an important topic that bears directly on the question of the response time to changes in sediment depositional fluxes. The question is: If the amount of POM reaching the sediment is changed – as the result of remedial actions, for example – when are these changes reflected in changed nutrient and oxygen fluxes?

16.2 STEADY STATE MODEL

We begin the presentation of the steady state model with a description of the method chosen to compute the diagenesis flux from the ammonia flux. This method is useful both for steady state modeling, and to estimate yearly average fluxes (Chapter 15). Of all the methods presented in previous chapters, this is found to be both convenient and accurate.

16.2 A   Computing the Diagenesis Flux

In order to apply either the steady state or time variable model, it is necessary to estimate the diagenesis flux J_N – or equivalently the depositional flux J_{PON} – in some way. For the Chesapeake Bay data set, a rather elaborate least squares fitting method using the results of the time variable model is employed. J_N is chosen by examining the model fit to both the ammonia flux and the SOD (Section 3.4 B). Since these data are used for the primary model calibration, the refined estimation methods are appropriate. For the other data sets, we used a more straightforward method that depends only on the ammonia flux and the overlying water dissolved oxygen concentration for the estimate of J_N.

The relationship between ammonia diagenesis J_N – the rate at which organic matter is mineralized and ammonia is released to the pore water – and ammonia flux to the overlying water $J [NH_4]$ is (Eq. 3.19)

$$J[NH_4] = J_N \frac{s^2}{s^2 + \kappa_{NH_4,1}^2} \tag{16.1}$$

where s is the surface mass transfer coefficient

$$s = \frac{SOD}{[O_2(0)]} \tag{16.2}$$

and $\kappa_{NH_4,1}$ is the reaction velocity for nitrification in the aerobic layer. The SOD is related to ammonia diagenesis since as particulate organic matter decays to produce ammonia at a rate J_N, it also produces reduced organic carbon at a rate J_C. This, in turn, is eventually oxidized and results in SOD.

The SOD, and the fraction of reduced endproducts that are buried, are determined by the full sediment model equations (Appendix B). However, an approximate analysis can be made as follows: The ratio of carbon to nitrogen produced by diagenesis is closely approximated by the Redfield ratio $a_{C,N}$, the stoichiometric composition of reactive organic matter

$$J_C = a_{C,N} J_N \tag{16.3}$$

If all the organic carbon were oxidized, the SOD would be

$$SOD = a_{O_2,C} J_C = a_{O_2,C} a_{C,N} J_N = a_{O_2,N} J_N \tag{16.4}$$

where $a_{O_2,C}$, $a_{C,N}$, and $a_{O_2,N}$ are the oxygen to carbon, carbon to nitrogen and oxygen to nitrogen Redfield ratios, respectively. It is the oxygen required to oxidize the organic carbon to CO_2. Hence the SOD can be computed from ammonia diagenesis J_N using Eqs. (16.3–16.4). The equation for ammonia diagenesis as a function of ammonia flux and overlying water oxygen concentration is found by solving Eqs. (16.1–16.4) simultaneously. The result is

$$J_N = \frac{J[NH_4]}{3} + \sqrt[3]{d_1} + \frac{J[NH_4]^2}{\sqrt[3]{d_1}} \tag{16.5a}$$

where

$$d_1 = \frac{2J[NH_4]^3 a_{O_2,N}^2 + 27J[NH_4]\kappa_{NH_4,1}^2 O_2(0)^2}{54 a_{O_2,N}^2}$$

$$+ \frac{J[NH_4]\kappa_{NH_4,1} O_2(0)\sqrt{4J[NH_4]^2 a_{O_2,N}^2 + 27J[NH_4]\kappa_{NH_4,1}^2 O_2(0)^2}}{6\sqrt{3} a_{O_2,N}^2} \qquad (16.5b)$$

and $\kappa_{NH_4,1}$ is temperature adjusted using $\theta_{NH_4,1}$. The MACSYMA solution is presented in Fig. 16.1. This approximation assumes that all carbon diagenesis eventually becomes SOD, namely that losses due to sulfide fluxes and burial are negligible. From an analysis of Chesapeake Bay model fluxes, these losses amount to no more than 18% (Fig. 14.16).

The second source of error is due to the time lags that occur between the production of oxygen equivalents by carbon diagenesis and their eventual oxidation. These are caused by the formation and the subsequent oxidation of FeS(s). From an analysis of the steady state version of the sediment model (Section 9.6 D), it is known that the time lag effect causes an error of approximately a factor of two between the carbon diagenesis estimated from SOD assuming steady state, and the actual flux (Fig. 9.5). Thus, although these errors are not negligible, the approximations can be used so long as the magnitude of the error involved is recognized.

Figure 16.2 presents the results of using Eqs. (16.5) to estimate J_N and then using the full steady state model to estimate all the fluxes, including $J[NH_4]$. Since this is a circular computation – $J[NH_4]$ is used to estimate J_N, which is used to estimate $J[NH_4]$ – we expect perfect agreement. However, it is clear that for low ammonia fluxes, the model significantly underestimates the observed ammonia fluxes for the three data sets.

There are a number of possible causes for this problem. Eq. (16.1) ignores the overlying water contribution to the ammonia flux. If it is included Eq. (16.1) is replaced by Eq.(3.19)

$$J[NH_4] = J_N \frac{s^2}{s^2 + \kappa_{NH_4,1}^2} - [NH_4(0)] \left(\frac{1}{s} + \frac{s}{\kappa_{NH_4,1}^2} \right)^{-1} \qquad (16.6)$$

Using this equation in place of Eq. (16.1) does not materially complicate the resulting solution for J_N, which is shown in Fig. 16.3. In fact, this modification does not improve the results, which are indistinguishable from Fig. 16.2.

The key is to modify Eq. 16.4 by making $a_{O_2,N}$ a function of temperature T using the usual formulation

$$SOD = a_{O_2,N}\theta_{a_{O_2,N}}^{(T-20)} J_C \qquad (16.7)$$

For $\theta_{a_{O_2,N}}^{(T-20)} = 1.068$, the agreement is almost perfect as shown in Fig. 16.4. The temperature dependency (Eq. 16.7) is needed to compensate for the temperature dependencies in the full sediment model. The implication to be drawn from Eq. (16.7)

(c1)   eq1:jn*s^2-jnh4*(s^2+kappa[nh4]^2)=0

(d1)
$$jn\, s^2 - jnh4 \left(s^2 + \kappa_{nh4}^{\ 2} \right) = 0$$

(c2)   eq2:s=sod/o2

(d2)
$$s = \frac{sod}{o2}$$

(c3)   eq3:sod=a[o2,n]*jn

(d3)
$$sod = jn\, a_{o2,\,n}$$

(c4)   eq4:ev(eq1,eq2,eq3,eval)

(d4)
$$\frac{jn^3\, a_{o2,\,n}^{\ 2}}{o2^2} - jnh4 \left(\frac{jn^2\, a_{o2,\,n}^{\ 2}}{o2^2} + \kappa_{nh4}^{\ 2} \right) = 0$$

(c5)   sol:solve(eq4,jn)[3]

(d5)
$$jn = \frac{\left(\dfrac{2\, jnh4^3\, a_{o2,\,n}^{\ 2} + 27\, jnh4\, \kappa_{nh4}^{\ 2}\, o2^2}{54\, a_{o2,\,n}^{\ 2}} + \dfrac{jnh4\, \kappa_{nh4}\, o2 \sqrt{4\, jnh4^2\, a_{o2,\,n}^{\ 2} + 27\, \kappa_{nh4}^{\ 2}\, o2^2}}{6\sqrt{3}\, a_{o2,\,n}^{\ 2}} \right)^{1/3}}{1}$$
$$+ \frac{jnh4^2}{9 \left(\dfrac{2\, jnh4^3\, a_{o2,\,n}^{\ 2} + 27\, jnh4\, \kappa_{nh4}^{\ 2}\, o2^2}{54\, a_{o2,\,n}^{\ 2}} + \dfrac{jnh4\, \kappa_{nh4}\, o2 \sqrt{4\, jnh4^2\, a_{o2,\,n}^{\ 2} + 27\, \kappa_{nh4}^{\ 2}\, o2^2}}{6\sqrt{3}\, a_{o2,\,n}^{\ 2}} \right)^{1/3}} + \frac{jnh4}{3}$$

Fig. 16.1   MACSYMA solution of Eqs. (16.1) and (16.4).

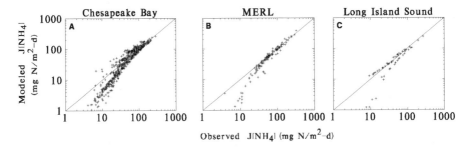

Fig. 16.2 Comparison of modeled and observed ammonia flux using Eqs. (16.4) and (16.5).

is that for cold temperatures, the steady state model converts only a portion of J_C into SOD. The remainder is buried.

The comparisons between data and the other fluxes computed by the steady state sediment model are shown in Fig. 16.5. The four data sets, analyzed using the time variable model in Chapters 14 and 15, are presented. The Chesapeake Bay parameter set is used (Tables 15.5 and 15.6). The computer program is listed in Appendix B. The point by point scatter is large for all the fluxes. For nitrate and SOD the model exhibits no apparent bias. However, for the phosphorus fluxes the model computes a significantly smaller range of fluxes than the observations. This is a direct result of the steady state assumption. The trapping mechanism for phosphate operates less effectively at steady state. For aerobic overlying water, the storage capacity of the sediment for phosphate reaches its maximum and the phosphate flux increases to its maximum, in comparison to the transient phase. A more detailed investigation of the steady state model is presented in the next section.

16.3 MODEL SENSITIVITY

The steady state model can be used to understand how the model responds to various changes in external parameters. These examples are chosen to illustrate certain points of interest.

16.3 A Depositional Flux

The principal driving force for fluxes is, of course, the depositional flux of POM. Fig. 16.6 presents the results for increasing J_{POC}. The resulting diagenesis and sediment-water fluxes are presented. The overlying water concentrations are zero for all constituents except dissolved oxygen $[O_2(0)] = 10$ mg O_2/L. The difference between diagenesis fluxes and fluxes to the overlying water at steady state is the amount either converted (ammonia to nitrate to nitrogen gas) or buried. Silica flux (D) and SOD (E) are essentially proportional to carbon diagenesis. Ammonia flux (A), and to a lesser extent, phosphate flux (C), exhibit some curvature in the response. The (c1) eq1:jn*s^2-jnh4*(s^2+kappa[nh4]^2)-kappa[nh4]^2*s*nh40=0

(d1)
$$- \text{jnh4}\left(s^2 + \kappa_{nh4}^2\right) + \text{jn}\, s^2 - \kappa_{nh4}^2 \, \text{nh40}\, s = 0$$

(c2) eq2:s=sod/o2

(d2)
$$s = \frac{\text{sod}}{\text{o2}}$$

(c3) eq3:sod=a[o2,n]*jn

(d3)
$$\text{sod} = \text{jn}\, a_{o2,n}$$

(c4) eq4:ev(eq1,eq2,eq3,eval)

(d4)
$$- \text{jnh4}\left(\frac{\text{jn}^2 a_{o2,n}^2}{\text{o2}^2} + \kappa_{nh4}^2\right) + \frac{\text{jn}^3 a_{o2,n}^2}{\text{o2}^2} - \frac{\text{jn}\, \kappa_{nh4}^2 \, \text{nh40}\, a_{o2,n}}{\text{o2}} = 0$$

(c5) sol:solve(eq4,jn)[3]

(d5)
$$\text{jn} = \frac{\kappa_{nh4}\, \text{o2}\sqrt{\left(\begin{array}{c}-\left(\text{jnh4}^2 \kappa_{nh4}^2 \, \text{nh40}^2 - 4\, \text{jnh4}^4\right) a_{o2,n}^2 \\ -\left(4\, \kappa_{nh4}^4 \, \text{nh40}^3 - 18\, \text{jnh4}^2 \kappa_{nh4}^2 \, \text{nh40}\right) \text{o2}\, a_{o2,n}^2 + 27 \\ *\, \text{jnh4}^2 \kappa_{nh4}^2 \, \text{o2}^2\end{array}\right)}}{6\sqrt{3}\, a_{o2,n}^2}$$

$$+ \frac{2\, \text{jnh4}^3 a_{o2,n}^2 + \text{jnh4}\, \kappa_{nh4}^2 \left(9\, \text{nh40}\, \text{o2}\, a_{o2,n} + 27\, \text{o2}^2\right)}{54\, a_{o2,n}^2}\Bigg)^{1/3}$$

$$+ \frac{\text{jnh4}^2 a_{o2,n} + 3\, \kappa_{nh4}^2 \, \text{nh40}\, \text{o2}}{9\, a_{o2,n}\left(\frac{\kappa_{nh4}\, \text{o2}\sqrt{\left(\begin{array}{c}-\left(\text{jnh4}^2 \kappa_{nh4}^2 \, \text{nh40}^2 - 4\, \text{jnh4}^4\right) a_{o2,n}^2 \\ -\left(4\, \kappa_{nh4}^4 \, \text{nh40}^3 - 18\, \text{jnh4}^2 \kappa_{nh4}^2 \, \text{nh40}\right) \text{o2}\, a_{o2,n}^2 + 27 \\ *\, \text{jnh4}^2 \kappa_{nh4}^2 \, \text{o2}^2\end{array}\right)}}{6\sqrt{3}\, a_{o2,n}^2} + \frac{2\, \text{jnh4}^3 a_{o2,n}^2 + \text{jnh4}\, \kappa_{nh4}^2 \left(9\, \text{nh40}\, \text{o2}\, a_{o2,n} + 27\, \text{o2}^2\right)}{54\, a_{o2,n}^2}\right)^{1/3}}$$

$$+ \frac{\text{jnh4}}{3}$$

Fig. 16.3 MACSYMA solution for J_N as a function of $J[NH_4]$ and $[NH_4(0)]$ using Eqs. (16.1, 16.2, and 16.6).

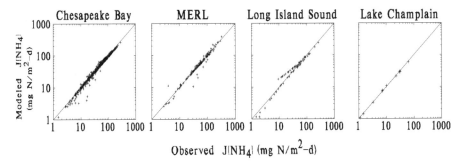

Fig. 16.4 Comparison of observed and modeled ammonia fluxes using Eqs. (16.5) and (16.7).

model computes a small sulfide flux (F) which increases nonlinearly as deposition increases. The nitrate flux (B) begins to increase sharply and then levels off.

These responses can be rationalized by examining the behavior of the model pore water and mass transfer coefficients (Fig. 16.7). The pore water concentrations are labeled with the layer number. The overlying water concentration (0) does not change. The aerobic (1) and anaerobic (2) layer concentrations vary as a result of increasing depositional flux. The changes are related to the variation in s, the surface mass transfer coefficient, and the aerobic layer depth H_1. Increasing the carbon depositional flux increases the SOD which decreases H_1. As a result nitrification decreases, increasing ammonia flux and decreasing nitrate flux. A smaller H_1 decreases the trapping efficiency, and the phosphate flux curves upward. Notice how difficult it is to explain the changes in pore water concentrations. The reason is that they are the result of changes in J_{POC} and H_1 rather than the causes of change. Fig. 16.7 also presents the solid phase concentrations which are varying linearly with the depositional flux. Note the preponderance of G_3 organic matter, and roughly the order of magnitude decrease in each of the other G classes.

16.3 B   Overlying Water Dissolved Oxygen

Fig. 16.8 presents the model response to varying overlying water dissolved oxygen. For anoxic conditions the overlying water dissolved oxygen is zero $[O_2(0)] = 0$ and so also is the aerobic layer depth $H_1 = 0$. As a result ammonia, sulfide, and phosphate fluxes are equal to the nitrogen, carbon (in oxygen equivalents), and phosphorus diagenesis fluxes respectively, and nitrate and SOD are zero. Silica flux is limited by the solubility constraint. As oxygen increases, the changes associated with increasing oxidation and phosphate partitioning occur. Ammonia flux decreases, nitrate flux increases, phosphate flux decreases, SOD increases – now that there is oxygen in the overlying water to oxidize the sulfide – and sulfide flux correspondingly decreases.

Pore water concentrations are presented in Fig. 16.9. The pore water concentrations are responding to the effect of increasing overlying water oxygen concentra-

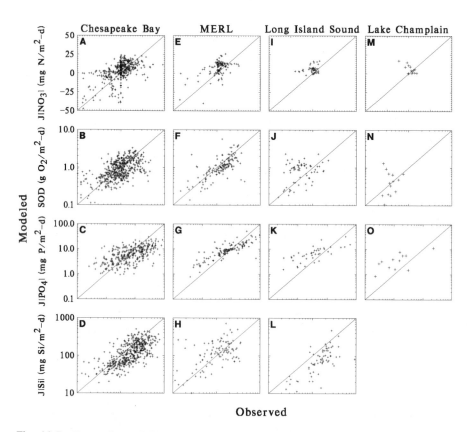

Fig. 16.5 Comparison of observed and modeled fluxes using the steady state model.

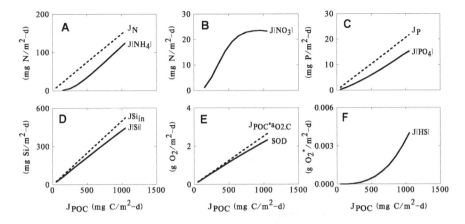

Fig. 16.6 Model response to changes in organic matter depositional flux.

tion, namely to increase the aerobic layer depth. The nitrification rate increases, so it might be expected that the aerobic layer ammonia concentration would decrease. However, since s, the surface mass transfer coefficient, also decreases as shown, the pore water concentration in layer 1 is forced to increase in order for the ammonia flux to escape. This is a good example of the fact that pore water concentrations adjust to circumstances rather than cause flux variations.

Nitrate concentrations increase as would be expected, due to the increasing nitrification. However, while the pore water concentrations are increasing, the nitrate flux levels off, reflecting the decrease in s.

Phosphate concentrations are increasing in both layers 1 and 2. The increase in layer 2 is the consequence of the reduction of the phosphate flux due to the increased trapping as H_1 increases. Since the phosphorus produced by diagenesis is a constant, the decreased flux causes an increase in storage of inorganic phosphorus in the sediment. This increase is reflected in both the pore water and solid phase concentrations, since they are linked by a linear partition coefficient.

The steady state silica flux is not materially affected by the overlying water dissolved oxygen since the extent of silica trapping is less due to its smaller partition coefficient. Therefore since s is decreasing, layer 1 silica concentrations must increase to supply the same flux to the overlying water. Again, pore water concentrations are adjusting to accommodate the changing mass transfer coefficient.

16.3 C Overlying Water Nitrate Concentrations

Changing the overlying water nitrate concentration results in a number of interesting changes due to the interrelationships in the model (Fig. 16.10). As expected, the nitrate flux from the sediment to the overlying water decreases and reverses to a negative flux from the overlying water to the sediment. What may not have been anticipated is the reduction in SOD that also occurs. This is a consequence of the

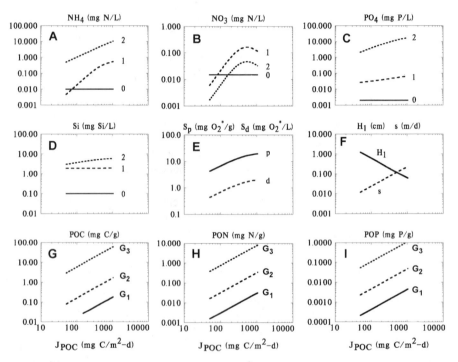

Fig. 16.7 Model pore water and particulate concentration responses to changes in organic matter depositional flux. Pore water concentration of (A) ammonia, (B) nitrate, (C) phosphate, and (D) silica. Numbers refer to the overlying water (0), aerobic (1) and anaerobic layer (2). Dissolved S_d and particulate S_p sulfide in the anaerobic layer (E). Aerobic layer depth H_1 and surface mass transfer s (F). G_1, G_2, and G_3 particulate organic carbon (G), nitrogen (H), and phosphorus (I) concentrations.

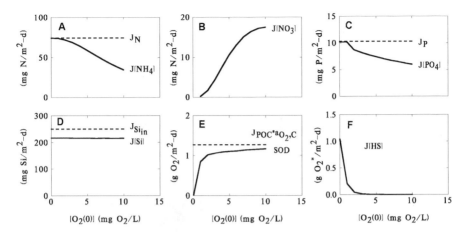

Fig. 16.8 Model flux responses to changes in overlying water dissolved oxygen concentration $[O_2(0)]$. (A) ammonia flux, (B) nitrate flux, (C) phosphate flux, (D) silica flux, (E) oxygen flux (SOD), and (F) sulfide flux. Diagenesis fluxes of ammonia and phosphate J_N, J_P, and the oxygen equivalents of the carbon diagenesis $a_{O_2,C}J_{POC}$ and the depositional flux of silica $J_{Si,in}$ also shown as dashed lines.

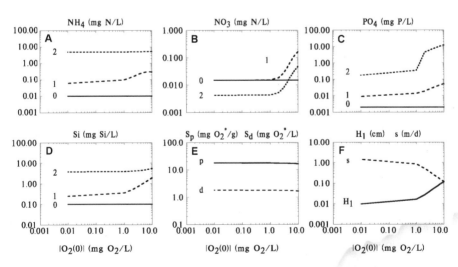

Fig. 16.9 Model pore water responses to changes in overlying water dissolved oxygen concentration. Pore water concentration of (A) ammonia, (B) nitrate, (C) phosphate, and (D) silica. Numbers refer to the overlying water (0), aerobic (1) and anaerobic layer (2). Dissolved S_d and particulate S_p sulfide in the anaerobic layer (E). Aerobic layer depth H_1 and surface mass transfer s (F).

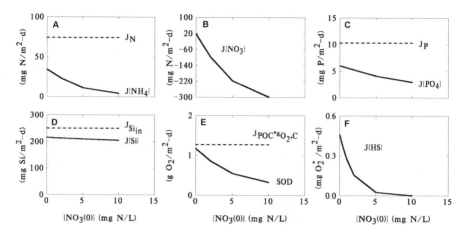

Fig. 16.10 Model flux responses to changes in overlying water nitrate concentration $[NO_3(0)]$. (A) ammonia flux, (B) nitrate flux, (C) phosphate flux, (D) silica flux, (E) oxygen flux (SOD), and (F) sulfide flux. Diagenesis fluxes of ammonia and phosphate J_N, J_P, and the oxygen equivalents of the carbon diagenesis $a_{O_2,C}J_{POC}$ and the depositional flux of silica $J_{Si,in}$ also shown as dashed lines.

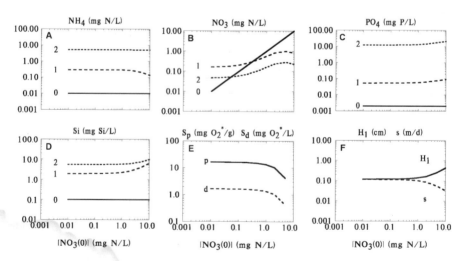

Fig. 16.11 Model pore water response to changes in overlying water nitrate concentration $[NO_3(0)]$. Pore water concentration of (A) ammonia, (B) nitrate, (C) phosphate, and (D) silica. Numbers refer to the overlying water (0), aerobic (1) and anaerobic layer (2). Dissolved S_d and particulate S_p sulfide in the anaerobic layer (E). Aerobic layer depth H_1 and surface mass transfer s (F).

denitrification reaction in which nitrate is the oxidizer of organic matter. Thus less organic matter is available for the reduction of sulfate to sulfide, and therefore, less oxygen is consumed by sulfide oxidation. The nitrate is replacing the oxygen as the oxidizer of organic matter. The rest of the changes are due to the reduction in SOD, which increases H_1 with consequences that are equivalent to increasing overlying water dissolved oxygen (Fig. 16.8).

The corresponding changes in pore water concentrations are shown in Fig. 16.11. The overlying water nitrate concentration is increasing and the reversal of the flux occurs when the layer 0 concentration exceeds that in layer 1. The consequence of the increasing consumption of organic matter by nitrate is seen in the decreasing concentration of particulate S_p and dissolved S_d sulfide concentrations. This, in turn lowers the SOD, with the consequences described above.

16.4   TIME TO STEADY STATE

The steady state version of the model is useful mainly as an instructional vehicle. It provides insight into the behavior of the model and, to the extent that the model reproduces reality, the behavior of sediments. However, for practical purposes the time variable model is necessary since it takes a relatively long time for the sediment to actually reach steady state. The purpose of this section is to determine the relevant time scales.

The method adopted is to examine the model response – which is termed its transient response – to an abrupt change in the input depositional fluxes. The transient response of the diagenesis model is considered first, followed by an analysis of the flux models. In particular, the response of the phosphate flux model is explored using an analytical investigation. Finally, the response of the full flux model is examined using numerical simulations. We begin with the response of the diagenesis kinetics.

16.4 A   Diagenesis

The transient responses for particulate organic carbon, nitrogen, and phosphorus are determined by the nature of the mass balance equations. Since these are similar, only the POC equations are explicitly analyzed. The conclusions apply to PON and POP as well. The mass balance equation for POC_i, the concentration of POC in the ith diagenesis class ($i = 1, 2,$ or 3) in the anaerobic layer at $T = 20°C$ is (Eq. 7.1)

$$H_2 \frac{dPOC_i}{dt} = -k_{POC,i} POC_i H_2 - w_2 POC_i + J_{POC,i} \qquad (16.8)$$

where: $k_{POC,i}$ is the first-order reaction rate coefficient, w_2 is the sedimentation velocity, and $J_{POC,i} = f_{POC,i} J_{POC}$ is the depositional flux of the ith G class of POC from the overlying water to the sediment. For constant coefficients, in particular, for constant temperature and depositional flux, this equation is easily solved

$$POC_i(t) = POC_i(0)e^{-\lambda t} + POC_i(\infty)(1 - e^{-\lambda t}) \qquad (16.9)$$

where

$$\lambda = k_{POC,i} + \frac{w_2}{H_2} \qquad (16.10)$$

and

$$POC_i(\infty) = \frac{J_{POC,i}}{k_{POC,i} H_2 + w_2} \qquad (16.11)$$

the final steady state concentration. The time it takes for the concentration of $POC_i(t)$ to change from the initial concentration $POC_i(0)$ to the final concentration $POC_i(\infty)$ is determined by the magnitude of the exponent λ. Its inverse τ is called the time constant of the equation

$$\tau \triangleq \frac{1}{\lambda} = \frac{1}{k_{POC,i} + w_2/H_2} \qquad (16.12)$$

When one, two and three time constants have elapsed, the model has reached 67%, 86%, and 95% of the new steady state value, respectively.

The time constant is related to the half-life, $t_{1/2}$, of POC_i in the sediment by the relationship

$$t_{1/2} = \frac{\ln(2)}{\lambda} = 0.693\tau \qquad (16.13)$$

For the three G classes, the time constants and half-lives are listed in Table 16.1. The

Table 16.1   Time Constants and Half Lives for Diagenesis[a]

	Units	G_1	G_2	G_3
τ	d	28.6	555.	14,600
	yr	0.078	1.52	40
$t_{1/2}$	d	19.8	385.	10,100
	yr	0.054	1.0	27.7

[a]Temperature = 20°C

magnitude of the reaction rates essentially determine the time constants for G_1 and G_2 as can be seen from Eq. (16.12). The sedimentation velocity w_2 and the active layer depth H_2 determine the time constant for G_3. Since G_3 is inert, its long time constant does not affect the response time of the diagenesis flux, which is controlled by the time constants of G_1 and G_2. Thus for ammonia and nitrate the time to 95% of steady state is approximately $3\tau \approx 4.5$ years, since no appreciable storage of these solutes occurs.

For the other fluxes, it is more difficult to determine the time constants because a significant amount of mass is in storage and this must be depleted in order to reach steady state. The case for phosphate is examined next.

16.4 B   Phosphate Flux

The equilibration time for the individual fluxes can be determined by an analysis of the governing equations. The method is simply to isolate λ. For phosphate, the mass balance equations for layers 1 and 2 are

$$0 = H_1 \frac{dC_{T1}}{dt} = s(C_{d0} - f_{d1}C_{T1}) + w_{12}(f_{p2}C_{T2} - f_{p1}C_{T1})$$

$$+ K_{L12}(f_{d2}C_{T2} - f_{d1}C_{T1}) - w_2 C_{T1} \tag{16.14a}$$

$$H_2 \frac{dC_{T2}}{dt} = -w_{12}(f_{p2}C_{T2} - f_{p1}C_{T1}) - K_{L12}(f_{d2}C_{T2} - f_{d1}C_{T1})$$

$$+ w_2(C_{T1} - C_{T2}) + J_P \tag{16.14b}$$

where C_{T1} and C_{T2} is the total (sorbed + dissolved) phosphate in layers 1 and 2. Adding equations (16.14) together yields

$$H_2 \frac{dC_{T2}}{dt} = s(C_{d0} - f_{d1}C_{T1}) - w_2 C_{T2} + J_P \tag{16.15}$$

which is an equation in both C_{T2} and C_{T1}. Solving the aerobic layer equation (16.14a) yields the relationship between the two concentrations

$$C_{T1} = \frac{sC_{d0} + (w_{12}f_{p2} + K_{L12}f_{d2})\, C_{T2}}{sf_{d1} + w_{12}f_{p1} + K_{L12}f_{d1} + w_2} \tag{16.16a}$$

$$C_{T1} = \frac{sC_{d0}}{sf_{d1} + w_{12}f_{p1} + K_{L12}f_{d1} + w_2} \tag{16.16b}$$

$$+ \frac{w_{12}f_{p2} + K_{L12}f_{d2}}{sf_{d1} + w_{12}f_{p1} + K_{L12}f_{d1} + w_2} C_{T2} \tag{16.16c}$$

$$= C_{OLW} + r_{12}C_{T2} \tag{16.16d}$$

Equation (16.16a) has two terms: The first, C_{OLW}, (Eq. 16.16b) involves the source of phosphate from the overlying water. The second term (Eq. 16.16c), which can be written in terms of r_{12} (Eq. 5.7), involves C_{T2} and, therefore, contributes to λ. Thus Eq. (16.15) becomes

$$H_2 \frac{dC_{T2}}{dt} = s(C_{d0} - f_{d1}r_{12}C_{T2}) - w_2 C_{T2} + J_P - sf_{d1}C_{OLW} \tag{16.17}$$

$$= (-sf_{d1}r_{12} - w_2)\, C_{T2} + sC_{d0} + J_P - sf_{d1}C_{OLW} \tag{16.18}$$

Collecting the terms that multiply C_{T2} yields

$$\frac{dC_{T2}}{dt} = -\lambda C_{T2} + S_T \tag{16.19}$$

where

$$\lambda = \frac{sf_{d1}r_{12} + w_2}{H_2} \qquad (16.20)$$

and S_T are the source terms that are independent of C_{T2}. Using the definition of the time constant τ (Eq. 16.12) yields

$$\tau = \frac{1}{\lambda} = \frac{H_2}{sf_{d1}r_{12} + w_2} \qquad (16.21)$$

Thus the time constant is determined by the magnitude of the loss terms in the denominator and the size of the active layer of the sediment in the numerator. The losses are the transfer of phosphate to the overlying water, and the loss to the deep sediment by burial.

Aerobic Overlying Water Fig. 16.12 presents an evaluation of Eq. (16.21) using the calibrated phosphate flux model parameters for both aerobic and anaerobic overlying water. Table 16.2 lists the values used. The time constants cannot be larger

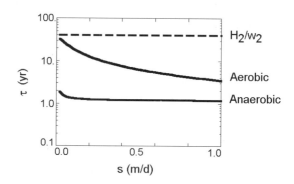

Fig. 16.12 Variation of the time constant τ as a function of the surface mass transfer coefficient s (Eq. 16.21).

than the time required for sedimentation to displace the sediment in the active layer: $\tau < H_2/w_2 = 10 \text{ cm} / 0.25 \text{ cm/yr} = 40 \text{ yr}$.

For the aerobic case, particulate mixing is larger than dissolved mixing and $r_{12} \approx 1$. Thus

$$\tau \approx \frac{H_2}{sf_{d1} + w_2} \qquad (16.22)$$

As the surface mass transfer coefficient s increases, the time constant decreases (Eq. 16.22) as shown in Fig. 16.12. The reduction in the time constant as s increases occurs because the stored sediment phosphorus can be lost at an increasing rate as a flux to the overlying water.

Table 16.2 Phosphate Flux Model Parameters for Time Constant Computation

Parameter[a]	Units	Aerobic Layer	Anaerobic Layer
$[O_2(0)]$	mg O_2/L	6	0.1
w_2	cm/yr	0.25	0.25
H_2	cm	10	10
D_p	m²/d	0.6×10^{-4}	1.5×10^{-6}
D_d	m²/d	5.0×10^{-4}	5.0×10^{-4}
m_1	kg/L	0.5	0.5
m_2	kg/L	0.5	0.5
π_1	L/kg	3×10^4	133
π_2	L/kg	100	100

[a] $w_{12} = D_p/(H_2/2)$, $K_{L12} = D_d/(H_2/2)$, $f_{di} = 1/(1 + m_i\pi_i)$ $i = 1, 2$, $f_{pi} = 1 - f_{di}$.

Since the overlying water is aerobic, the dissolved fraction f_{d1} is small and the phosphate flux is small as well. Using a typical annual average value for an aerobic station, $s = 0.2$ (m/d), yields a time constant of $\tau = 12.5$ years.

Anaerobic Overlying Water   For the anaerobic case the dissolved fraction in the aerobic layer is much larger due to the decrease in the aerobic layer partition coefficient. For this case surface mass transfer is no longer rate limiting. This can be seen from the expression for λ (Eq. 16.20) with r_{12} explicitly included (see Eq. 16.16d)

$$\lambda = \frac{sf_{d1}\left(\dfrac{w_{12}f_{p2} + K_{L12}f_{d2}}{sf_{d1} + w_{12}f_{p1} + K_{L12}f_{d1} + w_2}\right) + w_2}{H_2} \tag{16.23}$$

For sf_{d1} large relative to the particle and dissolved phase mixing coefficients, the expression simplifies to

$$\lambda \approx \frac{w_{12}f_{p2} + K_{L12}f_{d2} + w_2}{H_2} \tag{16.24}$$

Thus the rate limitation is the speed with which stored phosphate in the anaerobic layer can be transported to the aerobic layer by either particle mixing $w_{12}f_{p2}$ or interstitial water diffusion $K_{L12}f_{d2}$. Fig. 16.12 presents the results. Note that s no longer affects the time constant, which is now less than two years. For this calculation the minimum particle diffusion coefficient D_p is used (Table 16.2) because bioturbation would be suppressed by the low overlying water dissolved oxygen concentration. Since this time constant is much shorter than the aerobic case, the response is more rapid. The actual transient responses are examined in the next section where numerical results using the full model are presented.

16.4 C   Numerical Simulations

The transient response of the full time variable model is more complex than can be captured by a simple time constant analysis. The reason is that the various components of the model interact and affect the time variable behavior. In order to analyze a specific situation, the response to an abrupt decrease of the depositional flux to 1.0% of its value is examined. A decrease to zero is not used, since numerical difficulties can occur. Initially, the model is equilibrated to a constant depositional flux. All the overlying water concentrations are set to zero except oxygen. Two cases are presented: an aerobic and an anaerobic overlying water. Table 16.3 lists the parameter and input values specific to the transient response calculations.

Table 16.3   Transient Response Parameters

Parameter	Value	Units
J_{PON}	100.0	mg N/m^2-d
Aerobic $[O_2(0)]$	6.0	mg/L
Anaerobic $[O_2(0)]$	0.1	mg/L
Temperature	20.0	^oC

Aerobic Overlying Water   The results for aerobic overlying water are shown in Fig. 16.13. Particulate organic carbon (Fig. 16.13A) decreases exponentially following the time constant analysis given above. G_1 carbon decreases rapidly whereas G_2 carbon reacts more slowly. Although not shown, particulate organic nitrogen and phosphorus react similarly. Since G_1 and G_2 nitrogen are decreasing quickly, ammonia diagenesis, J_N, also decreases as shown in Fig. 16.13B. The ammonia flux, $J[NH_4]$, decreases even more rapidly because the depth of the aerobic layer is increasing, due to the reduction in sediment oxygen demand, as shown in Fig. 16.13C. SOD is decreasing, but slightly less slowly than J_C (in units of oxygen equivalents) because the stored sulfide is also being oxidized. The reason that SOD is slightly larger than J_C is that the oxygen consumed by nitrification is also included in the SOD. The decrease in surface mass transfer, s, as a result of the decrease in SOD is also shown.

The nitrate flux also decreases (Fig. 16.13D) but less slowly than the ammonia flux. The reason is that the initial increase in nitrification due to the increase in the depth of the aerobic zone provides additional nitrate. There is also an initial sharp decrease just after the abrupt drop in depositional flux. This is due to the initial sharp drop in surface mass transfer coefficient (Fig. 16.13C). However, the flux then increases. The reason is that the magnitude of the flux is related to the magnitude of the source of nitrate as well as the mass transfer coefficient. As shown in Fig. 16.14D, the aerobic layer nitrate concentration responds to the decrease in s by increasing in concentration, thereby increasing the flux.

This initial drop is much more apparent in the phosphate flux (Fig. 16.13E). Phosphate has a longer time constant than ammonia, nitrate, or SOD. After five years

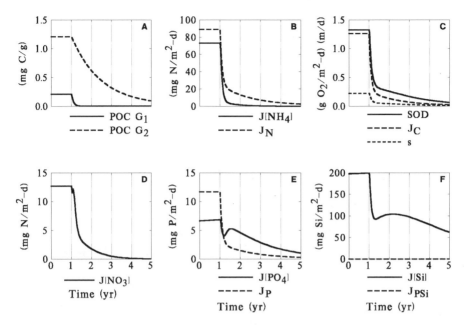

Fig. 16.13 Transient response to an abrupt reduction in depositional flux. Aerobic overlying water. (A) G_1 and G_2 carbon. (B) Ammonia diagenesis and ammonia flux. (C) SOD, carbon diagenesis J_C and surface mass transfer coefficient s. (D) Nitrate flux. (E) Phosphate diagenesis J_P and flux $J[PO_4]$. (F) Silica depositional flux $J_{P_{Si}}$ and flux to the overlying water $J[Si]$.

there is still a substantial phosphate flux even though phosphate diagenesis J_P has decreased to approximately 5% of its original value. The phosphate flux transient is projected to last for quite a long time since s has decreased (Fig. 16.13C) and the time constant for phosphate increases as s decreases (Fig. 16.12).

Silica (Fig. 16.13F) has the longest time constant as indicated from the results. Whereas the depositional flux of silica J_{PSi} drops abruptly, the silica flux remains elevated. The reason is that there is a substantial quantity of biogenic silica stored in the sediment and it provides the source for a continual supply. Additionally, the silica dissolution reaction is a function of the particulate silica concentration which is also decreasing.

The transient responses for the active layer solute concentrations are shown in Fig. 16.14. Ammonia (Fig. 16.14B) and nitrate (Fig. 16.14D) exhibit rapid declines characteristic of solutes that are not stored to a significant extent. Note the initial increase in nitrate concentration, due to the abrupt decrease in surface mass transfer coefficient. Sulfide (Fig. 16.14C) decreases more slowly, indicating further storage. Finally, phosphate (Fig. 16.14E) and silica (Fig. 16.14F) decrease only slightly during the first five years of the transient response.

Logarithmic plots for twenty years of simulation are shown in Fig. 16.15. Ammonia, nitrate, and SOD all reach their steady state values after ten years. However,

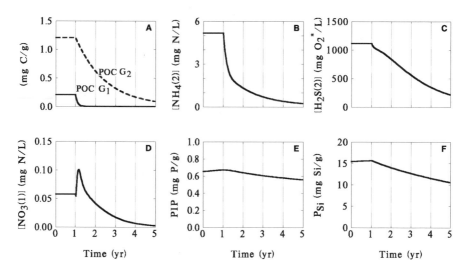

Fig. 16.14   Transient response to an abrupt reduction in depositional flux. Aerobic overlying water. (A) G_1 and G_2 carbon. (B) Pore water ammonia concentration [$NH_4(2)$]. (C) Total sulfide concentration [$\Sigma H_2S(2)$]. (D) Pore water nitrate concentration [$NO_3(1)$]. (E) Particulate inorganic phosphate concentration [PIP]. (F) Particulate silica concentration [P_{Si}].

both phosphate and silica are yet to reach their steady state values. The phosphate flux is still larger than the diagenesis flux, and both PIP and PSi are still declining after twenty years. Of course, the time constants of both phosphate and silica have as their upper bound the time required for sedimentation to replace the sediment in the active layer. After three time constants $3H_2/w_2$ have elapsed, virtually all memory of the previous depositional flux has been removed from the system and the sediment has equilibrated to the new depositional fluxes. At this point the transient response is over.

Anaerobic Overlying Water   The transient response for the situation where the overlying water is anaerobic is less complex then the preceding case. The oxygen concentration is set to 0.1 mg/L rather than zero to avoid numerical problems. The results are shown in Fig. 16.16 (top). Ammonia flux (B) is now equal to ammonia diagenesis since nitrification is limited by the low dissolved oxygen. It drops rapidly due to the G_1 decline and then more slowly, due to the slower G_2 decline (A). The situation for sediment oxygen demand is also more straightforward (C). Although carbon diagenesis decreases sharply (A), SOD, which is small to begin with due to the inhibition of sulfide oxidation by low dissolved oxygen, decreases slightly (C). The surface mass transfer coefficient (C), which is initially large, decreases somewhat as well. The nitrate flux (D), which is small to begin with, decreases further.

Phosphate flux (E) is almost equal to phosphate diagenesis since the trapping by the aerobic layer is very small due to the reduction of phosphate sorption in response to the low overlying water dissolved oxygen. However, the time constant is still in

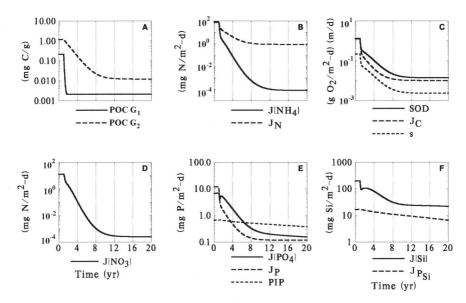

Fig. 16.15 Transient response to an abrupt reduction in depositional flux. Aerobic overlying water. (A) G_1 and G_2 carbon. (B) Ammonia diagenesis and ammonia flux. (C) SOD, carbon diagenesis J_C, and surface mass transfer coefficient s. (D) Nitrate flux. (E) Phosphate diagenesis J_P and flux $J[PO_4]$. (F) Silica depositional flux $J_{P_{Si}}$ and flux to the overlying water $J[Si]$.

excess of one year so that the transient extends beyond the first few years. Finally the silica flux (F) exhibits almost no reduction for the first five years, indicating that the transient for silica is much longer (Fig. 16.16F).

Logarithmic plots for twenty years of simulation are shown in Fig. 16.16 (bottom). Ammonia, nitrate, SOD, and phosphate all reach their steady state values after ten years . However, silica has yet to reach its steady state value although it is decreasing faster than the aerobic case. Again, the upper bound is set by the sedimentation rate and the depth of the active layer.

16.5 CONCLUSIONS

The transient response has a relatively short duration for ammonia, nitrate, and SOD. They are primarily determined by the time constant for G_2. The transient response for phosphate is of intermediate duration if the overlying water is aerobic, and is comparable to ammonia for anaerobic overlying water. For silica, the transient response is quite long, longer than the twenty years of simulated response time.

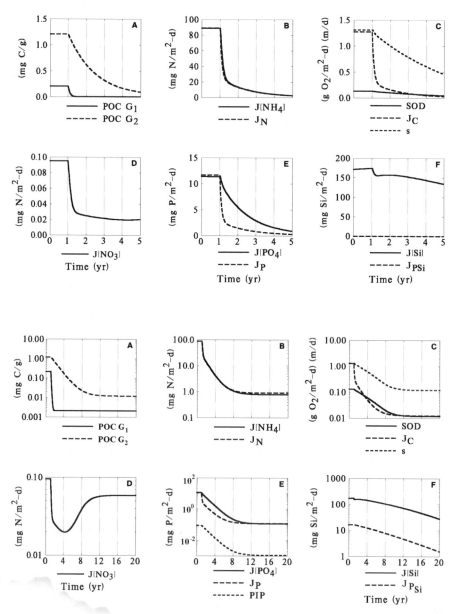

Fig. 16.16 Transient response to an abrupt reduction in depositional flux. Anaerobic overlying water. (A) G_1 and G_2 carbon. (B) Ammonia diagenesis and ammonia flux. (C) SOD, carbon diagenesis J_C, and surface mass transfer coefficient s. (D) Nitrate flux. (E) Phosphate diagenesis J_P and flux $J[PO_4]$. (F) Silica depositional flux $J_{P_{Si}}$ and flux to the overlying water $J[Si]$.

Appendix 16A: Model Equations

The model equations used in the steady state model are listed below. The solutions are found by numerically integrating the equations to steady state.

Diagenesis

$$H_2 \frac{d\text{POC}_i}{dt} = -k_{\text{POC},i}\theta_{\text{POC},i}^{(T-20)}\text{POC}_i H_2 - w_2\text{POC}_i + f_{\text{POC},i} J_{\text{POC}} \qquad (16\text{A}.1)$$

$$J_C = \sum_{i=1}^{2} k_{\text{POC},i}\theta_{\text{POC},i}^{(T-20)}\text{POC}_i H_2 \qquad (16\text{A}.2)$$

and similar equations for G_{PON} and G_{POP}, J_N and J_P.

Mass balance equations
These are the general equations for C_{T1} and C_{T2}.

$$H_1 \frac{dC_{T1}}{dt} = -\frac{\kappa_1^2}{s}C_{T1} + s(f_{d0}C_{T0} - f_{d1}C_{T1})$$

$$+ w_{12}(f_{p2}C_{T2} - f_{p1}C_{T1}) + K_{L12}(f_{d2}C_{T2} - f_{d1}C_{T1}) - w_2 C_{T1} + J_{T1} \qquad (16\text{A}.3)$$

$$H_2 \frac{dC_{T2}}{dt} = -\kappa_2 C_{T2} - w_{12}(f_{p2}C_{T2} - f_{p1}C_{T1})$$

$$- K_{L12}(f_{d2}C_{T2} - f_{d1}C_{T1}) + w_2(C_{T1} - C_{T2}) + J_{T2} \qquad (16\text{A}.4)$$

$$f_{di} = \frac{1}{1 + m_i \pi_i} \quad i = 1, 2 \qquad (16\text{A}.5)$$

$$f_{pi} = 1 - f_{di} \qquad (16\text{A}.6)$$

$$w_{12} = \frac{D_p \theta_{D_p}^{(T-20)}}{H_2} \frac{G_{\text{POC},1}}{G_{\text{POC,R}}} \frac{[O_2(0)]}{K_{M,D_p} + [O_2(0)]} \qquad (16\text{A}.7)$$

$$K_{L12} = \frac{D_d \theta_{D_d}^{(T-20)}}{H_2/2} \qquad (16\text{A}.8)$$

The kinetic and source terms for each solute are listed next. If the term is not listed, then it is zero.

Ammonia

$$\kappa_1^2 = \kappa_{NH4,1}^2 \theta_{NH4}^{(T-20)} \left(\frac{K_{M,NH4} \theta_{K_{M,NH4}}^{(T-20)}}{K_{M,NH4} \theta_{K_{M,NH4}}^{(T-20)} + [NH_4(1)]} \right) \left(\frac{[O_2(0)]}{2K_{M,NH4,O_2} + [O_2(0)]} \right)$$
(16A.9)

$$J_{T2} = J_N$$
(16A.10)

Nitrate

$$\kappa_1^2 = \kappa_{NO3,1}^2 \theta_{NO3}^{(T-20)}$$
(16A.11)

$$\kappa_2 = \kappa_{NO3,2} \theta_{NO3}^{(T-20)}$$
(16A.12)

$$J_{T1} = J_N - J[NH_4]$$
(16A.13)

Sulfide

$$\kappa_1^2 = \left(\kappa_{H_2S,dl}^2 f_{dl} + \kappa_{H_2S,pl}^2 f_{pl} \right) \theta_{H_2S}^{(T-20)} \frac{[O_2(0)]}{2K_{M,H_2S,O_2}}$$
(16A.14)

$$J_{T2} = a_{O_2,C} J_C - a_{O_2,NO_3} \left(\frac{\kappa_{NO3,1}^2}{s}[NO_3(1)] + \kappa_{NO3,2}[NO_3(2)] \right)$$
(16A.15)

Oxygen

$$CSOD = \frac{\left(\kappa_{H_2S,dl}^2 f_{dl} + \kappa_{H_2S,pl}^2 f_{pl} \right) \theta_{H_2S}^{(T-20)}}{s} \frac{[O_2(0)]}{2K_{M,H_2S,O_2}} [\Sigma H_2 S(1)]$$
(16A.16)

$$NSOD = a_{O_2,NH4} \frac{\kappa_{NH4,1}^2 \theta_{NH4}^{(T-20)}}{s} \left(\frac{K_{M,NH4} \theta_{K_{M,NH4}}^{(T-20)}}{K_{M,NH4} \theta_{K_{M,NH4}}^{(T-20)} + [NH_4(1)]} \right)$$

$$\left(\frac{[O_2(0)]}{2K_{M,NH4,O_2} + [O_2(0)]} \right) [NH_4(1)]$$
(16A.17)

Phosphate

$$\pi_1 = \pi_2(\Delta\pi_{PO_4,1}) \qquad\qquad [O_2(0)] > [O_2(0)]_{crit,PO_4} \quad (16A.18)$$

$$\pi_1 = \pi_2(\Delta\pi_{PO_4,1})^{\dfrac{[O_2(0)]}{[O_2(0)]_{crit,PO_4}}} \qquad [O_2(0)] \leqslant [O_2(0)]_{crit,PO_4} \quad (16A.19)$$

$$J_{T2} = J_P \qquad\qquad\qquad\qquad (16A.20)$$

Particulate Silica

$$H_2 \dfrac{dP_{Si}}{dt} = -S_{Si}H_2 - w_2 P_{Si} + J_{P_{Si}} + J_{DetrSi} \qquad (16A.21)$$

$$S_{Si} = k_{Si}\theta_{Si}^{(T-20)} \dfrac{P_{Si}}{P_{Si} + K_{M,P_{Si}}} \left([Si]_{sat} - f_{d2}[Si(2)]\right) \qquad (16A.22)$$

Silicate

$$\kappa_2 = k_{Si}\theta_{Si}^{(T-20)} \dfrac{P_{Si}}{K_{M,P_{Si}} + P_{Si}} f_{d2} H_2 \qquad (16A.23)$$

$$\pi_1 = \pi_2(\Delta\pi_{Si,1}) \qquad\qquad [O_2(0)] > [O_2(0)]_{crit,Si} \quad (16A.24)$$

$$\pi_1 = \pi_2(\Delta\pi_{Si,1})^{\dfrac{[O_2(0)]}{[O_2(0)]_{crit,Si}}} \qquad [O_2(0)] \leqslant [O_2(0)]_{crit,Si} \quad (16A.25)$$

$$J_{T2} = k_{Si}\theta_{Si}^{(T-20)} \dfrac{P_{Si}}{K_{M,P_{Si}} + P_{Si}} [Si]_{sat} H_2 \qquad (16A.26)$$

Part VI

Metals

The following chapters present models for metal fluxes. These are considerably more complex than the nutrient flux models because of the changing redox states for the metals considered: manganese and iron. Both oxidized and reduced manganese and iron need to be modeled explicitly. As a prelude, a simple model of calcium and alkalinity flux is considered in Chapter 17. The novel feature is explicitly considering calcium carbonate formation. This produces the first nonlinear set of algebraic equations for which MACSYMA is required. Manganese fluxes were also measured during the MERL experiments (Chapter 15), and a model is formulated and applied to these observations in Chapters 18 and 19. In Chapter 18 the normal progression is followed, a steady state model followed by a time variable analysis. Because of its importance, the influence of pH on the oxidation rate of Mn(II) is explicitly considered. In Chapter 19 a model of the overlying water is included with the same oxidation kinetics. This is the most interesting model, since the coupled behavior is instructive. This is followed by an iron flux model in Chapter 20, which is calibrated using field data from a lake and a reservoir. The final chapter presents a model for cadmium in sediments (Chapter 21). It focuses on the oxidation of cadmium sulfide and the liberation of dissolved cadmium to the pore water as well as the overlying water. The motivation for its construction is to aid in understanding the toxicity of metals in sediments. It uses a multilayered structure and it is the most complex and last of the models discussed in this book.

17

Calcium and Alkalinity

17.1 INTRODUCTION

This chapter presents the first of the models developed in this book that explicitly use the solubility of solid phases to determine the partitioning of species between dissolved and particulate forms. Models which explicitly consider both mass transport and chemical equilibrium are inherently complex and require numerical methods (Boudreau and Canfield, 1993, deRooij, 1991, Jahnke et al., 1994). In previous chapters, linear partition coefficients are employed to parameterize this important process – for solid and dissolved phase phosphate (Chapter 6) and sulfide (Chapter 9). In this chapter the solubility equations are explicitly included in the equation set, together with the usual mass balance equations. We present a model for calcium and alkalinity and include the possible formation of calcium carbonate.

What we find interesting about this model is the concept of excess calcium that emerges from the mathematical equations, which simplifies the understanding of calcium carbonate precipitation. Also there is an interesting relationship between the closed system chemical equations and their solutions, and the equations and solutions that result when the same chemistry is applied to the layered sediment model.

17.2 CALCIUM CARBONATE

Most freshwater and marine sediments contain large concentrations of calcium carbonate. Typical values are 10 to 100 mg $CaCO_3$/g or 1 to 10% of the dry weight (Green and Aller, 1998). This is in the same order of magnitude as the organic car-

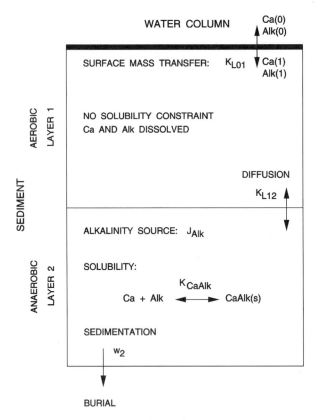

Fig. 17.1   Schematic diagram of the calcium and alkalinity flux model. The formation of $CaCO_3(s)$ is indicated by CaAlk(s), since this is the approximation used to simplify the model.

bon concentration in sediments. Calcium carbonate provides a buffer system for the pH in sediment pore water and also influences the concentration of bicarbonate and carbonate ions. The objective of the model formulated in this section is to reproduce the observations of pore water and solid phase concentrations of calcium, alkalinity, and calcium carbonate. The model structure is shown in Fig. 17.1.

The only chemical mechanism included in the model is the precipitation of calcium carbonate. Chemical speciation in the pore water is greatly simplified. Thus the model should be viewed more as a feasibility investigation into the methods and difficulties of including chemical equilibrium reactions into the sediment flux framework rather than a definitive model of calcium carbonate formation in sediments.

17.3   CHEMISTRY AND SIMPLIFICATIONS

A model of a chemical system is specified by the chemical components and the species formed by the components (Morel, 1983). This type of system is termed a closed system, since the concentrations of the components are thought of as being constant. This is the normal situation considered in typical chemical equilibrium calculations. The use of components in model formulations also applies to systems that are open (Di Toro, 1976), namely systems that have external sources and sinks of the components. This is the situation considered in this chapter.

The equilibrium chemistry of carbon dioxide, alkalinity, and calcium in natural waters is well understood.* It is specified by three components: calcium (Ca), alkalinity (Alk), and total inorganic carbon (TIC). However, using three components results in an equation set that is not conveniently solved. Therefore a simplified set of components and equations is used in the analysis presented below.

The equation that determines the solubility of calcium carbonate is

$$[Ca^{2+}][CO_3^{2-}] = K_{s1} \tag{17.1}$$

where the brackets [ ] denote molar concentrations and K_{s1} is the solubility product of calcium carbonate $CaCO_3(s)$. Ionic strength corrections are assumed to be included in K_{s1}. Since the objective of this exercise is to produce a tractable set of model equations, additional approximations are introduced. The concentration of Ca^{2+} is approximated with the concentration of total dissolved calcium denoted by [Ca]. This ignores the contributions of complexes of calcium with bicarbonate $CaHCO_3^+$ and other ligands to the total dissolved calcium. Thus

$$[Ca] = \left[Ca^{2+}\right] + \left[CaHCO_3^+\right] + \cdots \simeq \left[Ca^{2+}\right] \tag{17.2}$$

The concentration of carbonate can be obtained using the definition of alkalinity

$$[Alk] = [HCO_3^-] + 2[CO_3^{2-}] + [OH^-] - [H^+] \tag{17.3}$$

In the pH range that is typical of sediment pore waters, pH = 7–8, and for large enough alkalinity, [Alk] > 0.1 meq/L, the alkalinity is essentially all bicarbonate

$$[Alk] \simeq [HCO_3^-] \tag{17.4}$$

The deprotonation reaction for bicarbonate HCO_3^- which yields carbonate CO_3^{2-} is

$$HCO_3^- \leftrightarrow H^+ + CO_3^{2-} \tag{17.5}$$

and the mass action equation is

$$\frac{[H^+][CO_3^{2-}]}{[HCO_3^-]} = K_2 \tag{17.6}$$

where K_2 is the equilibrium constant for the reaction Eq. (17.5). Using Eq. (17.4) for HCO_3^- yields

$$\frac{[H^+][CO_3^{2-}]}{[Alk]} = K_2 \tag{17.7}$$

*Butler (1991), Loewenthal and Marais (1976), Morse and Mackenzie (1990), Stumm and Morgan (1981).

Thus the carbonate concentration becomes

$$[CO_3^{2-}] = \frac{K_2[Alk]}{[H^+]} \tag{17.8}$$

Substituting this equation in the solubility mass action equation (17.1) yields

$$[Ca][Alk] = \frac{K_{s1}}{K_2}[H^+] \triangleq K_{CaAlk} \tag{17.9}$$

where K_{CaAlk} is the apparent solubility constant of $CaCO_3(s)$ in terms of total dissolved calcium [Ca] and alkalinity [Alk]. It is an apparent or conditional solubility constant because it is a function of pH. The effect of decreasing pH, which increases H^+, increases the apparent solubility constant, as it should since $[CO_3^{2-}]$ is decreasing as pH decreases (Eq. 17.8).

This simplification reduces the number of components to two, [Ca] and [Alk], and requires only that the pH be specified. This is a significant reduction in the complexity of the equations.

17.4   CLOSED SYSTEM

Consider, initially, a closed system with no inflows or outflows. Let Ca_T and Alk_T be the total concentrations of calcium and alkalinity in the system. The mass balance equations are

$$[Ca] + [CaCO_3(s)] = Ca_T \tag{17.10}$$

and

$$[Alk] + 2[CaCO_3(s)] = Alk_T \tag{17.11}$$

where $[CaCO_3(s)]$ is the concentration of calcium carbonate that forms. Unlike the usual formulation*, the concentration of calcium carbonate is explicitly included in the mass balance equations. The reason is that the usual goal of the analysis is computing the pH of a saturated solution whereas for this model the concentration of calcium carbonate is required.

The term $2[CaCO_3(s)]$ in the alkalinity equation arises from the two equivalents of alkalinity in each mole of $CaCO_3(s)$ – see Eq. (17.3). Substituting these equations into the mass action equation (17.9), dropping the brackets and denoting $CaCO_3(s)$ by $CaCO_3$ for notational convenience yields

$$(Ca_T - CaCO_3)(Alk_T - 2CaCO_3) = K_{CaAlk} \tag{17.12}$$

which is a quadratic equation that can be solved for $CaCO_3$.

---

*Butler (1991), Loewenthal and Marais (1976), Stumm and Morgan (1970).

It is more convenient and instructive to solve, instead, for dissolved calcium Ca_d

$$Ca_d = Ca_T - CaCO_3 \qquad (17.13)$$

Finding the equation for Ca_d yields

$$Ca_d^2 - \left(Ca_T - \frac{1}{2}Alk_T\right)Ca_d - \frac{K_{CaAlk}}{2} = 0 \qquad (17.14)$$

The point is that the concentration of dissolved calcium is a function only of the difference $Ca_T - \frac{1}{2}Alk_T$ which we define as the excess calcium Ca_x

$$Ca_x \triangleq Ca_T - \frac{1}{2}Alk_T \qquad (17.15)$$

Hence, the equation for Ca_d is

$$Ca_d^2 - (Ca_x)Ca_d - \frac{K_{CaAlk}}{2} = 0 \qquad (17.16)$$

The reason that dissolved calcium concentration Ca_d is a function of only excess calcium Ca_x is a reflection of the chemical fact that changing the concentration of solid phase calcium carbonate in a solution that is already saturated – and therefore already in equilibrium with $CaCO_3(s)$ – does not affect the concentration of dissolved calcium. The system is already saturated with respect to calcium carbonate, so calcium carbonate is already present. Adding more calcium carbonate does not cause any chemical change.

To see that excess calcium is independent of $CaCO_3$, note that adding a quantity $\Delta CaCO_3$ of calcium carbonate to the system increases Ca_T by $\Delta CaCO_3$ and increases Alk_T by $2\Delta CaCO_3$, since each mole of calcium carbonate is two moles of alkalinity. However, the excess calcium Ca_x (Eq. 17.15) is unchanged and therefore, so is the dissolved calcium concentration since it depends only on the excess calcium concentration (Eq. 17.16).

Whereas the concentration of dissolved calcium is dependent on only excess calcium, the concentration of calcium carbonate does not have this property since adding calcium carbonate increases its concentration. Thus the concentration of calcium carbonate depends on both the concentration of total calcium and alkalinity.

The concentration of dissolved calcium is found by solving Eq. (17.16)

$$Ca_d = \frac{Ca_x}{2}\left(1 \pm \sqrt{1 + \frac{2K_{CaAlk}}{Ca_x^2}}\right) \qquad (17.17)$$

Once the dissolved calcium is known, the concentration of calcium carbonate follows from Eq. (17.10)

$$CaCO_3 = Ca_T - Ca_d \qquad (17.18)$$

We will see in Section 17.5 that the forms of the solutions for the equations that result from the sediment flux model are similar.

17.4 A Solution Behavior

The concentration of calcium carbonate is determined by the difference between the total and dissolved calcium, Eq. (17.10), which in turn is determined by the excess calcium (Eq. 17.17). While the idea of excess calcium is important to understanding the structure of the solution, it is nonetheless instructive to examine the solution in terms of the two independent variables: total calcium and alkalinity, Ca_T and Alk_T.

A simple way of approaching this problem is to obtain the relationship between the concentrations of Ca_T and Alk_T for a fixed concentration of $CaCO_3$. The solubility constraint is

$$[Ca][Alk] = (Ca_T - CaCO_3)(Alk_T - 2CaCO_3) = K_{CaAlk} \qquad (17.19)$$

For a fixed concentration of $CaCO_3$ this equation gives a relationship between Ca_T and Alk_T

$$Ca_T = \frac{K_{CaAlk}}{Alk_T - 2CaCO_3} + CaCO_3 \qquad (17.20)$$

For various concentrations of $CaCO_3$, the relationship between Ca_T and Alk_T is shown in Fig. 17.2.

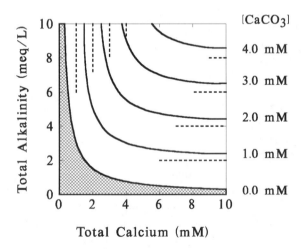

Total Calcium (mM)

Fig. 17.2 Concentration of $CaCO_3(s)$ as a function of total alkalinity and total calcium concentrations. $CaCO_3 = 0$ in the shaded region. The dotted lines are the stoichiometric limits to the quantity of $CaCO_3$ (s) that can form as Ca_T or $Alk_T \to \infty$.

The lack of symmetry in Fig. 17.2, which is implied by Eq. (17.9), is due to the requirement that two moles of alkalinity are required to form one mole of calcium carbonate, whereas only one mole of calcium suffices. If the graph were constructed with the axes: Ca_T and $2Alk_T$, then the graph would be symmetric in Ca_T and $2Alk_T$.

There is a range of Ca_T and Alk_T over which no precipitation occurs. This is signaled by the mathematics if either of the resulting dissolved concentrations are negative. That is, if either of the dissolved concentrations, computed from the relationships

$$Ca_d = Ca_T - CaCO_3 < 0 \tag{17.21a}$$

$$Alk_d = Alk_T - 2CaCO_3 < 0 \tag{17.21b}$$

is negative, that signals that the solubility constraint (Eq. 17.9) cannot be satisfied with chemically reasonable, namely with positive, concentrations and that the only chemically realistic solution is $CaCO_3 = 0$. The solubility constraint (Eq. 17.19) becomes

$$(Ca_T - CaCO_3)(Alk_T - 2CaCO_3) = (Ca_d)(Alk_d) \leqslant K_{CaAlk} \tag{17.22}$$

which means that the solution is undersaturated with respect to calcium carbonate. The shaded region in Fig. 17.2 indicates where no $CaCO_3$ can form. For larger concentrations of Ca_T and Alk_T, calcium carbonate forms as shown.

This example illustrates an important point. When computing the solutions of a chemical equilibrium problem from the algebraic equations that result from the mass balance and mass action equations, the solution must be tested to insure that it is physically realistic, namely that all the computed concentrations – the solid phase and dissolved concentrations – are positive. The reason is that the mathematical solutions of sets of equations where the inequalities that express the range of possibilities – saturated and unsaturated conditions, for example

$$(Ca_d)(Alk_d) \leqslant K_{CaAlk} \tag{17.23}$$

are replaced by equalities

$$(Ca_d)(Alk_d) = K_{CaAlk} \tag{17.24}$$

which are only true if the solid phase precipitates – do not necessarily result in positive concentrations. We will see that the need to test the realism of the solutions extends to the more complicated settings examined below.

17.5   SEDIMENT MODEL EQUATIONS AND SOLUTIONS

The model formulation is shown in Fig. 17.1. The dependent variables are Alk(1) and Alk(2), the total alkalinity in layer 1 and layer 2 respectively; Ca(1) and Ca(2), the total calcium in the same sequence; and $CaCO_3(2)$, the calcium carbonate in layer 2. Calcium carbonate precipitation in layer 1 is not considered since the layer is usually much thinner than layer 2. The mass balance and mass action equations are as follows

Layer 1 alkalinity

$$0 = s\big(\mathrm{Alk}(0) - \mathrm{Alk}(1)\big) + K_{L12}\big(\mathrm{Alk_d}(2) - \mathrm{Alk}(1)\big) \qquad (17.25a)$$

Layer 2 alkalinity

$$0 = -K\big(\mathrm{Alk_d}(2) - \mathrm{Alk}(1)\big) - w_2\big(2\mathrm{CaCO_3}(2)\big) + J_{\mathrm{Alk}} \qquad (17.25b)$$

Layer 1 calcium

$$0 = s\big(\mathrm{Ca}(0) - \mathrm{Ca}(1)\big) + K_{L12}\big(\mathrm{Ca_d}(2) - \mathrm{Ca}(1)\big) \qquad (17.25c)$$

Layer 2 calcium

$$0 = -K_{L12}\big(\mathrm{Ca_d}(2) - \mathrm{Ca}(1)\big) - w_2\big(\mathrm{CaCO_3}(2)\big) \qquad (17.25d)$$

CaCO$_3$(s) solubility

$$\mathrm{Ca_d}(2)\mathrm{Alk_d}(2) = K_{\mathrm{CaAlk}} \qquad (17.25e)$$

The dissolved calcium and alkalinity in layer 2 are

$$\mathrm{Ca_d}(2) = \mathrm{Ca}(2) - \mathrm{CaCO_3}(2) \qquad (17.26a)$$

$$\mathrm{Alk_d}(2) = \mathrm{Alk}(2) - 2\,\mathrm{CaCO_3}(2) \qquad (17.26b)$$

Since no calcium carbonate is allowed to precipitate in layer 1, the dissolved concentrations: $\mathrm{Ca_d}(1)$ and $\mathrm{Alk_d}(1)$ are equal to the total concentrations $\mathrm{Ca}(1)$ and $\mathrm{Alk}(1)$.

The solution to these equations can be found as follows. The variables $\mathrm{Alk}(1)$, $\mathrm{Alk}(2)$, $\mathrm{Ca}(1)$, and $\mathrm{Ca}(2)$ are eliminated from the five simultaneous equations (17.25a-17.25e). The resulting equation for $\mathrm{CaCO_3}(2)$ is a quadratic equation of the form

$$a\mathrm{CaCO_3^2} + b\mathrm{CaCO_3} + c = 0 \qquad (17.27)$$

where

$$a = 2\,(s + K_{L12})^2\, w_2^2 \qquad (17.28a)$$

$$b = -w_2\,(s + K_{L12})\left\{ \begin{array}{l} (2\mathrm{Ca}(0) + \mathrm{Alk}(0))s\,K_{L12} \\ +J_{\mathrm{Alk}}\,(s + K_{L12}) \end{array} \right\} \qquad (17.28b)$$

$$c = -s K_{L12}\left\{ \begin{array}{l} s K_{L12}\,[K_{\mathrm{CaAlk}} - \mathrm{Alk}(0)\mathrm{Ca}(0)] \\ -J_{\mathrm{Alk}}\mathrm{Ca}(0)\,(s + K_{L12}) \end{array} \right\} \qquad (17.28c)$$

The MACSYMA solution is presented in Fig. 17.3. It is remarkable that the solution to these five simultaneous equations is reduced to the solution of a quadratic equation, as it is for the closed system analyzed in the last section (Eq. 17.14). This suggests

Layer 1 Alkalinity:

(c2) $e1:0 = s*(Alk[0] - Alk[1]) + K[L12]*(Alk[2] - 2*CaAlk[2] - Alk[1])$

(d2) $$0 = \left(alk_0 - alk_1\right) s + \left(- 2\ caalk_2 + alk_2 - alk_1 \right) k_{l12}$$

Layer 2 Alkalinity:

(c3) $e2:0 = - K[L12]*(Alk[2] - 2*CaAlk[2] - Alk[1]) - w2*(2*CaAlk[2]) + J[Alk]$

(d3) $$0 = - 2\ caalk_2\ w2 - \left(- 2\ caalk_2 + alk_2 - alk_1 \right) k_{l12} + j_{alk}$$

Layer 1 Calcium:

(c4) $e3: s*(Ca[0] - Ca[1]) + K[L12]*(Ca[2] - CaAlk[2] - Ca[1])$

(d4) $$\left(ca_0 - ca_1\right) s + \left(- caalk_2 + ca_2 - ca_1 \right) k_{l12}$$

Layer 2 Calcium:

(c5) $e4: 0= - K[L12]*(Ca[2] - CaAlk[2] - Ca[1]) - w2*(CaAlk[2])$

(d5) $$0 = - caalk_2\ w2 - \left(- caalk_2 + ca_2 - ca_1 \right) k_{l12}$$

CaCO3 Solubility

(c6) $e5:(Ca[2] - CaAlk[2])*(Alk[2]-2*CaAlk[2]) = K[CaAlk]$

(d6) $$\left(alk_2 - 2\ caalk_2\right) \left(ca_2 - caalk_2\right) = k_{caalk}$$

Eliminate everything except CaAlk[2]

(c7) $eq1:facsum(eliminate([e1,e2,e3,e4,e5],[Alk[1],Alk[2],Ca[1],Ca[2]])[1],CaAlk[2])$

(d7)
$$2\ caalk_2^{\ 2} \left(s + k_{l12}\right)^2 w2^2 - caalk_2 \left(s + k_{l12}\right)$$
$$* \left(2\ ca_0\ k_{l12}\ s + alk_0\ k_{l12}\ s + j_{alk}\ s + j_{alk}\ k_{l12}\right) w2 - k_{l12}\ s$$
$$* \left(k_{caalk}\ k_{l12}\ s - alk_0\ ca_0\ k_{l12}\ s - ca_0\ j_{alk}\ s - ca_0\ j_{alk}\ k_{l12}\right)$$

Find the degree of the polynomial

(c8) $powers(ratexpand(eq1),CaAlk[2])$

(d8) $$[0, 1, 2]$$

Coefficients of the polynomial

(c9) $facsum(ratcoef(ratexpand(eq1),CaAlk[2],2))$

(d9) $$2 \left(s + k_{l12}\right)^2 w2^2$$

(c10) $facsum(ratcoef(ratexpand(eq1),CaAlk[2],1))$

(d10) $$- \left(s + k_{l12}\right) \left(2\ ca_0\ k_{l12}\ s + alk_0\ k_{l12}\ s + j_{alk}\ s + j_{alk}\ k_{l12}\right) w2$$

(c11) $facsum(ratcoef(ratexpand(eq1),CaAlk[2],0))$

(d11) $$- k_{l12}\ s \left(k_{caalk}\ k_{l12}\ s - alk_0\ ca_0\ k_{l12}\ s - ca_0\ j_{alk}\ s - ca_0\ j_{alk}\ k_{l12}\right)$$

Fig. 17.3 MACSYMA solution of Eq. (17.27).

that so long as the chemistry can be simplified to the point that the equations for a closed system reduce to a manageable result, then the open system equations can also be solved at the same level of algebraic difficulty. We will subsequently see that this observation continues to be true for more complex systems.

Once $CaCO_3$ has been found, the fluxes follow from solving Eqs. (17.26) for the dissolved concentrations in layer 2, and Eqs. (17.25a) and (17.25c) for the layer 1 concentrations. The fluxes of calcium and alkalinity to the sediment are found from the mass transfer relationships

$$J[Alk] = s\left(Alk\,(1) - Alk\,(0)\right) \tag{17.29}$$

$$J[Ca] = s\left(Ca\,(1) - Ca\,(0)\right) \tag{17.30}$$

17.5 A   Simplified Solution

Before examining the solution of Eq. (17.27), it is instructive to consider the limiting case where $K_{CaAlk} \to 0$ which corresponds to a completely insoluble calcium carbonate. In fact, this is a somewhat unrealistic assumption as can be seen in Fig. 17.2 where the dotted lines also represent the contours for the $K_{CaAlk} \to 0$ case. Nevertheless, the resulting solution is instructive.

For $K_{CaAlk} = 0$, Eq. (17.25e) requires that either Ca_d (2) or Alk_d (2) be zero or, using the definitions (Eqs. 17.26)

$$CaCO_3(2) = \frac{1}{2}Alk(2) \tag{17.31}$$

or

$$CaCO_3(2) = Ca(2) \tag{17.32}$$

The conditions are that either all the alkalinity (Eq. 17.31) or all the calcium (Eq. 17.32) becomes calcium carbonate, whichever is the least abundant. Assuming that alkalinity is in the shortest supply so that Eq. (17.31) is true, Eqs. (17.25a–17.25b) become

$$0 = s\left(Alk(0) - Alk(1)\right) + K_{L12}\left(-Alk(1)\right) \tag{17.33}$$

and

$$0 = K_{L12}\left(-Alk(1)\right) - w_2\left(2CaCO_3(2)\right) + J_{Alk} \tag{17.34}$$

Solving Eq. (17.33) for Alk(1) yields

$$Alk(1) = \frac{s}{s + K_{L12}}Alk(0) \tag{17.35}$$

and solving Eq. (17.34) for $CaCO_3(2)$ yields

$$CaCO_3(2) = \frac{1}{2w_2}\left[J_{Alk} + \left(\frac{1}{s} + \frac{1}{K_{L12}}\right)^{-1}Alk(0)\right] \tag{17.36}$$

To understand this solution, express this equation as the burial flux of calcium carbonate $J_{CaCO_3(s)} = w_2 CaCO_3$ (2)

$$J_{CaCO_3(s)} = w_2 CaCO_3(2) = \frac{1}{2}\left[J_{Alk} + \left(\frac{1}{s} + \frac{1}{K_{L12}} \right)^{-1} Alk(0) \right] \quad (17.37)$$

The concentration of calcium carbonate $CaCO_3(2)$ is determined by a balance between the loss by burial $w_2\, CaCO_3(2)$ – the left-hand side of the equation – and the sources of alkalinity to the sediment on the right-hand side of Eq. (17.37). For example, if the right-hand side of Eq. (17.37) increases, then the concentration $CaCO_3(2)$ must increase so that Eq. (17.37) remains true.

There are two sources of alkalinity. Alkalinity is transferred to the sediment via diffusion from the overlying water. The reciprocal of the sum of the reciprocals of the mass transfer coefficients $(1/s + 1/K_{L12})^{-1}$ is the total mass transfer resistance for transferring overlying water alkalinity $Alk(0)$ to the anaerobic layer. This source is added to the source of alkalinity due to sulfate reduction J_{Alk}. Since $CaCO_3$ is assumed to be completely insoluble, both of these sources of alkalinity are transformed entirely into calcium carbonate. The $1/2$ reflects the two moles of alkalinity that are required to make one mole of calcium carbonate, Eq. (17.11).

If the calcium is in shortest supply, then Eq. (17.32) is true. Using this result and solving Eqs. (17.25c-17.25d) for $J_{CaCO_3(s)}$ yields

$$J_{CaCO_3(s)} = w_2 CaCO_3(2) = \left(\frac{1}{s} + \frac{1}{K_{L12}} \right)^{-1} Ca(0) \quad (17.38)$$

where the calcium carbonate concentration is now determined by a balance between the loss of calcium via sedimentation of calcium carbonate $w_2\, CaCO_3(s)$ and the source of calcium from the overlying water $(1/s + 1/K_{L12})^{-1} Ca(0)$.

17.5 B   Parameter Estimates

The terms in the calcium and alkalinity model equations (Eqs. 17.27 and 17.28) that need to be estimated in order to obtain numerical results, in addition to the overlying water concentrations, are the surface mass transfer coefficient s, which controls the rate of mass transfer from the overlying water to the pore water, and the sediment alkalinity source J_{Alk}. Both of these can be related to the ammonia flux $J[NH_4]$ and the overlying water dissolved oxygen concentration. For a specified $J[NH_4]$ and $[O_2(0)]$, the ammonia diagenesis J_N, SOD and s can be computed. The equations are derived in Section 16.2 A.

In addition, there is also a sediment source of alkalinity J_{Alk}, which is related to the electron acceptor being consumed by carbon diagenesis. This can also be computed from ammonia diagenesis J_N and Redfield stoichiometry to obtain carbon diagenesis J_C. If it is assumed that all of the organic carbon diagenesis reduces sulfate to sulfide, then the following formula applies (Chapter 1)

$$CH_2O + \frac{1}{2}SO_4^{2-} \rightarrow HCO_3^- + \frac{1}{2}H_2S + H_2O \quad (17.39)$$

Therefore each mole of organic carbon reacted produces one equivalent of bicarbonate alkalinity.

However, this conclusion depends on which endproduct is assumed for the sulfide formed, in this case H_2S. If iron sulfide is the final repository of the sulfide, which is assumed in the sulfide and oxygen flux models (Chapter 9), then the reaction is

$$CH_2O + \frac{4}{9}FeOOH(s) + \frac{4}{9}SO_4^{2-} \rightarrow \frac{1}{9}CO_2 + \frac{8}{9}HCO_3^- + \frac{4}{9}FeS(s) + \frac{7}{9}H_2O$$

$$(17.40)$$

which is almost a 1 to 1 molar ratio between CH_2O mineralized and alkalinity HCO_3^- produced. The difference between Eq. (17.39) and Eq. (17.40) is the $1/9$th mole of organic matter required to reduce the Fe(III) to Fe(II) which does not produce alkalinity.

Finally if methane is the endproduct of diagenesis (Section 11.6 A)

$$CH_2O \rightarrow \frac{1}{2}CO_2 + \frac{1}{2}CH_4 \qquad (17.41)$$

then no alkalinity is produced if the methane escapes to the overlying water. However, if the dissolved methane is subsequently oxidized with sulfate as the oxidant, then the reaction is

$$\frac{1}{2}CO_2 + \frac{1}{2}CH_4 + \frac{1}{2}SO_4^{2-} \rightarrow HCO_3^- + \frac{1}{2}H_2S + H_2O \qquad (17.42)$$

and again the stoichiometric ratio is 1:1.

An analysis of the pore water relationships in Chesapeake Bay sediments (Fig. 12.1D) indicated that a 0.5 to 1 SO_4^{2-}/alkalinity stoichiometry ($= 1{:}1$ CH_2O/alkalinity) is actually observed. We adopt this stoichiometry for the calculations presented below.

17.6   APPLICATION TO LONG ISLAND SOUND

The data for this application come from observations of fluxes from three stations in Long Island Sound.* The parameters required for the calcium-alkalinity flux model are the usual transport parameters as well as those specific to these components (Table 17.1).

17.6 A   Chemical Parameters

The only chemical parameter required is the apparent solubility of $CaCO_3$, K_{CaAlk}. Fig. 17.4 presents the pore water data for dissolved calcium (Ca) and alkalinity (Alk) for three years from three Long Island Sound stations. The product: Ca × Alk =

---

*Aller (1980b), Aller and Yingst (1980).

Table 17.1 Parameter Values for the Calcium Carbonate Model

Parameter	Units	Value	Parameter	Units	Value
K_{L12}	(m/d)	0.001	Ca(0)	(mM)	9.0
w_2	(cm/yr)	0.5	Alk(0)	(mM)	2.0
$\kappa_{NH_4,1}$	(m/d)	0.15	$O_2(0)$	(mg O_2/L)	5.0
$a_{O_2,N}$†	(g O_2/g N)	2.54×5.68	K_{CaAlk}	(mM^2)	10–30

†$a_{O_2,N} = a_{O_2,C} \times a_{C,N}$

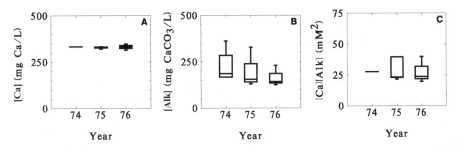

Fig. 17.4 Box plots of pore water concentration of (A) dissolved calcium Ca_d, (B) dissolved alkalinity Alk_d, and (C) the apparent solubility product $Ca_d \times Alk_d$ from three Long Island Sound stations. Years as indicated.

K_{CaAlk} is also presented. It ranges from approximately 20 mM^2 to 40 mM^2. This sets the range for K_{CaAlk}. With the parameters established, the model is evaluated by specifying an ammonia flux and computing the resulting concentrations.

17.6 B   Results

Fig. 17.5 presents the results for $K_{CaAlk} = 10$–30 mM^2. Alkalinity flux increases as ammonia flux increases (Fig. 17.5A). This is due to the increased production of alkalinity as a consequence of the increase in sulfate reduction (Eq. 17.39) in response to the increase in organic matter deposition that must have occurred to fuel the increase in ammonia flux. For $K_{CaAlk} = 30$ mM^2 the solubility of $CaCO_3$(s) is exceeded for $J[NH_4] > 10$ mg N/m^2-d and $CaCO_3$(s) starts to form (Fig 17.5E). At that point the calcium flux changes from zero to negative (Fig. 17.5B) which is the flux into the sediment that is required to support the burial of $CaCO_3$(s). Also the flux of alkalinity from the sediment to the overlying water increases less rapidly (Fig. 17.5A) since a portion is being buried. For smaller solubility products, $K_{CaAlk} = 10$ and 20 mM^2, $CaCO_3$(s) forms over the entire range of $J[NH_4]$ investigated (Fig. 17.5E) and therefore there is a corresponding calcium flux to the sediment (Fig. 17.5B).

Pore water concentrations are compared in Fig. 17.5C and D. The data are plotted at $J[NH_4] = 10$ mg N/m^2-d for convenience only. The annual average ammonia flux is actually somewhat larger. Alkalinity and calcium are reproduced for $K_{CaAlk} = 30$ mM^2. Calcium concentrations are essentially equal to the overlying water concen-

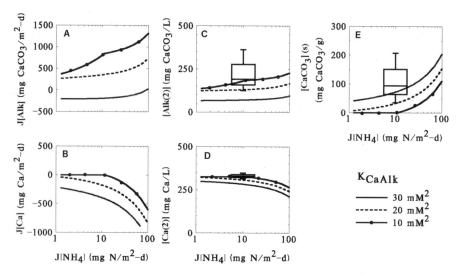

Fig. 17.5 Comparison of model results (lines) to Long Island Sound flux and pore water data (box plots) for three apparent solubility products K_{CaAlk}. (A) Alkalinity flux $J[Alk]$, (B) calcium flux $J[Ca]$, (C) pore water alkalinity $Alk_d(2)$, (D) pore water calcium $Ca_d(2)$, (E) calcium carbonate $CaCO_3(2)$. Data plotted at $J[NH_4] = 10$ mg N/m²-d for convenience.

trations (Table 17.1) whereas the alkalinity is larger, due to the additional source that results from sulfate reduction.

The concentration of $CaCO_3(s)$ is computed to increase from less than 10 mg $CaCO_3(s)$/g (1% of dry weight) to over 100 mg $CaCO_3(s)$/g (10%). Long Island Sound sediments contain between 25 and 200 mg/g. The model reproduces the observations for the lower values of the solubility constant. This is in contrast to the pore water results. Perhaps there is an additional source of calcium carbonate to the sediments, such as $CaCO_3(s)$ from calcareous animals (Green et al., 1993), which accounts for the additional $CaCO_3(s)$.

17.7   CONCLUSION

The model is reasonably successful in reproducing the general features of the data. Pore water calcium concentrations are predicted to be close to the overlying water value, 9 mM = 326 mg/L, whereas alkalinity is predicted to be larger than the overlying value of 2 meq = 100 mg/L. Calcium carbonate concentrations are predicted to be in the range of 10 to 100 mg/g which is the range of the observations. These results are obtained using transport parameters that are calibrated from Chesapeake Bay sediment flux data, suggesting that these parameters are also representative for these Long Island Sound sediments. In particular, the sedimentation velocity for these sediments is quite close to 0.5 cm/yr used for these calculations.

18

Manganese I: Sediment Flux

18.1   INTRODUCTION

The purpose of this chapter is to present a model of the manganese flux from sediments. Models have been constructed to describe the manganese cycling in sediments.* Their primary focus is calculating the pore water and, more recently, the solid phase vertical profiles. In principle, once the vertical pore water profile is known, the flux can be computed via Fick's Law (Eq. 2.47) applied to the sediment-water interface. However, the proper vertical scale is quite small near that interface (~1–5 mm) so models that are designed to predict nutrient, oxygen, and metal fluxes must properly represent the transport and reactions that occur in that surface layer where direct measurements are difficult. It is for this reason that the models constructed in this book rely primarily on measurements of the fluxes and only secondarily on pore water concentrations.

The model is applied to two manganese flux data sets, one of which is quite substantial (Hunt and Kelly, 1988). It is part of the MERL data set, analyzed in Chapter 15.† We begin with a discussion of the relevant chemistry of manganese in sediments.

18.1 A   Manganese Chemistry

The chemistry of manganese in surface waters and sediments has been studied for quite some time (Stumm and Morgan, 1996). Manganese exists in two valance states: the +4 state Mn(IV) in oxic waters and the +2 state Mn(II) in anoxic waters. Mn(IV) is very insoluble and precipitates to form manganese dioxide $MnO_2(s)$

$$Mn(IV)^{4+} + 2H_2O \rightarrow MnO_2(s) + 4H^+ \tag{18.1}$$

which is the predominant form of manganese in oxic surface waters. It usually exists as a coating on particles (Jenne, 1968). As the particles settle to the sediment, manganese is transported as well, providing a source of manganese to the sediments.

In the oxic layer of the sediment, $MnO_2(s)$ is stable. However, particle mixing causes manganese-containing particles to be transported to the anaerobic layer of the sediment where manganese dioxide is thermodynamically unstable and a reduction reaction occurs. Mn(IV) is reduced to Mn(II). For this to occur, two electrons are required as shown by the reduction half reaction

$$MnO_2(s) + 2e^- \rightarrow Mn(II)^{2+} - 4H^+ + 2H_2O \tag{18.2}$$

The primary source of electrons in sediments is decaying organic matter CH_2O (Section 1.2 B) and the oxidation half reaction is (Eq. 1.14)

$$CH_2O + H_2O \rightarrow CO_2 + 4H^+ + 4e^- \tag{18.3}$$

The overall reduction reaction can be constructed by supplying the electrons required in Eq. (18.2) from Eq. (18.3) to form the redox reaction

$$MnO_2(s) + \frac{1}{2}CH_2O + 2H^+ \rightarrow Mn(II)^{2+} + \frac{1}{2}CO_2 + \frac{3}{2}H_2O \tag{18.4}$$

In contrast to Mn(IV), Mn(II) is more soluble and exists in the mg/L range in sediment pore waters (e.g., Fig. 18.6C). As a consequence it can diffuse to the oxic layer of the sediment where it is subject to oxidation. The oxidation of Mn(II) to Mn(IV) occurs via the loss of two electrons

$$Mn(II)^{2+} \rightarrow Mn(IV)^{4+} + 2e^- \tag{18.5}$$

For oxygen as the electron acceptor, the overall reaction can be found using the reduction half reaction for oxygen (Eq. 1.15)

$$O_2 + 4H^+ + 4e^- \rightarrow 2H_2O \tag{18.6}$$

so that

$$Mn(II)^{2+} + \frac{1}{2}O_2 + 2H^+ \rightarrow Mn(IV)^{4+} + H_2O \tag{18.7}$$

The $Mn(IV)^{4+}$ that is formed precipitates as manganese dioxide (Eq. 18.1) and the overall redox reaction is

$$Mn(II)^{2+} + \frac{1}{2}O_2 + H_2O \rightarrow MnO_2(s) + 2H^+ \tag{18.8}$$

This is the reaction that occurs in the aerobic layer.

*Aller (1980b), Gratton et al. (1990), Holdren et al. (1975), van Cappellen and Wang (1995, 1996), Wang and van Cappellen (1996). †The MERL manganese data were collected and kindly provided to the author by Carlton D. Hunt (see the acknowledgments).

18.1 B   Reaction Kinetics

The kinetics of this reaction have been examined and can be described using the equation (Morgan, 1967)

$$\frac{d[\mathrm{Mn(II)}]}{dt} = -k'_{\mathrm{Mn}}[\mathrm{Mn(II)}] - k''_{\mathrm{MnO_2}}[\mathrm{Mn(II)}][\mathrm{MnO_2(s)}] \qquad (18.9)$$

where

$$k'_{\mathrm{Mn}} = k_{\mathrm{Mn}}[\mathrm{O_2}][\mathrm{OH^-}]^2 \qquad (18.10a)$$

$$k''_{\mathrm{MnO_2}} = k_{\mathrm{MnO_2}}[\mathrm{O_2}][\mathrm{OH^-}]^2 \qquad (18.10b)$$

An example is shown in Fig. 18.1. The reaction is initially first-order (Eq. 18.10a) and then becomes autocatalytic ((Eq. 18.10b) as the manganese dioxide participates in the oxidation. The first order reaction (k'_{Mn}) is slow in the normal pH ranges of surface waters, $k'^{*}_{\mathrm{Mn}} \simeq 0.003 \ \mathrm{d^{-1}}$ at pH = 8, and decreases or increases by 10^2 at pH = 7 or 9 respectively. However, the autocatalytic reaction can increase the rate. The oxidation can also be bacterially mediated and proceed more rapidly.[†] We will initially use a first-order reaction with a constant rate constant and then incorporate the additional complications.

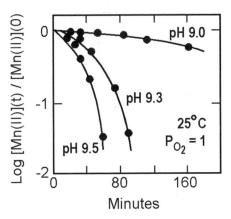

Fig. 18.1   Manganese (II) concentration versus time for various pHs from Morgan (1967).

18.1 C   Solubility

Fig. 18.2 presents a solubility diagram for manganese as a function of pH + pe where pe is the activity of the electron, which is a measure of the redox potential (Lindsay, 1979). The pH of sediment pore water is typically pH = 6 to 8. The oxic layer has a

---

[*]Davies and Morgan (1989), von Langen et al. (1997). [†]Dortch and Hamlin-Tillman (1995), Jaquet et al. (1982), Wehrli et al. (1995).

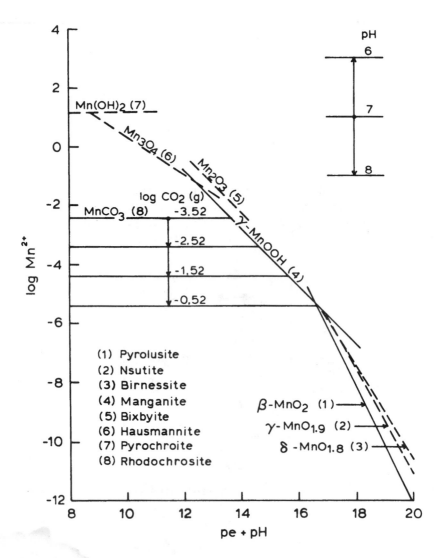

Fig. 18.2 Solubility of manganese minerals at pH = 7, from Lindsay (1979). The insert shows the effect of varying pH.

pe $\sim$ 13 in the presence of dissolved oxygen, whereas for the anoxic layer pe $\sim$ −3, the redox potential at which sulfate reduction takes place.

For the oxic conditions pH + pe $\sim$ 20, and manganese dioxide MnO_2(s) is the predominant species. The concentration of manganese is predicted to be very low $< 10^{-7}M$ (see Fig. 18.2). Thus there should be no dissolved manganese in the oxic layer at equilibrium. However, we will see below that equilibrium is not attained and it is the kinetics of the formation of manganese dioxide MnO_2(s) that controls the dissolved manganese concentration.

For anoxic conditions, pH + pe $<$ 10 and the concentration may be controlled by the solubility of manganese carbonate $MnCO_3$(s) (rhodochrosite). It is unlikely that manganese sulfide MnS(s) is present, since iron sulfide FeS(s) is present in most sediments and it is more insoluble than MnS(s) (Emerson et al., 1983). Typical pore water manganese concentrations are in the mg/L range or 10^{-4} M which is within the stability field of $MnCO_3$(s) (Fig. 18.2).

In addition Mn(II) will partition to sorption sites on the sediment particles as discussed below. Thus not all the Mn(II) that is formed by the reduction of MnO_2(s) remains in dissolved form even if manganese carbonate does not form. Some will sorb to sediment particles. Therefore the transfer of Mn(II) from the anoxic to the oxic layer occurs via particle mixing, which transports particulate Mn, and via the diffusion of soluble Mn(II).

18.2   STEADY STATE MODEL

The reactions that occur in the aerobic and anaerobic layer suggest a conceptual model for manganese fluxes from sediments that is illustrated in Fig. 18.3. This type of model has been suggested by a number of investigators, for example see Aller (1994), Davison (1985), Sundby et al. (1986). It is instructive to think of the processes in temporal sequence, corresponding to the numbers in Fig. 18.3.

1. Particulate manganese dioxide MnO_2(s) settles to the aerobic layer of the sediment.

2. Particle mixing moves the particle downward into the anaerobic layer of the sediment.

3. Manganese dioxide is unstable in a reducing environment, so it is reduced to soluble Mn(II) (Stone, 1987).

4. Mn(II) diffuses to the aerobic layer.

5. Oxidation of Mn(II) to manganese dioxide occurs in the aerobic layer which regenerates manganese dioxide MnO_2(s).

6. However, the oxidation reaction (5) converting Mn(II) to Mn(IV) which precipitates as MnO_2(s) is competing with the diffusion of Mn(II) to the overlying water.

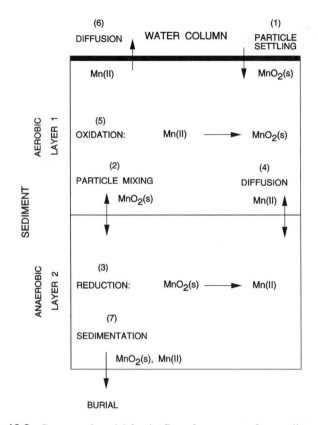

Fig. 18.3 Conceptual model for the flux of manganese from sediment.

7. If the oxidation of Mn(II) is fast relative to its diffusion to the overlying water, then the only other sink is via burial by sedimentation. If, on the other hand, the oxidation of Mn(II) is slow, then Mn(II) escapes to the overlying water as a flux of soluble manganese, completing the cycle.

This description is idealized since the processes all occur simultaneously. However, the main point is that the magnitude of the flux to the overlying water is determined by a competition between the rate of oxidation of Mn(II) to $MnO_2(s)$ in the aerobic layer, and the diffusion of Mn(II) from the aerobic layer to the overlying water.

18.2 A   Partitioning

In addition to the redox reactions discussed above, the chemistry of manganese (II) itself is also important. Since it is a divalent cation it can adsorb to sorption sites that are present in the sediment. In particular it can sorb to the freshly precipitated hydrous iron oxide (Dzombak and Morel, 1990) in the aerobic layer and to other sorption sites in the anaerobic layer.

Manganese (II) can also form solid species (see Fig. 18.2). For the model presented below, the formation of particulate manganese is modeled as a reversible partitioning reaction

$$Mn(II)_d \rightleftharpoons Mn(II)_p \qquad (18.11)$$

where $Mn(II)_d$ and $Mn(II)_p$ are dissolved and particulate manganese (II), respectively. The rationale for this choice is the same as that employed for the phosphate and sulfide flux models (Chapters 6 and 9). First, the resulting equations can be solved analytically, which is an important aid to understanding the model's behavior. Second, linear partitioning can sometimes be a realistic description of the relationship between dissolved and particulate chemical. It is the limiting case of the Langmuir model which is implicit in most surface complexation model formulations.* Finally, the general problem of computing the chemical composition of pore water would involve using a numerical chemical equilibrium model, e.g., Jahnke et al. (1994). Mass balance equations are required for the various chemicals that affect the pore water chemistry – hydrogen ion, carbon dioxide, and so on, e.g., Di Toro (1976). Thermodynamic data are required for the relevant aqueous complexes, and the stable and metastable phases, some of which are uncertain. Finally, sorption as well as precipitation reactions need to be considered. All this is necessary to compute the fraction of a chemical that is either dissolved or particulate.

The equivalent partitioning model employs only a partition coefficient π_{Mn}, the ratio of particulate to dissolved chemical concentration.

$$\pi_{Mn} = \frac{\left(Mn(II)_p / m\right)}{Mn(II)_d} \qquad (18.12)$$

where $Mn(II)_p$ is the concentration of particulate manganese per unit sediment volume and m is the concentration of solids per unit sediment volume. Thus $Mn(II)_p / m$ is the concentration of manganese per unit sediment solids (e.g., μg Mn/g solids). Since $Mn(II)_d$ is the concentration of dissolved manganese per unit volume of pore water (e.g., μg Mn/L), π_{Mn} has the units of L/kg.

It is possible to build some level of chemical realism into the partitioning formulation. If necessary, π_{Mn} can be varied as a function of other physical and chemical parameters such as pH in order to produce more realistic behavior. The practical question is: Does the added difficulty of including equilibrium chemistry into the model structure result in added realism? Whatever the answer, it is prudent to begin the modeling using linear partitioning and to examine the utility of the results.

18.2 B   Model Formulation

The model structure is shown in Fig. 18.4. It is an elaboration of the conceptual model presented in Fig. 18.3. There are four dependent variables: Mn(II) and

---

*Dzombak and Morel (1990), Westall and Hohl (1980).

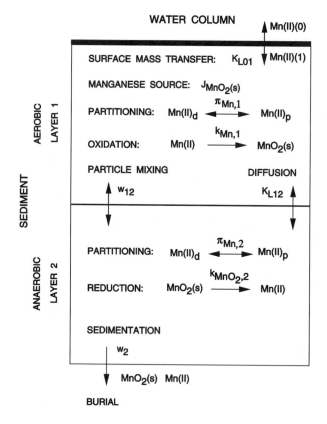

Fig. 18.4 Schematic of the model for the flux of manganese from sediment.

$MnO_2(s)$ in layers 1 and 2. These concentrations are denoted by $Mn(1)$ and $Mn(2)$, and $MnO_2(1)$ and $MnO_2(2)$, respectively. They correspond to the total $Mn(II)$ and $Mn(IV)$ in each layer. The source of manganese to the sediment is the settling of particulate manganese dioxide from the overlying water to the sediment. This depositional flux is denoted by J_{MnO_2}.

Two reactions occur in the aerobic layer. $Mn(II)$ partitions to form particulate manganese. This reaction is parameterized with a linear partition coefficient, $\pi_{Mn,1}$. In addition, dissolved $Mn(II)$ is oxidized to $MnO_2(s)$ following Eq. (18.8) with first-order rate constant $k'_{Mn,1}$. Thus the rate of $MnO_2(s)$ production in layer 1 per unit area is

$$S_{MnO_2} = k'_{Mn,1} H_1 f_{d1} Mn(1) \tag{18.13}$$

where H_1 is the depth of the aerobic layer and f_{d1} is the fraction of $Mn(1)$ that is dissolved. Following the usual procedure (Section 3.3 B), the depth of the aerobic layer H_1 is estimated from the ratio of the diffusion coefficient D_1 in layer 1 and the surface mass transfer coefficient s

$$H_1 = \frac{D_1}{s} \tag{18.14}$$

where the surface mass transfer coefficient is obtained from the sediment oxygen demand SOD computed by the oxygen portion of the sediment flux model, and the overlying water oxygen concentration $O_2(0)$

$$s = \frac{SOD}{O_2(0)} \tag{18.15}$$

Thus the oxidation rate becomes

$$k'_{Mn,1} H_1 = k'_{Mn,1} \frac{D_1}{s} = \frac{\kappa^2_{Mn,1}}{s} \tag{18.16}$$

where

$$\kappa_{Mn,1} = \sqrt{k'_{Mn,1} D_1} \tag{18.17}$$

which is the definition of the reaction velocity in layer 1. The rationale for this definition and Eq. (18.14) has been described in Section (3.3 C).

Two reactions also occur in the anaerobic layer. Mn(II) partitions to form particulate manganese which is parameterized with a linear partition coefficient, $\pi_{Mn,2}$. This may be different from the aerobic layer partition coefficient $\pi_{Mn,1}$ due to differences in pH, the nature of the particles, the absence of freshly precipitated iron oxyhydroxides, and whatever else is influencing the extent of the formation of Mn(II) particulates. In addition, $MnO_2(s)$ is reduced to Mn(II) following Eq. (18.4) with first-order rate, $k_{MnO_2,2}$. Hence the source of Mn(II) in the anaerobic layer is

$$S_{Mn(II)} = k_{MnO_2,2} H_2 MnO_2(2) = \kappa_{MnO_2,2} MnO_2(2) \tag{18.18}$$

where H_2 is the depth of the anaerobic layer and

$$\kappa_{MnO_2,2} = k_{MnO_2,2} H_2 \tag{18.19}$$

which is the usual definition of a reaction velocity in layer 2 (Chapter 4, Eq. 4.9).

The mass transport between the overlying water and layer 1 is quantified via the surface mass transfer coefficient K_{L01} which is set equal to s (Eq. 18.15)

$$K_{L01} = s \tag{18.20}$$

as in the models for oxygen and nutrient fluxes (Section 3.3 B). Particle mixing with mixing velocity w_{12} and diffusive exchange with mass transfer coefficient K_{L12} between layers 1 and 2 are included as before, as is the loss of manganese by burial with sedimentation velocity w_2.

18.2 C Equations and Solutions

The mass balance equations for the model follow from the reactions and transport processes discussed above. They are

Layer 1 Mn(II)

$$0 = H_1 \frac{d\text{Mn}(1)}{dt} = -s\big(f_{d1}\text{Mn}(1) - \text{Mn}(0)\big) + K_{L12}\big(f_{d2}\text{Mn}(2) - f_{d1}\text{Mn}(1)\big)$$

$$+w_{12}\big(f_{p2}\text{Mn}(2) - f_{p1}\text{Mn}(1)\big) - w_2\text{Mn}(1) - \frac{\kappa^2_{\text{Mn},1}}{s} f_{d1}\text{Mn}(1)$$

$$(18.21a)$$

Layer 2 Mn(II)

$$0 = H_2 \frac{d\text{Mn}(2)}{dt} = -K_{L12}\big(f_{d2}\text{Mn}(2) - f_{d1}\text{Mn}(1)\big)$$

$$- w_{12}\big(f_{p2}\text{Mn}(2) - f_{p1}\text{Mn}(1)\big) - w_2\big(\text{Mn}(2) - \text{Mn}(1)\big) + \kappa_{\text{MnO}_2,2}\text{MnO}_2(2)$$

$$(18.21b)$$

Layer 1 MnO$_2$(s)

$$0 = H_1 \frac{d\text{MnO}_2(1)}{dt} = \frac{\kappa^2_{\text{Mn},1}}{s} f_{d1}\text{Mn}(1) + w_{12}\big(\text{MnO}_2(2) - \text{MnO}_2(1)\big)$$

$$- w_2\text{MnO}_2(1) + J_{\text{MnO}_2}$$

$$(18.21c)$$

Layer 2 MnO$_2$(s)

$$0 = H_2 \frac{d\text{MnO}_2(2)}{dt} = -\kappa_{\text{MnO}_2,2}\text{MnO}_2(2)$$

$$- w_{12}\big(\text{MnO}_2(2) - \text{MnO}_2(1)\big) - w_2\big(\text{MnO}_2(2) - \text{MnO}_2(1)\big)$$

$$(18.21d)$$

The particulate f_p and dissolved f_d fractions are computed from the partition coefficients and the concentration of sediment solids m in each layer

$$f_{d1} = \frac{1}{1 + m\pi_{\text{Mn},1}} = 1 - f_{p1} \qquad (18.22a)$$

$$f_{d2} = \frac{1}{1 + m\pi_{\text{Mn},2}} = 1 - f_{p2} \qquad (18.22b)$$

The solutions to these steady state equations (18.21) are found using the symbolic computation program MACSYMA (1993). The procedure is presented in Appendix

18A. The solutions are extremely complicated if the dissolved concentration in the overlying water $Mn(0)$ is included. However, for $Mn(0) = 0$, the solutions are reasonably concise. The reason is that for this case only one equation (Eq. 18.21c) is nonhomogeneous* with an external source of manganese. The rest are homogeneous equations with terms that are restricted to linear functions of the four unknowns. It turns out that the equations in such a set have relatively concise solutions.

The solution of Eqs. (18.21) for $Mn(2)$ for the case $Mn(0) = 0$ is (Fig. 18A.2, d11)

$$Mn(2) = \frac{J_{MnO_2} \kappa_{MnO_2,2}}{\left(\kappa_{MnO_2,2} + w_2\right)\left(s f_{d1} r_{12} + w_2\right) + f_{d1} \dfrac{\kappa_{Mn,1}^2}{s} r_{12} w_2} \tag{18.23}$$

which differs from the MACSYMA solution by the factored denominator (d15). The ratio of the layer 2 to layer 1 concentrations r_{12} is found to be (Fig. 18A.3, d31)

$$r_{12} \triangleq \frac{Mn(1)}{Mn(2)} = \frac{f_{d2} K_{L12} + f_{p2} w_{12}}{f_{d1}\left(K_{L12} + s + \kappa_{Mn,1}^2/s\right) + f_{p1} w_{12} + w_2} \tag{18.24}$$

The layer 1 solution is obtained using this relationship

$$Mn(1) = r_{12} Mn(2) \tag{18.25}$$

The manganese flux follows from the relationship between surface mass transfer and dissolved layer 1 concentrations

$$J[Mn] = s f_{d1} Mn(1) \tag{18.26}$$

Substituting Eqs. (18.23 and 18.25) into this equation yields

$$J[Mn] = s f_{d1} \frac{J_{MnO_2} \kappa_{MnO_2,2} r_{12}}{\left(\kappa_{MnO_2,2} + w_2\right)\left(s f_{d1} r_{12} + w_2\right) + f_{d1} \dfrac{\kappa_{Mn,1}^2}{s} r_{12} w_2} \tag{18.27}$$

which is the flux of manganese to the overlying water. Various special cases of this equation are examined below.

Once the $Mn(II)$ solutions are known, the MnO_2 (s) concentrations are easily found. They are

$$MnO_2(1) = \frac{\left(f_{d1} Mn(1)\kappa_{Mn,1}^2/s + J_{MnO_2}\right)\left(\kappa_{MnO_2,2} + w_{12} + w_2\right)}{\left(w_{12} + w_2\right)\left(\kappa_{MnO_2,2} + w_2\right)} \tag{18.28}$$

---

*An equation is homogeneous if it has no forcing function. A nonhomogeneous equation has a forcing function, in this case, a constant.

and

$$MnO_2(2) = \frac{\begin{array}{c} Mn(2)\left(f_{d2}K_{L12} + f_{p2}w_{12} + w_2\right) \\ -Mn(1)\left(f_{d1}K_{l12} + f_{p1}w_{12} + w_2\right) \end{array}}{\kappa_{MnO_2,2}} \tag{18.29}$$

These solutions (Eqs. 18.23–18.29) can be checked by evaluating the global mass balance equation

$$J_{MnO_2} = J[Mn] + w_2\left(Mn(2) + MnO_2(2)\right) \tag{18.30}$$

which states that the input of particulate manganese to the sediment J_{MnO_2} is either returned to the overlying water as a flux of dissolved manganese, or is buried as either Mn(II) or $MnO_2(s)$. The calculation is performed using MACSYMA (Appendix 18A, Fig. 18A.3). The result, the last line of the output, is zero which verifies that Eq. (18.30) is satisfied.

18.2 D   Simplified Solutions

The steady state solution apportions the flux of manganese dioxide to the sediment J_{MnO_2} to either a flux of Mn(II) to the overlying water $J[Mn]$ or to the burial of $MnO_2(s)$ and Mn(II) (Eq. 18.30). The equations can be simplified considerably by examining a special case which, as shown subsequently, appears to be a reasonable representation of what actually occurs.

If the rate of reduction of $MnO_2(s)$ in the anaerobic layer is rapid, then $\kappa_{MnO_2} \rightarrow \infty$ and Eq. (18.27) becomes

$$J[Mn] = \frac{J_{MnO_2}}{1 + \beta} \tag{18.31}$$

where

$$\beta = \frac{w_2}{s f_{d1} r_{12}} \tag{18.32}$$

Thus β controls the fraction of incoming manganese that is recycled to the overlying water. It depends on the ratio of the burial velocity w_2 and the surface mass transfer coefficient s modified by the dissolved fraction f_{d1} in layer 1 and r_{12}.

The expression for r_{12} can be simplified slightly using the fact that most of the manganese (II) in the sediment is in particulate phases. Thus

$$f_{p1} \simeq 1$$

$$f_{p2} \simeq 1 \tag{18.33}$$

and r_{12} becomes

$$r_{12} = \frac{f_{d2}K_{L12} + w_{12}}{f_{d1}\left(K_{L12} + s + \dfrac{\kappa_{Mn,1}^2}{s}\right) + w_{12} + w_2} \tag{18.34}$$

If the dissolved fractions f_{d1} and f_{d2} are small enough and s is not small – note the $\kappa_{Mn,1}^2/s$ term – then

$$r_{12} \simeq \frac{w_{12}}{w_{12} + w_2} \tag{18.35}$$

And since the particle mixing velocity w_{12} is usually large relative to the sedimentation velocity, w_2, then $r_{12} \to 1$ as indicated in Eq. (18.35). Thus the fraction of incoming particulate $MnO_2(s)$ that is recycled to the overlying water is controlled by

$$\beta \simeq \frac{w_2}{s f_{d1}} \tag{18.36}$$

A large β corresponds to a small fraction recycled (Eq. 18.31). A large β occurs if the partitioning in the aerobic layer is large so that f_{d1} is small. To understand the magnitudes involved, consider a typical sedimentation velocity of $w_2 = 10^{-5}$ m/d (0.36 cm/yr.). A typical surface mass transfer coefficient is

$$s = SOD/O_2(0) = \left(\sim 1 \text{ g/m}^2\text{-d}\right) / \left(\sim 10 \text{ g/m}^3\right) \simeq 0.1 \text{ m/d}$$

Thus for $\beta \sim 1$, which corresponds to recycling one half of the incoming flux,

$$f_{d1} \simeq \frac{w_2}{\beta s} = \frac{10^{-5}}{(1)(0.1)} = 10^{-4}$$

and since (Eq. 18.22)

$$f_d \sim \frac{1}{m \pi_{Mn}} \tag{18.37}$$

and $m \sim 1$ kg/L, a partition coefficient of $\pi_{Mn} \sim 10^4$ L/kg is required if the fraction recycled is one half. If the partition coefficient is much larger, then none of the $MnO_2(s)$ flux to the sediment is recycled as a dissolved flux to the overlying water and $J[Mn] \to 0$. On the other hand, if π_{Mn} is much smaller, then the sediment traps very little Mn, all the incoming flux is recycled, and $J[Mn] \simeq J_{MnO_2}$.

Anoxic overlying water It is commonly observed that the largest manganese fluxes to the overlying water are observed if the overlying water concentration of dissolved oxygen approaches zero (Sundby et al., 1986). An example from a reservoir is presented in Chapter 20. Thus it is important to examine the solutions for this limiting case.

As $O_2(0) \to 0$, $s = SOD/O_2(0) \to \infty$ and Eqs. (18.23–18.26) approach

$$J[Mn] \to J_{MnO_2} \left(\frac{\kappa_{MnO_2,2}}{\kappa_{MnO_2,2} + w_2}\right) \left(\frac{f_{d2} K_{L12} + f_{p2} w_{12}}{f_{d2} K_{L12} + f_{p2} w_{12} + w_2}\right) \tag{18.38}$$

In this case, recycling will be complete if both the bracketed expressions approach one. The first necessary condition is that

$$\kappa_{MnO_2,2} \gg w_2 \tag{18.39}$$

or, using the definition of the reaction velocity (Eq. 18.18), if

$$k_{MnO_2,2} \gg \frac{w_2}{H_2} \tag{18.40}$$

The term w_2/H_2 is the reciprocal of the residence time of $MnO_2(s)$ in the anaerobic layer. Therefore the condition in Eq. (18.40) is that the reduction of $MnO_2(s)$ to Mn(II) occurs rapidly relative to the residence time of $MnO_2(s)$ in H_2.

The second condition

$$f_{d2}K_{L12} + f_{p2}w_{12} \gg w_2 \tag{18.41}$$

requires that the rate of transport of dissolved ($f_{d2}K_{L12}$) and particulate ($f_{p2}w_{12}$) manganese (II) to layer 1 is large relative to the rate at which it is buried w_2. This condition is necessary so that the manganese produced in layer 2 by the reduction of $MnO_2(s)$ to Mn(II) is transported to layer 1 rather than buried. Once it is in layer 1 it escapes to the overlying water without further modification, since the depth of the aerobic layer H_1 is approaching zero

$$\lim_{s \to \infty} H_1 = \lim_{s \to \infty} \frac{D_1}{s} = 0 \tag{18.42}$$

and no reoxidation of Mn(II) to $MnO_2(s)$ can occur.

Since both of these conditions are usually the case, the model produces the appropriate limiting condition namely

$$J[Mn] \to J_{MnO_2} \tag{18.43}$$

for $s \to \infty$ and no manganese is trapped by the sediment. We shall see subsequently that time variable effects come into play as well. In addition to the release of the incoming flux J_{MnO_2} there is also a release of stored manganese. The situation parallels that for phosphorus fluxes (Chapter 6).

Aerobic overlying water The other end of the spectrum is the behavior of $J[Mn]$ as $s \to 0$, corresponding to low SOD and high overlying water dissolved oxygen. To examine this case $J[Mn]$ (Eqs. 18.23–18.26) can be expanded in a Taylor series in s

$$J[Mn] = J_{MnO_2} \frac{\kappa_{MnO_2,2}}{\kappa^2_{Mn,1} w_2} s^2 + \cdots \tag{18.44}$$

The manganese flux decreases quadratically in s with a slope determined by the two reaction velocities and the sedimentation velocity.

The results of this section have shown that the model has reasonable limiting behavior at the extremes of the parameter values. Its ability to reproduce observations will be examined in the next section.

18.2 E   Manganese Flux Data

Two data sets will be examined. The first is a relatively small number of paired manganese and ammonia flux measurements made at three Long Island Sound stations (Aller, 1980a). The second is a large number of manganese (Hunt and Kelly, 1988) and nutrient flux measurements made at the MERL mesocosms (Nixon et al., 1986). The analysis technique is to attempt to reproduce the observed relationship between the ammonia $J[NH_4]$ and manganese $J[Mn]$ fluxes. As shown in Fig. 18.5 there is a proportional relationship between these two fluxes. The question is: What is the causal linkage?

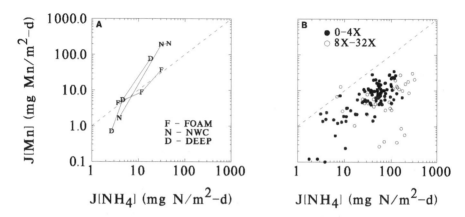

Fig. 18.5 Manganese versus ammonia flux. Data from (A) Long Island Sound and (B) the MERL mesocosm experiment (see Chapter 15 for a description of the experiment and the loading rates in the legend). A one-to-one relationship is indicated by the dashed line.

Relationship between Manganese and Ammonia Fluxes   The parameter in the manganese model equations that can ultimately be related to the ammonia flux is the surface mass transfer coefficient s. This parameter controls both the rate of mass transfer from the pore water to the overlying water (Eq. 18.21a) and the depth of the aerobic layer, H_1 (Eq. 18.16). As SOD increases or the overlying water dissolved oxygen decreases, one would expect that $J[Mn]$ would increase, since both the rate of surface mass transfer increases and the depth of the aerobic zone decreases. The latter effect decreases the residence time in the aerobic layer and makes the oxidation of Mn(II) to $MnO_2(s)$ less rapid (Eq. 18.16), enabling more Mn(II) to escape to the overlying water.

As shown in Eq. (18.44), this is indeed how the steady state solution behaves. The dependency of s on ammonia flux occurs because the ammonia flux and SOD are related. This has been worked out in Section 16.2 A. For a specified ammonia flux $J[NH_4]$, the ammonia diagenesis J_N is found (Eq. 16.5). The SOD is computed using Eq. (16.7) and Eq. (16.2) yields s. This is used in Eqs. (18.24–18.27) to compute $J[Mn]$. This establishes the relationship between ammonia and manganese fluxes.

Table 18.1   Parameter Values for the Steady State Model

Parameters	Small $\kappa_{Mn,1}$		Large $\kappa_{Mn,1}$		Units
	LIS	MERL	LIS	MERL	
J_{MnO_2}	150	20	50	10	mg Mn/m²-d
$\kappa_{Mn,1}$	0.25		2.0		m/d
$\kappa_{MnO_2,2}$	0.1		0.1		m/d
$\pi_{Mn,1}$	200		200		L/kg
$\pi_{Mn,2}$	100		100		L/kg
m_1	1.0		1.0		kg/L
m_2	1.0		1.0		kg/L
K_{L12}	0.05		0.05		m/d
w_{12}	0.0012		0.0012		m/d
w_2	0.5		0.5		cm/yr
$\kappa_{NH_4,1}$	0.15		0.15		m/d
$[O_2(0)]$	5.0		5.0		mg/L
$a_{O_2,C} \times a_{C,N}$	2.54 × 5.68		2.54 × 5.68		mg O_2/mg N

Comparison to Data   The model computations are compared to the Long Island Sound data in Fig. 18.6. The model parameters are listed in Table 18.1(LIS). Two cases are considered. The first (small $\kappa_{Mn,1}$) results from assigning a slow oxidation rate (Fig. 18.6A–D). The input flux of particulate manganese J_{MnO_2} is specified to be 150 mg Mn/m²-d for the Long Island Sound case, since the flux of manganese from the sediment $J[Mn]$ is approximately this magnitude at the high ammonia fluxes (Fig. 18.6A). The pore water and sediment data are plotted arbitrarily at an ammonia flux of 10 mg N/m²-d for convenience.

In order to reproduce the pore water (C) and sediment manganese concentrations (D), it is necessary to recycle most of the incoming manganese to the overlying water (B). Hence the oxidation reaction velocity $\kappa_{Mn,1}$ is set so that only a small fraction of the manganese flux to the aerobic layer J_{MnO_2} is oxidized to $MnO_2(s)$. As a result very little is trapped and the burial fraction is small (B). Since most of the flux is being recycled, the variation in s which occurs as $J[NH_4]$ varies has little effect on the manganese flux (A).

An alternate calibration is shown in Fig. 18.6E–H (Table 18.1 – large $\kappa_{Mn,1}$). The oxidation reaction velocity $\kappa_{Mn,1}$ is increased from 0.25 to 2.0 m/d so that a larger fraction of J_{MnO_2} is oxidized to $MnO_2(s)$ rather than released. As a consequence, the sediment manganese concentration is larger (H), and so also is the pore water concentration (G) because of the partitioning relationship (Eq. 18.12). The oxidation rate is now in the range where the variation in s is affecting the fraction that is buried versus that recycled (F). The burial fraction is predominant at low ammonia flux and the flux to the overlying water predominates at high ammonia flux.

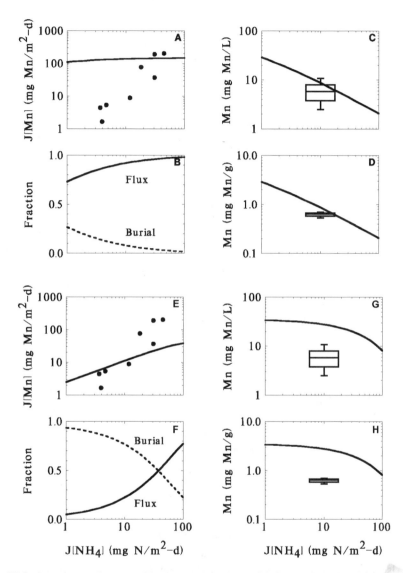

Fig. 18.6 Manganese linear partitioning model: slow oxidation rate (A−D), fast oxidation rate (E−H). Comparison of model predictions (line) to data (points) from Long Island Sound as a function of the ammonia flux. (A,E) Manganese flux, (B,F) Mn flux normalized to depositional flux $J[Mn]/J_{MnO_2}$, (C,G) pore water Mn concentration $Mn(2)_d$, (D,H) sediment solid phase concentration $Mn(2)_p/m$.

A similar analysis (Table 18.1) for the MERL flux data, shown previously in Fig. 18.5, is presented in Fig. 18.7 with results similar to LIS. The flux of particulate manganese to the sediments appears to be smaller $J_{MnO_2} = 10$–20 mg Mn/m^2-d than for Long Island Sound $J_{MnO_2} = 50$–150 mg Mn/m^2-d. As a consequence, the pore water and sediment manganese concentrations are also smaller. The same problem recurs. Either the pore water and solid phase data can be reproduced but not the variation in $J[Mn]$ (small $\kappa_{Mn,1}$, Fig. 18.7A–D) or the variation is reproduced (E) with pore water and sediment manganese concentrations that are too high (large $\kappa_{Mn,1}$, Fig. 18.7G and H).

18.2 F   Summary

The steady state linear partitioning model is reasonably successful in reproducing a major feature of the Long Island Sound and MERL data sets, namely the correlation between increasing ammonia and manganese fluxes. However, it cannot do so and be consistent with the pore water and sediment manganese concentrations at the same time.

For a steady state condition, there are only two possible exit pathways for the incoming manganese: either it escapes as a flux, or it is buried. Therefore, in order for the model to reproduce the observations it is necessary that it predicts a higher degree of burial when ammonia fluxes are smaller, and less burial at larger ammonia fluxes. Perhaps, instead of burial as the alternative, it is storage in the sediment that is occurring at low ammonia fluxes. Since this is inherently a time variable effect, it must be included explicitly.

18.3   TIME VARIABLE MODEL

In this section we present the time variable version of the model. It may be that the difficulties exhibited by the model are due to the steady state assumption.

18.3 A   Model Equations

The model as described in Section 18.2 is the basis for the time variable version. The mass balance equations for the model are based on Eqs. (18.21) with the addition of the entrainment terms (e.g., $\dot{H}_1$ etc., Section 13.4 A).

Also included is the dependency of the manganese (II) oxidation rate on the dissolved oxygen concentration (Morgan, 1967, Stumm and Morgan, 1970). The aerobic layer concentration $[O_2(1)]$ is related to the overlying water concentration $[O_2(0)]$ via $[O_2(1)] = [O_2(0)]/2$ (Eq. 3.33). The reaction rates are written using the rate constants $k_{Mn,1}$ and $k_{MnO_2,2}$ rather than the reaction velocities because the kinetics will subsequently be applied to the overlying water as well (Chapters 19 and 20) and the use of reaction velocities is no longer appropriate or useful.

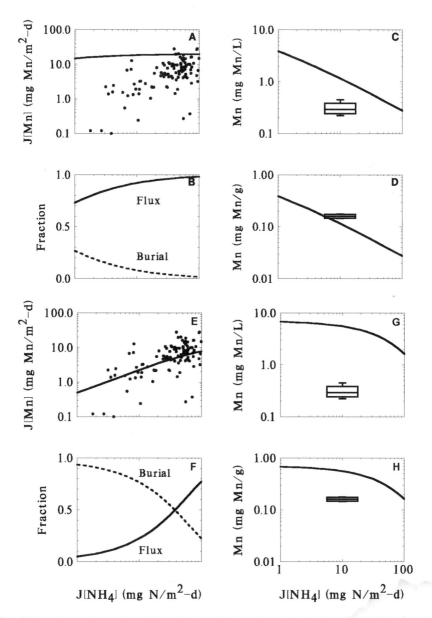

Fig. 18.7 Comparison of model predictions (line) to data (points) from the MERL experiment as a function of the ammonia flux: Slow oxidation rate (A–D), fast oxidation rate (E–H). (A,E) Manganese flux; (B,F) Mn flux normalized to depositional flux, (C,G) pore water Mn concentration $Mn(2)_d$, (D,H) sediment solid phase concentration $Mn(2)_p/m$.

The partitioning of Mn(II) to iron oxyhydroxide is included using the same formulation as employed for phosphorus (Eqs. 6.19–6.20)

$$\pi_{Mn,1} = \pi_{Mn,2}(\Delta\pi_{Mn,1}) \qquad\qquad [O_2(0)] > [O_2(0)]_{crit,Mn} \quad (18.45a)$$

$$\pi_{Mn,1} = \pi_{Mn,2}(\Delta\pi_{Mn,1})^{\beta_{Mn}} \qquad\qquad [O_2(0)] \leqslant [O_2(0)]_{crit,Mn}$$

where

$$\beta_{Mn} = \frac{[O_2(0)]}{[O_2(0)]_{crit,Mn}} \qquad\qquad (18.45b)$$

$\pi_{Mn,2}$ is the partition coefficient of dissolved Mn(II) in the anaerobic layer, and $\Delta\pi_{Mn,1}$ is the increase in the aerobic layer due to Mn(II) sorption to iron oxyhydroxide. Eqs. 18.45 provide a smooth decrease in partition coefficient from $\pi_{Mn,1}$ to $\pi_{Mn,2}$ as $O_2(0) \to 0$.

The equations are

Layer 1 Mn(II)

$$H_1\frac{d\text{Mn}(1)}{dt} = -s\left(f_{d1}\text{Mn}(1) - \text{Mn}(0)\right) + K_{L12}\left(f_{d2}\text{Mn}(2) - f_{d1}\text{Mn}(1)\right)$$

$$+w_{12}\left(f_{p2}\text{Mn}(2) - f_{p1}\text{Mn}(1)\right) - w_2\text{Mn}(1) - k_{Mn,1}[O_2(1)]f_{d1}\text{Mn}(1)H_1$$

$$+\text{Mn}(2)\,\dot{H}_1^+ - \text{Mn}(1)\left(\dot{H}_1 + \dot{H}_1^-\right) \qquad\qquad (18.46a)$$

Layer 2 Mn(II)

$$H_2\frac{d\text{Mn}(2)}{dt} = -K_{L12}\left(f_{d2}\text{Mn}(2) - f_{d1}\text{Mn}(1)\right) - w_{12}\left(f_{p2}\text{Mn}(2) - f_{p1}\text{Mn}(1)\right)$$

$$-w_2\left(\text{Mn}(2) - \text{Mn}(1)\right) + k_{MnO_2,2}\text{MnO}_2(2)H_2$$

$$-\text{Mn}(2)\left(\dot{H}_2 + \dot{H}_1^+\right) + \text{Mn}(1)\dot{H}_1^- \qquad\qquad (18.46b)$$

Layer 1 MnO$_2$(s)

$$H_1\frac{d\text{MnO}_2(1)}{dt} = k_{Mn,1}[O_2(1)]f_{d1}\text{Mn}(1)H_1 - w_2\text{MnO}_2(1)$$

$$+w_{12}\left(\text{MnO}_2(2) - \text{MnO}_2(1)\right) + J_{MnO_2}$$

$$+\text{MnO}_2(2)\dot{H}_1^+ - \text{MnO}_2(1)\left(\dot{H}_1 + \dot{H}_1^-\right) \qquad (18.46c)$$

Layer 2 MnO$_2$(s)

$$H_2 \frac{dMnO_2(2)}{dt} = -k_{MnO_2,2}MnO_2(2)H_2 - w_{12}(MnO_2(2) - MnO_2(1))$$

$$- w_2(MnO_2(2) - MnO_2(1))$$

$$- MnO_2(2)\left(\dot{H}_2 + \dot{H}_1^+\right) + MnO_2(1)\dot{H}_1^- \qquad (18.46d)$$

The numerical scheme used to solve these equations is the same implicit integration technique that is used for the nutrient model equations (Section 13.5 C). Since there are four equations matrix notation will be used. Equations (18.46) are written as

$$\mathbf{H}(t)\frac{d\mathbf{c}(t)}{dt} = \mathbf{A}(t)\mathbf{c}(t) + \mathbf{b}(t) \qquad (18.47)$$

where the boldface symbols denote vectors (lower case) and matrices (upper case). The manganese concentration state variable vector is

$$\mathbf{c}(t) = \begin{bmatrix} Mn(1) \\ Mn(2) \\ MnO_2(1) \\ MnO_2(2) \end{bmatrix} \qquad (18.48)$$

The matrix $\mathbf{A}(t) = [a_{ij}]$ are given in Table 18.2 where f_{12} and f_{21} are the effective mass transfer coefficients from layer 1 to layer 2 and from layer 2 to layer 1, respectively

$$f_{12} = K_{L12}f_{d1} + w_{12}f_{p1} \qquad (18.49a)$$

$$f_{21} = K_{L12}f_{d2} + w_{12}f_{p2} \qquad (18.49b)$$

and all other a_{ij}'s are zero. The forcing function is

$$\mathbf{b}(t) = \begin{bmatrix} sMn(0)(t) \\ 0 \\ J_{MnO_2}(t) \\ 0 \end{bmatrix} \qquad (18.50)$$

and $\mathbf{H}(t)$ is a diagonal matrix of layer depths

$$\mathbf{H}(t) = \text{diag}\left[H_1(t)\ H_2(t)\ H_1(t)\ H_2(t) \right] \qquad (18.51)$$

A finite difference equation is used to solve Eq. (18.47). An implicit forward in time scheme (Section 13.5 C) is used for the concentrations since some of the coefficients in $\mathbf{A}$ can become quite large, for example $\kappa_{Mn,1}^2/s$ if s is small, and an explicit scheme would require an unacceptably small Δt to remain stable. The $\mathbf{A}$ and $\mathbf{H}$ matrices are evaluated at time level t for the terms coupled to the other state

Table 18.2 Matrix elements of the $\mathbf{A}(t) = [a_{ij}]$ matrix.

$$a_{11} = -s f_{d1} - f_{12} - w_2 - k_{Mn,1}[O_2(1)] f_{d1} H_1 - \left(\dot{H}_1 + \dot{H}_1^- \right) \tag{18.54a}$$

$$a_{12} = f_{21} + \dot{H}_1^+ \tag{18.54b}$$

$$a_{21} = f_{12} + w_2 + \dot{H}_1^- \tag{18.54c}$$

$$a_{22} = -f_{21} - w_2 - \left(\dot{H}_2 + \dot{H}_1^+ \right) \tag{18.54d}$$

$$a_{24} = k_{MnO_2,2} H_2 \tag{18.54e}$$

$$a_{31} = k_{Mn,1}[O_2(1)] f_{d1} H_1 \tag{18.54f}$$

$$a_{33} = -w_{12} - w_2 - \left(\dot{H}_1 + \dot{H}_1^- \right) \tag{18.54g}$$

$$a_{34} = w_{12} + \dot{H}_1^+ \tag{18.54h}$$

$$a_{43} = w_{12} + w_2 + \dot{H}_1^- \tag{18.54i}$$

$$a_{44} = -k_{MnO_2,2} H_2 - w_{12} - w_2 - \left(\dot{H}_2 + \dot{H}_1^+ \right) \tag{18.54j}$$

variables such as s. Otherwise, the equation set would be nonlinear and more difficult to solve. The resulting finite difference equation is

$$\mathbf{H}(t) \frac{\mathbf{c}(t + \Delta t) - \mathbf{c}(t)}{\Delta t} = \mathbf{A}(t)\mathbf{c}(t + \Delta t) + \mathbf{b}(t) \tag{18.52}$$

which can be factored

$$[\mathbf{H}(t) - \Delta t \mathbf{A}(t)]\mathbf{c}(t + \Delta t) = \mathbf{H}(t)\mathbf{c}(t) + \Delta t \mathbf{b}(t) \tag{18.53}$$

and solved

$$\mathbf{c}(t + \Delta t) = [\mathbf{H}(t) - \Delta t \mathbf{A}(t)]^{-1} [\mathbf{H}(t)\mathbf{c}(t) + \Delta t \mathbf{b}(t)] \tag{18.55}$$

by inverting the matrix as shown, or, the more efficient procedure, using a linear equation solver (Press et al., 1989) applied to Eq. (18.50). It requires the solution of four simultaneous linear equations instead of the two required for the other variables such as ammonia or oxygen because the equations for Mn(II) and $MnO_2(s)$ in the two layers are coupled. This presents no essential difficulty, however, and the implementation is straightforward.

18.3 B   Manganese Model Application

Since the model is now time variable, the additional parameters required to execute the model, s and the entrainment terms, are needed as functions of time. The stratagem used in the steady state model – examining the relationship between ammonia and manganese fluxes and using approximate relationships to find s – is no longer feasible. It is necessary to execute the nutrient and oxygen model in parallel with the manganese model. This presents no difficulty, since the model has been applied to the MERL data set in Chapter 15. The manganese model equations are simply added to the model framework.

The manganese flux model is executed as follows. The overlying water concentrations of dissolved (A) and particulate (B) manganese are shown in Fig. 18.8 for all the tanks, together with the interpolated functions used for the model boundary conditions. The overlying water particulate concentration is assumed to be constant, for a reason that will become clear subsequently. For the equilibration phase the dissolved and particulate manganese concentrations for all the tanks are set to the overlying water concentrations of the control tanks. These conditions are cycled periodically until the sediment has come into equilibrium.

Once equilibrium has been reached, the nutrient loadings are increased in accordance with the various nutrient loading levels – 1X, 2X, and so on (Section 15.2 A) – and the model is run for the three years of the experiment. The overlying water dissolved manganese concentrations are set at the observed values for the three years (Fig. 18.8A). The particulate manganese concentrations are continued at the same constant value as used during the equilibration phase (Fig. 18.8B).

This procedure assumes that water column processes during the experiment do not change the depositional flux of manganese to the sediment. This is only an approximation, since the overlying water particulate manganese concentrations vary as dosing varies (Fig. 18.8B). However, overall the variation is not too large and the assumption of a constant flux from the overlying water appears to be a reasonable first approximation. The consequences of this assumption are evaluated below when the overlying water is explicitly included in the model.

18.3 C   Results

A comparison of the model and observed fluxes is shown in Fig. 18.9A. The sediment model parameters are listed in Table 18.4 located at the chapter's end and in Table 15.5. The general shape of the seasonal variation of the manganese fluxes is reproduced quite nicely. This variation is due solely to the variation in s since the flux of particulate manganese to the sediment is held constant (Figs. 18.9B).

The variation of manganese fluxes with respect to temperature, overlying water dissolved oxygen, and ammonia flux for both the data and the model are presented in Fig. 18.9B. The data exhibit a somewhat more pronounced variation with respect to ammonia flux than does the model. This is consistent with the findings of the steady state model (Figs. 18.6–18.7).

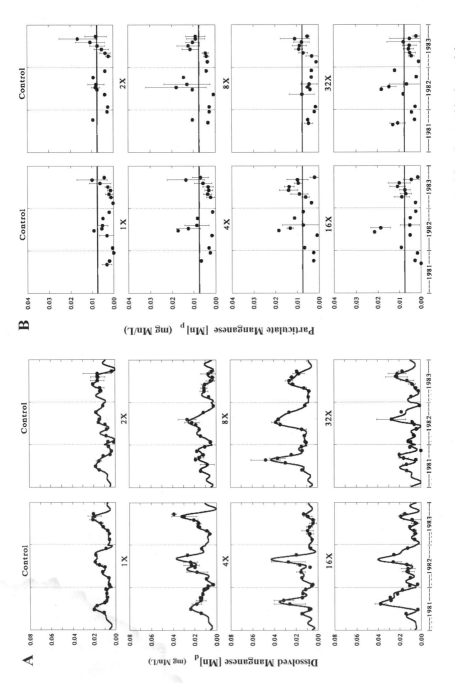

Fig. 18.8 (A) Dissolved manganese concentrations in the overlying water. Data (●). Line is the interpolated function used in the model computation. (B) Particulate concentrations in the overlying water. Line is the concentration used in the model for the depositional flux.

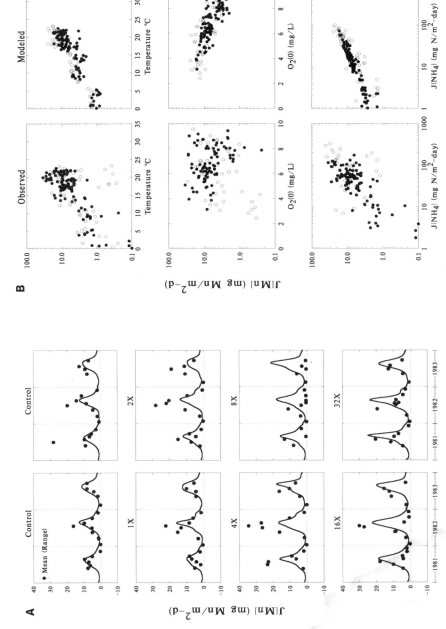

Fig. 18.9 (A) Comparison of observed (•) and modeled (lines) manganese fluxes. (B) Comparison of modeled and observed fluxes versus temperature, overlying water O_2 concentration, and ammonia flux $J[NH_4]$.

The variation in annual average manganese flux due to differing loading rates is examined in Fig. 18.10. The model captures the trend of increasing flux as nutrient loading increases for the first (A) and third (C) years. However, the sharp increases in the second year for the 0X to 4X tanks (B) are not reproduced.

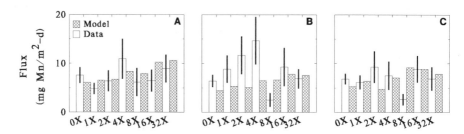

Fig. 18.10 Average annual manganese flux for the three years of the experiment. (A) 1981, (B) 1982, (C) 1983. Comparison of model and data.

The solid phase (A) and pore water (B) concentrations are compared in Fig. 18.11. The model concentrations are slightly higher than the observations, as was the case with the steady state model (Fig. 18.6–18.7). The lower pore water concentrations at the high loading rates are not reproduced by the model and cannot be reproduced by the partitioning model (Eq. 18.12) and be consistent with the solid phase concentrations, since these must be proportional.

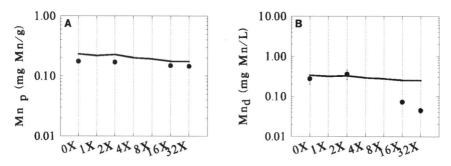

Fig. 18.11 Comparison of (A) solid phase and (B) pore water concentrations. Data (•) and model (line).

18.3 D Seasonal Variation

The manganese flux calculated using the time variable model exhibits a strong seasonal variation. What causes this variation? The flux is computed using the usual equation

$$J[\text{Mn}] = s\left(f_{d1}\text{Mn}(1) - f_{d0}\text{Mn}(0)\right) \tag{18.56}$$

where $f_{d1}Mn(1)$ and $f_{d0}Mn(0)$ are the dissolved manganese concentrations in layer 1 and the overlying water respectively and s is the surface mass transfer coefficient. The temporal variation in any of these can be the cause of the temporal variation in $J[Mn]$.

Fig. 18.12 presents the computed quantities. It is to be expected that the anaerobic

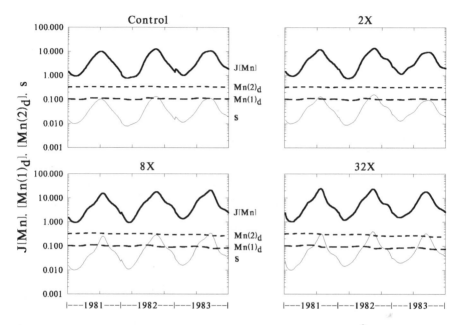

Fig. 18.12 Temporal variation of the manganese flux $J[Mn]$ (mg Mn/m^2-d), the dissolved manganese concentrations in layers 1 and 2: $Mn(1)_d = f_{d1}Mn(1)$ and $Mn(2)_d = f_{d2}Mn(2)$ (mg Mn/L), and the surface mass transfer coefficient s (m/d). $k_{Mn,1} = 5.0$ d^{-1}(mg O$_2$/L)$^{-1}$.

layer dissolved concentration $Mn(2)_d = f_{d2}Mn(2)$ is constant since its variation would require that the entire anaerobic layer manganese concentration $Mn(2)$ vary.

The reason that the aerobic layer concentration $Mn(1)_d$ is constant is due to the choice of the magnitude of the oxidation rate constant $k_{Mn,1} = 5.0$ d^{-1}(mg O$_2$/L)$^{-1}$ which for an overlying water oxygen concentration of $[O_2(0)] \simeq 10$ mg/L (Fig. 15.3) and $[O_2(1)] = [O_2(0)]/2$ is a first order rate constant of ~ 25 d^{-1}. Apparently this is not rapid enough to affect the resulting concentration.

The overlying water manganese concentration $Mn(0)_d$ also varies (Fig. 18.8), but it is small (~ 10 μg/L) relative to the aerobic layer concentration (~ 100 μg/L, Fig. 18.12).

Therefore, the variation in manganese flux (Eq. 18.56) is due entirely to the variation of the surface mass transfer coefficient s, which multiplies the term $\left(f_{d1}Mn(1) - f_{d0}Mn(0)\right) \simeq f_{d1}Mn(1) = Mn(1)_d$ which is not varying. This is clearly seen in Fig. 18.12 where $J[Mn]$ follows s exactly, differing only by a constant amount, which on a logarithmic scale corresponds to the constant factor $Mn(1)_d$.

Variation in s   The physical explanation for the dependency of the manganese flux on the surface mass transfer coefficient is illustrated in Fig. 18.13. The man-

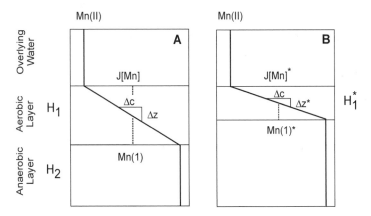

Fig. 18.13   Schematic diagram illustrating the effect of vaying H_1.

ganese flux $J[\text{Mn}]$ is proportional to the vertical gradient of the manganese concentration profile via Fick's law (Eq. 2.47)

$$J[\text{Mn}] = -D\frac{d\text{Mn}_d}{dz} \simeq -D\frac{\Delta\text{Mn}_d}{\Delta z} \tag{18.57}$$

As H_1 decreases $H_1^* < H_1$ the gradient *increases:* $\Delta\text{Mn}_d/\Delta z^* > \Delta\text{Mn}_d/\Delta z$ since $\Delta z^* < \Delta z$. Therefore the flux increases $J[\text{Mn}]^* > J[\text{Mn}]$. Note that the average concentration in the aerobic layer, identified in Fig. 18.13 as $\text{Mn}(1) = \text{Mn}(1)^*$, is the same in both cases.

The relationship between H_1 and the surface mass transfer coefficient s, is (Eq. 3.13)

$$s = \frac{D_1}{H_1} \tag{18.58}$$

Therefore, as H_1 decreases, s increases and so does $J[\text{Mn}]$. In the model calculation H_1 is actually not directly computed. Rather the sediment oxygen demand SOD is computed and s is obtained from (Eq. 3.14)

$$s = \frac{\text{SOD}}{[O_2(0)]} \tag{18.59}$$

where $[O_2(0)]$ is the overlying water dissolved oxygen concentration. However, once s is known, then so is H_1 via Eq. 18.58.

Variation in $k_{\text{Mn},1}$ The effect of an order of magnitude increase in $k_{\text{Mn},1} = 50.0\,\text{d}^{-1}(\text{mg O}_2/\text{L})^{-1}$ is shown in Fig. 18.14. The resulting fluxes are shown in Fig. 18.15. The variation in $\text{Mn}(1)_d$, which is in phase with the variation of s, increases

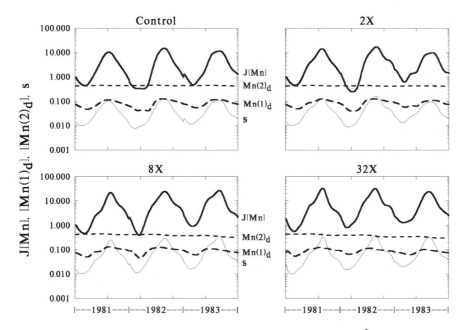

Fig. 18.14 Temporal variation of the manganese flux $J[\mathrm{Mn}]$ (mg Mn/m^2-d), the dissolved manganese concentrations in layers 1 and 2: $\mathrm{Mn}(1)_d = f_{d1}\mathrm{Mn}(1)$ and $\mathrm{Mn}(2)_d = f_{d2}\mathrm{Mn}(2)$ (mg Mn/L), and the surface mass transfer coefficient s (m/d). $k_{\mathrm{Mn},1} = 50.0\ \mathrm{d}^{-1}(\mathrm{mg\ O_2/L})^{-1}$.

Table 18.3 Relationship between $J[\mathrm{Mn}]$ and $J[\mathrm{NH_4}]$

	Data	Model	
		Fig. 18.9	Fig. 18.15
Reaction rate $k_{\mathrm{Mn},1}\left(\mathrm{d}^{-1}(\mathrm{mg\ O_2/L})^{-1}\right)$	–	5.0	50.0
Log-log slope*	0.661±0.069	0.454	0.618

*Slope m of the relationship $\log(J[\mathrm{Mn}]) = m\log(J[\mathrm{NH_4}]) + b$

the amplitude of the variation in $J[\mathrm{Mn}]$. Since s is in phase with $J[\mathrm{NH_4}]$, this also increases the slope of the relationship between $J[\mathrm{Mn}]$ and $J[\mathrm{NH_4}]$ – compare Fig. 18.9B and Fig. 18.15B.

A more quantitative comparison is made in Table 18.3 which lists the observed and model computed slopes between $\log(J[\mathrm{Mn}])$ and $\log(J[\mathrm{NH_4}])$. The larger aerobic layer oxidation rate $k_{\mathrm{Mn},1} = 50.0\ \mathrm{d}^{-1}(\mathrm{mg\ O_2/L})^{-1}$ reproduces the observed slope of the relationship.

An estimate of the magnitude at which $k_{\mathrm{Mn},1}$ becomes important can be made using the steady state solution for the approximation $k_{\mathrm{MnO_2},2} \to \infty$ (Eqs. 18.31,

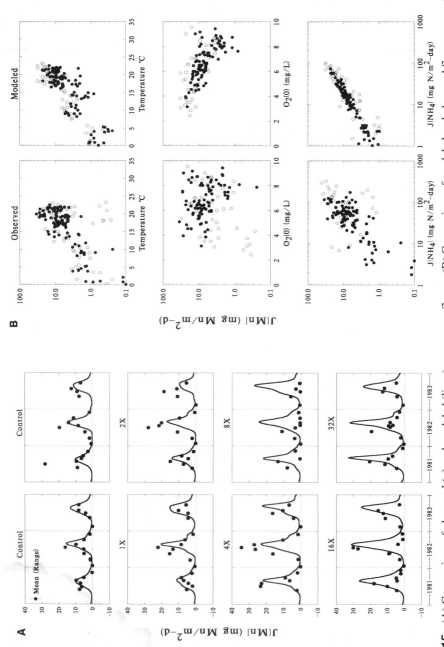

Fig. 18.15 (A) Comparison of observed ($\bullet$) and modeled (lines) manganese fluxes. (B) Comparison of modeled and observed fluxes versus temperature, overlying water O_2 concentration, and ammonia flux $J[NH_4]$. $k_{Mn,1} = 50.0$ d^{-1}(mg O_2/L)$^{-1}$.

18.32, 18.34)

$$J[\mathrm{Mn}] = J_{\mathrm{MnO_2}} \frac{s f_{d1}}{w_2} \frac{f_{d2} K_{L12} + w_{12}}{f_{d1}\left(K_{L12} + s + k'_{\mathrm{Mn},1} H_1\right) + w_{12} + w_2} \tag{18.60}$$

where $k'_{\mathrm{Mn},1}$ is the first order rate constant and the equality $k'_{\mathrm{Mn},1} H_1 = \kappa^2_{\mathrm{Mn},1}/s$ has been used (Eq. 18.16). Thus for $k'_{\mathrm{Mn},1}$ to be significant it must be larger than the other terms in the sum in which it appears

$$k'_{\mathrm{Mn},1} > \frac{s + K_{L12}}{H_1} \tag{18.61}$$

For the MERL experiment $K_{L12} = D_d/(H_2/2) = (25\ \mathrm{cm^2/d})/(5\ \mathrm{cm}) = 5\ \mathrm{cm/d}$. For $H_1 \sim 1\ \mathrm{mm} = 0.1\ \mathrm{cm}$, $k'_{\mathrm{Mn},1} > 50\ \mathrm{d^{-1}}$. But $k'_{\mathrm{Mn},1} = k_{\mathrm{Mn},1}[O_2(1)] = k_{\mathrm{Mn},1}[O_2(0)]/2$. With $[O_2(0)] \simeq 10\ \mathrm{mg/L}$, $k_{\mathrm{Mn},1} > 10\ \mathrm{d^{-1}}(\mathrm{mg}\ O_2/\mathrm{L})^{-1}$ which is in reasonable agreement with the results in Table 18.3. This calculation provides a nice illustration of the utility of analytical solutions.

18.3 E   Summary

The time variable version of the model captures many of the features exhibited by the data: a strong seasonal variation (Fig. 18.9A); an increasing manganese flux with increasing nutrient inputs (Fig. 18.10); and a slight depletion of the sediment manganese concentration, as a consequence of the increased manganese flux (Fig. 18.11). The results also point to the importance of the aerobic layer oxidation rate $k_{\mathrm{Mn},1}$. Therefore any process that would cause $k_{\mathrm{Mn},1}$ to change should be examined. The pH variation exhibited in the MERL experiments, which is known to significantly affect the oxidation rate of manganese (II) to manganese (IV) (Fig. 18.1), has not been taken into account. This is addressed in the next section.

18.4   EFFECT OF pH

The purpose of this section is to examine the effect of pH variations on the flux of manganese from the sediment. The rate of Mn(II) oxidation is strongly affected by the pH (Fig. 18.1), so it is reasonable to expect that this effect would be important. Since the oxidation reaction takes place in the aerobic layer of the sediment, this is the pH that is required.

18.4 A   pH in the Aerobic Layer

The pH in the aerobic layer is different from that in the overlying water, e.g., de Beer et al. (1997). It is determined by the alkalinity $\mathrm{Alk_1}$ and inorganic carbon $\mathrm{TIC_1}$ concentrations in that layer. These, in turn, are determined by the mass fluxes of alkalinity and inorganic carbon from the adjacent layers and internal sources and sinks. For the sake of simplicity, we assume that neither internal sources nor the sinks of Alk and TIC in the aerobic layer are significant. Any other assumption would require a consideration of the acid-base reactions in the layer as well as any TIC and alkalinity sources. Because the aerobic layer is quite thin, it is likely that their effects are not large relative to the mass fluxes from the adjoining layers. Therefore, alkalinity and TIC are assumed to be conservative in layer 1.

The schematic diagram is presented in Fig. 18.16. The mass balance equations

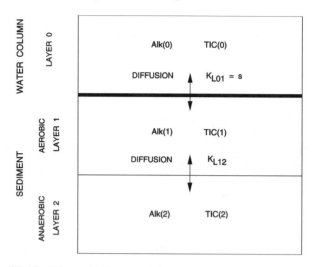

Fig. 18.16   Schematic diagram of the model for pH in the aerobic layer.

for this model are

$$s(\text{Alk}_0 - \text{Alk}_1) + K_{L12}(\text{Alk}_2 - \text{Alk}_1) = 0 \qquad (18.62a)$$

and

$$s(\text{TIC}_0 - \text{TIC}_1) + K_{L12}(\text{TIC}_2 - \text{TIC}_1) = 0 \qquad (18.62b)$$

where s and K_{L12} are the mass transfer coefficients for layer 0–1 and layer 1–2, respectively. Subscripts are used to identify the layer. Solving these Eqs. (18.62) for the layer 1 concentrations yields

$$\text{Alk}_1 = \alpha \text{Alk}_0 + (1 - \alpha)\text{Alk}_2 \qquad (18.63a)$$

and

$$\text{TIC}_1 = \alpha \text{TIC}_0 + (1 - \alpha)\text{TIC}_2 \qquad (18.63b)$$

where

$$\alpha = \frac{s}{s + K_{L12}} \qquad (18.64)$$

Once Alk_1 and TIC_1 are known, the pH is computed using the equations that describe the carbon dioxide system (Stumm and Morgan, 1996).

The mass balance equations for the carbon dioxide system are

$$[TIC] = [CO_2] + [HCO_3^-] + [CO_3^{2-}] \tag{18.65a}$$

$$[Alk] = [HCO_3^-] + 2[CO_3^{2-}] \tag{18.65b}$$

and the mass action equations are

$$\frac{[HCO_3^-][H^+]}{[CO_2]} = K_1 \tag{18.66a}$$

$$\frac{[CO_3^{2-}][H^+]}{[HCO_3^-]} = K_2 \tag{18.66b}$$

where

$$[CO_2] = \text{carbon dioxide concentration} \qquad \text{(mol/L)}$$
$$[HCO_3^-] = \text{bicarbonate concentration} \qquad \text{(mol/L)}$$
$$[CO_3^{2-}] = \text{carbonate concentration} \qquad \text{(mol/L)}$$
$$[H^+] = \text{hydrogen ion concentration} \qquad \text{(mol/L)}$$
$$K_1 = \text{carbon dioxide equilibrium constant (mol/L)}$$
$$K_2 = \text{bicarbonate equilibrium constant} \qquad \text{(mol/L)}$$

Eliminating $[H^+]$, $[HCO_3^-]$ and $[CO_3^{2-}]$ yields

$$K_1 K_2 (Alk - 2TIC) + K_1 H (Alk - TIC) + H^2 Alk = 0 \tag{18.67}$$

which relates the pH to the alkalinity and inorganic carbon concentrations. The square brackets and charges are dropped for notational convenience.

The aerobic layer alkalinity and TIC concentrations are given by Eqs. (18.63). For the alkalinity, both Alk_0 and Alk_2 are known by measurement. However, the TIC's are not measured. They can be computed from the respective pHs and alkalinities. Using Eq. (18.67) to solve for TIC yields

$$TIC = \frac{Alk\left(K_1(K_2 + H) + H^2\right)}{K_1(2K_2 + H)} \tag{18.68}$$

The relationship between pH and hydrogen ion concentration

$$H_0 = 10^{-pH_0} \tag{18.69}$$

together with Alk_0 determine TIC_0 using Eq. (18.68). A similar procedure is applied to layer 2 to obtain TIC_2 using pH_2 and Alk_2. Then Eq. (18.63b) yields TIC_1. Since the layer1 alkalinity and TIC are now known, Eq. (18.67) can be solved for H_1 to obtain the pH_1. The result is

$$H_1 = \frac{K_1 (\text{TIC}_1 - \text{Alk}_1) \pm \sqrt{4\text{Alk}_1 (2\text{TIC}_1 - \text{Alk}_1) K_1 K_2 + (\text{TIC}_1 - \text{Alk}_1)^2 K_1^2}}{2\text{Alk}_1}$$

(18.70)

and

$$\text{pH}_1 = -\log_{10} [H_1]$$

(18.71)

An example computation is given in Fig. 18.17 which presents the pH in the aerobic layer as a function of the pH in the overlying water, for various overlying water alkalinities and for two values of the surface mass transfer coefficient s.

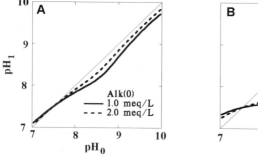
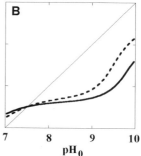

Fig. 18.17 Aerobic layer pH_1 as a function of overlying water pH_0. (A) $s = 0.1$ m/d. (B) $s = 0.01$ m/d. Assumed conditions: anaerobic layer $\text{pH}_2 = 7.5$; alkalinity $\text{Alk}_2 = 3$ meq/L; and layer 1–2 mass transfer coefficient $K_{L12} = 0.01$ m/d. Alk_0 is noted in the figure. Lines of equal pHs are also shown.

The question is: What is determining pH_1? A part of the behavior has a straight-forward rationalization. For a large surface mass transfer coefficient $s = 0.1$ m/d (A), the aerobic layer pH_1 tracks the overlying water pH_0 because the mixing from the overlying water is sufficiently large so aerobic layer Alk and TIC are determined by the overlying water alkalinity and TIC and the anaerobic layer Alk and TIC have little effect. For the smaller surface mass transfer coefficient $s = 0.01$ m/d (B), the aerobic layer pH_1 remains close to the anaerobic layer $\text{pH}_2 = 7.5$, since it is more isolated from the overlying water.

What is puzzling, however, is why the aerobic layer pH_1 varies as it does with varying overlying water alkalinity $\text{Alk}(0)$. Increasing the alkalinity from $\text{Alk}(0) = 1$ meq/L to 2 meq/L – going from the bold solid lines to the dashed lines in Fig. 18.17 – makes the overlying water alkalinity more like the alkalinity of the *anaerobic* layer

Alk (2) = 3 meq/L. One would expect that the same should happen to the pH. That is, the pH in the aerobic layer should move closer to the *anaerobic* layer pH = 7.5 as well. But the reverse happens. The aerobic layer pH_1 becomes more like the *overlying water* pH_0.

The key to understanding this result in particular, and the CO_2 system in general, is to form the ratio of Eq. (18.65b) to Eq. (18.65a)

$$\frac{[Alk]}{[TIC]} = \frac{[HCO_3^-] + 2[CO_3^{2-}]}{[CO_2] + [HCO_3^-] + [CO_3^{2-}]} \tag{18.72a}$$

$$= \frac{[HCO_3^-](1 + 2K_2/[H^+])}{[HCO_3^-]([H^+]/K_1 + 1 + K_2/[H^+])} \tag{18.72b}$$

$$= \frac{1 + 2K_2/[H^+]}{[H^+]/K_1 + 1 + K_2/[H^+]} \tag{18.72c}$$

This shows that the pH is determined by the ratio of alkalinity to inorganic carbon [Alk] / [TIC] and not the individual values. This relationship is shown in Fig. 18.18.

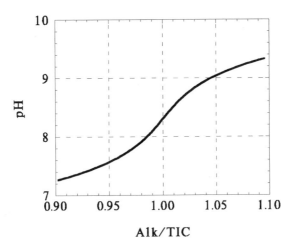

Fig. 18.18   pH as a function of [Alk] / [TIC]. Parameters: $K_1 = 10^{-6.3}$, $K_2 = 10^{-10.3}$.

Note that small changes in [Alk] / [TIC] result in large changes in pH due to the lack of buffering capacity in the pH = 7–9 range (Stumm and Morgan, 1996). As pH approaches pK_1 or pK_2, buffering is increased and the relationship flattens out. This relationship is the key to understanding what determines the pH.

The ratio of alkalinity to TIC in layer 1 is found using Eqs. (18.63)

$$\frac{\text{Alk}_1}{\text{TIC}_1} = \frac{\alpha\,\text{Alk}_0 + (1-\alpha)\,\text{Alk}_2}{\alpha\,\text{TIC}_0 + (1-\alpha)\,\text{TIC}_2} \tag{18.73a}$$

$$= \frac{\alpha\,\text{Alk}_0 + (1-\alpha)\,\text{Alk}_2}{\alpha r_0^{-1}\text{Alk}_0 + (1-\alpha)\,r_2^{-1}\text{Alk}_2} \tag{18.73b}$$

where $r_0 = \text{Alk}_0/\text{TIC}_0$ and $r_2 = \text{Alk}_2/\text{TIC}_2$. Factoring this expression and grouping the terms yields

$$\frac{\text{Alk}_1}{\text{TIC}_1} = \frac{1+\rho}{r_0^{-1} + r_2^{-1}\rho} \tag{18.74}$$

where

$$\rho = \frac{K_{L12}}{s}\frac{\text{Alk}_2}{\text{Alk}_0} \tag{18.75}$$

The limiting cases with respect to the variation of ρ are

$$\begin{aligned}\frac{\text{Alk}_1}{\text{TIC}_1} &\to r_0 = \frac{\text{Alk}_0}{\text{TIC}_0} \\ \text{pH}_1 &\to \text{pH}_0\end{aligned} \qquad \text{as} \qquad \rho \to 0 \tag{18.76}$$

$$\begin{aligned}\frac{\text{Alk}_1}{\text{TIC}_1} &\to r_2 = \frac{\text{Alk}_2}{\text{TIC}_2} \\ \text{pH}_1 &\to \text{pH}_2\end{aligned} \qquad \text{as} \qquad \rho \to \infty \tag{18.77}$$

Thus the behavior of the aerobic layer pH is determined by the magnitude of ρ.

We can now understand the results in Fig. 18.17. For a large s (Fig. 18.17A), $\rho = (K_{L12}/s)(\text{Alk}_2/\text{Alk}_0)$ is small, Eq. (18.76) applies, and $\text{pH}_1 \to \text{pH}_0$. This is the obvious case: ρ is controlled by the ratio of the mass transport coefficients.

But ρ is also influenced by the alkalinity ratio $\text{Alk}_2/\text{Alk}_0$ (Eq. 18.75). As the overlying water alkalinity $\text{Alk}_0 = 1.0 \to 2.0$ meq/L approaches the anaerobic layer alkalinity $\text{Alk}_2 = 3.0$ meq/L – going from the bold solid lines to the dashed lines in Fig. 18.17 – this causes ρ to decrease. Thus Eq. (18.76) applies, and again $\text{pH}_1 \to \text{pH}_0$. What is happening in this case is more subtle. Increasing Alk_0 causes an increase in TIC_0 since pH_0 is assumed to be fixed. Hence more TIC_0 is transported into layer 1, making the TIC in layer 1 more like layer 0. This is a larger effect than making the alkalinity in layer 1 more like layer 2 so that $\text{pH}_1 \to \text{pH}_0$. The behavior for any set of modifications can be understood by examining Eq. (18.74).

18.4 B   Effect of pH on the Manganese Flux

The pH in the aerobic layer can now be calculated using the equations developed in the previous section (Eqs. 18.63 and 18.70). The results are shown in Fig. 18.19. The

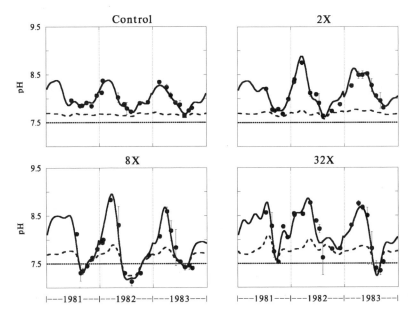

Fig. 18.19 Computed aerobic layer pH (dashed line). pH in the overlying water (solid line is interpolated to the data points). Anaerobic layer pH is assumed to be constant (dotted line).

solid line is an interpolation of the data in the overlying water, which are also shown. The dashed line is the computed pH in the aerobic layer. The dotted line is at the pH of the anaerobic layer, which is assumed to be $pH = 7.5$, consistent with pore water observations. The alkalinities are $Alk_0 = 2.0$ meq/L and $Alk_2 = 3.0$ meq/L, which are also estimated from observations. The remaining parameters s and K_{L12} are obtained from the simulation to be presented below.

For the control and 2X loading, there is little computed variation in pH_1. The reason is that the surface mass transfer coefficient s is small, corresponding to a comparatively small SOD (Fig. 15.9), so the aerobic layer resembles the anaerobic layer. However, for the more heavily loaded tanks, the SOD increases, s increases, and the aerobic layer starts to be influenced by the overlying water pH, which is varying. Hence the variation in pH_1 is more substantial for these tanks, sometimes approaching one half pH unit. As shown in the next section, this corresponds to one order of magnitude increase in oxidation reaction rate. It is for this reason – the expected importance of the pH variation in the aerobic layer – that this refinement is added to the model.

Reaction Kinetics The incorporation of pH dependent reaction kinetics is straightforward. The aerobic layer reaction rate R used previously (Eq. 18.46a)

$$R = -k_{Mn,1}[O_2(1)]f_{d1}Mn(1)H_1 \tag{18.78}$$

is replaced by a pH-dependent expression (Eq. 18.10a)

$$R = -k_{Mn,OH,1}[O_2(1)]\left[OH^-\right]^2 f_{d1}Mn(1)H_1 \tag{18.79}$$

where the dependency on $\left[OH^-\right]^2$ is from an analysis of experimental results. This is expressed as follows

$$R = -k_{Mn,OH,1}[O_2(1)]\left(10^{2(pH_1-7)}\right) f_{d1}Mn(1)H_1 \tag{18.80}$$

where the pH is normalized to pH $= 7$ so that the numerical value of the reaction rate constant velocity $k_{Mn,1}$ corresponds to $pH_1 = 7$. This procedure is used so the magnitude of $k_{Mn,OH,1}$ easily understood. The $+2$ in the exponent results from the dependency on $\left[OH^-\right]^2$

$$\left[OH^-\right]^2 = \left(\frac{K_W}{[H^+]}\right)^2 = \left(\frac{K_W}{10^{-pH}}\right)^2 = K_W^2 10^{2pH} \tag{18.81}$$

where $K_W = \left[H^+\right]\left[OH^-\right]$, the ionization constant for water. With the kinetics modified, it remains only to rerun the model using the computed pH_1 and examine the results.

Results   The consequences of this modification are shown in Fig. 18.21. The concentrations are shown in Fig. 18.20. There are almost no discernible differences when compared to Fig. 18.15. Perhaps, in hindsight, this should not have been too surprising if one considers the small pH variation in the aerobic layer (Fig. 18.19).

Actually the pH is important, as we will see in the next chapter. But it is important in the overlying water where the variation in pH is larger.

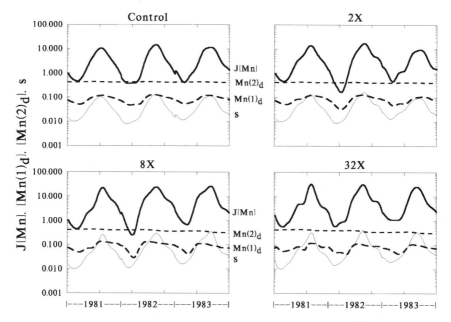

Fig. 18.20 Temporal variation of the manganese flux $J[Mn]$ (mg Mn/m^2-d), the dissolved manganese concentrations in layers 1 and 2: $Mn(1)_d = f_{d1}Mn(1)$ and $Mn(2)_d = f_{d2}Mn(2)$ (mg Mn/L), and the surface mass transfer coefficient s (m/d). $k_{Mn,1,OH} = 2.0$ d^{-1}(mg O$_2$/L)$^{-1}$.

Table 18.4 Sediment Model Parameters

Parameter*	Units	Fig. 18.9	Fig. 18.15	Fig. 18.21	Fig. 19.7
$\Delta\pi_{Mn,1}$	–	3.0	3.0	3.0	3.0
$\pi_{Mn,2}$	L/kg	700	700	700	700
$O_2(0)_{crit,Mn}$	mg O$_2$/L	2.0	2.0	2.0	2.0
$k_{Mn,1}$	(a)	5.0	50.0	0.0	0.0
$k_{Mn,OH,1}$	(a)	–	–	2.0	(c)
$k_{Mn,OH,MnO_2,1}$	(b)	–	–	–	(c)
$k_{MnO_2,2}$	d^{-1}	4.0	4.0	4.0	4.0
Alk$_0$	meq/L	–	–	2.0	2.0
pH$_2$	–	–	–	7.5	7.5
Alk$_2$	meq/L	–	–	3.0	3.0
$f_{colloidal}$	–	–	–	–	0.5
$w_{MnO_2,in}$	m/d	0.7	0.7	1.20	1.20

*$\theta_{Mn,1} = \theta_{Mn,OH,1} = \theta_{Mn,OH,MnO_2,1} = \theta_{MnO_2,2} = 1.08.$ (a)d^{-1}(mg O$_2$/L)$^{-1}$. (b)d^{-1}(mg O$_2$/L)$^{-1}$(mg Mn/L)$^{-1}$ $(c)k_{Mn,OH,MnO_2,1} = k_{Mn,OH,MnO_2,0} = 2 \times 10^{-4}$ (Fig. 19.9A) or $k_{Mn,OH,1} = k_{Mn,OH,0} = 0.003$ (Fig. 19.9B).

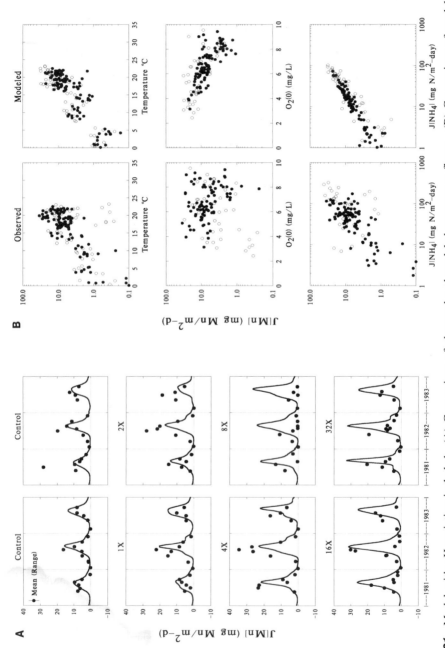

Fig. 18.21 Model with pH variation included. (A) Comparison of observed and modeled manganese fluxes. (B) Comparison of modeled and observed fluxes versus temperature, overlying water O_2 concentration, and ammonia flux $J[NH_4]$.

Appendix 18A: MACSYMA

The solution of the four simultaneous equations would not have been possible without MACSYMA. The output is listed in Figs. 18A.1 and 18A.2. The first four lines in Fig. 18A.1 (c2) to (c5) assign the mass balance equations to the variables eq1 to eq4. The term k[Mn,1] is used to represent $\kappa_{Mn,1}^2/s$. The next line (c6) checks the equations by adding them. The results (d6) are the source and sink terms: loss of dissolved manganese to the overlying water, source of MnO_2 to the aerobic layer, and burial of dissolved Mn(II) and MnO_2(s). (c7) is a string of commands in a block separated by commas (, , ⋯). The equation set eq is named. The variables mno2[1] and mno2[2] are eliminated, and the remaining equations are solved for mn[1]and mn[2]. The solution for mn[2] is displayed (d8).

The next set of instructions (Fig. 18A.2) evaluates r_{12} and substitutes it into the solutions. First r_{12} is found (c9). Then the numerator and denominator are found separately, and used to define r_{12}. They are expressed as equations that are then used to define the substitutions in scsimp. Once again it seems almost magical that the complicated expression (d8) can be reduced to (d11).

The flux is computed in (d13). MACSYMA needs some help to factor the denominator of (d11) and (d13). If the term $f_{d1}w_2k_{Mn,1}r_{12}$ is subtracted, the remaining expression factored, and the term added back, the result is (d15), the factored denominator.

The solution for MnO_2 in layer 1 is computed by eliminating MnO_2(2) from the equation set, and solving for MnO_2(1) in (c16). The idea is to leave Mn(1) and Mn(2) in the equation set so the solution can be expressed using these concentrations as well, since they are now known. The same procedure is applied to solve for MnO_2(2).

The mass balance equation is checked in Fig. 18A.3. The solutions are listed, the substitutions made and simplified. The result is zero as it should be.

Aerobic Sediment Layer

(c2)

eq1: s*(-fd[1]*Mn[1])+k[l12]*(fd[2]*Mn[2]-fd[1]*Mn[1])
+w[12]*(fp[2]*Mn[2]-fp[1]*Mn[1])
-(k[Mn,1])*Mn[1]*fd[1]-w[2]*Mn[1]

(d2)

$$- fd_1 \, mn_1 \, s - fd_1 \, mn_1 \, k_{mn,1} + \left(fd_2 \, mn_2 - fd_1 \, mn_1 \right) k_{l12} + \left(fp_2 \, mn_2 - fp_1 \, mn_1 \right) w_{12} - mn_1 \, w_2$$

(c3)

eq2: w[12]*(MnO2[2]-MnO2[1])+(k[Mn,1])*Mn[1]*fd[1]
-w[2]*MnO2[1]+j[MnO2]

(d3)

$$j_{mno2} + fd_1 \, mn_1 \, k_{mn,1} + \left(mno2_2 - mno2_1 \right) w_{12} - mno2_1 \, w_2$$

Anaerobic Sediment Layer

(c4)

eq3: -k[l12]*(fd[2]*Mn[2]-fd[1]*Mn[1])+k[MnO2,2]*MnO2[2]
-w[12]*(fp[2]*Mn[2]-fp[1]*Mn[1])
+w[2]*Mn[1] -w[2]*Mn[2]

(d4)

$$mno2_2 \, k_{mno2,2} - \left(fd_2 \, mn_2 - fd_1 \, mn_1 \right) k_{l12} - \left(fp_2 \, mn_2 - fp_1 \, mn_1 \right) w_{12} - mn_2 \, w_2 + mn_1 \, w_2$$

(c5)

eq4: -w[12]*(MnO2[2]-MnO2[1])-k[MnO2,2]*MnO2[2]
+w[2]*MnO2[1] -w[2]*MnO2[2]

(d5)

$$- mno2_2 \, k_{mno2,2} - \left(mno2_2 - mno2_1 \right) w_{12} - mno2_2 \, w_2 + mno2_1 \, w_2$$

Check Mass Balance

(c6)   ratsimp(eq1+eq2+eq3+eq4)

(d6)

$$- fd_1 \, mn_1 \, s + j_{mno2} + \left(- mno2_2 - mn_2 \right) w_2$$

Solve the Simultaneous Equations

(c7)   (eq:[eq1,eq2,eq3,eq4],s1:eliminate(eq,[MnO2[1],MnO2[2]]),s2:solve(s1,[Mn[1],Mn[2]]))$

(c8)   s2[1][2]

(d8)

$$mn_2 = \frac{\left(\begin{array}{c} fd_1 \, j_{mno2} \, k_{mno2,2} \, s + fd_1 \, k_{mn,1} \, j_{mno2} \, k_{mno2,2} + fd_1 \, k_{l12} \\ * \, j_{mno2} \, k_{mno2,2} + fp_1 \, w_{12} \, j_{mno2} \, k_{mno2,2} + w_2 \, j_{mno2} \\ * \, k_{mno2,2} \end{array} \right)}{\left(\begin{array}{c} \left(\begin{array}{c} w_{12} \left(fd_1 \, fp_2 \, k_{mno2,2} + fd_1 \, fp_2 \, w_2 \right) + k_{l12} \\ * \left(fd_1 \, fd_2 \, k_{mno2,2} + fd_1 \, fd_2 \, w_2 \right) + fd_1 \, w_2 \, k_{mno2,2} \\ + fd_1 \, w_2^2 \end{array} \right) s + w_{12} \\ * \left(fp_1 \, w_2 \, k_{mno2,2} + fp_1 \, w_2^2 \right) + k_{mn,1} \\ * \left(fd_1 \, w_2 \, k_{mno2,2} + fd_1 \, fd_2 \, w_2 \, k_{l12} + fd_1 \, fp_2 \, w_2 \, w_{12} + fd_1 \, w_2^2 \right) + k_{l12} \\ * \left(fd_1 \, w_2 \, k_{mno2,2} + fd_1 \, w_2^2 \right) + w_2^2 \, k_{mno2,2} + w_2^3 \end{array} \right)}$$

Fig. 18A.1 MACSYMA solution of Eqs. (18.21).

Substitute r12 into Mn[2]

(c9)   r12z:ratsimp(rhs(s2[1][1])/rhs(s2[1][2]))

(d9)
$$\frac{fd_2 \, k_{l12} + fp_2 \, w_{12}}{fd_1 \, s + fd_1 \, k_{mn,\,1} + fd_1 \, k_{l12} + fp_1 \, w_{12} + w_2}$$

(c10)   (r12eqnum:r12num=ratnumer(r12z),r12eqden:r12den=ratdenom(r12z),r12eq:r12=r12num/r12den)$

(c11)   Mn2eqz:ratsimp(scsimp(s2[1][2],r12eqnum,r12eqden,r12eq))

(d11)
$$mn_2 = \frac{j_{mno2} \, k_{mno2,\,2}}{\left(fd_1 \, k_{mno2,\,2} \, r12 + fd_1 \, w_2 \, r12\right) s + fd_1 \, w_2 \, k_{mn,\,1} \, r12 + w_2 \, k_{mno2,\,2} + w_2^{\,2}}$$

Compute the Flux

(c12)   Mn1eqz:ratsimp(scsimp(s2[1][1],r12eqnum,r12eqden,r12eq))$

(c13)   jMneqz:s*fd[1]*rhs(Mn1eqz)

(d13)
$$\frac{fd_1 \, j_{mno2} \, k_{mno2,\,2} \, r12 \, s}{\left(fd_1 \, k_{mno2,\,2} \, r12 + fd_1 \, w_2 \, r12\right) s + fd_1 \, w_2 \, k_{mn,\,1} \, r12 + w_2 \, k_{mno2,\,2} + w_2^{\,2}}$$

Factor the Denominator

(c14)   sden:ratdenom(rhs(Mn2eqz))

(d14)
/R/
$$\left(fd_1 \, k_{mno2,\,2} + fd_1 \, w_2\right) r12 \, s + fd_1 \, w_2 \, k_{mn,\,1} \, r12 + w_2 \, k_{mno2,\,2} + w_2^{\,2}$$

(c15)   factor(sden-(fd[1]*w[2]*k[Mn,1]* r12))+fd[1]*w[2]*k[Mn,1]* r12

(d15)
$$\left(k_{mno2,\,2} + w_2\right) \left(fd_1 \, r12 \, s + w_2\right) + fd_1 \, w_2 \, k_{mn,\,1} \, r12$$

Compute the MnO2 Solutions

(c16)   (s1:eliminate(eq,[MnO2[2]]),s21:solve(s1[1],MnO2[1]),facsum(s21[1],Mn[1],(k[MnO2,2] + w[12] + w

(c17)   facsum(%,(k[MnO2,2] + w[12] + w[2]))

(d17)
$$mno2_1 = \frac{\left(j_{mno2} + fd_1 \, mn_1 \, k_{mn,\,1}\right) \left(k_{mno2,\,2} + w_{12} + w_2\right)}{\left(w_{12} + w_2\right) \left(k_{mno2,\,2} + w_2\right)}$$

(c18)   (s1:eliminate(eq,[MnO2[1]]),s22:solve(s1[2],MnO2[2]))$

(c19)   facsum(s22[1],Mn[1])

(d19)
$$mno2_2 = \frac{mn_2 \left(fd_2 \, k_{l12} + fp_2 \, w_{12} + w_2\right) - mn_1 \left(fd_1 \, k_{l12} + fp_1 \, w_{12} + w_2\right)}{k_{mno2,\,2}}$$

Fig. 18A.2   MACSYMA solution of Eqs. (18.21), continued.

List the solutions

(d27)
$$jmn = \frac{fd_1\, j_{mno2}\, k_{mno2,\,2}\, r12\, s}{\left(fd_1\, k_{mno2,\,2}\, r12 + fd_1\, w_2\, r12\right) s + fd_1\, w_2\, k_{mn,\,1}\, r12 + w_2\, k_{mno2,\,2} + w_2^2}$$

(d28)
$$mn_1 = \frac{j_{mno2}\, k_{mno2,\,2}\, r12}{\left(fd_1\, k_{mno2,\,2}\, r12 + fd_1\, w_2\, r12\right) s + fd_1\, w_2\, k_{mn,\,1}\, r12 + w_2\, k_{mno2,\,2} + w_2^2}$$

(d29)
$$mn_2 = \frac{j_{mno2}\, k_{mno2,\,2}}{\left(fd_1\, k_{mno2,\,2}\, r12 + fd_1\, w_2\, r12\right) s + fd_1\, w_2\, k_{mn,\,1}\, r12 + w_2\, k_{mno2,\,2} + w_2^2}$$

(d30)
$$mno2_2 = \frac{mn_2 \left(fd_2\, k_{l12} + fp_2\, w_{12} + w_2\right) - mn_1 \left(fd_1\, k_{l12} + fp_1\, w_{12} + w_2\right)}{k_{mno2,\,2}}$$

(d31)
$$r12 = \frac{fd_2\, k_{l12} + fp_2\, w_{12}}{fd_1\, s + fd_1\, k_{mn,\,1} + fd_1\, k_{l12} + fp_1\, w_{12} + w_2}$$

Evaluate the Mass Balance

(c32)   mb:j[MnO2]-jmneqz-w[2]*(mn2eqz+mno22z)

(d32)
$$-w_2 \left(\frac{\dfrac{j_{mno2}\, k_{mno2,\,2}}{\left(fd_1\, k_{mno2,\,2}\, r12 + fd_1\, w_2\, r12\right) s + fd_1\, w_2\, k_{mn,\,1}\, r12 + w_2\, k_{mno2,\,2} + w_2^2}}{} + \frac{mn_2 \left(fd_2\, k_{l12} + fp_2\, w_{12} + w_2\right) - mn_1 \left(fd_1\, k_{l12} + fp_1\, w_{12} + w_2\right)}{k_{mno2,\,2}} \right)$$
$$- \frac{fd_1\, j_{mno2}\, k_{mno2,\,2}\, r12\, s}{\left(fd_1\, k_{mno2,\,2}\, r12 + fd_1\, w_2\, r12\right) s + fd_1\, w_2\, k_{mn,\,1}\, r12 + w_2\, k_{mno2,\,2} + w_2^2} + j_{mno2}$$

Substitute the Solutions for Mn(1) and Mn(2)

(c33)   ratsimp(ev(mb,mn[1]=mn1eqz,mn[2]=mn2eqz))

(d33)
$$\frac{\begin{pmatrix} fd_1\, w_2\, j_{mno2}\, r12\, s \\ + \left(fd_1\, w_2\, k_{mn,\,1} + fd_1\, w_2\, k_{l12} + fp_1\, w_2\, w_{12} + w_2^2 \right) j_{mno2} \\ * \, r12 + \left(- fd_2\, w_2\, k_{l12} - fp_2\, w_2\, w_{12} \right) j_{mno2} \end{pmatrix}}{\left(fd_1\, k_{mno2,\,2} + fd_1\, w_2\right) r12\, s + fd_1\, w_2\, k_{mn,\,1}\, r12 + w_2\, k_{mno2,\,2} + w_2^2}$$

Substitute for r12

(c34)   ratsimp(ev(%,r12=r12z))

(d34)   0

Fig. 18A.3   MACSYMA evaluation of the global mass balance.

19

Manganese II: Overlying Water-Sediment Interaction

19.1 INTRODUCTION

The flux of dissolved manganese from the sediment to the overlying water is fueled by the depositional flux of particulate manganese from the overlying water to the sediment J_{MnO_2}. In the previous chapter this flux was either specified as a constant input parameter or computed using a settling velocity w_0 applied to the observed overlying water particulate concentration Mn_p. This is a useful approach for model development and calibration. However, for actual applications it is necessary to be able to predict the water column manganese concentrations as well.

This chapter presents a version of the manganese flux model that includes the overlying water as an integral part of the model. This formulation is the prototype for the actual coupled water column-sediment model that is applied subsequently in Chapter 20.

The addition of a third layer to the model is conceptually and computationally straightforward. Since the application is initially to the MERL data set, the water column layer is modeled as one completely mixed volume. The transport terms for this layer, in addition to those from the aerobic layer of the sediment, are simply the advective inflow and outflow.

Some choices need to be made in the kinetic formulation. The question is: What controls the speciation between dissolved and particulate manganese in the water column? For the sediment aerobic layer, two reactions are considered: partitioning of Mn(II) to sediment solids, and the oxidation reaction which converts Mn(II) to $MnO_2(s)$. Clearly the oxidation reaction is important in the water column, since the overlying water is usually oxic and the oxidation reaction proceeds fairly rapidly. What about the partitioning reaction? The question is examined next.

19.1 A   Partitioning of Manganese (II)

The principal dissolved species of manganese (II) in water are the divalent cation Mn^{2+} and its hydrolysis product $Mn(OH)^+$ (see Section 18.1 A and Lindsay 1979). One expects, based on the partitioning behavior of other divalent metal cations, that manganese (II) should partition to particles in general and to hydrous iron oxides in particular. In principle there is no difficulty in allowing both the partition reaction

$$Mn_d \leftrightarrow Mn_p \tag{19.1}$$

and the oxidation reaction

$$Mn(II) \rightarrow MnO_2(s) \tag{19.2}$$

to occur simultaneously in the water column. After all, this is exactly the formulation that is used in the sediment aerobic layer. The issue is: Can the partition coefficient be determined in some independent way, or can it be shown that the partitioning reaction is insignificant relative to the oxidation reaction in the production of particulate manganese?

An estimate of the expected partition coefficients can be made using a linear free energy relationship (Brezonik, 1994). The idea is that the variation in the equilibrium constant for a series of chemical species reacting with one ligand can be used to predict the variation in the equilibrium constant for another ligand. The usual choice for sorption reactions is the equilibrium constant for metal hydrolysis (Baes and Mesmer, 1976)

$$K_{MOH} = \frac{[M^{2+}][OH^-]}{[MOH^+]} \tag{19.3}$$

which can be compared to the intrinsic sorption constants to hydrous iron oxide (Dzombak and Morel, 1990), and the sorption constants to sediment organic carbon (Mahony et al., 1996). The relationships are presented in Fig. 19.1A. Manganese (II) has the smallest hydrolysis constant of the cations listed, and therefore, it is expected to have a comparatively small partition coefficient to hydrous iron oxide and particulate organic carbon.

Fig. 19.1B presents the observed partition coefficient for manganese π_{Mn} from the MERL experiment, computed from the overlying water particulate $[Mn_p]$ and dissolved $[Mn_d]$ manganese concentrations and the concentration of suspended solids m

$$\pi_{Mn} = \frac{[Mn_p]/m}{[Mn_d]} \tag{19.4}$$

This computation assumes that all the particulate manganese is the result of sorption onto suspended solids and not a result of the oxidation of Mn(II). It can be thought

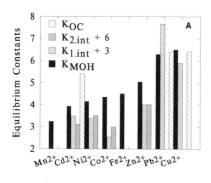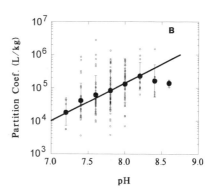

Fig. 19.1 (A) Linear free energy relationship between the first hydrolysis constants of metal ions, the intrinsic constants of sorption to hydrous iron oxide, and the partition coefficient to organic carbon at pH = 7. (B) Partition coefficients versus overlying water pH for the MERL data set.

Table 19.1 Manganese and Cadmium Partition Coefficients

pH	Observed π_{Mn} (L/kg)	Estimated $\pi_{Mn,OC}$ (L/kg OC)	Measured $\pi_{Cd,OC}$ (L/kg OC)
7	10^4	10^5–10^6	$10^{5.40}$
8	10^5	10^6–10^7	$10^{5.96}$

of as the partition coefficient that is necessary in order to explain the observed magnitude of particulate manganese. The results for pH = 7 and 8 are shown in Table 19.1. The magnitudes of π_{Mn} are quite large. At pH = 7, $\pi_{Mn} = 10^4$ L/kg (Fig. 19.1B). If the particles were between 1 and 10% organic carbon, the weight fraction of organic carbon would be in the range $f_{OC} = 0.01$–0.1. The range in the organic carbon normalized partition coefficient

$$\pi_{Mn,OC} = \frac{\pi_{Mn}}{f_{OC}} \tag{19.5}$$

would be 10^5–10^6 L/kg OC. For pH = 8 the range would be $\pi_{Mn,OC} = 10^6$–10^7 L/kg OC.

For cadmium the measured organic carbon normalized partition coefficients are $\pi_{Cd,OC} = 10^{5.4}$ and $10^{5.96}$ L/kg OC at pH = 7 and 8, respectively (Mahony et al., 1996). Thus $\pi_{Mn,OC}$ would need to be at least as large as $\pi_{Cd,OC}$, for $f_{OC} = 0.1$, and a factor of ten larger for $f_{OC} = 0.01$, in order to account for the partition coefficients shown in Fig. 19.1B. But manganese has a hydrolysis equilibrium constant, $10^{3.24}$, which is approximately one-half order of magnitude smaller than for cadmium, $10^{3.92}$. Thus it is unlikely that partitioning alone is the mechanism that generates particulate manganese. Of course, the partition coefficient estimated from

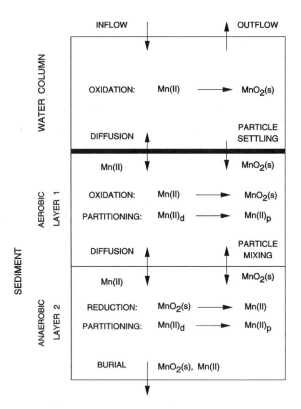

Fig. 19.2 Schematic of the manganese sediment flux model including the overlying water.

the linear free energy relationship could be used in the model, but its influence would be small as demonstrated by the results in Table 19.1.

19.2 MODEL FORMULATION

The model formulation is shown in Fig. 19.2. The water column segment is linked to the sediment aerobic layer by diffusive exchange and particle settling. The oxidation of Mn(II) is included. Transport in the water column is an advective flow, which is the conventional representation for a completely mixed lake model.* Similar models have been presented for metal radiotracer experiments conducted in lake and marine mesocosms.[†]

The model equations that result for the water column concentrations are

---

*Chapra (1997), Schnoor (1996), Thomann and Mueller (1987). [†]Diamond et al. (1990), Nyffeler et al. (1986), Santschi et al. (1987).

Layer 0 Mn(II)

$$H_0 \frac{d\text{Mn}(0)}{dt} = s\big(f_{d1}\text{Mn}(1) - \text{Mn}(0)\big) - \kappa_{\text{Mn},0}\text{Mn}(0) + q_0\big(\text{Mn}(0)_i - \text{Mn}(0)\big)$$

(19.6a)

Layer 0 MnO$_2$(s)

$$H_0 \frac{d\text{MnO}_2(0)}{dt} = \kappa_{\text{Mn},0}\text{Mn}(0) - w_1\text{MnO}_2(0) + q_0\big(\text{MnO}_2(0)_i - \text{MnO}_2(0)\big)$$

(19.6b)

where

$$\kappa_{\text{Mn},0} = k'_{\text{Mn},0} H_0$$

(19.7)

the usual expression for a reaction velocity. $k'_{\text{Mn},0}$ is the first order rate constant for Mn(II) oxidation and H_0 is the depth of the overlying water.

The terms in Eq. (19.6a) represent the flux from the sediment to the overlying water

$$J[\text{Mn}] = s\big(f_{d1}\text{Mn}(1) - \text{Mn}(0)\big)$$

(19.8)

the oxidation of Mn(II) to MnO$_2$(s), the inflowing source of dissolved manganese $q_0\text{Mn}(0)_i$, and the loss by outflow $q_0\text{Mn}(0)$. The flow rate q_0 is included as an overflow rate (m/d)

$$q_0 = \frac{Q}{A}$$

(19.9)

where Q is the volumetric flow rate (m^3/d) and A is the surface area of the water column (m^2) so that the units of q_0 (m/d) are consistent with the other mass transport coefficients in the equations. The loss of particulate manganese via settling $w_1\text{MnO}_2(0)$ is also included in Eq. (19.6b).

The layer 1 Mn(1) equation (18.46) is coupled to the layer 0 equation through Mn(0) which is now computed using Eq. (19.6a). In addition the source term in the layer 1 MnO$_2$ (1) Eq. (18.46c) becomes

$$J_{\text{MnO}_2} = w_1\text{MnO}_2(0)$$

(19.10)

which couples the sediment to the overlying water particulate concentration computed by Eq. (19.6b). Thus the sediment manganese mass balance equations become

Layer 1 Mn(II)

$$H_1 \frac{d\mathrm{Mn}(1)}{dt} = -s\big(f_{d1}\mathrm{Mn}(1) - \mathrm{Mn}(0)\big) + K_{L12}\big(f_{d2}\mathrm{Mn}(2) - f_{d1}\mathrm{Mn}(1)\big)$$

$$+w_{12}\big(f_{p2}\mathrm{Mn}(2) - f_{p1}\mathrm{Mn}(1)\big) - w_2\mathrm{Mn}(1)$$

$$-\frac{\kappa_{\mathrm{Mn},1}^2}{s} f_{d1}\mathrm{Mn}(1) + \mathrm{Mn}(2)\,\dot{H}_1^+ - \mathrm{Mn}(1)\left(\dot{H}_1 + \dot{H}_1^-\right)$$

$$(19.11\mathrm{a})$$

Layer 2 Mn(II)

$$H_2 \frac{d\mathrm{Mn}(2)}{dt} = -K_{L12}\big(f_{d2}\mathrm{Mn}(2) - f_{d1}\mathrm{Mn}(1)\big)$$

$$-w_{12}f\big(f_{p2}\mathrm{Mn}(2) - f_{p1}\mathrm{Mn}(1)\big) - w_2\big(\mathrm{Mn}(2) - \mathrm{Mn}(1)\big)$$

$$+\kappa_{\mathrm{MnO}_2,2}\mathrm{MnO}_2(2) - \mathrm{Mn}(2)\left(\dot{H}_2 + \dot{H}_1^+\right) + \mathrm{Mn}(1)\dot{H}_1^-$$

$$(19.11\mathrm{b})$$

Layer 1 MnO$_2$(s)

$$H_1 \frac{d\mathrm{MnO}_2(1)}{dt} = \frac{\kappa_{\mathrm{Mn},1}^2}{s} f_{d1}\mathrm{Mn}(1) - w_2\mathrm{MnO}_2(1)$$

$$+w_{12}\big(\mathrm{MnO}_2(2) - \mathrm{MnO}_2(1)\big) + w_1\mathrm{MnO}_2(0)$$

$$+\mathrm{MnO}_2(2)\dot{H}_1^+ - \mathrm{MnO}_2(1)\left(\dot{H}_1 + \dot{H}_1^-\right)$$

$$(19.11\mathrm{c})$$

Layer 2 MnO$_2$(s)

$$H_2 \frac{d\mathrm{MnO}_2(2)}{dt} = -\kappa_{\mathrm{MnO}_2,2}\mathrm{MnO}_2(2) - w_{12}\big(\mathrm{MnO}_2(2) - \mathrm{MnO}_2(1)\big)$$

$$-w_2\big(\mathrm{MnO}_2(2) - \mathrm{MnO}_2(1)\big)$$

$$-\mathrm{MnO}_2(2)\left(\dot{H}_2 + \dot{H}_1^+\right) + \mathrm{MnO}_2(1)\dot{H}_1^-$$

$$(19.11\mathrm{d})$$

The result is six coupled equations in the six unknown concentrations of Mn(II) and MnO$_2$(s) in the three layers.

19.2 A   Analytical Solutions

The steady state equations are Eqs. (19.6–19.11) with all the time derivatives set equal to zero: $d\text{Mn}(0)/dt = d\text{Mn}(1)/dt = d\text{Mn}(2)/dt = 0$ and $d\text{MnO}_2(0)/dt = d\text{MnO}_2(1)/dt = d\text{MnO}_2(2)/dt = 0$, as well as the entrainment terms $\dot{H}_1 = \dot{H}_1^+ = \dot{H}_1^- = 0$ and similarly for layer 2. These can be solved analytically for the general case, but the solutions are too unwieldy to be useful. For the special case where the loss by burial in layer 2 is small relative to the loss by outflow in layer 0, then $w_2 = 0$ is a reasonable approximation and the equations can be solved and reduced to simple expressions.

The steady state equations with $w_2 = 0$ are

Layer 0 Mn(II)

$$0 = s\big(f_{d1}\text{Mn}(1) - \text{Mn}(0)\big) - \kappa_{\text{Mn},0}\text{Mn}(0) + q_0\big(\text{Mn}(0)_i - \text{Mn}(0)\big)$$

(19.12a)

Layer 0 MnO$_2$(s)

$$0 = \kappa_{\text{Mn},0}\text{Mn}(0) - w_1\text{MnO}_2(0) + q_0\big(\text{MnO}_2(0)_i - \text{MnO}_2(0)\big)$$

(19.12b)

Layer 1 Mn(II)

$$0 = -s\big(f_{d1}\text{Mn}(1) - \text{Mn}(0)\big) + K_{L12}\big(f_{d2}\text{Mn}(2) - f_{d1}\text{Mn}(1)\big)$$

$$+w_{12}\big(f_{p2}\text{Mn}(2) - f_{p1}\text{Mn}(1)\big) - \frac{\kappa_{\text{Mn},1}^2}{s} f_{d1}\text{Mn}(1)$$

(19.12c)

Layer 1 MnO$_2$(s)

$$0 = \frac{\kappa_{\text{Mn},1}^2}{s} f_{d1}\text{Mn}(1) + w_{12}\big(\text{MnO}_2(2) - \text{MnO}_2(1)\big) + w_1\text{MnO}_2(0)$$

(19.12d)

Layer 2 Mn(II)

$$0 = -K_{L12}\big(f_{d2}\text{Mn}(2) - f_{d1}\text{Mn}(1)\big) - w_{12}\big(f_{p2}\text{Mn}(2) - f_{p1}\text{Mn}(1)\big)$$

$$+\kappa_{\text{MnO}_2,2}\text{MnO}_2(2)$$

(19.12e)

Layer 2 MnO$_2$(s)

$$0 = -\kappa_{\text{MnO}_2,2}\text{MnO}_2(2) - w_{12}\big(\text{MnO}_2(2) - \text{MnO}_2(1)\big)$$

(19.12f)

which can be solved analytically to yield a remarkably simple and instructive set of solutions (Appendix 19A). The layer 0 concentrations are

$$\text{Mn}(0) = \text{Mn}(0)_i \frac{w_1 + q_0}{w_1 + q_0 + \kappa_{\text{Mn},0}} + \text{MnO}_2(0)_i \frac{w_1}{w_1 + q_0 + \kappa_{\text{Mn},0}}$$

$$(19.13)$$

$$\text{MnO}_2(0) = \text{Mn}(0)_i \frac{\kappa_{\text{Mn},0}}{w_1 + q_0 + \kappa_{\text{Mn},0}} + \text{MnO}_2(0)_i \frac{\kappa_{\text{Mn},0} + q_0}{w_1 + q_0 + \kappa_{\text{Mn},0}}$$

$$(19.14)$$

Note that the total outflowing manganese equals the total inflowing manganese

$$\text{Mn}(0) + \text{MnO}_2(0) = \text{Mn}(0)_i + \text{MnO}_2(0)_i \qquad (19.15)$$

which follows from a mass balance around the entire water column-sediment system

$$q_0\big(\text{Mn}(0)_i + \text{MnO}_2(0)_i\big) = q_0\big(\text{Mn}(0) + \text{MnO}_2(0)\big) \qquad (19.16)$$

The mass flux of inflowing manganese must be equal to the only loss terms for manganese, namely the layer 0 outflows. This is why the assumption $w_2 = 0$ simplifies the situation dramatically.

Limiting Forms   The concentrations of Mn(II) and MnO$_2$(s) that result in the water column (Eqs. 19.13–19.14) are controlled by ratios involving $\kappa_{\text{Mn},0}$, q_0, and w_1 which are the parameters for the water column processes: oxidation, outflow, and settling to the sediment. No sediment process parameters are involved. This surprising result requires an explanation.

The water column concentrations are linear functions of the inflowing concentrations weighted by ratios with the sum $\kappa_{\text{Mn},0} + q_0 + w_1$ in the denominator. Consider first what happens if the flow rate through the water column q_0 is large. Then the solutions become

$$\text{Mn}(0) = \text{Mn}(0)_i \frac{w_1 + q_0}{w_1 + q_0 + \kappa_{\text{Mn},0}} + \text{MnO}_2(0)_i \frac{w_1}{w_1 + q_0 + \kappa_{\text{Mn},0}} \qquad (19.17)$$
$$\rightarrow \text{Mn}(0)_i$$

and

$$\text{MnO}_2(0) = \text{Mn}(0)_i \frac{\kappa_{\text{Mn},0}}{w_1 + q_0 + \kappa_{\text{Mn},0}} + \text{MnO}_2(0)_i \frac{\kappa_{\text{Mn},0} + q_0}{w_1 + q_0 + \kappa_{\text{Mn},0}} \qquad (19.18)$$
$$\rightarrow \text{MnO}_2(0)_i$$

as $q_0 \rightarrow \infty$ so there are no changes in the water column concentrations; they remain at the inflow values. This is entirely reasonable. The large inflow provides an infinite supply of Mn(II) and MnO$_2$(s) so that whatever happens in the sediment has no effect in the water column.

Consider what happens if the settling velocity w_1 is large

$$Mn(0) = Mn(0)_i \frac{w_1 + q_0}{w_1 + q_0 + \kappa_{Mn,0}} + MnO_2(0)_i \frac{w_1}{w_1 + q_0 + \kappa_{Mn,0}} \quad (19.19)$$
$$\rightarrow Mn(0)_i + MnO_2(0)_i$$

and

$$MnO_2(0) = Mn(0)_i \frac{\kappa_{Mn,0}}{w_1 + q_0 + \kappa_{Mn,0}} + MnO_2(0)_i \frac{\kappa_{Mn,0} + q_0}{w_1 + q_0 + \kappa_{Mn,0}} \quad (19.20)$$
$$\rightarrow 0$$

as $w_1 \rightarrow \infty$. This is a more interesting result: All the inflowing manganese appears as dissolved Mn(II). The explanation can be arrived at by considering the constraint imposed by mass balance. Since the settling velocity is becoming large, the particulate concentration in the water column is approaching zero. Therefore, all the inflowing manganese must escape via the dissolved concentration in the water column, since – and this is why the simplification $w_2 = 0$ is so powerful – there is no other loss pathway. So the dissolved concentration increases to what is necessary to satisfy mass balance, namely the total inflowing concentration of manganese, both particulate and dissolved.

The converse situation occurs if the oxidation rate $\kappa_{Mn,0}$ is large

$$Mn(0) = Mn(0)_i \frac{w_1 + q_0}{w_1 + q_0 + \kappa_{Mn,0}} + MnO_2(0)_i \frac{w_1}{w_1 + q_0 + \kappa_{Mn,0}} \quad (19.21)$$
$$\rightarrow 0$$

and

$$MnO_2(0) = Mn(0)_i \frac{\kappa_{Mn,0}}{w_1 + q_0 + \kappa_{Mn,0}} + MnO_2(0)_i \frac{\kappa_{Mn,0} + q_0}{w_1 + q_0 + \kappa_{Mn,0}} \quad (19.22)$$
$$\rightarrow Mn(0)_i + MnO_2(0)_i$$

as $\kappa_{Mn,0} \rightarrow \infty$. All the dissolved manganese in the water column oxidizes to particulate manganese which must increase in concentration to the total inflowing manganese concentration so that the outflowing flux equals the inflowing flux.

Other Layers   The sediment concentrations of Mn (II) and MnO_2(s) in layers 1 and 2 are also reasonably concise expressions (Appendix 19A)

$$Mn(1) = \frac{Mn(0)_i \left(s \left(w_1 + q_0 \right) + \kappa_{Mn,0} w_1 \right) + MnO_2(0)_i w_1}{s f_{d1} \left(w_1 + q_0 + \kappa_{Mn,0} \right)} \quad (19.23a)$$

$$MnO_2(1) = MnO_2(2) \frac{w_{12} + \kappa_{MnO_2,2}}{w_{12}} \quad (19.23b)$$

$$\text{Mn}(2) = \text{Mn}(1)\frac{f_{d1}K_{L12} + f_{p1}w_{12}}{f_{d2}K_{L12} + f_{p2}w_{12}} + \text{MnO}_2(2)\frac{\kappa_{\text{MnO}_2,2}}{f_{d2}K_{L12} + f_{p2}w_{12}} \qquad (19.23c)$$

$$\text{MnO}_2(2) = \frac{\left(\text{Mn}(0)_i + \text{MnO}_2(0)_i\right)\left\{\left(\kappa_{\text{Mn},1}^2/s + \kappa_{\text{Mn},0}\right)w_1 s + w_1\kappa_{\text{Mn},0}\kappa_{\text{Mn},1}^2/s\right\} \\ + q_0 s\left(\text{Mn}(0)_i\kappa_{\text{Mn},1}^2/s + \text{MnO}_2(0)_i w_1\right)}{s\kappa_{\text{MnO}_2,2}\left(w_1 + q_0 + \kappa_{\text{Mn},0}\right)}$$

$$(19.23d)$$

The importance of the various parameter groupings will be examined using the numerical results presented below as a guide.

Manganese Flux   The most unanticipated result – although in hindsight it is, perhaps, an obvious result – is the solution for the sediment flux of Mn(II). It is obtained by evaluating the defining equation

$$J[\text{Mn}] = s\left(f_{d1}\text{Mn}(1) - \text{Mn}(0)\right) \qquad (19.24)$$

using the solutions for Mn (1) and Mn(0) given above (Eqs. 19.13 and 19.23a) and using Eq. (19.14) to substitute for the resulting expression. The result is (Appendix 19A)

$$J[\text{Mn}] = w_1\text{MnO}_2(0) \qquad (19.25)$$

It is equal to the flux of particulate $\text{MnO}_2(s)$ that settles from the water column to the sediment. There is a complete recycling of the particulate flux to the sediment as a dissolved flux from the sediment.

This prediction appears to be substantiated by the MERL experimental observations. Fig. 19.3 compares the manganese flux $J[\text{Mn}]$ to the particulate manganese concentration in the overlying water $\text{Mn}(0)_p$. The point by point comparison (Fig. 19.3A) shows no apparent relationship, although the linear regression line included in the figure does indicate that a relationship exists. If the data are grouped (Fig. 19.3B), a clear relationship emerges. The line corresponds to a settling velocity of $w_1 = 1.0$ m/d.

We can understand the flux solution (Eq. 19.25) by examining what happens if no $\text{MnO}_2(s)$ forms in the water column. This happens if no particulate manganese flows into the water column, $\text{MnO}_2(0)_i = 0$, and no manganese dioxide forms by oxidation of Mn(II), so that $\kappa_{\text{Mn},0} = 0$. For this case the concentrations in the overlying water (Eqs. 19.13–19.14) become

$$\text{MnO}_2(0) = 0 \qquad (19.26)$$

$$\text{Mn}(0) = \text{Mn}(0)_i \qquad (19.27)$$

The pore water concentration in the aerobic layer equilibrates with the overlying water concentration (Eq. 19.23a)

$$\text{Mn}(1)_d = f_{d1}\text{Mn}(1) = \text{Mn}(0) \qquad (19.28)$$

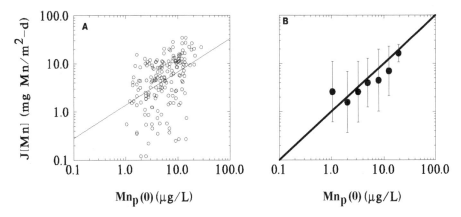

Fig. 19.3 (A) The observed manganese flux $J[\mathrm{Mn}]$ versus the particulate manganese concentration in the overlying water $\mathrm{Mn_p}(0)$. Only data for which $\mathrm{Mn_p}(0) > 1\ \mu g\ \mathrm{Mn/L}$ are included in order to reduce the influence of analytical errors at low concentrations. The linear regression line is shown. (B) The data are grouped in $0.2\ \log_{10}$ units. The mean $\pm$ the standard deviation is shown. The line corresponds to a settling velocity of $w_1 = 1$ m/d.

and all the other sediment concentrations adjust accordingly to this concentration. Since there are no mass balance consequences – these concentrations can be any value whatever without a need to remove manganese to maintain mass balance – no flux from the sediment is generated.

Note that no other solution is possible. If the pore water concentration in layer 1 were different from the layer 0 concentration, that would imply that a flux of manganese is present either from or to the overlying water. But under these circumstances the sediment concentrations would either be continually depleted or build up indefinitely, neither of which is a steady state solution. Therefore the zero flux case is the only allowable solution.

If $\mathrm{MnO_2}(0) \neq 0$ so that there is a flux $w_1\mathrm{MnO_2}(0)$ to the sediment, then a flux from the sediment is necessary for a mass balance to be maintained. Since the only possible flux is a dissolved flux $J[\mathrm{Mn}]$ to the overlying water – a burial flux has been eliminated as a possibility by assuming $w_2 = 0$ – the concentrations adjust to exactly return this incoming flux (Eq. 19.25).

This remarkable result, the sediment flux is controlled by water column processes and, in particular, the particulate concentration and settling velocity, was eventually noticed after many numerical simulations were performed. It provided the motivation for seeking analytical solutions to establish the precise form of the dependence. The solutions for the general case, $w_2 \neq 0$, are too complex to be useful, but removing the possibility of a burial flux simplifies the solutions so that the relationship between the particulate flux to the sediment and the recycled dissolved flux emerges. The utility of these analytical solutions to help in the understanding of the time variable numerical results is examined next.

19.3   TIME VARIABLE MODEL

The complete model includes the overlying water and the sediment mass balance equations. The equations are discussed first, followed by the a sequence of numerical calculations of increasing complexity and realism.

19.3 A   Vector-Matrix Equation

The equations used for the numerical calculations are expressed in vector-matrix form. They comprise the two-layer sediment model equations (Eqs. 18.46) with the addition of the water column as the third layer. The manganese concentration state variable vector is

$$\mathbf{c}(t) = \begin{bmatrix} \mathrm{Mn}(1) \\ \mathrm{Mn}(2) \\ \mathrm{MnO_2}(1) \\ \mathrm{MnO_2}(2) \\ \mathrm{Mn}(0) \\ \mathrm{MnO_2}(0) \end{bmatrix} \tag{19.29}$$

where two additional bottom rows have been added for the water column concentrations. The matrix $\mathbf{A}(t) = [a_{ij}]$ has elements listed in Table 19.2 where f_{12} and f_{21} are the effective mass transfer coefficients from layer 1 to layer 2 and from layer 2 to layer 1, respectively, as before (Eqs. 18.49) and all other a_{ij}'s are zero. The forcing function is

$$\mathbf{b}(t) = \begin{bmatrix} 0 \\ 0 \\ 0 \\ 0 \\ q_0 \mathrm{Mn}(0)_i \\ q_0 \mathrm{MnO_2}(0)_i \end{bmatrix} \tag{19.30}$$

The only sources are Mn(II) and $\mathrm{MnO_2}$(s) flowing into layer 0 at concentrations $\mathrm{Mn}(0)_i$ and $\mathrm{MnO_2}(0)_i$. The diagonal matrix of layer depths $\mathbf{H}(t)$ now includes the overlying water segment depth

$$\mathbf{H}(t) = \mathrm{diag}\left[H_1(t),\ H_2(t),\ H_1(t),\ H_2(t),\ H_0,\ H_0 \right] \tag{19.31}$$

An implicit finite difference equation (Eq. 18.55) is used to solve these equations.

19.3 B   Comparison to Steady State Results

How useful are the steady state solutions in understanding the behavior of the time variable model? The first step is to check the results of the time variable model when it is configured to be a steady state computation. This is accomplished by disabling

Table 19.2 Matrix elements of the $\mathbf{A}(t) = [a_{ij}]$ matrix.

$$a_{11} = -sf_{d1} - f_{12} - w_2 - k_{Mn,OH,1} H_1 [O_2(1)] \left(10^{2(pH_1-7)}\right) f_{d1}$$
$$- \left(\dot{H}_1 + \dot{H}_1^-\right)$$

$$a_{12} = f_{21} + \dot{H}_1^+$$

$$a_{15} = s$$

$$a_{21} = f_{12} + w_2 + \dot{H}_1^-$$

$$a_{22} = -f_{21} - w_2 - \left(\dot{H}_2 + \dot{H}_1^+\right)$$

$$a_{24} = k_{MnO_2,2} H_2$$

$$a_{31} = k_{Mn,OH,1} H_1 [O_2(1)] \left(10^{2(pH_1-7)}\right) f_{d1}$$

$$a_{33} = -w_{12} - w_2 - \left(\dot{H}_1 + \dot{H}_1^-\right)$$

$$a_{34} = w_{12} + \dot{H}_1^+$$

$$a_{36} = w_1$$

$$a_{43} = w_{12} + w_2 + \dot{H}_1^-$$

$$a_{44} = -k_{MnO_2,2} H_2 - w_{12} - w_2 - \left(\dot{H}_2 + \dot{H}_1^+\right) + \dot{H}_1^-$$

$$a_{51} = sf_{d1}$$

$$a_{55} = -s - H_0/t_0 - k_{Mn,OH,0} H_0 [O_2(0)] \left(10^{2(pH_0-7)}\right)$$

$$a_{65} = k_{Mn,OH,0} H_0 [O_2(0)] \left(10^{2(pH_0-7)}\right)$$

$$a_{66} = -H_0/t_0 - w_1$$

all the time varying elements. Thus all the parameters that cause the reaction rates to vary with temperature (the θ's) are set equal to one. This includes the nutrient and oxygen model parameters as well. The inflowing concentrations $Mn(0)_i$ and $MnO_2(0)_i$ are set to constant average values. The only time variation that is allowed to remain is that induced in $s = SOD/[O_2(0)]$ by the varying overlying water dissolved oxygen concentration $[O_2(0)]$. The reason is to ascertain the effect of just this source of variation. The burial velocity is set to zero as well to correspond to the steady state solutions and H_2 is set to 1.0 cm to reduce the time to steady state (see Chapter 16, Eqs. 16.21–16.22 for the analogous situation for phosphate).

The time variable model is initialized by cycling the inputs for the first year until a periodic steady state is achieved. Then average annual concentrations are computed. In addition, annual average parameters and forcing functions are computed, and the

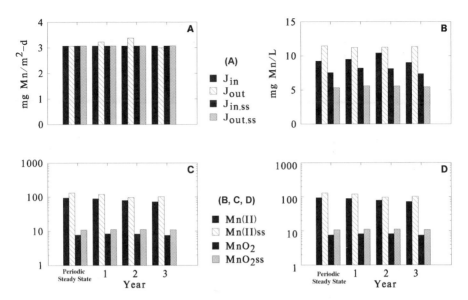

Fig. 19.4 Comparison of steady state and time variable solutions. The subscript ss indicates the steady state analytical solution. The time variable solution is for periodic steady state conditions, and the three years of the MERL simulation. (A) Mass balances: flux in (J_{in}) and flux out (J_{out}). Concentrations in (B) overlying water, (C) aerobic layer, and (D) anaerobic layer.

steady state equations are evaluated using these values. This procedure is followed for all the time variable-steady state comparisons.

The steady state results are compared to the annual average fluxes and concentrations in Fig. 19.4, labeled as "periodic steady state." The subsequent years 1 to 3 represent the results of the simulation of the three year MERL experiment. The mass balances are checked in Fig. 19.4A. Since the periodic steady state results are for an equilibrated model, the inflowing flux balances the out-flowing flux perfectly. However, the year 1 and 2 time variable model results show an imbalance with the outflow being larger than the inflow. This is occurring because the sediment is releasing more manganese than is settling to the sediment during this period. The cause is the change in overlying water conditions at the onset of the experiment. Of course the steady state solutions are perfectly in balance since they are the result that would be expected after the system has adjusted to the changed conditions.

The steady state and annual average concentrations for (B) the overlying water, (C) aerobic layer, and (D) anaerobic layer are somewhat different, reflecting the lack of steady state conditions in the time variable results. However the magnitudes are correct, which suggests that the steady state solutions can be used to guide calibration and to explain the time variable results.

19.4 CALIBRATION

The calibration of the three-layer model to the MERL data set is presented in this section. The particulate and dissolved concentrations $MnO_2(0)_i$ and $Mn(0)_i$ in the inflow to the water column for the three years are obtained from the average particulate and dissolved concentrations in the control tanks' water columns (Fig. 19.5). The idea is that the control tanks are replicating the bay and therefore the concentrations in the control tanks are similar to what is in the inflow to the other tanks. For the latter part of the third year the actual inflowing concentrations were determined and these are included in Fig. 19.5. They are similar to the control tank concentrations, thus validating the use of the control tank concentrations for the inflow.

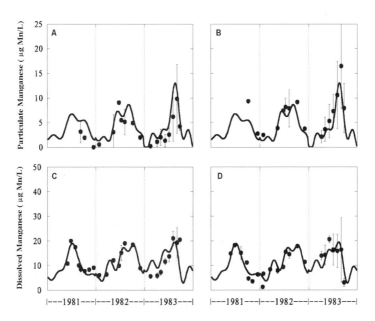

Fig. 19.5 Observed (•) concentrations of particulate (A, B) and dissolved (C, D) manganese in two (A, C and B, D) control mesocosms. The lines are fitted to both data sets and they are used as the inflowing concentrations $Mn(0)_i$ and $MnO_2(0)_i$.

The pH in the three layers of the model has been presented in Section 18.4 B, Fig. 18.19. The manganese sediment model and remaining parameter values are listed in Tables 18.4 and 15.5, respectively. The sediment model parameters are not changed from those established in Chapter 18. The calibration involves choosing the oxidation rate of Mn(II) in the overlying water that reproduces the particulate concentration. We know from the analytical results that if the particulate flux to the sediment is properly computed, then the sediment flux will also be correct. The resulting parameters are listed in Table 19.3. Also included are the parameters used to fit the manganese data from the Croton Reservoir (Chapter 20). The Croton Reservoir water column oxidation rate is within the same order of magnitude as the MERL calibration result.

Table 19.3 Sediment Model Parameters

Parameter*	Units	MERL Figs. 19.7, 19.8, 19.9B	Croton Fig. 20.13	MERL Fig. 19.9A
$k_{Mn,OH,0}$	(a)	0.003	0.001	
$k_{Mn,OH,MnO_2,0}$	(b)			2×10^{-4}

[a] $d^{-1} (\text{mg } O_2/L)^{-1}$. [b] $d^{-1} (\text{mg } O_2/L)^{-1} (\text{mg Mn/L})^{-1}$.

19.4 A   Model Results

The model is equilibrated to a periodic steady state by using the average water column concentrations of the control tanks for the inflowing concentrations and cycling the model. The same input is used for each cycle, with the ending condition for the previous year used as the initial condition. The idea is to simulate the state of the sediments before the nutrient enrichment experiment commenced. The presumption is that the behavior of the first year of the controls is characteristic of the average situation in Narragansett Bay in general. Once the computation is initialized, the simulation period is started, the treatment loadings are initiated at the appropriate time, and the model computation is continued to the end of the three years.

The annual average data and model results for manganese fluxes, and the water column particulate and dissolved concentrations, are shown in Fig. 19.6.

The manganese fluxes (A–C) increase with increased nutrient loading from the control (0X) to the most heavily loaded (32X). The magnitude and pattern are reproduced for the first and third year. The large flux increases in the 1X to 4X treatments for the second year (B) are not reproduced. These correspond to the large increase in particulate concentration (E). Overall, more manganese is in the tanks at this time than can be accounted for by the model. We have no good explanation for this observation.

The time variable computed and observed concentrations in the overlying water are shown in Figs. 19.7. Two sets of model computations are presented. The dashed lines are the model computed dissolved Mn(0) and particulate MnO_2(0) concentrations. These results correspond to the assumption that all the particulate manganese is filterable and would be quantified as particulate manganese. The solid lines are the concentrations that are computed if a fraction $f_{colloidal}$ of the particulate manganese is not filterable. The reason for this modification is discussed in the next section.

The model correctly computes the seasonal variation of dissolved manganese and the pattern of increasing concentrations to the 8X treatment and decreasing thereafter (Fig. 19.7A). The differences between the dashed and solid lines are not large

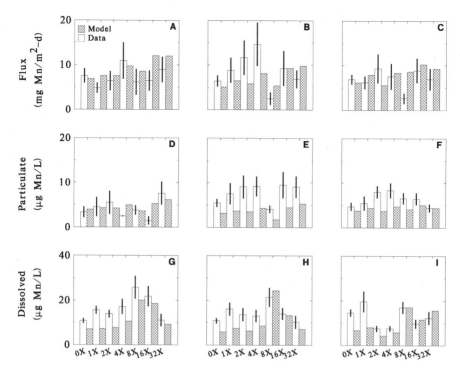

Fig. 19.6 Comparison of average annual data (± standard deviation) and model results: 1981 (A, D, G); 1982 (B, E, H); 1983 (C, F, I). Manganese flux (A, B, C). Water column particulate (D, E, F) and dissolved (G, H, I) manganese concentrations.

and the data do not clearly favor one over the other. There is a systematic under-prediction of the dissolved manganese during the winter and spring of each year. We have no explanation for this discrepancy. The increase in pH_0 that occurs (Fig. 18.19) causes a rapid increase in the rate at which the dissolved $Mn(0)$ oxidizes to particulate $MnO_2(0)$. The particulate manganese (Fig. 19.7B) settles from the water column to the sediment where it remains for a period of time and is then recycled to the overlying water as a flux.

The manganese fluxes which result are shown in Fig. 19.8A. The model reproduces the observed seasonal variation as well as the increase in flux as the tanks are enriched from 1X to 32X. The flux to the water column causes the dissolved concentration to increase which produces the seasonal pattern. As a result of the formation of particulate manganese in the water column, its settling to the sediment, and its release to the water column, there is a seasonal variation in the total manganese concentration in the water column as shown in Fig. 19.8B.

It might be supposed that the variation in total manganese concentration is also influenced by the variation in input concentrations, Fig. 19.5. However, this is not the case. It is possible to suppress the seasonal variation in the inflowing water by using constant input concentrations, which are the yearly averages of the concentrations

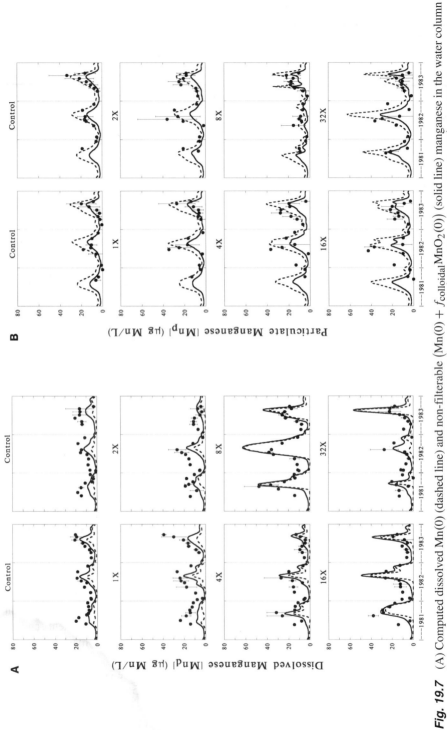

Fig. 19.7 (A) Computed dissolved Mn(0) (dashed line) and non-filterable (Mn(0) + $f_{colloidal}MnO_2(0)$) (solid line) manganese in the water column compared to observations. (B) Computed particulate $MnO_2(0)$ (dashed line) and filterable ($MnO_2(0) - f_{colloidal}MnO_2(0)$) (solid line) manganese in the water column compared to observations.

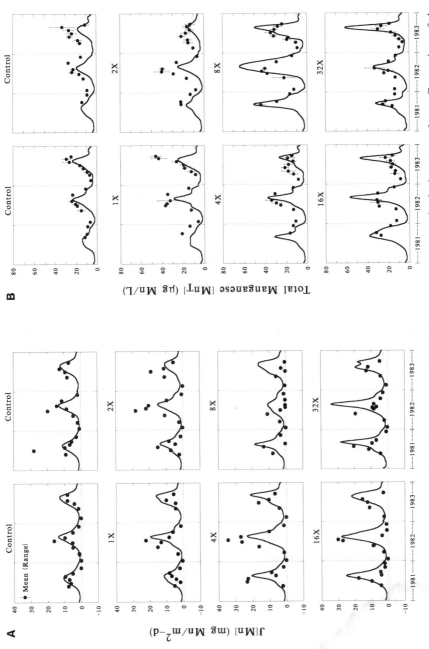

Fig. 19.8 (A) Manganese flux versus time for all the treatments. (B) Total manganese concentration in the water column. Comparison of observations (●) and model output (lines).

used previously (Fig. 19.5). The resulting total manganese concentrations are almost indistinguishable from the results using the time variable inputs. This is also true for all the other model outputs. Thus the temporal variation in total manganese in the water column is generated by the migration of manganese from the sediment to the water column and back again during the annual cycle.

19.4 B   Colloidal Fraction

A fraction of the particulate manganese that is formed by oxidation may be non-filterable and would be quantified as "dissolved" manganese (Buffle et al., 1988). Whether this phenomena is important in this experimental data is examined in this section. The suggestion that it may be important is not based on the model's ability to reproduce the observed particulate and dissolved manganese concentrations. In fact, the dashed lines (no colloidal manganese) capture the magnitudes of the peak particulate concentrations better than the solid lines (Figs. 19.7B). The use of a colloidal fraction does not appear to significantly change the fit to the dissolved data either (Fig. 19.7A). However if the data are analyzed in another way then a problem appears.

The fraction of manganese in the water column that is particulate is given by

$$f_p = \frac{\mathrm{MnO_2(0)}}{\mathrm{MnO_2(0)} + \mathrm{Mn(0)}} \tag{19.32}$$

The variation in f_p as a function of the pH is shown in Fig. 19.9. The upper solid

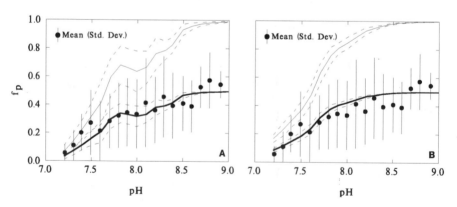

Fig. 19.9 Observed (•) and calculated (lines) particulate fraction versus pH in the water column. Model results assuming $f_{\mathrm{colloidal}} = 0$ (upper lines) and $f_{\mathrm{colloidal}} = \frac{1}{2}$ (lower lines). The dashed lines are the range in model results for that pH interval. (A) Computed using autocatalytic kinetics (Eq. 19.35). (B) Computed using linear kinetics (Eq. 19.36).

and dashed lines are the model average and range over each pH interval, assuming that all the particulate manganese $\mathrm{MnO_2(0)}$ is filterable and would be measured as particulate manganese. The model kinetics predict that all the manganese should be particulate $f_p \simeq 1$ above pH = 8.5. However the observations rarely exceed a particulate fraction of $f_p = 0.5$. Since the pH dependency of the oxidation kinetics is experimentally observed, another explanation for this discrepancy is required.

The supposition that a colloidal fraction is erroneously measured as a dissolved species is commonly used as an explanation (Gschwend and Wu, 1985) for the observed decrease in partition coefficient as the concentration of particles increase (O'Connor and Connolly, 1980). To be consistent in the model formulation, the colloidal fraction is assumed not to settle to the sediment so w_1 is replaced by $w_1 (1 - f_{colloidal})$ in Eqs. (19.6, 19.11, 19.12) and Table 19.2.

The lower solid and dashed lines in Fig. 19.9 are the model results assuming that 50% of the particulate manganese is colloidal ($f_{colloidal} = 0.5$) and, therefore, would not be filterable. They are computed from the truly particulate and dissolved concentrations using

$$Mn_d(0) = Mn(0) + f_{colloidal} MnO_2(0) \tag{19.33}$$

$$Mn_p(0) = MnO_2(0) - f_{colloidal} MnO_2(0) \tag{19.34}$$

19.4 C   Autocatalytic Oxidation Kinetics

The kinetics of Mn(II) oxidation display an autocatalytic behavior as is clear from Fig. 18.1. Depending on the magnitudes of the reaction rate constants and particulate manganese concentration, the reaction is either first order ($k'_{Mn} \gg k''_{MnO_2}[MnO_2(s)]$) or autocatalytic. The results in Fig. 19.9A are computed with a purely autocatalytic rate expression for the oxidation of Mn(II) (Eq. 18.10b)

$$R = -k_{Mn,OH,MnO_2,0}[O_2(0)] \left(10^{2(pH_0-7)}\right) [MnO_2(s)] f_{d0}[Mn(0)] H_0 \tag{19.35}$$

whereas the results in Fig. 19.9B use the first order reaction rate formulation (Eq. 18.10a)

$$R = -k_{Mn,OH,0}[O_2(0)] \left(10^{2(pH_0-7)}\right) f_{d0}[Mn(0)] H_0 \tag{19.36}$$

The reaction rates used are tabulated in Table 19.3. The fit is, perhaps, slightly better with the autocatalytic expression if one is inclined to believe that the reaction should be autocatalytic. The model in Fig. 19.9A has the proper shape in the vicinity of pH = 8. However, there is actually no strong reason to prefer one over the other.

19.4 D   Effect of Anaerobic Layer Depth

The depth of the anaerobic layer H_2 is an important parameter in the sediment model. It controls the time constant – the characteristic time for the model to equilibrate to new conditions – for conservative species such as phosphorus (Chapter 16). This is also true for the manganese model. The reason is that the anaerobic layer provides a large storage volume for manganese, and achieving steady state requires that the concentration Mn(2) in H_2 be equilibrated. This can happen if either the flux to or from the sediment $J[\text{Mn}]$, or the burial flux $w_2(\text{Mn}(2)+\text{MnO}_2(2))$, adjusts accordingly. The model results presented in Section 19.3 B (Fig. 19.4) were made with $H_2 = 1$ cm. As a result steady state is achieved relatively rapidly, even without a loss by burial, since the comparisons to the analytical model steady state solutions are made with $w_2 = 0$. The question remains: What is the appropriate H_2 for the MERL nutrient exchange experiment?

The manganese concentration in the anaerobic layer near the end of the third year of the simulation is compared to observations in Fig. 19.10A, B with $H_2 = 10$ cm. There is a decline in observed pore water manganese concentration at the higher

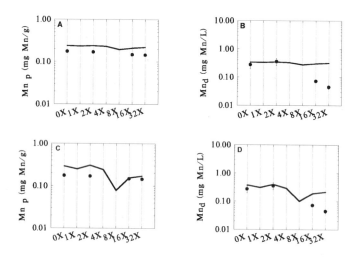

Fig. 19.10   Sediment (A and C) and pore water (B and D) observed (•) and computed (lines) manganese concentrations at the end of year 3. (A and B) $H_2 = 10$ cm. (C and D) $H_2 = 1$ cm.

enrichment levels which the model does not capture. The same results for $H_2 = 1$ cm are shown in Fig. 19.10C, D. The decline in Mn(2) with increasing nutrient inputs is the result of increasing fluxes of Mn to the overlying water in the first year of the simulation (Fig. 19.6). Since the anaerobic layer is thinner than in Fig. 19.10A and B, the sediment concentrations respond more rapidly. From these results it appears that the thinner layer $H_2 = 1$ cm is more representative of the short-term response observed in these experiments.

It should be pointed out that the depth of the anaerobic layer is an artificial construct that is the consequence of the assumed two-layer structure of the model. If the model had been constructed using multiple layers in the vertical (e.g., Chapter 21), then the parameters that control the long-term response are the mixing coefficient for particles and the depth of particle mixing. Although this formulation is preferable, it can add an unacceptably large computational burden if the model is to be applied as a part of a large water quality model with many individual sediment segments. The depth of the active layer H_2 is a surrogate for these parameters and plays a similar role in the model.

Appendix 19A: MACSYMA

Solving the six simultaneous equations, and finding the substituted forms, is greatly aided using MACSYMA. The output is listed in Figs. 19A.1 to 19A.3. The first six lines in Fig. 19A.1 (c3) to (c8) assign the mass balance equations to the variables eq1 to eq6. The term k[Mn,1] is used to represent $\kappa_{Mn,1}^2/s$. The next line (c9) checks the equations by adding them. The results (d9) are the source and sink terms: burial of dissolved Mn(II) and $MnO_2(s)$, inflow of Mn(II) and $MnO_2(s)$, and outflow of Mn(II) and $MnO_2(s)$ from the overlying water.

The solution (Fig. 19A.2) begins with (c10) that consists of two commands in a block separated by a comma. The equation set eqa is named and the equation set eq is defined with $w_2 = 0$. The block (c11) eliminates the sediment Mn(II) and $MnO_2(s)$ concentrations and solves for the water column Mn(II) and $MnO_2(s)$ concentrations, which are displayed (c12) and (c13). The next step (c14) is to substitute the water column solutions into the sediment equations forming the equation set eqb which is then solved for the sediment Mn(II) and $MnO_2(s)$ concentrations. They are displayed (Figs. 19A.2 and 19A.3) in (d15) through (d18). The manganese flux to the overlying water is computed in (c19).

The substituted solutions – solutions expressed in terms of other concentrations rather than in terms of the inflowing concentrations – are found using scsimp in (c20) for $J[Mn]$, and in (c21) for $MnO_2(1)$. Occasionally scsimp fails to find the substitution. In this case the procedure in (c22) and (c23) can be used. The idea is to invert the solutions for the concentrations which are to be substituted, in this case Mn(1) and $MnO_2(2)$. That is, use these equations to solve for the source terms $Mn(0)_i$ and $MnO_2(0)_i$. This produces a set of equations for $Mn(0)_i$ and $MnO_2(0)_i$ in terms of Mn(1) and $MnO_2(2)$. Then use these solutions, saved in ss1, to evaluate the solution for Mn(2) using ev and simplify the result. This produces a solution for Mn(2) in terms of Mn(1) and $MnO_2(2)$. If the substitution is useful, a simplified expression is produced as in this case (d23).

Water column equations

(c3)   eq1: q[0]*mn[0,i]-q[0]*mn[0]+s*(fd[1]*mn[1]-mn[0])-k[Mn,0]*mn[0]

(d3) $$\left(fd_1\ mn_1 - mn_0 \right) s - mn_0\ k_{mn,\ 0} + q_0\ mn_{0,\ i} - mn_0\ q_0$$

(c4)   eq2: q[0]*mno2[0,i]-q[0]*mno2[0]+k[Mn,0]*mn[0]-w[01]*mno2[0]

(d4) $$mn_0\ k_{mn,\ 0} - mno2_0\ w_1 + q_0\ mno2_{0,\ i} - mno2_0\ q_0$$

Aerobic sediment layer

(c5)   eq3: s*(mn[0]-fd[1]*mn[1])+k[l12]*(fd[2]*mn[2]-fd[1]*mn[1])
        +w[12]*(fp2*mn[2]-fp[1]*mn[1])-k[Mn,1]*mn[1]*fd[1]-w[2]*mn[1]

(d5) $$\left(mn_0 - fd_1\ mn_1 \right) s - fd_1\ mn_1\ k_{mn,\ 1} + \left(fd_2\ mn_2 - fd_1\ mn_1 \right) k_{l12} + w_{12} \left(mn_2\ fp2 - fp_1\ mn_1 \right) - mn_1\ w_2$$

(c6)   eq4: w[01]*mno2[0]+w[12]*(mno2[2]-mno2[1])+k[Mn,1]*mn[1]*fd[1]-w[2]*mno2[1]

(d6) $$fd_1\ mn_1\ k_{mn,\ 1} + \left(mno2_2 - mno2_1 \right) w_{12} - mno2_1\ w_2 + mno2_0\ w_1$$

Anaerobic sediment layer

(c7)   eq5: -k[l12]*(fd[2]*mn[2]-fd[1]*mn[1])+k[MnO2,2]*mno2[2]
        -w[12]*(fp2*mn[2]-fp[1]*mn[1])+w[2]*mn[1] -w[2]*mn[2]

(d7) $$mno2_2\ k_{mno2,\ 2} - \left(fd_2\ mn_2 - fd_1\ mn_1 \right) k_{l12} - w_{12} \left(mn_2\ fp2 - fp_1\ mn_1 \right) - mn_2\ w_2 + mn_1\ w_2$$

(c8)   eq6: -w[12]*(mno2[2]-mno2[1])-k[MnO2,2]*mno2[2]+w[2]*mno2[1] -w[2]*mno2[2]

(d8) $$- mno2_2\ k_{mno2,\ 2} - \left(mno2_2 - mno2_1 \right) w_{12} - mno2_2\ w_2 + mno2_1\ w_2$$

Check mass balance

(c9)   mb:ratsimp(eq1+eq2+eq3+eq4+eq5+eq6)

(d9) $$\left(- mno2_2 - mn_2 \right) w_2 + q_0\ mno2_{0,\ i} + q_0\ mn_{0,\ i} + \left(- mno2_0 - mn_0 \right) q_0$$

Fig. 19A.1 MACSYMA solution of Eqs. (19.12).

Solve the equations for the w[2]=0 case

(c10)   (eqa:[eq1,eq2,eq3,eq4,eq5,eq6],eq:ev(eqa,w[2]=0))$

(c11)   (s1:eliminate(eq,[mn[1],mno2[1]]),s2:eliminate(s1,[mn[2],mno2[2]]),s3:solve(s2,[mn[0],mno2[0]]))$

Water column solutions

(c12)   mn0sol:lhs(s3[1][1])=mn0:facsum(rhs(s3[1][1]),mn[0,i],mno2[0,i])

(d12)
$$mn_0 = \frac{mn_{0,i}\left(w_1 + q_0\right) + mno2_{0,i}\,w_1}{k_{mn,0} + w_1 + q_0}$$

(c13)   mno20sol:lhs(s3[1][2])=mno20:facsum(rhs(s3[1][2]),mn[0,i],mno2[0,i])

(d13)
$$mno2_0 = \frac{mno2_{0,i}\left(k_{mn,0} + q_0\right) + mn_{0,i}\,k_{mn,0}}{k_{mn,0} + w_1 + q_0}$$

Substitute the water column solutions and solve the sediment equations

(c14)   (eqb:ev(eq,mn[0]=mn0,mno2[0]=mno20),s4:solve(eqb,[mn[1],mn[2],mno2[1],mno2[2]]))$

Mn solutions

(c15)   mn1sol:lhs(s4[1][1])= mn1:facsum(rhs(s4[1][1]),mn[0,i],mno2[0,i])

(d15)
$$mn_1 = \frac{mn_{0,i}\left(w_1 s + q_0 s + w_1 k_{mn,0}\right) + mno2_{0,i}\,w_1\left(s + k_{mn,0} + q_0\right)}{fd_1\left(k_{mn,0} + w_1 + q_0\right) s}$$

(c16)   mn2sol:lhs(s4[1][2])= mn2:facsum(rhs(s4[1][2]),mn[0,i],mno2[0,i])

(d16)
$$mn_2 = \frac{\begin{pmatrix} mn_{0,i}\begin{pmatrix} fd_1 w_1 k_{mn,1} s + q_0 fd_1 k_{mn,1} s + fd_1 w_1 k_{mn,0} s + fd_1 w_1 k_{l12} s + q_0 fd_1 k_{l12} s + fp_1 \\ * w_1 w_{12} s + q_0 fp_1 w_{12} s + fd_1 w_1 k_{mn,0} k_{mn,1} + fd_1 w_1 k_{l12} k_{mn,0} + fp_1 w_1 w_{12} k_{mn,0} \end{pmatrix} \\ + mno2_{0,i}\,w_1\begin{pmatrix} fd_1 k_{mn,1} s + fd_1 k_{mn,0} s + fd_1 k_{l12} s + fp_1 w_{12} s + q_0 fd_1 s + fd_1 k_{mn,0} \\ * k_{mn,1} + q_0 fd_1 k_{mn,1} + fd_1 k_{l12} k_{mn,0} + fp_1 w_{12} k_{mn,0} + q_0 fd_1 k_{l12} + q_0 fp_1 w_1 \end{pmatrix} \end{pmatrix}}{fd_1\left(fd_2 k_{l12} + w_{12} fp2\right)\left(k_{mn,0} + w_1 + q_0\right) s}$$

Fig. 19A.2   MACSYMA solution of Eqs. (19.12), continued.

MnO2 solutions

(c17)   mno21sol:lhs(s4[1][3])= mno21:facsum(rhs(s4[1][3]),mn[0,i],mno2[0,i])

(d17)

$$
mno2_1 = \frac{\left(\begin{array}{c} mn_{0,i}\left(k_{mno2,2}+w_{12}\right)\left(w_1 k_{mn,1}s+q_0 k_{mn,1}s+w_1 k_{mn,0}s+w_1 k_{mn,0}k_{mn,1}\right) \\ + mno2_{0,i}\,w_1\left(k_{mno2,2}+w_{12}\right) \\ *\left(k_{mn,1}s+k_{mn,0}s+q_0 s+k_{mn,0}k_{mn,1}+q_0 k_{mn,1}\right) \end{array} \right)}{w_{12}\left(k_{mn,0}+w_1+q_0\right)k_{mno2,2}\,s}
$$

(c18)   mno22sol:lhs(s4[1][4])= mno22:facsum(rhs(s4[1][4]),mn[0,i],mno2[0,i])

(d18)

$$
mno2_2 = \frac{\left(\begin{array}{c} mn_{0,i}\left(w_1 k_{mn,1}s+q_0 k_{mn,1}s+w_1 k_{mn,0}s+w_1 k_{mn,0}k_{mn,1}\right) \\ + mno2_{0,i}\,w_1\left(k_{mn,1}s+k_{mn,0}s+q_0 s+k_{mn,0}k_{mn,1}+q_0 k_{mn,1}\right) \end{array} \right)}{\left(k_{mn,0}+w_1+q_0\right)k_{mno2,2}\,s}
$$

Mn Flux

(c19)   jmn:facsum(j[Mn]=-s*(mn0-fd[1]*mn1),mno2[0,i])

(d19)

$$
j_{mn} = \frac{mno2_{0,i}\,w_1\left(k_{mn,0}+q_0\right)+mn_{0,i}\,w_1 k_{mn,0}}{k_{mn,0}+w_1+q_0}
$$

Substituted solutions

(c20)   ratsimp(scsimp(jmn,mno20sol))

(d20)

$$
j_{mn} = mno2_0\,w_1
$$

(c21)   ratsimp(scsimp(mno21sol,mno22sol))

(d21)

$$
mno2_1 = \frac{mno2_2\,k_{mno2,2}+mno2_2\,w_{12}}{w_{12}}
$$

Use direct solution method

(c22)   ss1:solve([mn1sol,mno22sol],[mn[0,i],mno2[0,i]])$

(c23)   ratsimp(ev(mn2sol,ss1))

(d23)

$$
mn_2 = \frac{mno2_2\,k_{mno2,2}+fd_1\,mn_1\,k_{112}+fp_1\,mn_1\,w_{12}}{fd_2\,k_{112}+w_{12}\,fp2}
$$

Fig. 19A.3   MACSYMA solution of Eqs. (19.12), continued.

20

Iron Flux Model

20.1 INTRODUCTION

This chapter presents a model for iron flux from sediment. The model is constructed in exactly the same way as the manganese sediment and water column flux model described in the previous chapters. The model is applied to data from two lakes with seasonally anaerobic hypolimnia. The initial application is to Onondaga Lake. It was chosen because a complete eutrophication water column data set is available and has been modeled using a coupled water column-sediment model, and because iron data are also available. The hypolimnion of the lake becomes anaerobic each year, releasing significant quantities of iron to the water column. Thus Onondaga Lake provides a very good setting to test the validity of the iron sediment flux model.

The second lake is the Croton Reservoir. For this lake both iron and manganese data are available. Thus this application provides a test for both the iron and manganese models, together with the coupled water column-sediment eutrophication model.

20.2 IRON CHEMISTRY

Like manganese, the chemistry of iron in natural waters and sediments has been extensively studied (Stumm and Morgan, 1970). Iron exists in two valance states: the +3 state Fe(III) in oxic waters and the +2 state Fe(II) in anoxic waters. Fe(III) is very insoluble and forms iron oxyhydroxide, FeOOH(s) (goethite), and eventually more insoluble iron oxides. Fig. 20.1 presents a solubility diagram for iron as a function of

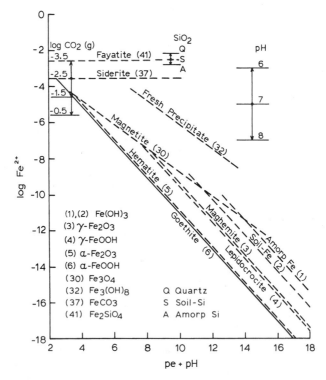

Fig. 20.1 Solubility of iron as a function of pH + pe for pH = 7 (Lindsay, 1979).

pH+ pe (Lindsay, 1979). At equilibrium iron is more insoluble in oxic waters than manganese (Fig. 18.2). However, iron carbonate (siderite) and manganese carbonate have approximately the same solubility. Thus pore water iron concentrations should be comparable to manganese.

Like manganese, iron usually exists as a coating on particles. However, unlike manganese, there are other forms of particulate iron that exist in natural waters. Since the earth's crust is approximately 2% iron by weight (Lindsay, 1979), particles that run off into natural waters contain a large amount of iron. As the particles settle to the sediment, iron is transported as well. This is the source of iron to the sediments.

Not all iron in sediments is reactive. It is convenient to denote the reactive portion of oxic iron as FeOOH(s) (goethite), and to assume that it includes iron hydroxide, $Fe(OH)_3(s)$ as well, since it differs from FeOOH(s) in water content only

$$FeOOH(s) + H_2O \rightarrow Fe(OH)_3 (s) \tag{20.1}$$

The term iron oxyhydroxide is meant to denote the sum of FeOOH(s), $Fe(OH)_3(s)$, and any other reactive iron solid phases that are present (Canfield, 1989).

In the oxic layer of the sediment, FeOOH(s) is stable. However, as particle mixing transports iron containing particles to the anaerobic layer of the sediment, iron oxyhydroxide is thermodynamically unstable and a reduction reaction occurs. Fe(III) is reduced to Fe(II). For this to occur, one electron is required as shown by the following reduction half-reaction

$$\text{FeOOH(s)} + e^- \rightarrow \text{Fe(II)}^{2+} - 3\text{H}^+ + 2\text{H}_2\text{O} \tag{20.2}$$

The source of electrons in sediments is organic matter mineralization, CH_2O (Section 1.2 B) and the overall reduction reaction can be written

$$\text{FeOOH(s)} + \frac{1}{4}\text{CH}_2\text{O} + 2\text{H}^+ \rightarrow \text{Fe(II)}^{2+} + \frac{1}{4}\text{CO}_2 + \frac{7}{4}\text{H}_2\text{O} \tag{20.3}$$

By contrast to Fe(III), Fe(II) is more soluble and exists in the low mg/L range in sediment pore waters. As a consequence it can diffuse to the oxic layer of the sediment where it is subject to oxidation. The oxidation of Fe(II) to Fe(III) occurs via the loss of one electron

$$\text{Fe(II)}^{2+} - e^- \rightarrow \text{Fe(III)}^{3+} \tag{20.4}$$

With oxygen as the electron acceptor, the overall reaction can be found using the half-reaction for oxygen (Eq. 1.15) followed by the precipitation of iron oxyhydroxide

$$\text{Fe(III)}^{3+} + 2\text{H}_2\text{O} \rightarrow \text{FeOOH(s)} + 3H^+ \tag{20.5}$$

to yield the overall redox reaction

$$\text{Fe(II)}^{2+} + \frac{1}{4}\text{O}_2 + \frac{3}{2}\text{H}_2\text{O} \rightarrow \text{FeOOH(s)} + 2\text{H}^+ \tag{20.6}$$

This is the reaction that occurs in the aerobic layer.

20.2 A   Reaction Kinetics

The kinetics of this reaction have been examined and can be described using the equation (Morgan, 1967)

$$\frac{d[\text{Fe(II)}]}{dt} = -k'_{\text{Fe}}[\text{Fe(II)}] \tag{20.7}$$

where

$$k'_{\text{Fe}} = k_{\text{Fe}}[\text{O}_2][\text{OH}^-]^2 \tag{20.8}$$

in the pH range of 5 to 7 (Stumm and Lee, 1961). An example is shown in Fig. 20.2. The reaction is first-order in Fe(II) concentration, and the rate constant is linear in oxygen and quadratic in hydroxide ion concentration (Eq. 20.8). The pH dependence

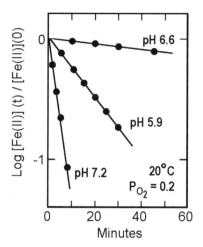

Fig. 20.2 Oxidation of Fe(II). Semilogarithmic plot of normalized iron concentration versus time (Stumm and Morgan, 1970).

of the reaction can be explained by examining the speciation of the Fe(II) and assuming that each species, e.g., Fe^{2+}, $FeOH^+$, $Fe(OH)_2$ (Millero, 1990), $Fe(CO_3)_2^{2-}$ (King, 1998), oxidizes at a different rate. A somewhat more rapid oxidation rate in lake water at pH = 6.8–7.3 has been observed (Emmemegger et al., 1998). What is important to note for the application presented below is that the oxidation reaction is rapid in the normal pH range of surface and aerobic layer pore waters (Fig. 20.2).

Thus the chemistry of manganese and iron are quite similar. The oxidized forms are both insoluble and form oxides. The reduced forms are soluble in the mg/L range. Their concentrations in pore water are regulated by solid phases and sorption to particles. Their flux to the overlying water is controlled by the rate at which the reduced forms are oxidized in the aerobic layer. A rapid oxidation rate prevents escape, since insoluble particles form. A reduced oxidation rate, or anoxic conditions in the overlying water, allows dissolved metal to escape as fluxes to the overlying water. Responses to lowered dissolved oxygen appear to be similar (Sundby et al., 1986). Hence it is expected that the same sediment flux model framework for can be used for both manganese and iron.

20.3 MODEL CONFIGURATION

The formulation of the sediment flux model for iron exactly parallels that developed for manganese. Iron is tracked both as dissolved Fe^{2+} and particulate FeOOH(s) iron. A schematic is presented in Fig. 20.3. The equations for the aerobic (1) and anaerobic (2) sediment layers are

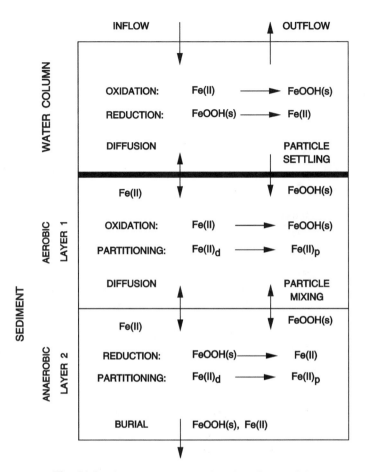

Fig. 20.3 Schematic diagram of the iron flux model.

Layer 1 Fe(II)

$$H_1 \frac{d\text{Fe}(1)}{dt} = -s\left(f_{d1}\text{Fe}(1) - \text{Fe}(0)\right) + K_{L12}\left(f_{d2}\text{Fe}(2) - f_{d1}\text{Fe}(1)\right)$$

$$+ w_{12}\left(f_{p2}\text{Fe}(2) - f_{p1}\text{Fe}(1)\right) - w_2\text{Fe}(1)$$

$$- k_{\text{Fe,OH,1}} H_1 O_2(1) \left(10^{2(\text{pH}_1 - 7)}\right) f_{d1}\text{Fe}(1)$$

$$+ \text{Fe}(2)\,\dot{H}_1^+ - \text{Fe}(1)\left(\dot{H}_1 + \dot{H}_1^-\right)$$

$$(20.9a)$$

Layer 2 Fe(II)

$$H_2 \frac{d\text{Fe}(2)}{dt} = -K_{L12}\big(f_{d2}\text{Fe}(2) - f_{d1}\text{Fe}(1)\big)$$

$$-w_{12}\big(f_{p2}\text{Fe}(2) - f_{p1}\text{Fe}(1)\big) - w_2\big(\text{Fe}(2) - \text{Fe}(1)\big)$$

$$+k_{\text{Fe},2}\text{FeOOH}(2)H_2 - \text{Fe}(2)\left(\dot{H}_2 + \dot{H}_1^+\right) + \text{Fe}(1)\dot{H}_1^-$$

$$(20.9\text{b})$$

Layer 1 FeOOH(s)

$$H_1 \frac{d\text{FeOOH}(1)}{dt} = k_{\text{Fe,OH},1} H_1 O_2(1)\left(10^{2(\text{pH}_1 - 7)}\right) f_{d1}\text{Fe}(1) - w_2\text{FeOOH}(1)$$

$$+w_{12}\big(\text{FeOOH}(2) - \text{FeOOH}(1)\big) + w_1\text{FeOOH}(0)$$

$$+\text{FeOOH}(2)\dot{H}_1^+ - \text{FeOOH}(1)\left(\dot{H}_1 + \dot{H}_1^-\right) \qquad (20.9\text{c})$$

Layer 2 FeOOH(s)

$$H_2 \frac{d\text{FeOOH}(2)}{dt} = -k_{\text{Fe},2}\text{FeOOH}(2)H_2 - w_{12}\big(\text{FeOOH}(2) - \text{FeOOH}(1)\big)$$

$$-w_2\big(\text{FeOOH}(2) - \text{FeOOH}(1)\big)$$

$$-\text{FeOOH}(2)\left(\dot{H}_2 + \dot{H}_1^+\right) + \text{FeOOH}(1)\dot{H}_1^- \qquad (20.9\text{d})$$

where FeOOH(0) and Fe(0) refer to the particulate and dissolved iron concentrations in the bottommost water column segment, designated as segment 0, which overlays the sediment.

The water column equations for iron for this segment are

Layer 0 Fe(II)

$$H_0 \frac{d\text{Fe}(0)}{dt} = s\big(f_{d1}\text{Fe}(1) - \text{Fe}(0)\big) - k'_{\text{Fe,OH},0} H_0 \left(10^{2(\text{pH}_0 - 7)}\right)\text{Fe}(0)$$

$$+k_{\text{FeOOH},0} H_0 \frac{K_{\text{M,FeOOH},0}}{K_{\text{M,FeOOH},0} + [O_2(0)]}\text{FeOOH}(0)$$

$$+E_{0,-1}\big(\text{Fe}(-1) - \text{Fe}(0)\big) \qquad (20.10\text{a})$$

Layer 0 FeOOH(s)

$$H_0 \frac{d\text{FeOOH}(0)}{dt} = k'_{\text{Fe,OH},0} H_0 \left(10^{2(\text{pH}_0 - 7)}\right) \text{Fe}(0)$$

$$+ w_1 \left(\text{FeOOH}(-1) - \text{FeOOH}(0)\right)$$

$$- k_{\text{FeOOH},0} H_0 \frac{K_{\text{M,FeOOH},0}}{K_{\text{M,FeOOH},0} + [\text{O}_2(0)]} \text{FeOOH}(0)$$

$$+ E_{0,-1} \left(\text{FeOOH}(-1) - \text{FeOOH}(0)\right) \qquad (20.10b)$$

where $E_{0,-1}$ represents the vertical mixing coefficient between the bottom layer and the layer immediately above. The concentrations in the water column are numbered in reverse order to preserve the notation used previously. Thus layer -1 refers to the layer just above layer 0 in the water column.

An additional term has been added to the water column equations (20.10) to include the reduction of FeOOH if the dissolved oxygen concentration approaches zero. The rate of reduction of FeOOH to Fe(II) is given by

$$k_{\text{FeOOH},0} H_0 \frac{K_{\text{M,FeOOH},0}}{K_{\text{M,FeOOH},0} + [\text{O}_2(0)]} \text{FeOOH}(0) \qquad (20.11)$$

where $k_{\text{FeOOH},0}$ is the rate at which FeOOH(s) is reduced to Fe(II), $K_{\text{M,FeOOH},0}$ is the half-saturation constant for oxygen for this reaction which is set to a small number so that reduction occurs only if the dissolved oxygen concentration is almost zero. An analogous term

$$k_{\text{MnO}_2,0} H_0 \frac{K_{\text{M,MnO}_2,0}}{K_{\text{M,MnO}_2,0} + [\text{O}_2(0)]} \text{MnO}_2(0) \qquad (20.12)$$

is required for the manganese model when it is applied to situations where anoxia in the water column is occurring. The nitrogen, oxygen, phosphorus, and silica portions of the sediment model are identical to those presented previously.

20.4 APPLICATION TO ONONDAGA LAKE

Onondaga Lake is located immediately north of the city of Syracuse, in Onondaga County, Central New York State (Effler and Whitehead, 1996b). The lake has a length of 7.6 km and a width of 1.8 km. The outflow from the lake exits through a single outlet at the northern end and enters the Seneca River. A bathymetric map is shown in Fig. 20.4 and relevant physical data are listed in Table 20.1. The lake is comprised of two basins, commonly referred to as the south and north basins, which are separated by a slight saddle region that is located approximately 3.6 km from the outlet. The north basin has a maximum depth of 18.9 m and the south basin has a maximum depth of 19.9 m.

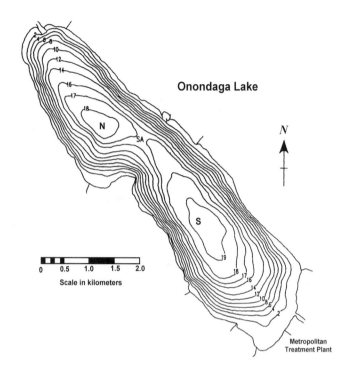

Fig. 20.4 Bathymetric map of Onondaga Lake. South (S) and North (N) sampling stations. Contours in meters.

Onondaga Lake receives surface runoff from a drainage basin of 689 km² which is almost entirely located in Onondaga County. The major freshwater inputs are Nine-mile Creek, Onondaga Creek, the Metropolitan Syracuse Sewage Treatment Plant (METRO), Ley Creek, Bloody Brook, and Harbor Brook.

As a result of urban and industrial development, Onondaga Lake has been used for both domestic and industrial waste disposal for over 100 years. Major industrial loadings include calcium, alkalinity, inorganic carbon, and chlorides that were as-sociated with the production of soda ash (Na_2CO_3), which began on the shores of Onondaga Lake in 1884 and continued until 1986. The METRO wastewater treat-ment plant was upgraded in 1979 to secondary treatment. Tertiary treatment for phosphorus removal was added in 1981. By the early '80s the plant capacity was 80 MGD.

The water quality in Onondaga Lake during the period simulated below is char-acteristic of a highly eutrophic lake. Due to the large nutrient loadings, summer chlorophyll *a* concentrations can exceed 100 μg/L. The lake is stratified in the sum-mer months from June to September with temperature differences exceeding 18 °C. Hypolimnetic hypoxia and anoxia exist for 9 to 10 months per year. Epilimnion dissolved oxygen concentrations are generally above 6 mg/L throughout the year.

20.4 A   Data Sets

A substantial data set is available for Onondaga Lake dating back to the late 1960s. A compilation and review of all data collected between 1968–1989 is available (Walker, 1991). Data sources include Onondaga County Department of Drainage and Sanitation (D&S), Onondaga County Department of Health (DOH), New York State Department of Environmental Conservation (DEC), Upstate Freshwater Institute (UFI), and the United States Geological Survey (USGS). The data have been assembled onto a single database containing 186,000 water quality observations.

Dissolved iron data were collected by the Department of Drainage and Sanitation in the years 1968–1975. Dissolved iron data were also collected by the Upstate Freshwater Institute in 1980–1981, 1985–1986, and 1988–1989. Data were collected at station 41, in the southern basin of the lake. It is the data from this station that will be used subsequently. Data were collected at 1 m depth intervals on a weekly basis. In 1985 iron data was recorded from May to November. In 1986 iron data for March through November are available. In 1987 no iron data were recorded. In 1988 data were recorded from May to October. In 1989 iron data are available from June to August. A comprehensive review of the water quality data is given in Effler and Harnett (1996).

20.4 B   Water Quality Model

A eutrophication model (RCA) for Onondaga Lake has been constructed and a detailed description is available (HydroQual, 1992). The model is a direct descendent of the WASP model (Di Toro et al., 1981). For Onondaga Lake the model has 32 state variables, including the two iron state variables Fe(II) and FeOOH(s). A list of the systems used are given in Table 20.2. The model is configured to simulate the annual cycle of phytoplankton production, nutrient cycling, and the resulting dissolved oxygen distribution. The sediment flux model provides the fluxes of nutrients and oxygen, and is identical to that used for the Chesapeake Bay and MERL simulations discussed previously.

Segmentation for Onondaga Lake   Onondaga Lake is segmented vertically into 10 water column segments. Each segment has a depth of 2 meters and a surface

Table 20.1 Physical characteristics of Onondaga Lake

Characteristic	Value
Surface Area (km^2)	12.0
Volume (m^3)	131. $\times$ 10^6
Mean Depth (m)	10.9
Maximum Depth (m)	19.9
Watershed Drained (km^2)	689.
Detention Time (days)	90.0

Table 20.2   Onondaga Lake water quality model systems

Number	System	Number	System
1	Salinity	17	Refractory POC
2	Phytoplankton – diatoms	18	Labile POC
3	Phytoplankton – summer group	19	Refractory DOC
4	Refractory POP	20	Labile DOC
5	Labile POP	21	Reactive DOC
6	Refractory DOP	22	Algal exudate DOC
7	Labile DOP	23	Oxygen equivalents
8	Total DIP	24	Dissolved oxygen
9	Refractory PON	25	Total active metal
10	Labile PON	26	Total inorganic carbon
11	Refractory DON	27	Alkalinity
12	Labile DON	28	Calcite
13	Total Ammonia	29	Calcium
14	Nitrate + Nitrate	30	Temperature
15	Biogenic Silica	31	Dissolved iron
16	Total Silica	32	Particulate iron

area chosen to match the hypsometry of the lake. The lake is assumed to be laterally well mixed, consistent with observations (Walker, 1991).

Iron Loading   Data for the composition of particles entering Onondaga Lake in 1981 have been reported (Effler and Whitehead, 1996a). The loading rate in 1981 for a "Fe-Mn-rich" class of solids is reported to contain $Fe(OH)_3(s)$, $Fe_2O_3(s)$, $FeOOH(s)$, $MnO_2(s)$, and other Fe and Mn compounds. Since the concentration of manganese is much less than iron in typical soil particles (Lindsay, 1979) it is assumed that these loading rates represent only iron inputs. Using the average tributary flows an estimate of the influent concentration can be obtained. This concentration was then used to estimate daily loads for the calibration years 1985–1989.

This loading represents all particulate iron compounds. In order to obtain the reactive iron load it is necessary to reduce the total load to account for the unreactive iron compounds. As there is no simple way to do this without sequential extraction information (Canfield, 1989), the load was reduced as part of the calibration procedure. A reduction of 40% was required in order to match the observations. All influent iron was assumed to be particulate since inflows to the lake were aerobic and the oxidation rate of ferrous iron is very rapid (Fig. 20.2).

Dissolved Oxygen Calibration   The original development of the Onondaga Lake model focused on calcite precipitation and deposition, with secondary emphasis on eutrophication. Although the model produced a reasonable fit of the dissolved oxygen profiles, it did not always reproduce the exact timing of the anoxia in Onondaga Lake. An example of model-data comparison for 1986 is shown in Figure 20.5. The model is unable to reproduce the observed hypolimnetic anoxia from February through May. Because of the strong relationship between iron flux from the

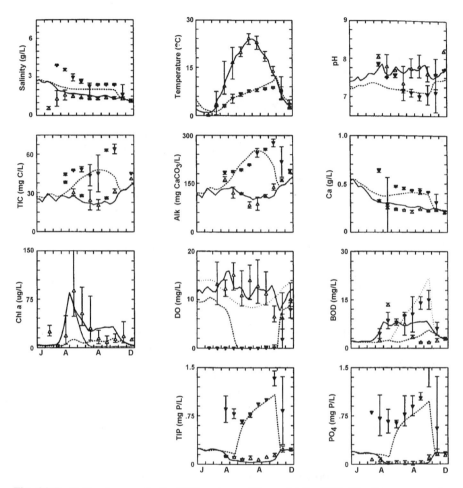

Fig. 20.5 Calibration results for RCA water quality model for 1986. Monthly averages for surface layer: model (solid line), data (▲) ± range; and bottom layer: model (dotted line), data (▼) ± range. Dotted line in the DO plot is the dissolved oxygen saturation concentration.

sediment and the overlying water dissolved oxygen, it was decided to use the measured dissolved oxygen data rather than the model results to specify when anoxia was present. This was done for three reasons. First, the dissolved oxygen data were very complete and thorough, with weekly dissolved oxygen profiles measured at 1 m depth intervals. Second, because dissolved oxygen is coupled to so many other systems, achieving a better dissolved oxygen fit to the data would have essentially involved recalibrating the existing water quality model, which would have been time consuming. Third, because the iron only responds to the dissolved oxygen but does not affect it, at least at these iron concentrations for which the oxygen consumed by iron oxidation is small, there is no loss of a feedback mechanism by specifying the dissolved oxygen directly.

20.4 C   Coupled Model Calibration

The model was run for five years, beginning January 1, 1985, to December 31, 1989. Iron loading was supplied via the tributaries based on the calculation procedure described above. The values of the constants used for the iron model in the water column and sediment are given below in Table 20.3. Note that the same oxidation and reduction rates are used for both the sediment and water column. This parallels the procedure followed in the manganese model. The model was cycled through 20 years of simulations – the 5-year sequence repeated four times – in order to generate equilibrium initial conditions. These were then used as the initial conditions for the final five-year calibration run.

Table 20.3   Parameters for the Iron Flux Models

Parameter*	Units	Onondaga Lake	Croton Reservoir
$k'_{Fe,OH,0}$	d^{-1}	10.0	10.0
$k_{FeOOH,0}$	d^{-1}	4.0	4.0
$K_{M,FeOOH,0}$	mg O_2/L	0.2	0.2
$k_{Fe,OH,1}$	$d^{-1}(mg\ O_2/L)^{-1}$	1.0	1.0
$k_{FeOOH,2}$	d^{-1}	4.0	4.0
$\Delta\pi_{Fe,1}$	–	15.0	15.0
$\pi_{Fe,2}$	L/kg	3000.0	100.0
$[O_2(0)]_{crit,Fe}$	mg O_2/L	2.0	2.0

$*\theta_{Fe,OH,0} = \theta_{FeOOH,0} = \theta_{Fe,OH,1} = \theta_{FeOOH,2} = 1.08.$

Figure 20.6 shows the iron calibration results for March through November 1986, when iron data were recorded, using the fitted dissolved oxygen for the months. Also shown are temperature and dissolved oxygen. The measured data average (open symbols) and range (bars) and the model computed monthly averages (solid line) and range (shading) are compared. It is interesting to note how early in the year anoxia started (B) and how long it persisted. In fact, in 1986 the bottom of the lake was essentially anoxic or hypoxic (DO < 2 mg/L) for the whole year.

The dissolved iron concentrations (Fig. 20.6C) increase as a consequence of the iron flux being generated by the sediment. The data for March were actually collected on March 30 and so the right edge of the shading should be compared to the data, to which it compares reasonably well. There is a continual buildup of dissolved iron due to an iron flux from the sediment. The model overestimates the amount of Fe^{2+} slightly in August but more noticeably in September and October. One reason for this may be the formation and subsequent precipitation of iron sulfide, FeS(s). This mechanism is discussed in more detail in Section 20.4 E. Note that the Fe^{2+} data for November 1986 are only for one day, November 3.

The computed particulate iron concentrations are shown in Fig. 20.6D. Unfortunately, there are no particulate iron data with which to compare the model results. When anoxia is not present the particulate iron will exhibit an increasing concentration with depth, due to the settling of particulate iron into the deeper parts of the lake where the volume is decreasing. As the particulate iron settles into the anoxic zone, it is reduced to form Fe^{2+}. Thus the particulate concentration will initially increase and then decrease in the anoxic zone.

In 1987 no iron data were collected. Fig. 20.7 shows the calibration results for 1988 for the months of May through October. It is interesting to note that the data for both 1988 and 1989 exhibit much more short-term variation than in 1986. It is not clear why this happens, and although the model fits the monthly concentration averages, it cannot reproduce the large variations. The observed dissolved oxygen data does not suggest any reaeration events occurring, which would cause large variations in Fe^{2+}.

Effect of Dissolved Oxygen It is of interest to determine the extent to which the model predictions are dependent on an accurate representation of the dissolved oxygen concentrations. This is examined in Fig. 20.8 where the results of using the observed (I) and modeled (II) dissolved oxygen concentration are compared. The onset of iron generation from the sediment is delayed until the dissolved oxygen is zero at the sediment-water interface. However, once anoxia occurs, the iron concentrations increase (IC and IIC) to almost the same extent. It is interesting that the maximum iron concentration during anoxia is not very sensitive to the duration of anoxia.

20.4 D Analysis of the Total Iron Distribution

The distribution of iron in the water column can be understood using the results of an analysis of a simple two-layer model of the lake, representing the epilimnion and hypolimnion, the portions of the water column above and below the thermocline. We will assume for the sake of simplicity that the epilimnion is always aerobic and the hypolimnion is always anaerobic. The equations are

Layer 1 Fe(II)

$$H_1 \frac{d\mathrm{Fe}(1)}{dt} = -k'_{\mathrm{Fe},1} H_1 \mathrm{Fe}(1) + K_{L12}(\mathrm{Fe}(2) - \mathrm{Fe}(1)) \tag{20.13a}$$

Layer 2 Fe(II)

$$H_2 \frac{d\mathrm{Fe}(2)}{dt} = k'_{\mathrm{Fe},2} H_2 \mathrm{FeOOH}(2) - K_{L12}(\mathrm{Fe}(2) - \mathrm{Fe}(1)) + J[\mathrm{Fe}] \tag{20.13b}$$

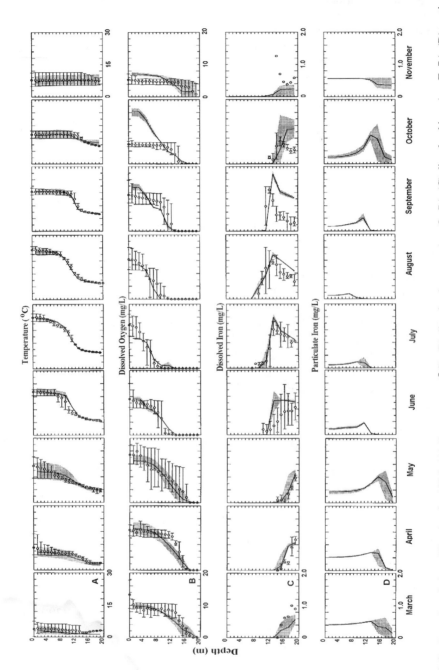

Fig. 20.6   Onondaga Lake calibration for 1986. (A) Temperature (°C), (B) dissolved oxygen (mg O$_2$/L), (C) dissolved iron (mg Fe/L), (D) particulate iron (mg Fe/L). Model average (line) and range (shading) and data average (•) and range are shown.

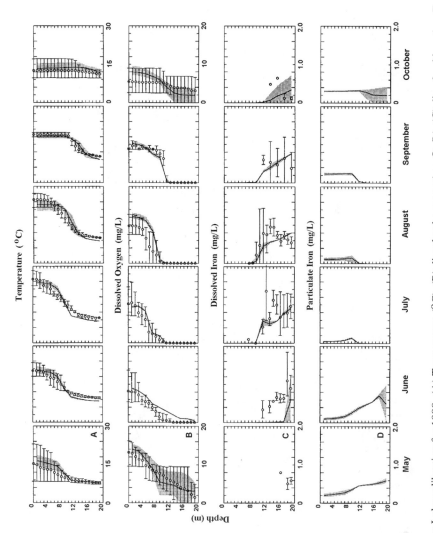

Fig. 20.7 Onondaga Lake calibration for 1988. (A) Temperature (°C), (B) dissolved oxygen (mg O$_2$/L), (C) dissolved iron (mg Fe/L), (D) particulate iron (mg Fe/L).

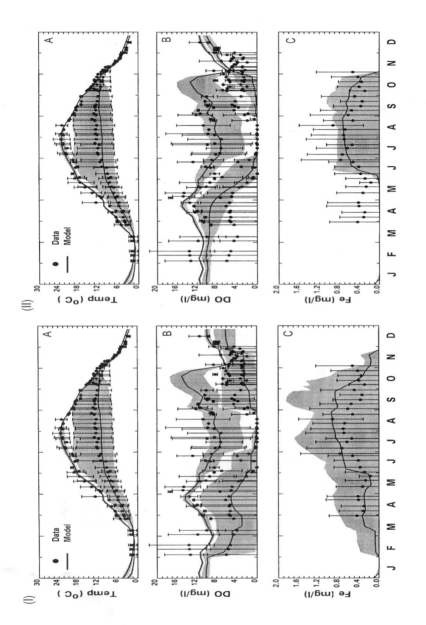

Fig. 20.8 Epilimnion and hypolimnion average concentrations and ranges: Model (lines and shading), data (●) and bars, for 1986. (I) Using measured DO, (II) using modeled DO for the iron calculation.

Layer 1 FeOOH(s)

$$H_1 \frac{d\text{FeOOH}(1)}{dt} = k'_{\text{Fe},1} H_1 \text{Fe}(1) + K_{L12}\big(\text{FeOOH}(2) - \text{FeOOH}(1)\big)$$

$$-w_1 \text{FeOOH}(1) \tag{20.13c}$$

Layer 2 FeOOH(s)

$$H_1 \frac{d\text{FeOOH}(2)}{dt} = -k'_{\text{Fe},2} H_2 \text{FeOOH}(2) - K_{L12}\big(\text{FeOOH}(2) - \text{FeOOH}(1)\big)$$

$$+w_1 \text{FeOOH}(1) - w_2 \text{FeOOH}(2) \tag{20.13d}$$

where $\text{Fe}(1)$, $\text{Fe}(2)$ and $\text{FeOOH}(1)$, $\text{FeOOH}(2)$ refer to the dissolved and particulate iron (Fe(II) and FeOOH) in the epilimnion (1) and hypolimnion (2) of the lake. The analogy to the sediment is almost perfect. The epilimnion corresponds to the aerobic layer H_1 and the hypolimnion to the anaerobic layer H_2. The first-order oxidation rate of Fe(II) in the epilimnion is $k'_{\text{Fe},1}$ and the first-order reduction rate of FeOOH in the hypolimnion is $k'_{\text{Fe},2}$. The only differences are in the transport terms. The diffusive mixing K_{L12} affects both the dissolved and particulate species, since the particles are suspended in the water column and are subjected to the mixing motions of the water. However, only the particles settle from the epilimnion to the hypolimnion w_1 and from the hypolimnion to the sediment w_2.

The steady state solutions, computed using MACSYMA, are listed in Figs. 20.9 and 20.10. Note that the relationship between the particulate concentration in the hypolimnion and sediment flux (d6)

$$J[\text{Fe}] = w_2 \text{FeOOH}(2) \tag{20.14}$$

is the same as found from the interactive water column-sediment model for manganese (Eq. 19.25), and for essentially the same reason. The deposited iron must be returned if burial in the sediment is ignored.

An interesting feature of the solutions has to do with cycling of particulate iron FeOOH that settles into the hypolimnion where it is reduced to soluble ferrous iron Fe(II). It then mixes into the epilimnion where it oxidizes to form particulate iron and the cycle repeats. The solutions reflect this cycling. The concentration of Fe(II) in the hypolimnion is always larger than the epilimnion concentration, Fig. 20.10 (d10)

$$\text{Fe}(2) = \text{Fe}(1)\left(1 + \frac{k'_{\text{Fe},1} H_1}{K_{L12}}\right)$$

$$\geqslant \text{Fe}(1) \tag{20.15}$$

since the term in parentheses is always greater than one. Conversely the epilimnetic particulate iron concentration (d11) is greater than the hypolimnetic concentration if

$w_1 = w_2$

$$\text{FeOOH}(1) = \text{FeOOH}(2) \frac{K_{L12} + k'_{Fe,2} H_2 + w_2}{K_{L12} + w_1}$$

$$= \text{FeOOH}(2) \frac{K_{L12} + k'_{Fe,2} H_2 + w_2}{K_{L12} + w_2} \quad \text{if } w_1 = w_2$$

$$= \text{FeOOH}(2) \left(1 + \frac{k'_{Fe,2} H_2}{K_{L12} + w_2} \right)$$

$$\geqslant \text{FeOOH}(2) \tag{20.16}$$

Consider, finally the distribution of total iron, the sum of dissolved and particulate iron

$$\text{Fe}_T(1) = \text{Fe}(1) + \text{FeOOH}(1) \tag{20.17a}$$

$$\text{Fe}_T(2) = \text{Fe}(2) + \text{FeOOH}(2) \tag{20.17b}$$

The ratio $\text{Fe}_T(2)/\text{Fe}_T(1)$

$$r_{21} = \text{Fe}_T(2)/\text{Fe}_T(1) \tag{20.18}$$

is always greater than one if the oxidation rate $k'_{Fe,1}$ of Fe(II) to FeOOH, or the reduction rate of FeOOH to Fe(II) $k'_{Fe,2}$, is fast enough (Fig. 20.10, d15,d16)

$$\lim_{k'_{Fe,1} \to \infty} r_{21} = 1 + \frac{w_1}{K_{L12}}$$

$$\geqslant 1 \tag{20.19}$$

$$\lim_{k'_{Fe,2} \to \infty} r_{21} = 1 + \frac{k'_{Fe,1} H_1 w_1}{K_{L12}^2 + (w_1 + k'_{Fe,1} H_1) K_{L12}}$$

$$\geqslant 1 \tag{20.20}$$

It is this mechanism – the downward advection of particulate iron and the upward diffusion of dissolved iron – that produces the maximum concentration of total iron at the boundary of the epilimnion and hypolimnion.

20.4 E   Iron Sulfide Precipitation

The suggestion was made in Section 20.4 C that the decline in dissolved iron in the latter portion of 1986, which was observed but not modeled (Fig. 20.6C), might be due to the precipitation of iron sulfide. The sediment model computes sulfide fluxes, and the water column model tracks the oxygen equivalents of dissolved sulfide. It is

Fe(II) Layer 1

(c2) eq1:k[l12]*(fe[2]-fe[1])-k[1]*fe[1]=0

(d2)
$$\left(fe_2 - fe_1\right) k_{l12} - fe_1 k_1 = 0$$

Fe(II) Layer 2

(c3) eq2:-k[l12]*(fe[2]-fe[1])+k[2]*feooh[2]+j[fe]=0

(d3)
$$- \left(fe_2 - fe_1\right) k_{l12} + j_{fe} + feooh_2 k_2 = 0$$

FeOOH Layer 1

(c4) eq3: k[1]*fe[1]+k[l12]*(feooh[2]-feooh[1])-w[1]*feooh[1]=0

(d4)
$$\left(feooh_2 - feooh_1\right) k_{l12} - feooh_1 w_1 + fe_1 k_1 = 0$$

FeOOH Layer 2

(c5) eq4:-k[2]*feooh[2]-k[l12]*(feooh[2]-feooh[1])+w[1]*feooh[1]-w[2]*feooh[2]=0

(d5)
$$- \left(feooh_2 - feooh_1\right) k_{l12} - feooh_2 w_2 - feooh_2 k_2 + feooh_1 w_1 = 0$$

Mass balance check

(c6) eq1+eq2+eq3+eq4

(d6)
$$j_{fe} - feooh_2 w_2 = 0$$

Solve the equations

(c7) c:factorsum(solve([eq1,eq2,eq3,eq4],[fe[1],fe[2],feooh[1],feooh[2]])[1])

(d7)
$$\left[fe_1 = \frac{\left(w_2 + k_2\right) j_{fe}}{k_1 w_2}, \quad fe_2 = \frac{\left(w_2 + k_2\right) j_{fe} \left(k_{l12} + k_1\right)}{k_1 w_2 k_{l12}}, \right.$$

$$\left. feooh_1 = \frac{j_{fe} \left(k_{l12} + w_2 + k_2\right)}{w_2 \left(k_{l12} + w_1\right)}, \quad feooh_2 = \frac{j_{fe}}{w_2} \right]$$

(c8) (fe1:rhs(c[1]),fe2:rhs(c[2]),feooh1:rhs(c[3]),feooh2:rhs(c[4]))$

Fig. 20.9 MACSYMA solution for a 2-layer model of iron in the water column. $k_1 = k_{Fe,1} H_1$, $k_2 = k_{Fe,2} H_2$.

Substitutions

(c9)   (rule1:fe[1]=fe1,rule2:feooh[2]=feooh2)$

(c10)   fe[2]=facsum(scsimp(fe2,rule1))

(d10)
$$fe_2 = \frac{fe_1 \left(k_{l12} + k_1\right)}{k_{l12}}$$

(c11)   feooh[1]=facsum(scsimp(feooh1,rule2))

(d11)
$$feooh_1 = \frac{feooh_2 \left(k_{l12} + w_2 + k_2\right)}{k_{l12} + w_1}$$

Analyze the total Fe distribution

(c12)   fet1:fe1+feooh1

(d12)
$$\frac{j_{fe} \left(k_{l12} + w_2 + k_2\right)}{w_2 \left(k_{l12} + w_1\right)} + \frac{\left(w_2 + k_2\right) j_{fe}}{k_1 \, w_2}$$

(c13)   fet2:fe2+feooh2

(d13)
$$\frac{\left(w_2 + k_2\right) j_{fe} \left(k_{l12} + k_1\right)}{k_1 \, w_2 \, k_{l12}} + \frac{j_{fe}}{w_2}$$

(c14)   ratio21:r21=ratsimp(fet2/fet1)$

Find the limit of r21 as k[1] -> infinity

(c15)   limit(ratio21,k[1],inf)

(d15)
$$r21 = \frac{k_{l12} + w_1}{k_{l12}}$$

Find the limit of r21 as k[2] -> infinity

(c16)   limit(ratio21,k[2],inf)

(d16)
$$r21 = \frac{k_{l12}^{\,2} + \left(w_1 + k_1\right) k_{l12} + k_1 \, w_1}{k_{l12}^{\,2} + \left(w_1 + k_1\right) k_{l12}}$$

Fig. 20.10   MACSYMA analysis of the total iron in a 2-layer water column model, continued from Fig. 20.9.

possible, therefore, to see if the sulfide concentrations are increasing to a point where precipitation is possible.

Figure 20.11 presents the vertical distributions of sulfide and dissolved iron for the months of May through October for 1986 and 1988. For 1986 (Fig. 20.11A, B) the buildup is reproduced until October when premature destratification in the model reduces the sulfide concentration. The computed sulfide concentrations for 1988 (Fig. 20.11C) increase more rapidly than is observed. The magnitude of the sediment flux of sulfide during anoxia is influenced by the magnitude of the sulfide partition coefficients $\pi_{H_2S,1}$ and $\pi_{H_2S,2}$. These are empirical parameters that were calibrated using the Chesapeake Bay data set (Chapter 14). Since these empirical parameters have not been calibrated for Onondaga Lake, it is somewhat unexpected that the results are generally reasonable.

The sulfide and iron data can be used to make a qualitative analysis of the likelihood that sulfide precipitation is the cause of the model overprediction of dissolved iron and sulfide. The excessive iron concentrations calculated by the model occur primarily in September and October of 1986 (Fig. 20.11B). There is no indication that the sulfide concentrations are increasing rapidly enough to initiate precipitation of iron sulfide. It is possible, of course, that FeS(s) is being formed continuously and that its addition as a mechanism would improve the model's predictive ability. However, a qualitative examination of the model-data discrepancies does not suggest that this is a first-order effect.

20.4 F   Sediment Concentrations

Some limited data are available for both pore water and sediment iron concentrations in Onondaga Lake. Sediment iron data are available for 1986 (Yin and Johnson, 1984). Pore water dissolved iron data are available for August and November 1991 (Cornwell, 1993). The data are summarized in Table 20.4, which also lists the model's calculated annual average concentrations. The computed and observed pore water concentrations are in the same order of magnitude and the range is similar. The solid phase comparison is also reasonable when the observed concentrations are corrected for the reactive fraction, as shown.

Table 20.4 Comparison of Observed and Computed Sediment and Pore Water Iron Concentrations: Annual Average (Range)

Phase	Model	Data
Pore water (mg Fe/L)	0.88	0.58
	(0.02–2.78)	(0.13–3.34)
Solid phase (mg Fe/g)	4.0	5.6[a]
	(3.0–5.5)	(3.6–7.8)[a]

[a]Reactive iron is estimated as 40% of total iron, consistent with the reactive fraction assumption used to estimate the loading.

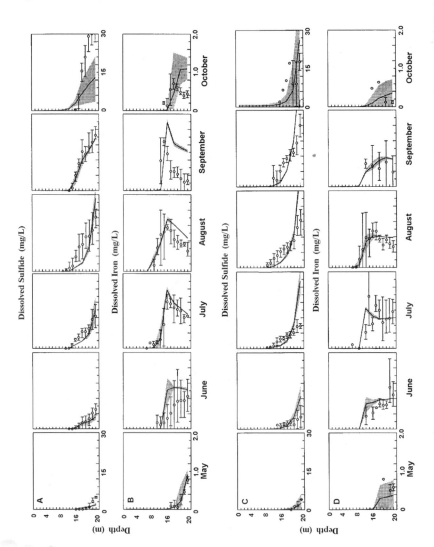

Fig. 20.11 Onondaga Lake calibration for dissolved sulfide (mg O_2^*/L) and dissolved iron (mg Fe/L). (A) 1986, (B) 1988.

20.4 G   Summary

The iron sediment and water column model formulated in this chapter can reproduce the available observations of water column and sediment concentrations in Onondaga Lake with reasonable fidelity. The model is initialized to a periodic steady state and then run for five years, beginning in 1985 and ending in 1989. The same oxidation and reduction kinetic rate constants are used in both the water column and sediment, consistent with the procedure followed for the manganese model. The seasonal development of the vertical distribution of iron in the water column is correctly reproduced. An initial increase occurs at the sediment-water interface at the beginning of anoxia (March–May 1986), followed by an almost uniform concentration in the hypolimnion (June), the development of a maximum concentration at the thermocline (July–September), and the subsequent decline at overturn.

The predicted sulfide concentrations are also in reasonable agreement with observations without any further calibration of the partitioning parameters for sulfide. There does not appear to be any strong evidence for the precipitation of iron sulfide if the criterion used is the discrepancy between model and data. This does not rule out the possibility, however. Therefore, the addition of this mechanism would be a logical next step. It would also remove the need to specify the partition coefficients for sulfide and iron, thereby removing a degree of freedom from the model.

20.5   APPLICATION TO THE CROTON RESERVOIR

As a part of a study of the water supply system for the City of New York (Metcalf et al., 1995), a eutrophication model for the Croton Reservoir was constructed. The iron and manganese models described in this chapter were incorporated in that model, and the results are presented below.

The Croton Reservoir is located approximately 40 miles northeast of New York City, in Westchester and Putnam Counties, in lower New York State. The reservoir represents part of the drinking water supply system for New York City and supplies on average approximately 140 million gallons per day of drinking water to the city. The Croton Reservoir was created by the impoundment of the Croton River by the New Croton Dam. At the northwest end of the reservoir is the Muscoot Dam which delineates the Muscoot Reservoir. Approximately 4 km upstream from the Croton Dam, in the Croton Reservoir (sometimes called the New Croton Reservoir), is a submerged dam that formed the original Croton Reservoir at this site.

The reservoir is oriented along a southeast-northwest axis, with the New Croton Dam located at the southeast end. A bathymetric map of the Croton Reservoir, which outlines the model surface segmentation, is shown in Fig. 20.12. The length of the Croton Reservoir is approximately 14.6 km. The maximum depth when full is 36 m. The surface area when full is 92×10^6 ft^2 and the volume when full is 4.14×10^9 ft^3. The average residence time of the Croton Reservoir is 91 days. The direct drainage area is 56 square miles.

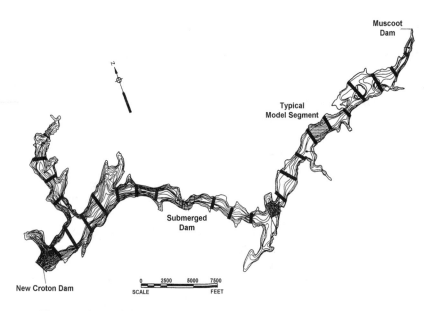

Fig. 20.12 New Croton Resevoir segmentation. Contour interval = 10 ft.

At the Muscoot Dam the depth of the Croton Reservoir is approximately 10 m depending on the water level in the reservoir. The depth increases to 24 m just after the submerged dam. At the New Croton Dam the maximum depth of the reservoir is 40 m.

Most of the inflow to the Croton Reservoir comes from the adjoining Muscoot Reservoir. The rest is inflow from tributaries, groundwater recharge, and surface runoff. The largest inflows occur in the months January through April. The smallest inflows occur in the months June through September. In order to maintain adequate stream conditions in the Croton River downstream of the reservoir, a minimum out-flow of 5.5 MGD is required to be maintained.

The nutrient loadings to the Croton Reservoir are from runoff, wastewater treat-ment plants, septic systems, atmospheric deposition, and the Muscoot Reservoir in-flow. Phosphorus loading is of most interest because it is the limiting nutrient. Phos-phorus loading from the Muscoot Reservoir for 1993 was estimated to be 42 kg/d. The two tributaries, the Kisco River and Hunter Brook, together contribute an esti-mated 6 kg/d to the Croton Reservoir. Atmospheric deposition is estimated to be 0.6 kg/d.

A summary of water quality data is presented in Table 20.5. The Croton reservoir is mesotrophic with intermediate levels of chlorophyll. It is highly stratified in the summer months from June to September with temperature differences exceeding 18°C. In 1993 hypolimnion hypoxia and anoxia lasted from June through November. Epilimnion dissolved oxygen concentrations were above 8 mg/L throughout the year.

Table 20.5   Croton Reservoir Water Quality Data for 1993

Variable	Units	Epilimnion		Hypolimnion	
		Summer	Winter	Summer	Winter
Temperature	°C	24.0	6.0	7.0	6.0
Chlorophyll *a*	μg/L	6.0	3.0		
pH		8.5	7.2	6.7	7.2
Dissolved Oxygen	mg O$_2$/L	9.0	11.0	0.0	11.0

20.5 A   Data Sets

Extensive water quality monitoring of the Croton and Muscoot Reservoirs has been carried out by the New York City Department of Environmental Conservation since 1976. The parameters that have been measured include both iron and manganese. Two sites representative stations are Station 5 which is located just upstream of the submerged original Croton Dam, and Station 6 which is located approximately 4.5 km downstream of the Muscoot Reservoir. For 1993, the data at Stations 5 and 6 were collected monthly from April to December. While many parameters such as dissolved oxygen and temperature were measured at 1 m depths, the sampling protocol for iron and manganese was different. Two grab samples were taken at a site: one sample above the thermocline and one sample below, near the reservoir bottom. Only dissolved iron and manganese were measured.

Monthly measurements of total iron and total manganese were made at the tributary inflows to the Muscoot and Croton Reservoirs. These data are used to generate the daily iron and manganese loads to the system. As all tributaries are aerobic, the influent iron and manganese loads are assumed to be entirely particulate. At present there are no sediment iron or manganese data available for the Croton Reservoir.

20.6   MODEL FRAMEWORK

The model of the Croton and Muscoot Reservoirs the water quality model has 34 state variables. The first 32 variables are the same as used in the Onondaga Lake model and are listed in Table 20.2. An additional 2 variables were added for manganese. Like iron, manganese is modeled as dissolved Mn(II) and particulate MnO$_2$(s) manganese as described in Chapters 18 and 19. Both the manganese and iron sediment models are included. This application provides a test of both models.

20.6 A   Manganese Flux Model Equations

The manganese flux model is virtually identical to that applied to the MERL mesocosm data set (Chapter 19, Eqs. 19.11 and 19.6). The only change is the addition of the reduction reaction of MnO$_2$(s) to Mn(II) in the water column if the hypolimnion becomes anoxic (Eq. 20.12). The water column equations for manganese are

Layer 0 Mn(II)

$$H_0 \frac{d\text{Mn}(0)}{dt} = s\big(f_{d1}\text{Mn}(1) - \text{Mn}(0)\big) - k'_{\text{Mn,OH},0} H_0 \left(10^{2(\text{pH}_0 - 7)}\right) \text{Mn}(0)$$

$$+ k_{\text{MnO}_2,0} H_0 \frac{K_{\text{M,MnO}_2,0}}{K_{\text{M,MnO}_2,0} + [\text{O}_2(0)]} \text{MnO}_2(0)$$

$$+ E_{0,-1}\big(\text{Mn}(-1) - \text{Mn}(0)\big) \tag{20.21a}$$

Layer 0 MnO$_2$(s)

$$H_0 \frac{d\text{MnO}_2(0)}{dt} = k'_{\text{Mn,OH},0} H_0 \left(10^{2(\text{pH}_0 - 7)}\right) \text{Mn}(0) + w_1 \big(\text{MnO}_2(-1) - \text{MnO}_2(0)\big)$$

$$- k_{\text{MnO}_2,0} H_0 \frac{K_{\text{M,MnO}_2,0}}{K_{\text{M,MnO}_2,0} + [\text{O}_2(0)]} \text{MnO}_2(0)$$

$$+ E_{0,-1}\big(\text{MnO}_2(-1) - \text{MnO}_2(0)\big) \tag{20.21b}$$

where $k_{\text{MnO}_2,0}$ is the rate at which MnO_2(s) is reduced to Mn(II), $K_{\text{M,MnO}_2,0}$ is the half-saturation constant for oxygen for this reaction, and $E_{0,-1}$ represents the vertical mixing coefficient between the bottom layer and the layer immediately above. The concentrations in the water column are numbered in reverse order to preserve the notation used previously. Thus layer -1 refers to the layer just above layer 0 in the water column. The remaining portions of the sediment model – the nutrient and oxygen components – are identical to those used previously.

Model Segmentation for Croton and Muscoot Reservoirs   The surface segmentation for the Croton and Muscoot model is shown in Figure 20.12. The model has 55 surface segments, 32 in the Croton Reservoir and 23 in the Muscoot. Vertically the model has 10 layers and uses a stretching coordinate system to adjust the layer depth to fit the varying bathymetry.

20.6 B   Model Results

Dissolved Oxygen   It is critical that the timing of the onset of anoxia be calculated correctly since this event triggers the release of iron and manganese from the sediment. As with the Onondaga Lake application, it was decided that actual dissolved oxygen data would be used to specify when anoxia was present. This allows us to examine how the iron and manganese model would perform if the eutrophication-dissolved oxygen model were perfectly calibrated to the dissolved oxygen data.

Calibration Scenario   The model was run for 1993. In order to ensure that the sediment had reached a steady state, the model was cycled for several years before using the results as the calibration output. The values of the constants used for the iron and manganese models in the water column and sediment are given Tables 20.3

Table 20.6 Parameters for the Manganese Flux Model

Parameter*	Units	Croton Reservoir
$k'_{Mn,OH,0}$	d^{-1}	10.0
$k_{MnO_2,0}$	d^{-1}	4.0
$K_{M,MnO_2,0}$	mg O_2/L	0.2
$k_{Mn,OH,1}$	$d^{-1}(mg\ O_2/L)^{-1}$	0.001
$k_{MnO_2,2}$	d^{-1}	4.0
$\Delta\pi_{Mn,1}$	–	200.0
$\pi_{Mn,2}$	L/kg	8000.0
$[O_2(0)]_{crit,Mn}$	mg O_2/L	2.0

*$\theta_{Mn,OH,0} = \theta_{MnO_2,0} = \theta_{Mn,OH,1} = \theta_{MnO_2,2} = 1.08$.

and 20.6. The iron model values are identical to those used for Onondaga Lake with one exception as noted below.

Calibration Results   As mentioned previously, the iron and manganese data for Croton Reservoir consisted of samples at two depths, which were collected approximately once a month. One sample was taken in the epilimnion, which was always oxic and hence yielded negligible dissolved iron and manganese. The hypolimnion sample was taken near the bottom. Therefore no comparison can be made to the computed vertical profile.

Figure 20.13 presents the calibration results for sites 5 and 6. The surface and bottom model (lines) and data (symbols) are shown for temperature and dissolved oxygen. The bottom dissolved iron and dissolved manganese concentrations are compared to the model computations for the bottommost layer of the model. The extended period of anoxia is evident, from early June until early November. For these simulations the observed time for the both onset and end of anoxia was used to correct the bottom layer dissolved oxygen computed by the model.

The iron data for station 5 (Fig. 20.13I) shows very large Fe^{2+} concentrations during anoxia, between 4 and 6 mg/L. The results for site 6 (II) show markedly lower iron concentrations, $\sim$ 1 mg/L for July and August, but $\sim$ 4 mg/L for early September. By comparison, Onondaga Lake bottom iron concentrations were $\lesssim$ 1 mg/L.

In order to reproduce this difference, it is necessary to understand what controls the maximum hypolimnion iron concentration. The following observation is useful. Within $\sim$ 14 days from the onset of anoxia, the bottom concentration will be approximately the same as the pore water iron concentration. For Onondaga Lake this was $\sim$ 1 mg/L. Hence, in order for the model to calculate 4 mg/L in the bottom layer of the Croton reservoir, the sediment partition coefficient for iron must be reduced in order to increase the pore water iron concentration. The results, which are shown in Fig. 20.13, utilize an anaerobic layer partition coefficient of $\pi_{Fe,2} = 100$ L/kg, whereas the value for Onondaga Lake is $\pi_{Fe,2} = 3000$ L/kg.

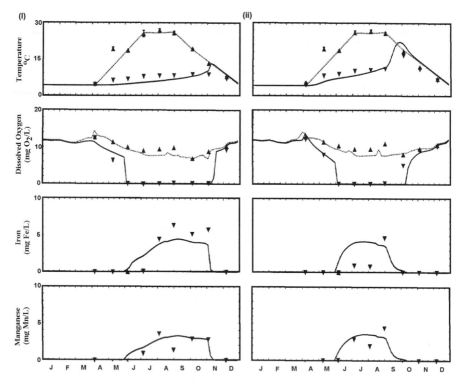

Fig. 20.13 Comparison of temperature, dissolved oxygen, iron, and manganese for (I) site 5 and (II) site 6, 1993. Model surface (dashed line) and bottom (solid line). Data for the epiliminion (▲) and hypolimnion (▼).

It is not clear why the pore water iron concentration in the Croton reservoir is larger, and the partition coefficient is lower, than in Onondaga Lake. This is the principal drawback associated with using an empirical partition coefficient in the sediment flux model to determine the pore water-solid phase speciation. There is no guarantee that it, or any other of the partition coefficients used in the model, are transferable from one application to another. We have seen this problem before in the phosphorus flux model (Chapter 15).

The manganese concentrations, by contrast, are approximately the same at site 5 and site 6. As can be seen, the model produces a good fit to the data with no need for further adjustments.

20.7 SUMMARY

The application of the iron flux model to the Croton reservoir highlights both the strengths and weaknesses of the model. Both the Onondaga Lake and Croton water column iron concentration profiles are reproduced with reasonable fidelity. Only the anaerobic layer sediment partition coefficient needed changing from the value used for Onondaga Lake in order to compute the higher iron concentrations observed in the Croton reservoir. The calibrated manganese model was able to reproduce the observed data without any further adjustments.

21

Cadmium and Iron

21.1 INTRODUCTION

The model presented in this chapter calculates the temporal and vertical variation of iron and cadmium sulfide, and sorbed and pore water cadmium (Di Toro et al., 1996b). It is based on the one-dimensional advective-dispersive mass balance equation. Unlike the models in the previous chapters, this model focuses, initially, on the vertical distributions of solid phase and pore water species. Transport by particle and pore water mixing is considered. The state variables are iron sulfide FeS, cadmium sulfide CdS, and the sorbed + pore water cadmium Cd_T. In addition, particulate organic carbon POC, iron oxyhydroxide FeOOH, and dissolved oxygen O_2, are explicitly modeled. The oxidation of the metal sulfides, the displacement of iron by cadmium in FeS to form CdS, and partitioning reactions to mimic solubility, are included.

The model is applied to three experimental data sets. Two of these are marine sediments with typical organic carbon and iron sulfide concentrations. The third employed a freshwater sediment with high POC and low FeS concentrations. The calibration produced similar partitioning parameters on an organic carbon basis for both sediments. However, the oxidation and partitioning behavior of the FeS in the freshwater sediment is distinctly different from that of the marine sediments.

21.2 TOXICITY OF METALS

The motivation for constructing this model comes from a need to understand the long-term fate of the sulfides of potentially toxic metals in sediments. The toxicity of certain cationic metals in sediments is strongly influenced by the quantities of the amorphous metal sulfides that are present.[*] The available sulfide is measured using a cold acid extraction that is called *acid volatile sulfide* (AVS). The AVS of a sediment is made up of the molar sum of the concentrations of iron sulfide (FeS) and the extracted portion of the other metal sulfides. The measure of potentially bioavailable metal in sediments is the total simultaneously extracted metal (ΣSEM). The ΣSEM is the sum of the metal concentrations associated with the *same* metal sulfides extracted by the AVS and any other metal bearing phase that is extracted in the cold HCl extraction used for AVS[†]. For example, metal sorbed to iron oxides and particulate organic carbon will also be extracted[‡]. It has been demonstrated that if the molar concentration of acid volatile sulfide (AVS) that is extracted from a sediment exceeds the molar sum of the simultaneously extracted metals (ΣSEM) that form more insoluble sulfides than iron sulfide – that is NiS, ZnS, CdS, PbS, and CuS, any one of which is denoted by MS – then those sediment metals will not be toxic to sediment dwelling organisms.[§]

This observation is quite important. The bulk concentration of metal in sediment bears no relationship to the toxicity observed in sediment toxicity tests. Fig. 21.1A presents the relationship of amphipod mortality – the organism used in the toxicity test – to the concentration of metal in the sediment on a dry weight basis. However, the ratio of SEM/AVS can predict the region of no organism mortality quite reliably, as shown in Fig. 21.1B. The reason is that for SEM/AVS $<$ 1, the excess AVS in the sediment ensures that there is no free metal available to the organism. All the metal is present as metal sulfide. Since the AVS and SEM of sediments are of critical importance in determining the toxicity of sediments, processes that create or destroy iron and metal sulfides are particularly important in determining the toxicity and fate of metals in sediments.

It is the seasonal cycle of SEM and AVS in natural sediments that is ultimately of interest. The practical question is: During the seasonal variation of AVS and SEM, can a sediment which is non-toxic (SEM $\leqslant$ AVS) during the period of high AVS concentration – usually the summer – become potentially toxic (SEM $>$ AVS) during periods of low AVS – usually the winter? This depends on the processes that control the quantities of FeS and MS in sediments. FeS is formed as a by-product of sulfate reduction – the oxidation of organic carbon with sulfate as the electron acceptor (Chapter 9). It is destroyed by oxidation with oxygen as the terminal oxidant. The quantity of MS that is present is controlled by a balance between its formation via the displacement reaction – which is discussed below – and its loss by oxidation.

---

[*]Di Toro et al. (1990a). [†]Allen et al. (1993). [‡]Tessier et al. (1979). [§]Berry et al. (1996), Di Toro et al. (1991a), Hansen et al. (1996a).

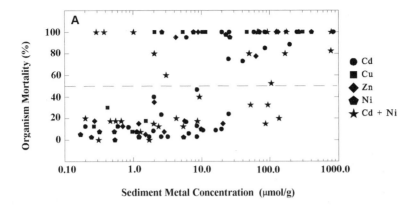

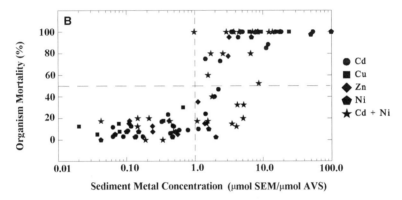

Fig. 21.1 Amphipod mortality versus (A) metal concentration and (B) SEM/AVS ratio for spiked sediment tests. Data from Berry et al. (1996), Di Toro et al. (1991a), Hansen et al. (1996a).

The purpose of this chapter is to present a model of the AVS and cadmium SEM in a sediment that includes these reactions. It is intended ultimately to be used in predicting the seasonal cycle of AVS and SEM as an aid in interpreting sediment quality criteria. The model, with some refinements, has been used to predict metal fluxes to the overlying water (Carbanaro, 1999).

21.3 MODEL STRUCTURE

The model structure is presented in Fig. 21.2. The reactions are indexed by (a), (b), and so on. Two diagrams are used, and the aerobic and anaerobic zones of the sediment are delineated as separate layers for expository purposes only. The model employs multiple vertical layers. The flux of particulate organic carbon (POC) to the sediment provides POC which can either oxidize aerobically using oxygen (a)

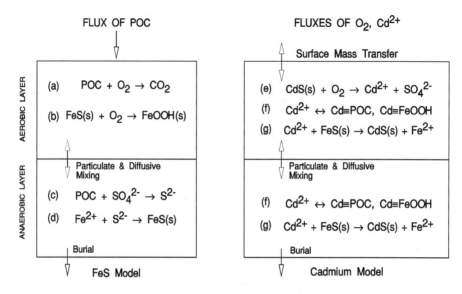

Fig. 21.2 Model schematic for the iron sulfide FeS, particulate organic carbon POC, and cadmium models.

or anaerobically using sulfate (c). The sulfide produced by sulfate reduction reacts with iron in the sediment (d) to produce iron sulfide FeS. Mixing of particulate and dissolved sulfide into the aerobic zone and the oxidation using oxygen (b) depletes FeS and produces iron oxyhydroxide FeOOH.

Cadmium sulfide is formed by the displacement reaction (g) in both the aerobic and anaerobic zones. It is oxidized in the aerobic zone (e) to produce dissolved cadmium Cd^{2+} in the pore water. Cadmium can sorb (f) to both the particulate organic carbon and the iron oxyhydroxide in the sediment – the complexes are denoted by Cd≡POC and Cd≡FeOOH respectively – to form particulate cadmium. Dissolved cadmium Cd^{2+} can also be lost to the overlying water by diffusion.

21.4   MODEL FRAMEWORK

The model is based on a one-dimensional vertical representation of a sediment. Mass balance equations are formulated that include the transport of particles and pore water by bioturbation, bioirrigation, and diffusion; and the production and destruction of the chemical species of interest by reactions or external sources. Models of this sort are quite commonly used to describe the vertical distribution of various sediment solid phase and pore water constituents.[†] The general form of the mass balance

[†]Boudreau (1991), Jahnke et al. (1994).

equation is

$$\frac{\partial [C(z)]}{\partial t} - \frac{\partial}{\partial z}\left((f_p D_p(z) + f_d D_d) \frac{\partial [C(z)]}{\partial z} \right) + w_s \frac{\partial [C(z)]}{\partial z} = \sum_j R_j \quad (21.1)$$

where $[C(z)]$ is the concentration of chemical per unit bulk volume of the sediment at depth z, f_p is the fraction of the chemical that is particulate, $D_p(z)$ is the diffusion coefficient for particle mixing, f_d is the dissolved fraction, and D_d is the diffusion coefficient for dissolved phase mixing. The loss by burial is represented by the sedimentation velocity w_s. For the applications presented below, the loss by burial is small and can be neglected so that $w_s = 0$. The term $\sum_j R_j$ is the sum of the sources $(+)$ and sinks $(-)$ of $C(z)$ due to reactions R_j. In order to simplify the notation, Eq. (21.1) will be represented by

$$\mathfrak{D}(f_p D_p + f_d D_d)[C] = \sum_j R_j \quad (21.2)$$

where the symbol $\mathfrak{D}(f_p D_p(z) + f_d D_d)$ represents the time and transport derivative terms with overall diffusion coefficient $D = f_p D_p(z) + f_d D_d$ applied to the concentration $[C]$

$$\mathfrak{D}(D)[C] \triangleq \frac{\partial [C(z)]}{\partial t} - \frac{\partial}{\partial z}\left(D \frac{\partial [C(z)]}{\partial z} \right) + w_s \frac{\partial [C(z)]}{\partial z} \quad (21.3)$$

Equations of this form will be used for each of the chemical species in the model which are listed in Table 21.1. The parameters used in the model are all listed in the Nomenclature at the end of the book.

Table 21.1 Model State Variables

State Variable	Name	State Variable	Name
POC$_1$	G$_1$ Carbon	FeS	Iron Sulfide
POC$_2$	G$_2$ Carbon	FeOOH	Iron Oxyhydroxide
O$_2$	Dissolved Oxygen	CdS	Cadmium Sulfide
		Cd$_T$	Dissolved + Sorbed Cadmium

21.4 A Particle and Pore Water Mixing

The model considers both particle and pore water mixing. Particle mixing by benthic animals is important because the oxidation of particulate iron and cadmium sulfide can only occur in the aerobic layer. Therefore, mixing of particles into that layer exposes more particles to oxidation and increases the rate of depletion.

The mixing is simulated using a conventional approach.[†] A diffusion coefficient is applied to particles that decreases exponentially in depth

$$D_p(z, t) = D_p(t)\theta_{D_p}^{(T-20)} \exp\left(-\frac{z}{z_{D_p}}\right) \tag{21.4}$$

Temperature dependency[‡] can be represented by the Arrhenius relationship using the θ approximation (Section 13.2 A). The e-folding depth z_{D_p} specifies the depth of bioturbation. The magnitude of the mixing is specified using a time dependent diffusion coefficient $D_p(t)$ which varies due to changes in benthic biomass (Fig. 13.1A[‡]). A logistic formulation

$$D_p(t) = D_p \frac{1}{1 + \exp\left(\ln(3)\dfrac{t_{D_p1/2} - t}{t_{D_p1/2} - t_{D_p1/4}}\right)} \tag{21.5}$$

is used to simulate the development of biomass in the colonization experiments to which this model is applied. The parameters are the times t to one-fourth $(t_{D_p1/4})$ and one-half $(t_{D_p1/2})$ of the maximum mixing, which are chosen from estimates of the biomass increase in the colonization experiments. The magnitude of the particle diffusion coefficient D_p can be related to the biomass of benthic animals present.[§] It can be as large as one-tenth of the pore water diffusion coefficient.

Pore water mixing is specified using a temperature dependent diffusion coefficient that is constant in depth

$$D_d(T) = D_d\theta_{D_d}^{(T-20)} \tag{21.6}$$

Bioirrigation would be included by adding a term that is proportional to the particle mixing formulation (Eq. 21.4). For the applications discussed below, no enhancement over molecular diffusion is observed.

21.4 B   Sediment Organic Carbon

The oxidation of organic carbon CH_2O with sulfate as the electron acceptor (Section 1.2 B)

$$2CH_2O + SO_4 \rightarrow S^{2-} + 2CO_2 + 2H_2O \tag{21.7}$$

is the source of sulfide S^{2-} that reacts with iron to produce iron sulfide

$$Fe^{2+} + S^{2-} \rightarrow FeS(s) \tag{21.8}$$

[†]Boudreau (1986a), Officer and Lynch (1989) . See Thoms et al. (1995) for a review of available models.
[‡]McCall and Tevesz (1982). [§]Matisoff (1982).

Organic carbon can also oxidize in the aerobic layer using oxygen

$$CH_2O + O_2 \rightarrow CO_2 + H_2O \tag{21.9}$$

Both of these reactions need to be considered, since the fraction of organic carbon that oxidizes anaerobically (Eq. 21.7) or aerobically (Eq. 21.9) affects the quantity of sulfide that is produced.

The rate at which organic carbon is oxidized depends on the electron acceptor (O_2 or SO_4) being used in the reaction. Also, with either electron acceptor, it has been found that the rate of decay is rapid at first, after which it proceeds much more slowly. A model that describes this behavior is the multiple G model (Chapter 12). The idea is that particulate organic carbon POC is made up of component parts that decay at differing first-order rates. The first fraction G_1 is the most reactive component and decays at a temperature dependent first-order rate $k_{POC,1}(T)$. The second fraction G_2 is a more slowly reacting component that decays at first-order rate $k_{POC,2}(T)$ and so on. For the purposes of this model, the final fraction G_3 is assumed to be conservative in the time scales being considered. Thus POC is composed of three fractions

$$POC = f_{POC,1}POC_{G_1} + f_{POC,2}POC_{G_2} + f_{POC,3}POC_{G_3} \tag{21.10}$$

where $f_{POC,1} + f_{POC,2} + f_{POC,3} = 1$. Modeling the decay of organic carbon involves modeling each of these G components.

The rate at which POC_{G_1} reacts depends on which electron acceptor (O_2 or SO_4) is being used and whether it is present in sufficient concentration for the reaction to occur. The normal way to formulate this dependency is to use a Michaelis-Menton expression for the oxygen dependency

$$R_{G_1,O_2} = k_{POC,1}(T)\frac{[O_2]}{[O_2] + K_{M,O_2}}[POC_1] \tag{21.11}$$

The square brackets [ ] denote concentrations. The Arrhenius temperature dependence is given by

$$k_{POC,1}(T) = k_{POC,1}\theta_{POC_1}^{T-20} \tag{21.12}$$

The oxidation rate R_{G_1,O_2} decreases if the concentration of oxygen falls below the half-saturation constant K_{M,O_2}. When this happens, organic matter is oxidized using sulfate as the electron acceptor at a first-order rate $k_{POC,1,SO_4}$. This reaction can be modeled using $1 - [O_2]/([O_2] + K_{M,O_2})$ which approaches 1 as $[O_2]/([O_2] + K_{M,O_2})$ approaches 0. That is

$$R_{G_1,SO_4} = k_{POC,1,SO_4}(T)\left(1 - \frac{[O_2]}{[O_2] + K_{M,O_2}}\right)[POC_1]$$

$$= k_{POC,1,SO_4}(T)\frac{K_{M,O_2}}{[O_2] + K_{M,O_2}}[POC_1] \tag{21.13}$$

where the second equality follows from an algebraic simplification of the term in parentheses. Thus sulfate reduction becomes important only as the Michaelis-Menton expression in Eq. (21.11) becomes small when the oxygen concentration falls below the half-saturation constant K_{M,O_2}.

The mass balance equation for POC_1 that includes both of these reactions (Eqs. 21.11 and 21.13) is

$$\mathfrak{D}(D_p)[POC_1] = -R_{G_1,O_2} - R_{G_1,SO_4} \qquad (21.14)$$

The source of POC_1 entering the sediment at $z = 0$ is the G_1 fraction $f_{POC,1}$ of the flux J_{POC}, that is $f_{POC,1}J_{POC}$. This is introduced into the surface layer of the model.

For POC_2, the G_2 component of particulate organic carbon, an analogous set of equations is written for the oxidation kinetics

$$R_{G_2,O_2} = k_{POC,2}(T)\frac{[O_2]}{[O_2] + K_{M,O_2}}[POC_2] \qquad (21.15)$$

$$R_{G_2,SO_4} = k_{POC,2,SO_4}(T)\frac{K_{M,O_2}}{[O_2] + K_{M,O_2}}[POC_2] \qquad (21.16)$$

and the mass balance equation

$$\mathfrak{D}(D_p)[POC_2] = -R_{G_2,O_2} - R_{G_2,SO_4} \qquad (21.17)$$

The source of POC_2 entering the sediment at $z = 0$ is $f_{POC,2}J_{POC}$. These equations model the POC dynamics in the sediment.

21.4 C Iron Monosulfide

The modeling of iron sulfide in sediments is complicated by the fact that, unlike particulate organic carbon, the dissolved species, Fe^{2+} and S^{2-}, are also important. The solubility of FeS is such that the interstitial water concentration of sulfide can approach millimolar concentrations (Morse et al., 1987). The oxidation of dissolved sulfide $\Sigma H_2S = H_2S + HS^- + S^{2-}$ will lower the interstitial water concentration of S^{2-} below that established by the solubility equilibrium

$$[Fe^{2+}][S^{2-}] = K_{sp} \qquad (21.18)$$

This will cause a dissolution of FeS, the reverse of Eq. 21.8 and, therefore, a potential loss of FeS by the further oxidation of ΣH_2S. Hence this mechanism is included in the model.

Partitioning Approximation   The method that is employed in order to simulate FeS dissolution is to partition total FeS (μmol/L-bulk) into particulate FeS_p (μmol/g-solids) and dissolved FeS_d (μmol/L-water) concentrations using a partition coefficient π_{FeS}. The equations are presented in Appendix 21A.1. The justification for the expedient of using a partition coefficient is simply that it is a simple and straightforward way to represent the fact that dissolved and particulate sulfides are both present. The alternative is to model the solution and solid phase chemistry in some detail. At minimum a model of the pH would be required, which is known to be a difficult problem in sediments (Boudreau, 1991, Marinelli and Boudreau, 1996). This, in turn, would necessitate the inclusion of the state variables required for its calculation which would further complicate an already complicated model.

The partitioning approximation has been used for both sulfide and phosphorus in Chapters 6 and 9. It is by no means a rigorous representation of the phenomenon, but it is easy to implement and replaces what would be a complicated computation involving many parameters – some of which may not be known very well – with a single empirical parameter π_{FeS}. A step in that direction has been made by prescribing the pH, using the equivalent solubility equation

$$[Fe^{2+}][HS^-] = K_{HS^-,S^{2-}}[H^+]K_{sp} \tag{21.19}$$

where $[HS^-]/[H^+][S^{2-}] = K_{HS^-,S^{2-}}$, and assuming $\Sigma H_2S = [HS^-]$ (Carbanaro, 1999).

Sulfide Oxidation The oxidation of FeS

$$FeS(s) + \frac{9}{4}O_2 + \frac{3}{2}H_2O \rightarrow FeOOH(s) + 2H^+ + SO_4^{2-} \tag{21.20}$$

destroys FeS and produces iron oxyhydroxide FeOOH which sorbs Cd^{2+} (Dzombak and Morel, 1990). The rates of oxidation of iron and other metal sulfides have been studied extensively at low pHs in the mining literature (Forssberg, 1985). However, the oxidation of FeS in more natural settings has not received much attention with one notable exception. Nelson (1978) studied the oxidation of synthetic FeS under a wide variety of conditions. Variations in pH, oxygen concentration, ionic strength, temperature, and the presence of catalytic metals were examined. His focus was on the initial rate of reaction. He proposed a surface complexation model that fit his experimental results quite successfully.

A model for the reaction kinetics of the entire time course of the reaction for synthetic FeS using Nelson's data, and for AVS in sediments from other experiments, is available (Di Toro et al., 1996a). Under certain conditions the oxidation is essentially first-order with respect to FeS concentration. The results indicate that the reaction rate is virtually the same for sediments and synthetic FeS at the same pH. This suggests that the experimental information that has been generated for the pH, temperature, and O_2 dependence of the oxidation rate is applicable to sediment AVS (Di Toro et al., 1996a). In particular, the iron oxidation rate is linear with respect to oxygen and iron sulfide concentration, and it has an Arrhenius temperature dependence

$$R_{FeS_p,O_2} = k_{FeS_p}(T)[O_2][FeS] \tag{21.21}$$

where

$$k_{FeS_p}(T) = k_{FeS}\theta_{FeS}^{T-20} \tag{21.22}$$

The reaction rate for dissolved sulfide is also known to be linear in the oxygen concentration and to have an Arrhenius temperature dependence (Boudreau, 1991, Morse et al., 1987)

$$R_{FeS_d,O_2} = k_{FeS_d}(T)[O_2][FeS] \qquad (21.23)$$

where the same formulation for temperature dependency (Eq. 21.22) is used.

Displacement Reaction   The displacement reaction

$$Cd^{2+} + FeS(s) \rightarrow CdS(s) + Fe^{2+} \qquad (21.24)$$

consumes FeS. Dissolved cadmium Cd^{2+} with concentration $f_{d,Cd_T}Cd_T$ titrates any available sulfide to form cadmium sulfide. The reaction occurs rapidly (Di Toro et al., 1990a). Thus the reaction rate expression needs only to depend on the concentration of the reactants

$$R_{Disp} = k_{Disp}\left(f_{d,Cd_T}[Cd_T]\right)[FeS] \qquad (21.25)$$

The reaction can proceed only if the concentrations of dissolved cadmium $f_{d,Cd_T}[Cd_T]$ and iron sulfide [FeS] are greater than zero. Therefore the reaction term is formulated as a product of these two concentrations and a reaction rate constant, k_{Disp}, which is set large enough to complete the reaction (Eq. 21.24).

Mass Balance Equation   The mass balance equation for total FeS that results using the partitioning approximation and the kinetics discussed above is

$$\mathcal{D}\left(f_{d,FeS}D_d + f_{p,FeS}D_p\right)[FeS] =$$
$$-R_{FeS_d,O_2} - R_{FeS_p,O_2} - R_{Disp} + a_{FeS,POC}\left(R_{G_1,SO_4} + R_{G_2,SO_4}\right)$$

$$(21.26)$$

The first two terms on the right-hand side represent the oxidation of dissolved and particulate sulfide (Eqs. 21.23 and 21.21). The third term is the loss via the displacement reaction (Eq. 21.25). The final term is the production of FeS via the oxidation of G_1 and G_2 carbon using sulfate as the electron acceptor (Eqs. 21.13 and 21.16). The stoichiometric coefficient ($a_{FeS,POC}$) is the amount of FeS produced per unit POC oxidized.

21.4D   Iron Oxyhydroxide

The result of the oxidation of FeS is the production of iron oxyhydroxide which we represent as FeOOH (goethite). The chemistry of iron is discussed in Section 20.2. The mass balance equation is

$$\mathcal{D}\left(D_p\right)[FeOOH] = R_{FeS_d,O_2} + R_{FeS_p,O_2} \qquad (21.27)$$

The source terms for the oxidation of dissolved and particulate FeS are from Eq. (21.26). The concentration of FeOOH is used to quantify the sorption of cadmium to this phase. This is discussed in more detail below.

21.4 E   Cadmium Sulfide

Since CdS is quite insoluble, the partitioning approximation employed for FeS is not necessary and only particulate CdS is considered. Cadmium sulfide is produced by the displacement reaction and destroyed by oxidation. We assume that the form of the oxidation kinetics is the same as for FeS (Eq. 21.21)

$$R_{CdS,O_2} = k_{CdS}(T)[O_2][CdS] \tag{21.28}$$

The mass balance equation for CdS is

$$\mathfrak{D}\left(D_p(z)\right)[CdS] = -R_{CdS,O_2} + R_{Disp} \tag{21.29}$$

The terms correspond to particle mixing (left hand side), oxidation of CdS (Eq. 21.28), and production of CdS by the displacement reaction (Eq. 21.25), respectively.

21.4 F   Dissolved and Sorbed Cadmium

The cadmium which is not present as CdS is assumed to be present as either dissolved cadmium in the pore water or as sorbed cadmium. The sum of these cadmium species is denoted by Cd_T. The fractions of dissolved (f_{d,Cd_T}) and particulate (f_{p,Cd_T}) cadmium are computed from a partitioning model described below. The production of Cd_T is via the oxidation of CdS. The loss is via the displacement reaction Eq. (21.24). The resulting mass balance equation is

$$\mathfrak{D}\left(f_{d,Cd_T}D_d + f_{p,Cd_T}D_p\right)[Cd_T] = R_{CdS,O_2} - R_{Disp} \tag{21.30}$$

Cadmium Partitioning   Cadmium partitioning is modeled in detail because it is important – the partitioning reactions are reasonably well understood – and it is not that difficult to do. Cadmium is assumed to partition to two solid phases: iron oxide and particulate organic carbon. The partitioning between the free sites and the aqueous phase is assumed to be linear. The result from these assumptions is a Langmuir model for each phase. The equations are presented in Appendix 21A.2.

21.4 G   Dissolved Oxygen

The reactions that influence the concentration of dissolved oxygen have all been formulated. It remains to gather the appropriate terms for the sinks of O_2

$$\mathfrak{D}(D_d)[O_2] = -a_{O_2,CdS}R_{CdS,O_2}$$

$$-a_{O_2,FeS}(R_{FeS_d,O_2} + R_{FeS_p,O_2}) - a_{O_2,POC}\left(R_{G_1,O_2} + R_{G_2,O_2}\right)$$

$$\tag{21.31}$$

They are the consumption of oxygen due to the oxidation of CdS (Eq. 21.28), dissolved and particulate FeS (Eq. 21.23 and 21.21), and the oxidation of G_1 and G_2 carbon (Eq. 21.11 and 21.15). The a terms are the stoichiometric coefficients: the quantity of oxygen consumed per unit oxidation of CdS, FeS, and POC, respectively.

21.5 SOLUTION METHOD

The solutions to these equations are obtained using an implicit finite difference approximation for each partial differential equation (Carnahan et al., 1969). The concentrations of the nonlinear terms and the other state variables in the kinetic expressions are lagged by one time step. For example, the resulting finite difference equation for POC_{G_1} is

$$\frac{[POC_1]_z^{t+\Delta t} - [POC_1]_z^t}{\Delta t}$$

$$= D_p\left(z + \frac{\Delta z}{2}\right)\frac{[POC_1]_{z+\Delta z}^{t+\Delta t} - [POC_1]_z^{t+\Delta t}}{\Delta z}$$

$$- D_p\left(z - \frac{\Delta z}{2}\right)\frac{[POC_1]_z^{t+\Delta t} - [POC_1]_{z-\Delta z}^{t+\Delta t}}{\Delta z}$$

$$- k_{POC,1}(T)\frac{[O_2]_z^t}{[O_2]_z^t + K_{M,O_2}}[POC_1]_z^{t+\Delta t}$$

$$- k_{POC,1,SO_4}(T)\frac{[K_{M,O_2}]}{[O_2]_z^t + K_{M,O_2}}[POC_1]_z^{t+\Delta t} \tag{21.32}$$

Note that in the kinetic terms the oxygen concentration $[O_2]_z^t$ is at time level t rather than at the implicit level $t + \Delta t$ for the state variable $[POC_1]_z^{t+\Delta t}$. The other finite difference equations are similar. The model vertical resolution Δz used for the computation is 1 mm, which is sufficient to resolve the sharp gradient of O_2 near the surface.

The integration time step for the calculation $\Delta t = 0.005 - 0.05$ d is chosen so that mass balance errors are typically less than 1%. The errors arise from the lagged terms in the numerical scheme (Eq. 21.32). The mass balances are computed as follows. The quantity of cadmium lost to the overlying water is computed from the gradient at the sediment water interface. This is compared to the quantity lost from the sediment itself, computed by summing the concentrations in each model layer. Similarly the change in oxygen equivalents in the sediment, computed by summation, is compared to the amount of oxygen that enters the sediment from the overlying water.

21.5 A   Parameter Values

The parameters and other inputs used in this model are either directly measured (e.g., temperature), abstracted from other sources, or obtained from fitting the model to the data presented below. The measured and abstracted parameters are listed in Table 21.2 for the three data sets. The primary source is the nutrient flux model which employs the same formulation for parts of the model. The other sources are listed.

The calibrated parameters are listed in Table 21.3. In principle the flux of organic carbon to the sediment J_{POC} can be measured using sediment traps (see discussion in Valiela 1995) or estimated from primary production measurements in the water column (Hargrave, 1984). The initial amount of G_1 and G_2 carbon, POC_1 and POC_2, are more difficult to measure. The former can be related to the chlorophyll in the sediment, whereas the latter can only be measured using long-term decomposition experiments, since it decays so slowly. As a rule of thumb, the ratios of G_1, G_2, and G_3 carbon in sediments are roughly 1:10:100 (see discussion in Chapter 12 and Table 12.5).

The oxidation rate for FeS has been measured for synthetic FeS (Nelson, 1978) and sediment AVS (Di Toro et al., 1996a). It is in the range of $k_{FeS} \approx 24$ d^{-1} or 3 d^{-1} $(mg\ O_2/L)^{-1}$. The value obtained from calibration is 1.0 d^{-1} $(mg\ O_2/L)^{-1}$. An initial experiment for synthetic CdS (Di Toro et al., 1996a) gave $k_{CdS} \approx 0.01$ d^{-1} or ~ 0.001 d^{-1} $(mg\ O_2/L)^{-1}$. The calibrated value for the marine sediments is 0.01 $(mg\ O_2/L)^{-1}$ and for the lake sediments is 0.0001 $(mg\ O_2/L)^{-1}$ which straddles the experimental value. The reaction rate k_{Disp} for the displacement reaction is set to a large number so the reaction is essentially complete.

The partition coefficient for FeS for the marine sediment is $\pi_{FeS} = 100$ L/kg which is the value used in the sulfide model in Chapter 9. A much larger value $\pi_{FeS} = 10^4$ L/kg is found for the lake sediment. Possible reasons are discussed below.

The calibrated sorption parameters for cadmium-organic carbon are similar for the three applications. The organic carbon normalized partition coefficients $\pi_{Cd,POC} = 3 \times 10^5$ and 7×10^4 (L/kg OC) and sorption capacity $\sigma_{Cd,POC} = 1.7 \times 10^3$ (μmol/g OC) are both within the range of values found for soils and sediments ($\pi_{Cd,POC} = 10^4 - 10^5$ L/kg OC, $\sigma_{Cd,POC} = 10^3 - 10^4$ μmol/g OC (Mahony et al., 1996). Detailed descriptions of the model applications are given next.

21.6   APPLICATIONS

The model is applied to three experimental data sets: a laboratory colonization experiment, a laboratory aquaria study, and a field colonization experiment. The designs are similar. Sediments are collected from a field site and cadmium is added in various amounts to produce ratios of Cd to AVS from below to above one. Since the additions are based on a previous estimate of the AVS, the actual and nominal ratios differ (Table 21.4). At various times during the experiment, AVS, SEM, and pore water cadmium are measured. The model is calibrated to these data.

Table 21.2 Parameter Values

Parameter	Experiment 1[a]	Experiment 2[b]	Experiment 3[c]	Units	References
D_d	0.75×10^{-4}	0.75×10^{-4}	0.75×10^{-4}	$m^2 d^{-1}$	(d)
θ_{D_d}	1.08	1.08	1.08	—	(d)
D_p	1.0×10^{-5}	1.0×10^{-7}	1.0×10^{-5}	$m^2 d^{-1}$	(d)
θ_{D_p}	1.117	1.117	1.117	—	(d)
$t_{Dp1/4}$	50.0	—	250.	d	(e)
$t_{Dp1/2}$	90.0	—	290.	d	(e)
z_{Dp}	50.0	50.0	20.0	mm	(f)
$k_{POC,1}$	0.35	0.35	0.35	d^{-1}	(d)
$\theta_{POC,1}$	1.08	1.08	1.08	—	(d)
$k_{POC,1,SO_4}$	0.035	0.035	0.035	d^{-1}	(d)
$\theta_{POC,1,SO_4}$	1.08	1.08	1.08	—	(d)
$k_{POC,2}$	0.018	0.018	0.018	d^{-1}	(d)
$\theta_{POC,2}$	1.08	1.08	1.08	—	(d)
$k_{POC,2,SO_4}$	0.0018	0.0018	0.0018	d^{-1}	(d)
$\theta_{POC,2,SO_4}$	1.15	1.15	1.15	—	(d)
K_{M,O_2}	0.1	0.1	0.1	mg O_2 L^{-1}	(d)
$f_{POC,1}$	0.65	0.65	0.65	—	(d)
$f_{POC,2}$	0.25	0.25	0.25	—	(d)
$f_{POC,3}$	0.10	0.10	0.10	—	(d)
ρ	2.5	2.5	2.5	kg/L	(d)
ϕ	0.9	0.9	0.9	L/L-bulk	(d)
θ_{FeS}	1.08	1.08	1.08	—	(g)
θ_{CdS}	1.08	1.08	1.08	—	(i)
$POC_3(z, 0)$	10.0	10.0	144.0	mg C g^{-1}	(i)
$AVS(z, 0)$	17.0	25.0	0.5	μmol g^{-1}	
$T(z, t)$	18.4	21.5	5.0	°C	(j)
$\pi_{Cd,FeOOH}$	10^3	10^3	10^3	L/mol	(k)
$\sigma_{Cd,FeOOH}$	0.2	0.2	0.2	mol/mol	(l)

[a]Hansen et al. (1996b), [b]Di Toro et al. (1996b), [c]Hare et al. (1994), [d] Nutrient flux model, [e]Estimated from the biomass buildup during the experiment, [f]Estimated from the benthic biomass (McCall and Tevesz, 1982), [g]Boudreau (1986a), Krishnaswami and Lal (1978), [h]Di Toro et al. (1996b), [i]Assumed $= \theta_{FeS}$, [j]Measured (Hansen et al., 1996b, Tessier et al., 1993) , [k]Measured (Hansen et al., 1996b, Hare et al., 1994), [l]Dzombak and Morel (1990).

Table 21.3 Calibrated Parameter Values

Parameter	Experiment 1	Experiment 2	Experiment 3	Units
J_{POC}	0.1	0.0^a	0.015	gC m^{-2}d^{-1}
$POC_1(z, 0)$	0.01	0.01	0.0015	mg C g^{-1}
$POC_2(z, 0)$	0.10	0.10	0.015	mg C g^{-1}
k_{FeS}	1.0	1.0	0.0032	d^{-1}(mg O$_2$ L^{-1})$^{-1}$
k_{CdS}	0.01	0.01	0.001	d^{-1}(mg O$_2$ L^{-1})$^{-1}$
k_{Disp}	10.0	10.0	10.0	d^{-1}(mmol/L)$^{-1}$
π_{FeS}	10^2	10^2	10^4	L/kg
$\pi_{Cd,OC}$	3×10^5	3×10^5	7×10^4	L/kg OC
$\sigma_{Cd,OC}$	1.7×10^3	1.7×10^3	1.7×10^3	μmol/g OC

^{a}Known from experimental design, no POC added.

21.6 A   Laboratory Colonization Experiment

The first application of the model is to a laboratory colonization experiment (Hansen et al., 1996b). Contaminated sediments in aquaria are continuously supplied with raw seawater. Larval forms of organisms progressively colonize the sediment. Replicate sediment cores were retrieved at 28, 57, and 118 days at the termination. These were sliced into 6 mm or 12 mm sections for the surface and deeper layers respectively and analyzed for SEM$_{Cd}$ and AVS. Pore water concentrations were measured using diffusional samplers at depths of 2 and 8 cm.

Comparison to Observations   The profiles of AVS are shown in Fig. 21.3. The symbols are the slice average concentrations. The solid lines are the model-computed AVS (= FeS + CdS), and the dashed lines are the model-computed CdS concentrations. The difference between these two lines is the model-computed FeS. The longer dashed lines near the sediment water interface are the model-computed dissolved oxygen profiles. The model overestimates the amount of AVS in the controls and the 0.1 treatment at 28 days. However, the remainder of the results are in general agreement with observations.

It is interesting to note that the reduction in surface FeS occurs quickly. By the 28th day the FeS (the difference between the AVS and CdS) in the top 1 to 2 cm. has decreased dramatically. By contrast, the AVS at depth is increasing slowly as the result of continuing POC mineralization and sulfate reduction. The cadmium sulfide in the surface layers has also decreased as can be seen from the dashed line in the 0.8 treatment and the solid line in the 3.0 treatment, where the solid and the dashed lines are colinear. Both of these reductions occur as a result of oxidation reactions in the oxic layer, the small surface layer where O$_2$ is positive.

The profiles for SEM$_{Cd}$ are shown in Fig. 21.4I. The solid and dashed lines are the model-computed SEM and CdS concentrations, respectively. The difference is Cd$_T$ the cadmium that has either been released as a result of the oxidation of CdS (0.1 and 0.8 treatments) or that, in addition, was there initially (3.0 treatment). The

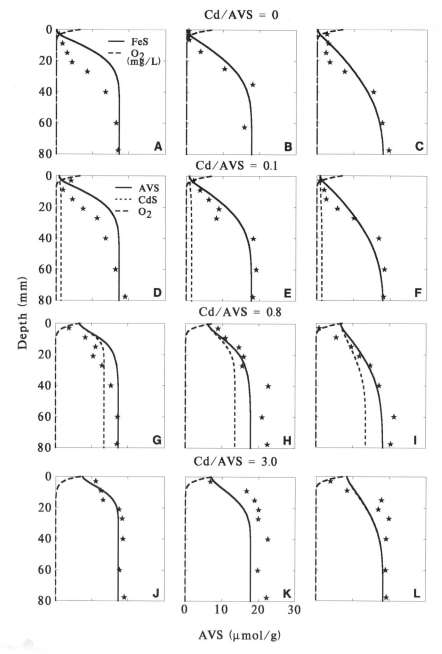

Fig. 21.3 Vertical distributions of acid volatile sulfide (AVS): Data (★) and model results. Line types are identified in (A) and (D). For (A)-(C) model results are for AVS = FeS concentrations. For the rest, the model results are for AVS = FeS + CdS. The concentrations of CdS are shown separately. Rows represent varying Cd/AVS ratios. Columns represent different times: (A, D, G, J) $t = 28$ d, (B, E, H, K) $t = 57$ d, (C, F, I, L) $t = 118$ d.

Table 21.4 Boundary and Initial Conditions

Variable	Experiment 1	Experiment 2		Experiment 3	Units
$O_2(0, t)$	7.88	7.5		12.8	mg O_2 L^{-1}
$Cd_d(0, t)$	0.0	t^*	$Cd_d(0, t)^*$	0.0	d, $\mu g\,L^{-1}$
		0.0	0.0		
		4.5	36.0		
		8.0	98.0		
		13.5	160.0		

SEM/AVS Ratios[†]

Experiment 1		Experiment 2		Experiment 3	
Nominal	Actual	Nominal	Actual	Nominal	Actual
0.0	0.0	0.4	0.4	0.0	0.0127
0.1	0.0823			0.1	0.0848
0.8	0.793			0.5	0.365
3.0	2.77			2.0	1.28
				10.0	6.96

[†] mol/mol. *Variable overlying water concentration $Cd_d(t)$ as a function of time t (d).

model reproduces the observations with the exception of the top 6 mm slice. These measurements are the most prone to distortions due to an uneven surface layer in the core. Therefore, some variation is expected. The discrepancy in Fig. 21.4D is inexplicable since the SEM appears to increase between day 28 (D) and day 57 (E).

The difference between SEM (the solid line) and CdS (the dashed line) is the total sorbed + pore water Cd_T. For the 0.1 and 0.8 ratios this is the cadmium released by oxidation. It increases as time progresses. For Cd/AVS = 3 it is initially the cadmium in excess of the initial CdS. In this case, the difference decreases, since the flux of cadmium to the overlying water is larger than the rate of production of Cd_T by oxidation of CdS.

The computed SEM/AVS ratio is compared to the observations in Fig. 21.4II. With the exception of certain instances in the top layer, the model correctly reproduces the increases of SEM/AVS in the top 1 to 2 cm. The initial SEM/AVS is constant from the top to the bottom of the sediment at the start of the experiment. Thus the extent of increase is most easily seen by comparing the values near the surface of the sediment, where oxidation of the metal sulfides is occurring, to the values at depth that are only slightly affected by the production of AVS due to organic carbon mineralization.

The SEM/AVS increases for the 0.1 and 0.8 ratios as CdS oxidizes. With the exception of the first slice, the data also increase. For 0.8 the SEM/AVS increases to greater than one, indicating that toxicity is a possibility. In fact an analysis of the fauna that colonized the sediment indicated that it was statistically different from the control and 0.1 treatments (Hansen et al., 1996b).

The measured pore water distributions are compared to observations in Fig. 21.5. Data were collected at four sampling times at depths of 2 and 8 cm using diffusional samplers. The data are compared to the model results for the surface, the depth average, and the bottom concentrations. The data are only roughly comparable to the model calculations. For the Cd/AVS = 0.1 treatment (Fig. 21.5A) the model is computing much lower concentrations than are observed. Concentrations of 10 μg/L are computed only in the surface layer of the model. The large observed concentrations, which were also observed in the controls (Fig. 21.5A), are not consistent with pore water data from the other two experiments discussed below. They are likely the result of contamination. For the Cd/AVS = 0.8 treatment, the data are roughly comparable with the sediment average concentration. For the Cd/AVS = 3.0 treatment, the bottom and average concentrations are consistent with the observations, although the model computes a more rapid loss of pore water cadmium via diffusion to the overlying water than is observed (Fig. 21.5C).

State Variables The time-depth profiles of the model state variables which produce these results are examined next. The concentrations of POC_{G_1} and POC_{G_2} as a function of depth and time are shown in Fig. 21.6A, B. The flux of POC to the sediment J_{POC} results in an increase in the surface layers for both G_1 and G_2. Particle mixing causes a gradual increase in the lower layers. Since the mineralization of POC_{G_1} is more rapid than POC_{G_2} the latter increases more rapidly in the lower layers than POC_{G_1}.

Figure 21.7I presents the FeS, FeOOH, and CdS time-depth profiles for the 0.8 and 3.0 treatments. The FeS distributions are shown in Fig. 21.7IA and B. Note the reversal of the depth axes used in Fig. 21.6. Initially the concentration in the 0.8 treatment is constant in depth as shown. Then the FeS in the surface layers decreases due to oxidation and it increases in the bottom layer due to POC mineralization and FeS precipitation.

The situation for the Cd/AVS = 3.0 treatment is quite different. Initially there is no FeS present since cadmium was added in excess of the FeS and all the FeS was converted to CdS by the displacement reaction (Eq. 21.24). As time passes FeS is produced in the lower layers by POC mineralization and precipitation. It is then converted to CdS by the displacement reaction (Eq. 21.24) so that the concentration of FeS is essentially zero. However, near the surface, a small amount of FeS forms that escapes conversion due to the lower concentrations of dissolved cadmium (Fig. 21.7II F).

The FeOOH that forms as a result of FeS oxidation is shown in Fig. 21.7I C, D (depth axes as in Fig. 21.6). More FeOOH forms in the 0.8 treatment than in the 3.0 treatment since more FeS oxidizes in the 0.8 treatment. The reason is that there is more initially there – compare Fig. 21.7I C to D – than in the 3.0 treatment.

The distributions of CdS are shown in Fig. 21.7I E and F. The initially constant distribution is modified as CdS is oxidized in the surface layer. Since the depth distribution of O_2 is roughly the same in both cases (Fig. 21.3), the oxidation produces

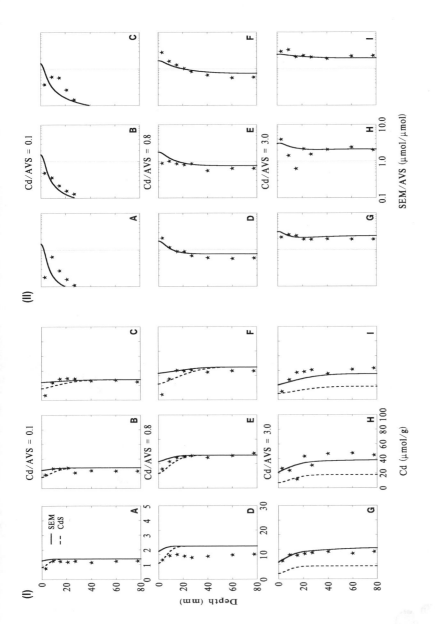

Fig. 21.4 Vertical distributions of (I) simultaneously extracted metal (SEM), and (II) SEM to AVS ratio: Data (★) and model results. Line types are identified in (IA). Rows represent varying Cd/AVS ratios. Columns represent different times: (A, D, I) $t = 28$ d, (B, E, H) $t = 57$ d, (C, F, I) $t = 118$ d.

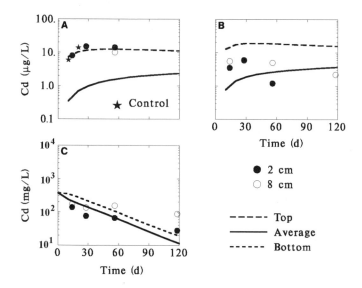

Fig. 21.5 Time course of pore water concentrations: Data from the control (★), 2 cm (●), and 8 cm (○) depth. Model results as lines identified in the legend. (A) Cd/AVS = 0.1, (B) Cd/AVS = 0.8, (C) Cd/AVS = 3.0.

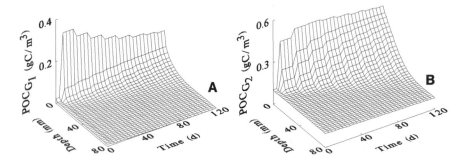

Fig. 21.6 Time-depth-concentration plots. (A) Particulate organic carbon POC_{G_1}, (B) POC_{G_2} for the control treatment.

a similar proportional reduction in CdS concentration in the surface layers of the sediments. Particle mixing brings CdS to the oxic interface and causes it to decrease in depth as time passes.

The distributions of dissolved and sorbed cadmium are presented in Fig. 21.7II. More cadmium is sorbed to the organic carbon in the 3.0 treatment (Fig. 21.7II B) than the 0.8 treatment (II A) because initially cadmium exceeded FeS in the 3.0 treatment. The similarity of the sorbed (II A) and (II C) concentrations and the dissolved concentration (II E) is because the partitioning is in the linear range of the Langmuir sorption model. By contrast in the 3.0 treatment, the Cd sorbed to

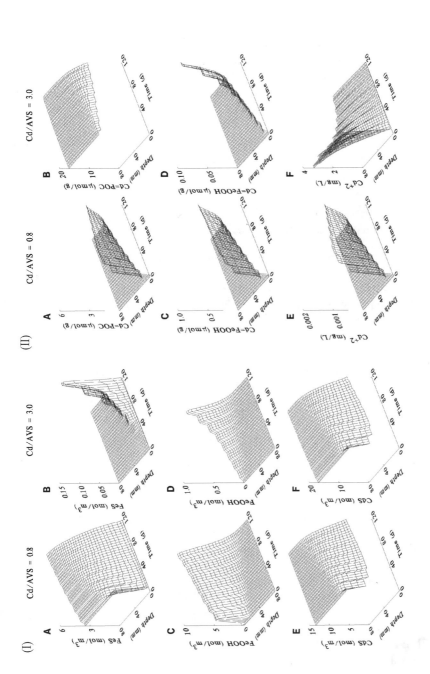

Fig. 21.7 Time-depth-concentration plots. Columns for Cd/AVS = 0.8 and 3.0. (I)(A, B) concentrations of iron sulfide FeS, (C, D) iron oxyhydroxide FeOOH, and (E, F) cadmium sulfide CdS. (II) (A, B) cadmium sorbed onto particulate organic carbon, (C, D) and onto iron oxyhydroxide. (E, F) cadmium pore water concentration.

POC (II B) is at the sorption capacity (17 μmol/g) until the dissolved concentration decreases in the surface layers, at which time the sorbed concentration decreases. The pore water cadmium concentration (II F) is initially constant in depth, and then decreases as Cd diffuses to the overlying water.

The cadmium sorbed to iron oxyhydroxide (Fig. 21.7II C and D) is smaller than that sorbed to POC (compare to II A and B). The shapes of the profiles are similar to the FeOOH profiles (Fig. 21.7I C and D – note the depth scale reversal).

The pore water profiles are shown in Fig. 21.7II E and F. Concentrations are highest in the surface layers for the 0.8 treatment since cadmium is produced there by oxidation. For the 3.0 treatment the reverse is true since the pore water cadmium is initially present in the pore water at the same concentration. It is then progressively lost from the surface layer to the overlying water so that it decreases both in the surface layer and progressively with depth as shown.

21.6 B   Laboratory Bioturbation Experiment

The second experiment employed a marine sediment (Long Island Sound) held in laboratory aquaria. The overlying water was not changed during the experiment. The experiment was designed to emphasize the effect of bioturbation. The treatments consisted of unmanipulated sediments (controls) and spiked sediments at SEM/AVS = 0.4 with and without the polychaete *Neanthes arenaceodentata*. The organism density was 60 juveniles/aquarium (1 individual/cm^2 = 1.24 g dry wt/m^2) a value likely to occur in the field. The experiment was terminated at $t = 30$ d. The measurements were made as in the laboratory colonization experiment described above.

The vertical profiles for the controls and the treatments, with (filled symbols) and without (unfilled symbols) animals are shown in Fig. 21.8 for the solid phase and Fig. 21.9 for the pore water. The lines are computed using the same oxidation coefficients as in the colonization experiment. However, the particle mixing coefficient corresponds to the known polychaete biomass and its relationship to diffusion coefficient (Chapter 13, Fig. 13.1A). Particle mixing is set to zero for the case without polychaetes.

The effect of this magnitude of bioturbation is small as can be seen from the differences between the filled and unfilled symbols. The model computations are virtually identical with the exception of the pore water result as discussed below. The AVS distributions for the unspiked (Fig. 21.8A) and the spiked (B) sediments are quite similar to each other. The SEM distribution for the spiked sediment (C) is nearly constant. The reason is that the overlying water in this experiment was static and accumulated the cadmium that diffused out of the sediment (Table 21.4). This can be seen in the distribution of the pore water (Fig. 21.9A), which has increased to 160 μg/L. This is a different result from the laboratory colonization experiment (Fig. 21.7II E and F) where the overlying water was rapidly replaced.

The pore water profile (Fig. 21.9A) shows a rapid decline to essentially zero by $z = 3$ mm, in striking contrast to the lab colonization experiment (Fig. 21.5A, B). The model results with and without bioturbation are shown for two values of the displacement reaction rate coefficient k_{Disp} (Eq. 21.25) in Fig. 21.9. The larger value

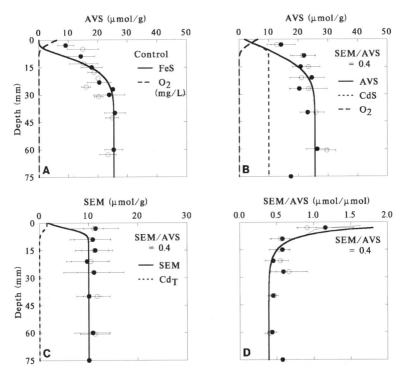

Fig. 21.8 Vertical distribution of the AVS in the control (A), the treatment AVS (B), the SEM (C), the SEM to AVS ratio (D). Data for the treatments with (•) and without (○) bioturbation: Mean ± standard deviation. Model results as the lines. Model results are virtually identical for the bioturbation or no bioturbation cases.

(B) causes the reaction (Eq. 21.24) to go to completion and nearly reproduces the observed results. It may be fortuitous, but the slightly sharpened profile computed without bioturbation (dashed line) corresponds to the slightly higher pore water concentration observed at the interface. This pore water profile, collected with a diffusion sampler with more resolution (Carignan et al., 1985), is more consistent with model computations and with the results found in the laboratory colonization experiment (Fig. 21.5).

21.6 C   Freshwater Field Colonization Experiment

The third experiment is a freshwater field colonization experiment (Hare et al., 1994). It was conducted in an oligotrophic lake where the bottom water temperature remains at 5°C throughout the year. Trays of spiked sediments were placed into a lake from which the sediments were initially obtained. The experiment began in August 1990. Cores and diffusional samplers were retrieved in May, August, and October of the following year. Pore water cadmium profiles and solid phase SEM and AVS were measured.

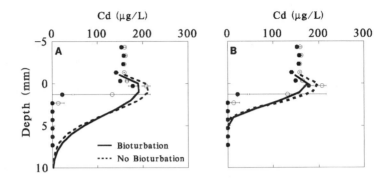

Fig. 21.9 Vertical distribution of the dissolved cadmium in the pore water. Data for the treatments with (•) and without (○) bioturbation: Mean ± standard deviation. Model results are lines with line types identified in A. (A) $k_{Disp} = 10$ (d-mmol/L)$^{-1}$. (B) $k_{Disp} = 10^6$ (d-mmol/L)$^{-1}$.

The initial AVS concentration is quite low (0.5 μmol/g) and decreases to approximately 0.2 μmol/g during the year-long experiment. The AVS and SEM results are shown in Fig. 21.10. Cadmium SEM, AVS, and SEM/AVS ratio are presented for the three sampling periods. The data for all the treatments SEM/AVS = 0.0, 0.1, 0.5, 2.0, and 10.0 are presented in each figure with alternating filled and unfilled symbols.

In order to reproduce the observations, it is necessary to lower the CdS and FeS oxidation coefficients k_{FeS} and k_{CdS} by two and three orders of magnitude respectively (Table 21.3). For FeS, there is an indication that at low concentration, the oxidation rate decreases (see Fig. 5 in Di Toro et al., 1996a). This may represent the oxidation of another, more resistant, component of AVS. That is, AVS is known to consist of amorphous FeS, mackinawite, and, possibly, greigite (Morse et al., 1987), which may be the more slowly oxidized component.

As a consequence of the low oxidation rate and low organic carbon flux – corresponding to the oligotrophic state of the lake – the model predicts oxygen penetration to 80 mm (the dotted lines in Fig. 21.10B).

Cadmium SEM concentrations (Fig. 21.10A) are observed and computed to remain constant except in the top centimeter. The decrease reflects the loss of cadmium via a diffusive flux to the overlying water. AVS concentrations (Fig. 21.10B) increase with respect to depth. However the model is unable to compute the observed changes at depth for the different treatments – the spread of the data points at 75 mm – and still maintain observed concentrations near the surface. The oxidation rate for FeS is chosen as a compromise to approximate the surface and bottom observations. SEM/AVS concentrations (Fig. 21.10C) increase for the smallest treatments but not for the largest, and the model reproduces this observation.

Pore water concentrations are presented in Fig. 21.11. Each row represents a treatment; the columns are for the May, August, and October sampling. The data are the triplicate measurements. The partitioning model roughly reproduces the data from the top 20 mm for all the treatments. But it does not reproduce the sharply lower

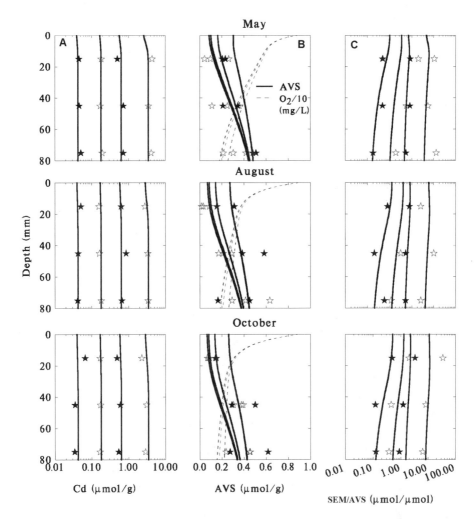

Fig. 21.10 Vertical distributions of (A) SEM, (B) AVS and dissolved oxygen, and (C) SEM to AVS ratio (columns) for the three sampling times (rows). Data (★) alternately filled and unfilled for the sequence of treatments: Cd/AVS = 0.0, 0.1, 0.5, 2.0, 10.0. Lines are the model results, corresponding to each of the treatments. The dotted line (second column) is the modeled O_2 concentration divided by 10 to conform to the plotting scale.

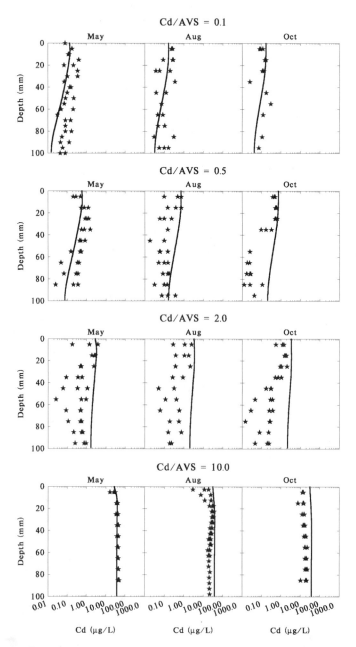

Fig. 21.11 Vertical distributions of the pore water cadmium concentration: data (★) for each of the three triplicates and model results (line). Columns represent different times. Rows represents the different Cd/AVS ratios.

concentrations at depth ($>$ 40 mm) for the intermediate treatments (0.5 and 2.0). The situation is especially mysterious for the 2.0 treatment. With SEM exceeding AVS, free Cd should exist for all depths. Therefore, for partition coefficients that are consistent with the remaining data, the model predicts between 1 and 10 μg/L of dissolved cadmium. The data are all below the predictions especially at depth. There appears to be strong binding sites that are active at low concentrations ($<$ 1 μg/L) which are not modeled by the Langmuir sorption model. It is encouraging that the surface pore water data are reasonably well reproduced, which is the zone of likely benthic organism exposure.

21.7 CONCLUSIONS

A model has been presented for the temporal and vertical variation of AVS, and cadmium SEM and interstitial water concentrations in sediments. It is based on the one-dimensional advective-dispersive mass balance equation. Transport by particle and pore water mixing is considered. The diffusive flux to the overlying water is also included. The state variables include the components of AVS and SEM, namely FeS, CdS, and sorbed + pore water cadmium Cd_T. In addition, particulate organic carbon, iron oxyhydroxide, and dissolved oxygen are explicitly modeled. The oxidation, displacement, and partitioning reactions which are sources and sinks of the state variables are included.

A model of this complexity has a large number of parameters for which values must be specified. Most of these are obtained from the previously calibrated sediment flux model or other reported values. The remaining parameters are evaluated by calibration to three experimental data sets. Two of these used a marine sediment of representative organic carbon and AVS concentrations. The third was a freshwater sediment with a high POC and low AVS.

For these sediments, similar partitioning parameters on an organic carbon basis are suitable for all the data sets. However, the oxidation and partitioning behavior of the FeS in the low AVS freshwater sediment was distinctly different from the high AVS cases, and from the reaction rates determined from suspensions of both synthetic and natural sediments. The experiments displayed a small fraction of AVS which did not oxidize rapidly. This suggests that AVS is made up of at least two distinct components and that the lake sediment was composed of this resistant component.

The model successfully reproduces the major features of the experiments: the decrease in AVS and SEM and an increase in SEM/AVS ratio. It is less successful in reproducing the pore water cadmium concentrations. For the freshwater experiment, the model simulates the surface layer concentrations to within an order of magnitude.

The computed partitioning to the iron oxyhydroxide generated by the oxidation of FeS is not as significant as that computed partitioning to particulate organic carbon. This follows from using laboratory sorption constants for the iron phase and calibrating the POC sorption constants. Since the latter are in the range of experimentally determined POC sorption constants, they appear to be reasonable. However, since no measurements of reactive iron oxides are available for comparison to model computations, the conclusion as to the importance of the iron phase and, in fact, this entire portion of the model, is necessarily speculative.

An additional feature described by this model – the increase in AVS as time passes – is observed to some extent in the bottom layers of the sediment in the lab colonization experiment (Fig. 21.3 – compare G, H, I and J,K,L). It is the result of continuing sulfate reduction. In other field and laboratory experiments, this phenomenon is seen more clearly.*

Finally, the reverse phenomenon to that which motivated the development of this model has also been seen: the SEM/AVS progressively *decreasing* from a ratio greater than one to essentially one for cadmium[†] and zinc.* The production of FeS (Eqs. 21.7 and 21.8) and the formation of the metal sulfide by the displacement reaction (Eq. 21.24) causes this to occur.

The further development of the model requires a set of experiments which includes measurements of the organic carbon input and initial conditions, the reactive iron concentrations, and pore water chemical species that are necessary to improving the modeling of iron sulfide partitioning. A seasonal study in which the cycles of the state variables and the fluxes to and from the sediment are monitored would be desirable.

Appendix 21A: Partitioning Equations

A.1   FeS PARTITIONING

The equations for iron sulfide partitioning are presented below. The mass balance equation is

$$[\text{FeS}] = \phi\text{FeS}_d + \rho_s(1 - \phi)\text{FeS}_p \qquad (21\text{A}.1)$$

where ϕ (L-water/L-bulk) is the porosity and ρ_s (g/L-solids) is the density of the solid particles. The porosity ϕ and density ρ_s convert the units to mmol/L-bulk volume from mmol/L of pore water for [FeS$_d$] and mmol/kg-solid phase for [FeS$_p$]. The partitioning relationship is

$$\pi_{\text{FeS}} = \frac{[\text{FeS}_p]}{[\text{FeS}_d]} \qquad (21\text{A}.2)$$

The partition coefficient is used to compute the fraction of the total FeS that is dissolved

$$f_{d,\text{FeS}} = \frac{\phi[\text{FeS}_d]}{[\text{FeS}]} = \frac{\phi}{\phi + \rho_s(1 - \phi)\pi_{\text{FeS}}} \qquad (21\text{A}.3)$$

---

*DeWitt et al. (1996), Liber et al. (1996). [†]DeWitt et al. (1996).

and particulate

$$f_{p,FeS} = \frac{\rho_s(1-\phi)[FeS_p]}{[FeS]} = \frac{\rho_s(1-\phi)\pi_{FeS}}{\phi + \rho_s(1-\phi)\pi_{FeS}} = 1 - f_{d,FeS} \qquad (21A.4)$$

The porosity ϕ and density ρ_s are included to conform to the usual partitioning formulation. However, since π_{FeS} is a purely empirical quantity, the inclusion of these terms is of no real importance other than to maintain dimensional consistency.

A.2   CADMIUM PARTITIONING

Three mass balance equations describe the situation. The mass balance equation for total cadmium is the sum of dissolved cadmium $[Cd^{2+}]$, cadmium sorbed to iron oxide $[Cd{\equiv}FeOOH]$, and cadmium sorbed to particulate organic carbon $[Cd{\equiv}POC]$

$$[Cd_T] = \phi[Cd^{2+}] + \rho_s(1-\phi)[Cd{\equiv}FeOOH] + \rho_s(1-\phi)[Cd{\equiv}POC] \quad (21A.5a)$$

The porosity ϕ and density ρ_s convert the units to mmol/L-bulk volume from mmol/L of pore water for $[Cd^{2+}]$ and mmol/kg-solid phase for $[Cd{\equiv}FeOOH]$ and $[Cd{\equiv}POC]$. The mass balance equation for the total sorption sites on the particulate organic carbon $[{\equiv}POC]_T$ is

$$[{\equiv}POC]_T = [{\equiv}POC] + [Cd{\equiv}POC] \qquad (21A.5b)$$

the sum of the free $[{\equiv}POC]$ and cadmium bound $[Cd{\equiv}POC]$ sites. The analogous equation for total sorption sites on the iron oxide is

$$[{\equiv}FeOOH]_T = [{\equiv}FeOOH] + [Cd{\equiv}FeOOH] \qquad (21A.5c)$$

The distribution between the dissolved and bound cadmium components is obtained from the mass action law applied to the sorption reaction

$$Cd^{2+} + {\equiv}POC \rightleftarrows Cd{\equiv}POC \qquad (21A.5d)$$

so that

$$\frac{[Cd{\equiv}POC]}{[Cd^{2+}][{\equiv}POC]} = \pi_{Cd,POC} \qquad (21A.5e)$$

where $\pi_{Cd,POC}$ is the partition coefficient to POC. The result for iron oxide is the same

$$\frac{[Cd{\equiv}FeOOH]}{[Cd^{2+}][{\equiv}FeOOH]} = \pi_{Cd,FeOOH} \qquad (21A.5f)$$

where $\pi_{Cd,FeOOH}$ is the partition coefficient to FeOOH.

The site density $[\equiv POC]_T$ for POC is computed from the total $[POC] = [POC_{G_1}] + [POC_{G_2}] + [POC_{G_3}]$, and the specific binding capacity ($\sigma_{Cd,POC}$) for POC

$$[\equiv POC]_T = \sigma_{Cd,POC}[POC] \tag{21A.6}$$

where $\sigma_{Cd,POC}$ is the concentration of sorption sites per unit organic carbon. Similarly the site density $[\equiv FeOOH]_T$ for FeOOH is computed from the total $[FeOOH]$ and the specific binding capacity ($\sigma_{Cd,FeOOH}$) for FeOOH

$$[\equiv FeOOH]_T = \sigma_{Cd,FeOOH}[FeOOH] \tag{21A.7}$$

Equations (21A.5a–21A.7) can be solved for the concentration of sorbed cadmium as a function of dissolved cadmium to yield the Langmuir isotherm equations

$$[Cd\equiv POC] = \frac{\pi_{Cd,POC}[Cd^{2+}]\sigma_{Cd,POC}[POC]}{1 + \pi_{Cd,POC}[Cd^{2+}]} \tag{21A.8}$$

and

$$[Cd\equiv FeOOH] = \frac{\pi_{Cd,FeOOH}[Cd^{2+}]\sigma_{Cd,FeOOH}[FeOOH]}{1 + \pi_{Cd,FeOOH}[Cd^{2+}]} \tag{21A.9}$$

For the model computations, however, it is necessary to solve for the dissolved cadmium concentration $[Cd^{2+}]$ from the total cadmium concentration $[Cd]_T$. A cubic equation in $x = [Cd^{2+}]$ results (Appendix 21B)

$$q_3 x^3 + q_2 x^2 + q_1 x + q_0 = 0 \tag{21A.10}$$

where

$$q_3 = -\pi_{Cd,FeOOH}\pi_{Cd,POC}\phi \tag{21A.11a}$$

$$q_2 = \pi_{Cd,FeOOH}\pi_{Cd,POC}(\phi - 1)\rho_s([\equiv FeOOH]_T + [\equiv POC]_T) \tag{21A.11b}$$
$$-\phi(\pi_{Cd,FeOOH} + \pi_{Cd,POC}) + \pi_{Cd,FeOOH}\pi_{Cd,POC}[Cd]_T$$

$$q_1 = \pi_{Cd,POC}(\phi - 1)\rho_s([\equiv POC]_T) \tag{21A.11c}$$
$$+\pi_{Cd,FeOOH}(\phi - 1)\rho_s([\equiv FeOOH]_T)$$
$$+(\pi_{Cd,FeOOH} + \pi_{Cd,POC})[Cd]_T - \phi$$

$$q_0 = [Cd]_T \tag{21A.11d}$$

Note that Eqs. (21A.11a–21A.11c) are symmetric with respect to an interchange between the POC and FeOOH terms, the reason being that the sorption equations for these two phases have the same form. The cubic Eq. (21A.10) is solved using the known analytical solution (Press et al., 1989).

Once the dissolved concentration is computed, the dissolved f_{d,Cd_T} and particulate f_{p,Cd_T} fractions can be calculated

$$f_{d,Cd_T} = \frac{[Cd^{2+}]}{[Cd]_T} \tag{21A.12}$$

$$f_{p,Cd_T} = 1 - f_{d,Cd_T} \tag{21A.13}$$

These are used in the mass balance Eqs. (21.30).

Appendix 21B: MACSYMA

The solution of the five simultaneous equations (21A.5), which are nonlinear due to the mass action equations (Eq. 21A.5e and 21A.5f), is a straightforward exercise with MACSYMA. The output is listed in Fig. 21B.1. The first five lines (c3) to (c7) assign the mass balance equations to the variables eq1 to eq3, and the mass action equations to eq4 and eq5. (c8) is a string of commands in a block separated by commas (, ,). The equation set eqt is named. The variables Cd[POC], Cd[FeOOH], POC[f], FeOOH[f], are eliminated sequentially. This leaves a final equation in Cd[d], the dissolved cadmium concentration, which the powers command identifies it as a cubic. The coefficients of each of the powers of Cd[d] are found using the ratcoef command. These are listed in Eqs. (21A.11).

Total Cadmium, POC, and FeOOH Mass Balance

(c3)   eq1:Cd[T]=Cd[d]*phi+Cd[POC]*rho*(1-phi)+Cd[FeOOH]*rho*(1-phi)

(d3) $$cd_t = (1 - \phi)\, cd_{poc}\, \rho + cd_{feooh}\, (1 - \phi)\, \rho + cd_d\, \phi$$

(c4)   eq2:POC[T]=POC[f]+Cd[POC]

(d4) $$poc_t = cd_{poc} + poc_f$$

(c5)   eq3:FeOOH[T]=FeOOH[f]+Cd[FeOOH]

(d5) $$feooh_t = cd_{feooh} + feooh_f$$

(c6)   eq4:Cd[POC]=pi[Cd,POC]*Cd[d]*POC[f]

(d6) $$cd_{poc} = pi_{cd,\,poc}\, cd_d\, poc_f$$

(c7)   eq5:Cd[FeOOH]=pi[Cd,FeOOH]*Cd[d]*FeOOH[f]

(d7) $$cd_{feooh} = pi_{cd,\,feooh}\, cd_d\, feooh_f$$

Solve the Equations Simultaneously

(c8)   (eqt:[eq1,eq2,eq3,eq4,eq5],mdeq1:eliminate(eqt,[Cd[POC]]),
mdeq2:eliminate(ratexpand(mdeq1),[Cd[FeOOH]]),
mdeq3:eliminate(ratexpand(mdeq2),[POC[f]]),mdeq3[1]:mdeq3[1]/(phi-1)/rho,
mdeq4:eliminate(mdeq3,[FeOOH[f]]),mdeq4x:ratexpand(mdeq4)[1],print(mdeq4[1]=0))\$

$$
\left(\left(pi_{cd,\,feooh}\, pi_{cd,\,poc}\, cd_d^{\,2} + pi_{cd,\,poc}\, cd_d \right) \phi - pi_{cd,\,feooh}\, pi_{cd,\,poc}\, cd_d^{\,2} - pi_{cd,\,poc}\, cd_d \right)
$$
$$
* \rho\, poc_t + \rho \left(\phi \left(pi_{cd,\,feooh}\, pi_{cd,\,poc}\, cd_d^{\,2}\, feooh_t + pi_{cd,\,feooh}\, cd_d\, feooh_t \right) \right.
$$
$$
\left. - pi_{cd,\,feooh}\, pi_{cd,\,poc}\, cd_d^{\,2}\, feooh_t - pi_{cd,\,feooh}\, cd_d\, feooh_t \right)
$$
$$
+ \left(pi_{cd,\,feooh}\, pi_{cd,\,poc}\, cd_d^{\,2} + \left(pi_{cd,\,poc} + pi_{cd,\,feooh} \right) cd_d + 1 \right) cd_t
$$
$$
+ \left(- pi_{cd,\,feooh}\, pi_{cd,\,poc}\, cd_d^{\,3} + \left(- pi_{cd,\,poc} - pi_{cd,\,feooh} \right) cd_d^{\,2} - cd_d \right) \phi = 0
$$

(c9)   powers(mdeq4x,Cd[d])

(d9) $$[0, 1, 2, 3]$$

(c10) q0=facsum(ratcoef(mdeq4x,Cd[d],0),pi[Cd,FeOOH],pi[Cd,POC])

(d10) $$q0 = cd_t$$

(c11) q1=facsum(ratcoef(mdeq4x,Cd[d],1),pi[Cd,FeOOH],pi[Cd,POC])

(d11) $$q1 = pi_{cd,\,poc} \left(\phi\, \rho\, poc_t - \rho\, poc_t + cd_t \right) + pi_{cd,\,feooh} \left(\phi\, \rho\, feooh_t - \rho\, feooh_t + cd_t \right) - \phi$$

(c12) q2=facsum(ratcoef(mdeq4x,Cd[d],2),pi[Cd,FeOOH],pi[Cd,POC])

(d12) $$q2 = pi_{cd,\,feooh}\, pi_{cd,\,poc} \left(\phi\, \rho\, poc_t - \rho\, poc_t + \phi\, \rho\, feooh_t - \rho\, feooh_t + cd_t \right) - pi_{cd,\,poc}\, \phi - pi_{cd,\,feooh}\, \phi$$

(c13) q3=facsum(ratcoef(mdeq4x,Cd[d],3),pi[Cd,FeOOH],pi[Cd,POC])

(d13) $$q3 = - pi_{cd,\,feooh}\, pi_{cd,\,poc}\, \phi$$

Fig. 21B.1   MACSYMA solution of Eqs. (21A.5).

Appendix A: Data Tables

The calibration of comprehensive and interactive nutrient, oxygen, and metals flux models require, above all, high-quality and comprehensive data sets. These data sets are the result of efforts by scientists who developed the methods for reliably measuring sediment fluxes and applied these techniques in a systematic and comprehensive way. Their efforts are specifically acknowledged and appreciated.

It is likely that the data will ultimately prove to be the more enduring part of this book, as the models presented herein are superceded. No small effort is required to assemble large data sets—not to mention the effort to collect them—so these tables should prove to be useful in themselves.

They are not absolutely complete, in the sense that *every* number is tabulated. But they provide the bulk of the information used, in a form that should provide a very good start for anyone who is interested in reproducing the calculations presented herein, or extending the work.

Directions for downloading these tables are also available on the authors page at the Manhattan College (manhattan.edu) and HydroQual, Inc. (hydroqual.com).

A.1 CHESAPEAKE BAY

The SONE data set[1] listed in Table A1 consists of nutrient and oxygen fluxes and overlying water concentrations and temperature measured four times a year from 1985 through 1988 in Chesapeake Bay. Four main bay stations, two stations in the Potomac estuary, two in the Patuxent estuary are included (see Fig.14.1 for station locations). Fluxes and overlying water concentrations of NH_4, NO_3, O_2, PO_4, and Si are measured in triplicate from subcores taken from a large box core obtained from each station. In addition, solid phase data POC, PON, POP, and chlorophyll are determined. While only four years of data are included in Table A1, this sampling program has continued through the 1990s. The methods employed for collecting these data are available in Cowan and Boynton (1996).

The BEST data set[2] is an expanded set of measurements taken in 1988 that extended the sampling stations into the southern bay and the lower tributaries. The same sampling techniques were employed and some additional parameters were measured. These data are not tabulated, but are available in the downloadable version.

The interstitial water data set (Bricker et al., 1977) was developed during the years 1971 to 1976. Stations throughout the main bay were sampled for pore water vertical profiles of pH, Eh, pS, and concentrations of SO_4, HCO_3, Fe, Mn, PO_4, NH_4, and SiO_2. The data have been reported and analyzed in a number of dissertations and papers.[3] Table A2 presents the 0–10 cm average concentrations at various stations as a function of distance from the mouth of the bay.

A.2 MERL

The MERL data set is from the Nutrient Addition Experiment (Nixon et al., 1986, Oviatt et al., 1986) conducted at the University of Rhode Island mesocosm experimental facility (Pilson et al., 1980). The details are presented in Section 15.2 A. Complete data reports are available,[4] a practice for which the investigators are to be applauded. In addition, manganese flux, overlying water, and sediment compositional data were also collected.[5] These data were generously provided by Carlton Hunt.

Table A3 presents the nutrient and dissolved oxygen overlying water concentrations and fluxes, and the manganese flux and temperature. The three controls are C1, C5, and C8. The various treatments are labeled 1X, 2X, and so on (Section 15.2 A).

---

[1]Boynton et al. (1985a, 1986, 1988a,b, 1985b,c,d), Boynton and Kemp (1985), Garber et al. (1988)
[2]Boynton et al., 1989; Burdige (1989)
[3]Bray (1973), Bray et al. (1973), Bricker and Troup (1975), Holdren (1977), Holdren et al. (1975), Matisoff (1977), Matisoff et al. (1975), Troup (1974), Troup and Bricker (1975), Troup et al. (1974)
[4]Frithsen et al. (1985a,b,c)
[5]Hunt and Kelly (1988)

The corresponding overlying water dissolved and particulate manganese concentrations, pH, and O_2 data are listed in Table A4.

A.3 LAKE CHAMPLAIN

The Lake Champlain data[6] were collected in July and October of 1994 and April 1995 at the stations shown in Fig. 15.20I. The measurements were made using the same techniques employed in the Chesapeake Bay data collection programs. Table A5 presents the nutrient and dissolved oxygen fluxes and overlying water concentrations and temperature. A more complete time series of overlying water concentrations are listed in Table A6. Pore water (Table A7) and solid phase (Table A8) 0–10 cm average concentrations are also included.

---

[6]Cornwell and Owens (1998a)

Chesapeake Bay Nutrient Flux Data

Sta- tion	Date	SOD g O$_2$/ m^2-d	J[NH$_4$] mg N/ m^2-d	J[NO$_3$] mg N/ m^2-d	J[PO$_4$] mg P/ m^2-d	J[Si] mg Si/ m^2-d	O$_2$ mg O$_2$/ L	NH$_4$ mg N/ L	NO$_3$ mg N/ L	PO$_4$ mg P/ L	Si mg Si/ L	T °C
SP	05/08/85	2.407	90.9	-33.7	8.2	222.2	7.77	0.086	0.776	0.011	0.305	17.8
SP	06/26/85	1.870	20.9	0.0	4.5	0.0	4.91	0.075	0.535	0.018	1.130	23.6
SP	08/21/85	1.163	39.2	2.5	4.0	372.1	4.59	0.060	0.249	0.030	1.113	26.2
SP	10/15/85	0.780	14.7	-15.3	0.9	93.7	6.41	0.104	0.508	0.024	0.925	18.5
SP	05/04/86	1.433	41.2	-27.5	1.2	202.5	7.71	0.110	1.015	0.012	1.333	14.4
SP	06/25/86	2.133	34.8	-1.2	4.4	122.5	5.42	0.081	0.767	0.024	1.594	24.2
SP	08/21/86	0.800	103.6	-21.2	8.7	251.5	4.52	0.205	0.272	0.034	0.980	25.1
SP	11/11/86	0.797	5.8	2.3	0.0	87.0	7.51	0.196	0.461	0.022	0.867	12.8
SP	04/20/87	0.867	24.2	-8.7	0.5	179.0	6.83	0.196	0.832	0.017	1.266	11.1
SP	06/11/87	1.427	0.0	-12.5	2.9	119.1	3.36	0.145	0.436	0.016	0.701	21.6
SP	08/17/87	1.007	20.6	6.6	7.1	153.3	4.91	0.109	0.285	0.024	1.747	27.6
SP	11/10/87	0.567	-0.3	-22.8	2.8	68.3	9.11	0.157	0.770	0.017	0.717	11.5
SP	04/17/88	0.670	17.9	-9.9	-3.5	0.0	9.37	0.158	1.126	0.019	1.450	11.5
SP	06/03/88	0.430	34.0	-36.9	1.2	28.3	4.72	0.189	0.805	0.020	1.693	
SP	08/16/88	0.810	40.3	17.7	5.7	76.2	5.79	0.111	0.466	0.013	1.065	29.1
SP	11/03/88	0.240	0.0	5.9	0.0	0.0	8.62	0.104	0.434	0.019	0.908	11.4
SP	12/12/88	0.257	0.0	-7.2	-1.1	7.0	9.69	0.177	0.866	0.016	1.286	
R64	05/06/85	0.870	62.2	1.9	0.9	192.1	5.95	0.117	0.126	0.005	0.317	13.8
R64	06/25/85	1.353	146.2	-3.1	64.8	679.1	0.91	0.371	0.014	0.023	1.573	21.6
R64	08/20/85	2.280	128.9	-1.8	20.0	370.6	1.99	0.395	0.023	0.063	1.310	25.6
R64	10/14/85	1.347	115.8	-1.1	5.2	404.5	4.14	0.295	0.035	0.020	0.857	21.1
R64	05/05/86	0.817	65.5	-19.8	0.2	267.9	5.52	0.174	0.254	0.007	0.493	11.8
R64	06/24/86	0.377	115.7	-16.0	68.5	477.3	0.62	0.532	0.019	0.082	1.555	20.4
R64	08/20/86	0.040	235.0	2.2	52.7	317.1	0.07	0.541	0.009	0.125	1.743	26.1
R64	11/10/86	0.647	30.0	0.6	1.9	176.9	6.39	0.190	0.049	0.019	0.722	16.0
R64	04/20/87	0.593	13.1	-4.0	0.8	64.7	7.12	0.111	0.187	0.004	0.150	9.4

Chesapeake Bay Nutrient Flux Data

Sta-tion	Date	SOD g O$_2$/m^2-d	J[NH$_4$] mg N/m^2-d	J[NO$_3$] mg N/m^2-d	J[PO$_4$] mg P/m^2-d	J[Si] mg Si/m^2-d	O$_2$ mg O$_2$/L	NH$_4$ mg N/L	NO$_3$ mg N/L	PO$_4$ mg P/L	Si mg Si/L	T °C
R64	06/12/87	0.470	60.1	0.0	28.1	264.9	0.29	0.354	0.005	0.072	0.925	16.9
R64	08/17/87	0.157	76.3	-0.4	15.9	314.6	0.11	0.278	0.002	0.068	1.590	25.7
R64	11/16/87	0.453	14.0	10.7	-0.4	124.7	8.07	0.123	0.042	0.012	0.393	12.8
R64	04/17/88	0.523	16.5	-1.3	0.0	96.0	7.46	0.053	0.123	0.005	0.095	10.9
R64	06/01/88	0.143	94.2	0.0	42.1	218.9	0.50	0.379	0.005	0.137	0.925	15.6
R64	08/11/88	0.510	186.3	1.3	33.7	231.0	0.35	0.440	0.004	0.084	1.319	25.8
R64	11/01/88	0.530	63.8	-4.0	0.0	159.6	6.34	0.144	0.052	0.009	0.540	11.6
R64	12/13/88	0.323	15.3	3.2	-2.7	70.6	8.72	0.096	0.074	0.006	0.415	
R78	05/07/85	1.053	62.4	-14.5	0.0	120.7	5.78	0.106	0.183	0.007	0.178	13.8
R78	06/27/85	0.805	56.4	2.7	7.8	277.5	4.91	0.075	0.536	0.018	1.130	21.5
R78	08/20/85	1.010	27.8	19.0	-3.3	231.0	2.15	0.385	0.014	0.055	1.471	25.7
R78	10/14/85	0.740	28.5	17.6	-1.2	186.1	3.42	0.201	0.148	0.021	1.103	20.2
R78	05/04/86	0.790	9.5	-19.3	0.0	67.6	2.47	0.183	0.242	0.004	0.339	10.4
R78	06/24/86	1.150	113.0	-22.2	38.5	275.7	1.39	0.364	0.042	0.038	1.419	21.0
R78	08/21/86	0.003	37.7	-1.7	29.9	114.6	0.06	0.403	0.006	0.147	1.813	25.3
R78	11/11/86	0.500	-9.7	5.2	1.4	93.9	4.14	0.228	0.073	0.031	0.986	16.9
R78	04/20/87	0.500	7.0	5.1	0.2	50.1	4.99	0.180	0.265	0.006	0.241	9.1
R78	06/11/87	0.035	44.2	3.1	15.9	96.4	0.28	0.418	0.006	0.060	0.815	15.4
R78	08/17/87	0.117	25.6	0.2	-2.4	117.5	0.71	0.277	0.003	0.042	1.707	26.5
R78	11/09/87	0.353	-17.8	4.1	18.2	47.4	6.72	0.189	0.195	0.023	0.757	13.7
R78	04/17/88	0.843	11.7	-6.8	-0.5	47.9	6.86	0.105	0.253	0.011	0.168	10.2
R78	06/01/88	-0.047	32.9	-10.4	7.3	84.5	0.61	0.218	0.030	0.025	0.491	15.3
R78	08/15/88	0.730	52.9	2.8	10.9	149.9	2.59	0.221	0.027	0.030	1.529	28.4
R78	11/03/88	0.503	37.4	9.1	6.3	57.5	5.34	0.175	0.089	0.020	0.669	14.9
PP	05/08/85	1.247	50.3	-13.9	1.6	274.4	5.72	0.120	0.107	0.007	0.377	15.8
PP	06/24/85	2.863	111.2	0.0	22.2	639.8	0.75	0.366	0.012	0.043	1.585	22.6

Chesapeake Bay Nutrient Flux Data

Sta-tion	Date	SOD g O₂/ m²-d	J[NH₄] mg N/ m²-d	J[NO₃] mg N/ m²-d	J[PO₄] mg P/ m²-d	J[Si] mg Si/ m²-d	O₂ mg O₂/ L	NH₄ mg N/ L	NO₃ mg N/ L	PO₄ mg P/ L	Si mg Si/ L	T °C
PP	08/20/85	1.073	74.1	-5.7	-2.3	355.3	2.93	0.239	0.025	0.022	1.183	26.1
PP	10/14/85	1.040	24.1	4.2	1.6	250.3	6.77	0.089	0.026	0.005	0.219	20.4
PP	05/05/86	1.030	21.8	-12.4	0.0	133.0	7.97	0.057	0.351	0.004	0.214	13.1
PP	06/24/86	0.687	49.1	-0.2	3.2	312.1	2.66	0.214	0.031	0.008	1.124	22.6
PP	08/18/86	0.740	79.0	7.5	0.0	170.8	5.82	0.066	0.004	0.011	0.725	26.3
PP	11/12/86	0.817	15.4	5.2	1.9	179.8	8.68	0.091	0.046	0.007	0.366	14.8
PP	04/22/87	0.900	0.5	0.0	0.3	138.0	10.61	0.018	0.185	0.002	0.110	12.4
PP	06/13/87	0.090	37.3	-1.4	9.8	198.2	0.50	0.205	0.003	0.045	0.878	18.1
PP	08/20/87	0.363	40.0	0.0	6.6	208.8	0.89	0.169	0.003	0.028	0.863	27.0
PP	11/12/87	0.660	13.6	1.0	-7.1	189.7	8.75	0.089	0.026	0.009	0.303	12.4
PP	04/21/88	0.553	2.8	0.0	0.0	78.0	9.37	0.007	0.121	0.004	0.082	11.2
PP	06/05/88	0.677	47.9	-4.7	0.5	267.6	4.42	0.140	0.065	0.007	0.576	18.9
PP	08/17/88	-0.123	44.8	0.4	16.0	135.1	0.01	0.234	0.002	0.040	1.274	26.0
PP	11/04/88	0.523	33.3	0.7		174.2	8.10	0.075	0.029	0.008	0.223	13.0
BV	05/06/85	1.503	18.9	9.3	0.0	165.4	6.94	0.035	0.029	0.046	0.914	19.8
BV	06/25/85	1.690	93.6	0.0	16.5	0.0	4.21	0.044	0.008	0.056	2.610	25.7
BV	08/22/85	0.697	78.6		2.4	56.0	3.14	0.266	0.144	0.111	2.390	26.1
BV	10/16/85	1.797	63.8	7.0	19.2	190.1	6.95	0.037	0.027	0.086	1.785	21.6
BV	05/08/86	3.230	31.1	14.0	6.2	407.0	6.37	0.026	0.013	0.015	0.801	19.1
BV	06/26/86	2.567	138.4	19.2	26.8	400.9	4.74	0.087	0.020	0.050	2.449	24.9
BV	08/18/86	1.593	138.9	4.3	11.7	177.5	5.48	0.186	0.019	0.215	2.618	28.5
BV	11/13/86	0.960	11.7	0.0	3.4	117.3	8.28	0.135	0.184	0.086	1.639	12.8
BV	04/23/87	1.443	66.0	-2.4	6.6	91.7	9.44	0.066	0.278	0.011	1.118	16.1
BV	06/15/87	2.783	110.1	18.0	33.5	340.7	3.75	0.087	0.016	0.075	2.532	25.7
BV	08/19/87	0.930	58.6	-1.9	8.8	238.4	2.81	0.242	0.042	0.100	3.086	28.7
BV	11/11/87	0.847	2.1	9.6	10.0	60.2	9.03	0.132	0.109	0.053	0.728	14.0

Chesapeake Bay Nutrient Flux Data

Station	Date	SOD g O₂/m²-d	J[NH₄] mg N/m²-d	J[NO₃] mg N/m²-d	J[PO₄] mg P/m²-d	J[Si] mg Si/m²-d	O₂ mg O₂/L	NH₄ mg N/L	NO₃ mg N/L	PO₄ mg P/L	Si mg Si/L	T °C
BV	04/21/88	1.640	42.4	-7.7	5.2	168.5	7.72	0.141	0.242	0.015	0.910	12.4
BV	06/04/88	1.470	97.9	-0.5	11.3	226.3	3.74	0.140	0.018	0.031	2.025	20.4
BV	08/21/88	0.807	72.0	-0.7	37.4	352.6	1.78	0.407	0.029	0.146	2.025	27.6
BV	11/02/88	1.200	8.6	0.0	4.2	96.2	8.02	0.085	0.134	0.040	0.819	11.9
SL	05/06/85	2.717	41.4	0.0	2.3	482.7	5.01	0.098	0.087	0.008	0.584	17.0
SL	06/25/85	3.540	27.6	29.7	8.9	512.8	2.95	0.136	0.039	0.012	1.796	24.5
SL	08/22/85	1.020	53.2	14.7	-1.2	0.0	4.66	0.177	0.094	0.031	1.544	26.3
SL	10/16/85	0.793	26.1	4.9	5.6	180.0	6.81	0.094	0.052	0.023	0.708	20.4
SL	05/08/86	1.137	12.3	-15.0	1.5	155.1	7.14	0.032	0.321	0.002	0.169	14.4
SL	06/26/86	2.963	57.7	7.3	0.8	260.0	5.23	0.098	0.022	0.004	1.933	24.5
SL	08/18/86		51.5	3.8	9.1	109.0	4.50	0.294	0.029	0.059	2.054	27.0
SL	11/13/86	0.510	-1.9	9.5	1.5	62.5	8.19	0.120	0.069	0.043	0.999	14.0
SL	04/23/87	1.290	8.4	13.2	0.6	110.1	12.10	0.022	0.048	0.002	0.320	14.2
SL	06/15/87	0.777	59.3	0.6	15.0	368.1	1.94	0.225	0.011	0.047	1.390	21.3
SL	08/19/87	0.757	73.7	3.5	16.1	200.9	3.78	0.227	0.015	0.046	1.905	27.7
SL	11/11/87	0.573	8.6	12.2	3.4	65.6	8.64	0.078	0.042	0.024	0.330	12.3
SL	04/21/88	1.043	6.4	2.6	0.0	75.5	10.84	0.026	0.133	0.003	0.068	12.0
SL	06/04/88	0.793	49.5	3.7	4.4	318.4	5.50	0.188	0.103	0.017	1.296	20.6
SL	08/21/88	0.850	77.3	4.4	13.3	302.0	2.84	0.289	0.022	0.075	2.145	27.7
SL	11/02/88	1.917	1.2	6.7	2.5	172.8	7.29	0.048	0.053	0.008	0.525	13.3
MP	05/09/85	0.170	19.6	-10.0	6.2	181.3	6.09	0.100	0.593	0.036	0.968	19.4
MP	06/24/85	0.367	19.0	0.0	5.7	63.6	2.09	0.228	0.012	0.022	1.405	24.7
MP	08/19/85	2.377	98.1	7.4	24.7	185.4	4.60	0.212	0.097	0.093	2.004	26.7
MP	10/17/85	2.063	136.0	-14.9	3.0	405.8	7.36	0.056	0.234	0.055	2.007	20.0
MP	05/07/86	1.357	30.1	-1.3	7.6	118.7	6.72	0.078	1.114	0.008	0.420	15.7
MP	06/23/86	1.903	36.7	-16.5	1.9	223.4	3.79	0.091	0.323	0.044	1.328	25.3

Chesapeake Bay Nutrient Flux Data

Station	Date	SOD g O$_2$/ m^2-d	J[NH$_4$] mg N/ m^2-d	J[NO$_3$] mg N/ m^2-d	J[PO$_4$] mg P/ m^2-d	J[Si] mg Si/ m^2-d	O$_2$ mg O$_2$/ L	NH$_4$ mg N/ L	NO$_3$ mg N/ L	PO$_4$ mg P/ L	Si mg Si/ L	T °C
MP	08/19/86	1.323	124.8	-17.8	8.4	188.8	5.03	0.099	0.381	0.081	1.743	26.8
MP	11/12/86	0.913	59.2	-14.6	-11.9	93.7	8.28	0.148	0.543	0.046	1.611	13.8
MP	04/22/87	0.810	0.0	15.5	4.2	60.7	8.30	0.182	1.240	0.037	2.740	14.3
MP	06/14/87	0.550	39.8	-9.9	6.8	174.6	4.33	0.207	0.570	0.055	1.307	22.2
MP	08/20/87	1.357	142.7	-24.8	11.4	331.5	4.27	0.124	0.290	0.050	1.706	29.9
MP	11/13/87	0.353	-3.8	-17.9	-17.3	-117.5	9.17	0.147	0.476	0.041	0.895	10.7
MP	04/19/88	0.783	38.4	-91.6	-1.4	122.5	8.86	0.195	1.534	0.037	0.844	12.6
MP	06/06/88	0.470	-9.3	0.0	1.8	215.5	6.22	0.118	1.058	0.053	3.013	20.4
MP	08/20/88	0.927	56.6	-41.0	-13.0	108.8	3.64	0.091	0.255	0.052	2.061	28.8
MP	11/05/88	0.707	47.5	-0.6	-4.7	39.8	8.77	0.146	0.431	0.035	1.355	11.0
RP	05/09/85	1.750	123.2	-21.4	10.5	316.4	3.20	0.197	0.067	0.017	0.536	15.6
RP	06/24/85	1.517	117.2	2.2	24.0	236.6	4.24	0.079	0.427	0.045	1.236	22.3
RP	08/19/85	2.285	107.2	9.8	0.0	26.5	5.40	0.219	0.009	0.018	1.390	26.0
RP	10/17/85	0.857	137.7	0.9	1.3	243.6	5.00	0.236	0.041	0.013	0.383	20.2
RP	05/06/86	2.343	130.9	-37.3	2.1	278.2	5.86	0.146	0.358	0.005	0.235	12.8
RP	06/23/86	0.937	198.8	-1.7	18.6	271.5	2.94	0.325	0.006	0.026	1.373	22.6
RP	08/19/86	0.000	224.8	0.7	48.2	154.6	0.05	0.721	0.007	0.166	1.876	25.9
RP	11/12/86	0.953	36.7	8.1	2.7	113.3	7.04	0.136	0.046	0.011	0.431	15.8
RP	04/22/87	0.643	33.4	23.2	0.0	107.6	6.95	0.117	0.304	0.007	0.182	11.2
RP	06/13/87	0.000	35.5	-0.7	16.1	110.3	0.19	0.345	0.006	0.061	0.917	18.4
RP	08/20/87	0.000	144.5	0.3	30.9	145.2	0.01	0.518	0.002	0.129	1.624	26.7
RP	11/13/87	0.627	14.9	9.1	6.6	92.8	8.87	0.111	0.047	0.013	0.243	12.0
RP	04/18/88	0.940	0.0	0.9	0.3	62.5	10.50	0.005	0.197	0.004	0.066	11.5
RP	06/06/88	0.150	105.8	0.0	36.5	199.3	0.54	0.453	0.005	0.085	0.899	18.3
RP	08/20/88	-0.010	37.3	1.0	16.4	152.0	0.02	0.335	0.002	0.078	1.254	26.7
RP	11/05/88	0.720	11.8	6.7	0.9	61.1	7.46	0.082	0.035	0.005	0.279	12.1

MERL Nutrient and Manganese Flux Data

Treat-ment	Julian day	SOD g O₂/m²·d	$J[NH_4]$ mg N/ m²·d	$J[NO_3]$ mg N/ m²·d	$J[PO_4]$ mg P/ m²·d	$J[Si]$ mg Si m²·d	$J[Mn]$ mg Mn/ m²·d	O_2 mg O_2/ L	NH_4 mg N/ m³	NO_3 mg N/ m³	PO_4 mg P/ m³	Si mg Si/ m³	T °C
C1	146	0.708	39.4	0.7	6.7	76.4	7.66	8.01	9.0	2.7	23.2	403.	17.3
C1	173	1.406	47.5	1.4		245.9	6.29	7.28	8.3	6.0	25.6	509.	19.3
C1	208	0.912	59.1	6.1		139.2	7.16	6.19	16.0	7.0	30.3	772.	20.8
C1	236	1.025	40.1		0.8		9.62	6.79	10.4	9.5	41.4	449.	17.0
C1	264	0.974	31.2	10.2	5.6	8.6	5.16	6.59	59.4	25.1	55.7	292.	16.6
C1	314	0.552	10.5	3.2	2.4	34.2	0.24	7.87	46.9	79.3	46.1	472.	10.0
C1	383	0.240	3.9	1.4	0.8	26.9	0.10	11.79					0.1
C1	439	0.480	7.2	2.4	1.3	33.2	1.22	11.58	6.2	4.2	4.7	30.	3.8
C1	447	0.475	8.7	2.4	1.2	27.8	1.32	11.50	4.9	2.5	1.5	21.	5.3
C1	502	1.147	29.2	10.1	4.1	143.0	5.04	7.54	14.0	12.5	15.3	446.	13.2
C1	546	1.663	47.7	12.3	4.2	191.2	5.26	6.61	17.5	10.5	19.7	911.	19.0
C1	565	1.327	49.6	7.0	7.6	147.0	16.25	4.52	16.4	6.4	27.0	1517	21.3
C1	594	2.232	50.8	17.0	8.5		9.77	5.36	46.8	36.7	39.9	1913	21.5
C1	642	1.390	26.4	12.6	4.0	184.1	5.09	5.92	43.3	43.3	40.3	1353	16.1
C1	698	0.610	4.0	5.3	0.5	56.4	1.57	6.97	73.8	123.4	40.0	853.	9.0
C1	775	0.430	2.7	3.8	0.5	17.4	0.12	11.90	3.5	2.5	9.9	422.	2.1
C1	863						1.66						
C1	908	1.793	64.1	20.0	11.2	167.3	4.09	6.83	14.3	15.4	7.4	761.	19.2
C1	936	1.872	52.0	24.2	8.6	159.4	8.20	5.70	53.0	41.9	28.8	1160	21.1
C1	992	1.361	54.0	10.4	10.9	181.6	8.20	6.08	25.9	23.0	33.1	1193	21.0
C5	146	0.845	39.0	4.1	6.5	-36.9	8.97	8.01	11.2	4.9	27.0	518.	17.5
C5	180	1.219	39.6	-1.6	8.0	151.7	28.02	7.48	15.4	6.7	39.9	242.	20.3
C5	208	1.291	44.2	2.3	-1.3	-249.	9.58	6.48	6.2	2.1	36.4	652.	19.7
C5	236	1.193	55.7		8.3		6.63	6.14	20.6	16.1	51.5	897.	18.1
C5	264	1.094	27.9	5.1	5.7	78.0	5.55	6.66	48.3	27.6	56.2	1115	17.0
C5	307	0.554	13.7	-16.7	1.4	58.9	2.76	8.15	59.5	66.1	48.6	817.	11.4

MERL Nutrient and Manganese Flux Data

Treat-ment	Julian day	SOD g O₂/ m²-d	J[NH₄] mg N/ m²-d	J[NO₃] mg N/ m²-d	J[PO₄] mg P/ m²-d	J[Si] mg Si m²-d	J[Mn] mg Mn/ m²-d	O₂ mg O₂/ L	NH₄ mg N/ m³	NO₃ mg N/ m³	PO₄ mg P/ m³	Si mg Si/ m³	T °C
C5	384	0.155	0.2	0.1	0.4	2.5	0.86	14.98	4.9	2.2	4.2	18.	0.8
C5	439	0.218	11.0	1.9	0.7	43.0	1.88	11.53	6.9	3.2	2.1	34.	4.6
C5	502	1.202	35.4	7.5	3.9	52.2	4.57	7.62	14.6	11.1	19.7	544.	14.6
C5	546	1.555	42.2	5.6	4.4	179.6	19.73	6.05	24.1	11.8	20.9	888.	19.3
C5	565	1.486	60.5	5.7	6.9		8.46	4.93	17.5	10.2	22.8	1442	21.8
C5	594	1.414	52.7	3.7	6.1		14.30	5.78	22.6	26.3	34.5	1858	22.5
C5	642	0.883	32.6	6.6	7.4	117.5	11.27	6.39	47.2	27.0	43.4	1342	17.4
C5	698	0.403	9.6	1.9	2.5	47.1	1.91	8.03	77.3	94.0	40.6	654.	9.0
C5	859	1.138	20.8	2.6	6.2	166.0	8.02	10.10	3.9	2.8	0.9	17.	14.0
C5	908	1.562	61.7	4.2	13.4	227.4	9.19	6.19	36.6	9.9	25.7	881.	19.2
C5	936	1.334	55.6	7.2	11.9	228.9	12.57	6.18	46.4	17.1	43.4	1364	21.5
C5	994	0.965	51.9	3.3	8.3	102.1	7.14	6.43	32.2	7.1	49.6	414.	22.0
C8	146	0.569	19.1	0.9	3.7	231.4	5.53	8.68	9.1	3.2	21.1	165.	16.8
C8	180	1.392	27.1	2.3	17.2	136.5		7.32	8.7	1.8	27.2	92.	19.2
C8	208	1.378	31.5	2.8		91.3		7.27	11.2	2.4	34.8	200.	19.3
C8	236	1.505	76.0				8.39	5.98	32.4	12.3	43.5	568.	17.8
C8	264	1.094	47.1	4.1	5.8	124.7	6.31	6.09	113.6	39.4	55.0	954.	17.6
C8	307	0.547	11.5	0.3	5.3	137.2	3.25	8.46	107.2	106.9	51.2	691.	10.6
C8	376	0.149	3.5	1.3	0.8	13.6	1.38	11.64	7.4	91.9	19.3	147.	0.9
C8	439	0.530	3.1	0.7	0.8	21.4	2.23	12.27	7.6	2.2	6.6	29.	4.0
C8	502	1.284	37.3	6.1	4.7	294.1	4.03	7.90	9.2	4.3	14.1	476.	13.6
C8	546	1.970	70.7	5.6			10.00	6.58	46.0	15.3	21.8	790.	18.7
C8	565	1.879	60.0	3.9	6.5		14.56	5.67	9.1	10.8	20.2	1172	21.8
C8	594	1.879	60.3	5.6	3.0	195.2		4.83	270.7	32.1	72.4	1856	22.0
C8	642	0.950	33.4	9.4	-3.3	124.9	10.55	6.15	108.9	53.2	44.0	982.	16.3
C8	698	0.454	4.2	7.3	0.7	43.1	2.43	8.02	72.6	111.8	38.1	691.	8.2

MERL Nutrient and Manganese Flux Data

Treat-ment	Julian day	SOD g O$_2$/ m^2-d	J[NH$_4$] mg N/ m^2-d	J[NO$_3$] mg N/ m^2-d	J[PO$_4$] mg P/ m^2-d	J[Si] mg Si m^2-d	J[Mn] mg Mn/ m^2-d	O$_2$ mg O$_2$/ L	NH$_4$ mg N/ m^3	NO$_3$ mg N/ m^3	PO$_4$ mg P/ m^3	Si mg Si/ m^3	T °C
C8	859	1.099	24.3	8.2	3.6	115.1	4.66	8.28	3.9	3.2	5.9	463.	19.6
C8	908	2.198	60.7	5.6	9.7	280.3	5.19	6.35	16.0	25.8	13.3	808.	19.6
C8	936	1.884	61.6	15.4	10.3	238.9	9.17	5.80	36.8	12.0	25.4	1321	20.5
C8	992	1.486	81.0	0.6	24.6	30.8	10.75	6.48	38.7	15.8	37.8	1256	21.2
1X	146	0.792	31.8	1.1	6.0	41.4	5.21	7.76	10.9	5.7	24.0	408.	17.3
1X	180	1.171	45.1	1.1	7.2	118.3	1.88	7.47	72.7	18.2	40.4	379.	19.2
1X	208	1.471	61.8	0.6	3.3	19.1	4.32	7.34	5.9	2.7	36.4	377.	19.1
1X	236	1.315	72.9		-0.1		6.99	6.84	31.4	13.4	48.9	769.	17.6
1X	264	2.246	131.8	4.1	33.3	239.9	8.32	8.18	30.1	22.0	49.1	83.	17.3
1X	307	0.970	18.6	4.4	3.8	42.7	2.17	9.42	128.1	97.6	57.3	72.	11.0
1X	376	0.278	6.4	2.6	0.8	11.3	0.54	11.59	107.6	200.3	53.6	196.	1.0
1X	439	0.370	11.1	0.7	1.2	12.5	3.21	17.17	5.6	1.7	4.7	17.	3.8
1X	502	1.831	55.5	2.3	7.6		15.26	8.65	9.0	6.4	25.8	62.	14.2
1X	546	2.143	100.6	9.1	10.3	347.5	13.29	6.72	154.8	42.2	46.8	526.	18.8
1X	565	2.657	117.2	3.0	18.1		22.25	6.76	6.7	2.5	22.3	822.	20.8
1X	594	1.781	58.3	4.1	15.5		8.45	3.14	210.9	81.5	81.8	2263	21.1
1X	642	1.476	53.0	-4.4	8.8	211.8	6.20	5.03	205.5	94.4	78.4	1699	17.0
1X	698	0.732	5.9	15.3	1.1	77.5	1.53	7.16	192.2	159.4	61.9	949.	8.5
1X	859	1.618	36.3	11.7	5.4	168.5	4.45	8.44	62.3	42.5	22.9	435.	18.4
1X	908	2.441	58.5	25.2	5.8		4.50	5.91	60.1	50.4	28.8	1094	18.4
1X	936	1.740	67.0	13.2	13.2	293.8	10.03	5.57	88.1	56.2	50.5	1340	20.5
1X	992	1.699	113.4	-2.5	20.9	360.0	5.63	4.44	114.9	23.3	74.3	1556	21.8
2X	140	0.499	17.6	1.4	4.2	257.0	2.15	8.81	10.4	4.2	26.1	334.	14.0
2X	180		8.1	-4.2	0.3	-16.7	7.18	10.55	13.4	2.8	48.2	29.	19.0
2X	208	1.843	98.7	0.2	5.2	79.5	14.95	9.16	7.0	4.2	63.0	39.	20.5
2X	236	1.874	100.9		9.3		4.50	5.99	57.7	16.5	66.5	850.	17.8

MERL Nutrient and Manganese Flux Data

Treat-ment	Julian day	SOD g O₂/ m²-d	J[NH₄] mg N/ m²-d	J[NO₃] mg N/ m²-d	J[PO₄] mg P/ m²-d	J[Si] mg Si m²-d	J[Mn] mg Mn/ m²-d	O₂ mg O₂/ L	NH₄ mg N/ m³	NO₃ mg N/ m³	PO₄ mg P/ m³	Si mg Si/ m³	T °C
2X	264	1.450	57.7	22.4	4.1	61.0	8.27	6.38	353.3	64.9	87.2	1067	16.8
2X	307	0.703	10.8	3.3	1.9	-36.0	1.39	7.85	382.2	175.3	96.4	915.	11.8
2X	390	0.178	0.7	3.1	0.7	1.7	0.22	13.65	90.2	326.3	71.7	71.	0.0
2X	439	0.665	30.1	4.7	5.4	34.3	2.23	15.39	87.1	33.3	34.8	55.	4.2
2X	502	1.651	50.6	8.7	9.5	169.7	10.72	8.73	419.3	72.9	57.3	364.	14.9
2X	546	2.244	92.5	-0.4	10.5	66.0	28.04	7.55	197.4	50.4	46.1	649.	19.0
2X	565	2.006	100.4	-8.3			21.78	4.84	217.4	73.0	75.9	1032	20.8
2X	594	0.343	46.8	0.8	8.3	79.7	20.27	4.53	349.7	164.8	111.2	999.	19.9
2X	642	0.996	27.6	10.7	2.3	54.2	9.53	5.51	351.7	248.3	146.8	547.	18.0
2X	698	0.442	-3.1	12.8	3.0	23.7	0.74	8.08	252.7	257.8	105.6	453.	9.1
2X	775	0.259	-0.2	5.0	1.9	12.6	0.07	12.21	240.1	205.8	91.4	308.	4.0
2X	859	1.090	26.3	1.6	5.8	41.9	10.76	13.91	5.3	2.8	9.3	12.	13.5
2X	908	1.596	55.8	3.6	18.4	113.3	18.83	8.46	6.4	3.4	21.1	34.	18.3
2X	936	1.243	59.3	-0.1	16.1	127.9	11.14	8.72	43.7	15.6	49.9	318.	20.8
2X	994	1.860	73.7		23.9	185.1	5.54	8.00	49.9	71.5	115.8	342.	21.9
4X	140	0.593	17.3	2.1	-0.8	-51.7	2.22	8.89	7.7	3.2	23.7	297.	14.3
4X	180	1.001	109.5	2.6	54.9	13.2	23.35	11.48	12.5	2.9	41.4	35.	19.5
4X	208	1.819	102.4	5.6	19.8	166.1	22.81	6.44	247.4	17.8	91.2	597.	19.3
4X	236	1.822	173.5		-29.0		6.40	3.86	515.8	118.5	129.1	840.	18.2
4X	264	1.421	36.7	-14.6	7.3	110.1	8.99	4.88	524.8	238.0	148.8	1300	16.5
4X	314	0.787	-9.1	13.4	6.3	76.4	1.85	7.24	632.4	388.9	169.0	994.	9.5
4X	376	0.229	1.8	5.2	0.9	8.0	0.12	14.82	320.5	514.2	164.6	26.	0.8
4X	447	0.667	41.0	2.1	6.8	47.5	1.43	19.44	172.7	379.8	128.2	79.	5.2
4X	502	1.901	73.9	0.5	41.2	84.3	16.21	8.25	435.9	110.8	137.5	238.	15.0
4X	546	2.506	124.7	6.5	23.9		26.35	9.01	300.8	149.1	91.9	822.	19.5
4X	565	2.554	114.9	-14.0	27.4	182.8	34.51	5.08	330.4	128.5	148.8	1006	19.9

MERL Nutrient and Manganese Flux Data

Treatment	Julian day	SOD g O$_2$/ m^2-d	J[NH$_4$] mg N/ m^2-d	J[NO$_3$] mg N/ m^2-d	J[PO$_4$] mg P/ m^2-d	J[Si] mg Si m^2-d	J[Mn] mg Mn/ m^2-d	O$_2$ mg O$_2$/ L	NH$_4$ mg N/ m^3	NO$_3$ mg N/ m^3	PO$_4$ mg P/ m^3	Si mg Si/ m^3	T °C
4X	594	2.909	91.3	-8.0	23.4		27.02	3.95	500.9	271.5	211.1	1245	21.3
4X	642	1.282	37.2	1.4	6.4	95.7	10.32	5.33	618.4	330.4	235.1	813.	17.0
4X	698	0.622	-13.1	51.4	-0.5	26.2	1.53	7.48	546.8	448.6	90.1	596.	9.0
4X	775	0.210	1.0	6.3	1.6	4.9	0.02	14.41	366.2	284.8	123.6	129.	3.0
4X	859	1.022	55.2	4.6	14.0	52.4	4.14	12.98	48.8	28.3	72.5	301.	21.8
4X	908	1.337	75.3	4.2		216.6	16.38	6.34	308.8	131.8	149.3	403.	21.7
4X	936	1.956	104.9	5.5	26.0	127.0	10.39	11.06	197.0	53.8	129.1	109.	21.8
4X	992	3.055	96.0	2.6	28.3	193.3	6.70	4.27	113.3	153.3	173.7	802.	21.4
8X	146	0.677	31.0	-0.6	1.2	75.0	7.79	8.47	9.2	4.5	19.0	377.	17.8
8X	180	2.170	90.0	9.9	11.7	20.0		16.23	19.9	18.9	77.3	25.	20.0
8X	208	2.366	131.5	1.9	6.2	-71.5	13.21	7.01	745.6	59.4	137.8	849.	20.5
8X	236	4.229			28.8		0.45	3.75	911.8	172.9	163.1	1355	18.2
8X	264	1.762	19.6	-41.4	1.7	79.2		4.66	880.1	433.7	213.0	1564	16.5
8X	314	1.006	-20.3	5.6	-6.3	149.3	3.15	5.97	1088.3	685.2	275.3	1080	9.7
8X	384	0.324	-19.4	18.2	-1.8	35.5	0.05	10.04	1103.1	859.7	363.9	649.	0.6
8X	439	0.353	42.8	6.6	10.2	48.3	3.57	17.82	750.1	933.3	396.9	-6.	4.6
8X	509	3.643	147.7	-1.2	34.2	80.4	10.92	10.75	776.0	431.9	272.2	973.	15.4
8X	546	2.856	94.8	-8.2	26.0	165.1	3.20	5.02	1149.4	460.1	273.2	1278	19.3
8X	572	2.734	131.9		31.0		0.35	2.43	1330.7	487.5	371.7	1892	20.9
8X	594	2.227	71.2	-36.3		-18.2	0.37	3.28	1252.9	787.6	430.0	1364	23.2
8X	642	2.222	-59.6	-1.4	-8.5	80.6	0.45	3.13	928.0	805.0	425.5	1235	18.0
8X	698	0.900	-40.8	-19.5	9.8	85.7	1.24	6.15	882.3	811.7	310.3	1032	10.5
8X	775	0.720	-27.4	35.7	2.6	24.2	0.55	10.67	880.1	818.3	313.1	574.	3.4
8X	859	1.001	24.7	10.4	2.7	61.9	5.76	8.09	4.6	2.8	4.6	517.	12.3
8X	859	2.196	42.3	-1.2	18.7	103.6	5.76	8.09	712.5	424.5	263.9	432.	13.0
8X	908	2.098	56.0	-2.8	24.2	118.2	0.27	4.31	896.4	487.5	312.5	1063	18.3

MERL Nutrient and Manganese Flux Data

Treat-ment	Julian day	SOD g O$_2$/ m^2-d	J[NH$_4$] mg N/ m^2-d	J[NO$_3$] mg N/ m^2-d	J[PO$_4$] mg P/ m^2-d	J[Si] mg Si m^2-d	J[Mn] mg Mn/ m^2-d	O$_2$ mg O$_2$/ L	NH$_4$ mg N/ m^3	NO$_3$ mg N/ m^3	PO$_4$ mg P/ m^3	Si mg Si/ m^3	T °C
8X	936	2.251	75.1	-3.0	26.1	120.1	2.83	4.53	902.7	456.9	332.3	1218	22.0
8X	992	2.652	196.8	-21.8	73.1	148.4	0.57	2.84	805.4	359.4	339.1	1113	22.2
16X	140	0.643	32.3	0.8	6.4	250.5	4.38	8.65	9.5	4.9	25.8	298.	14.0
16X	180	2.042	175.7	5.4	15.9	-25.9	9.96	13.02	951.0	56.2	220.3	440.	19.5
16X	208	3.017	235.6	-5.8	48.6	307.9	17.84	6.55	1882.5	214.9	414.3	671.	21.2
16X	236	2.321	46.0		-78.2		4.21	5.27	1867.5	438.8	483.4	1493	18.0
16X	264	1.524		-0.4	-6.6	154.0	4.68	3.51	1877.8	970.5	498.4	1975	18.0
16X	307	1.133	-48.9	20.0	-14.3	-10.6	2.04	6.10	2056.0	1195.5	549.7	1554	10.6
16X	342	0.761	-83.9	6.3	-3.9	35.5	2.12	7.78	1873.0	1356.2	567.8	1294	7.7
16X	384	0.470	-48.4	31.4	0.8	0.0	0.25	13.86	1564.6	1554.1	561.0	20.	1.4
16X	447	0.814	9.1	-15.7	4.6	51.0	2.67	14.12	2616.6	2206.3	939.0	0.	5.0
16X	506						9.02						
16X	509	2.474	245.0	-61.1	97.8	131.5	8.59	15.30	1985.8	1073.2	1013.6	1457	15.0
16X	553	2.959	157.9	128.4	52.8	225.4	27.25	13.50	1962.0	787.4	756.6	173.	19.2
16X	565	3.134	293.5	-456.3			30.09	5.36	1943.9	536.9	822.4	87.	19.3
16X	600	3.230		-12.6			3.71	3.07	3786.3	283.6	1153.3	997.	21.8
16X	650	1.349	165.1	-19.5	2.7	108.1	0.87	4.55	3891.8	453.8	929.1	1024	17.2
16X	705	0.854	-50.9	-210.1	6.9	40.8	1.10	7.45	3165.3	735.4	626.5	1031	9.0
16X	775	0.437	-2.9	73.0	7.7	7.9	2.49	15.79	2042.0	1000.2	579.4	11.	1.5
16X	859	1.327	111.7	-7.3	32.8	81.6	11.16	13.70	1619.4	515.0	469.8	15.	12.1
16X	908	2.172	136.1	296.5	41.0	174.9	12.33	7.00	1915.9	533.4	558.7	25.	18.9
16X	936	2.186	204.2	-21.6	70.2	204.7	15.42	6.56	1409.5	749.8	541.7	35.	22.1
16X	996	1.901	101.7	5.6	8.8	125.3	2.86	3.76	1334.2	619.7	600.5	1112	20.2
32X	140	0.600	27.1	1.2	3.9	324.0	4.42	8.68	10.2	36.1	25.1	383.	14.1
32X	180	0.864	-161.4	75.4		-50.6	11.42	14.93	2757.4	41.3	602.6	26.	18.6
32X	208	1.332	325.3	-0.3	34.3	164.4	20.43	9.54	3540.0	111.2	765.1	47.	21.5

MERL Nutrient and Manganese Flux Data

Treatment	Julian day	SOD g O$_2$/ m^2-d	J[NH$_4$] mg N/ m^2-d	J[NO$_3$] mg N/ m^2-d	J[PO$_4$] mg P/ m^2-d	J[Si] mg Si m^2-d	J[Mn] mg Mn/ m^2-d	O$_2$ mg O$_2$/ L	NH$_4$ mg N/ m^3	NO$_3$ mg N/ m^3	PO$_4$ mg P/ m^3	Si mg Si/ m^3	T °C
32X	236	3.331			85.6		9.48	3.84	3760.3	285.9	789.3	122.	18.2
32X	264	1.906	-104.5		15.6	144.1	6.90	3.51	3905.1	902.9	908.0	1598	17.3
32X	314	1.212	43.2	2.3	-21.3	281.7		13.21	3276.8	817.3	638.8	1266	10.8
32X	342	0.562	16.6	1.6	-21.6	-12.7	1.04	8.33	3814.2	828.0	920.5	1411	7.5
32X	384	0.533	-22.6	13.3	0.0	7.1	0.35	15.61	4441.7	1495.8	1038.8	238.	0.4
32X	447	0.787	30.9	-42.5	-4.9	71.7	2.16	15.37	6633.2	1779.4	1588.8	1.	5.6
32X	509	0.262	122.4	-33.8	49.9	141.6	19.39	15.38	4327.3	876.6	1309.6	31.	15.6
32X	553	2.246	-76.5	-8.6		288.9	8.29	7.06	3653.8	1091.1	992.7	158.	20.4
32X	566						9.23						
32X	572	4.358	6.9			342.4	6.78	9.02	1614.5	2671.1	1414.5	136.	21.1
32X	600	0.151		136.6			8.39	5.84	3909.2	710.3	1516.5	467.	22.5
32X	650	2.342	101.8	-7.2	28.5	201.8	4.37	6.69	4171.5	975.4	1080.5	117.	18.0
32X	705	1.238	-114.0	70.7	2.4	67.8	2.93	7.12	2365.7	2031.0	734.6	1057	9.4
32X	775	0.588	-33.4	-3.5	7.0	27.6	0.84	12.49	2603.6	2334.2	923.5	1115	3.5
32X	859	2.570	-93.2	11.3	35.4	119.8	4.37	13.63	1252.5	1548.1	780.4	10.	13.0
32X	908	3.588	8.2	-194.8	72.0		12.78	8.44	823.5	1324.6	780.8	15.	20.0
32X	943	4.934	-61.4	-302.8	69.9	260.8	11.77	5.92	1317.6	991.6	1009.9	1346	23.0
32X	996	3.458	91.1	-123.8			4.77	4.38	1632.0	1047.9	1080.9	1733	21.2

Lake Champlain Nutrient Flux Data

Station	Date	SOD g O$_2$/m^2-d	J[NH$_4$] mg N/m^2-d	J[NO$_3$] mg N/m^2-d	J[PO$_4$] mg P/m^2-d	O$_2$ g O$_2$/L	NH$_4$ mg N/L	NO$_3$ mg N/L	PO$_4$ mg P/L	T °C
2	04/21/94	0.272	0.12	0.00	0.22	8.06	0.018	0.051	0.005	15.93
2	07/19/94	0.427	30.98	−3.93	2.01	3.85	0.089	0.363	0.028	20.94
2	10/06/94	0.346	29.21	1.95	0.00	6.87	0.036	0.217	0.011	14.30
7	04/10/94	0.246	0.47	3.52	3.10	12.94	0.018	0.315	0.011	10.35
7	07/23/94	0.094	0.00	−4.22	2.30	10.50	0.015	0.272	0.006	20.94
7	10/04/94	0.383	1.12	−0.59	0.00	9.03	0.020	0.260	0.006	8.42
37	04/25/94	0.334	−2.19	9.27	1.09	10.87	0.031	0.044	0.002	7.57
37	07/20/94	0.402	0.00	0.00	2.90	5.89	0.015	0.015	0.011	20.94
37	10/17/94	0.438	6.98	5.46	1.54	10.48	0.021	0.127	0.005	11.57
41	04/25/94	0.358	7.37	2.71	0.37	10.87	0.031	0.044	0.002	7.57
41	07/20/94	0.754	22.12	0.00	2.60	5.89	0.015	0.015	0.011	20.94
41	10/17/94	0.381	3.28	4.35	0.30	10.48	0.021	0.127	0.005	1.57
50	04/11/94	0.254	−2.13	0.58	0.15	10.68	0.015	0.424	0.007	10.35
50	07/21/94	0.260	61.45	−11.21	23.08	7.97	0.019	0.091	0.006	20.94
50	10/12/94	0.502	1.47	6.70	0.67	10.20	0.015	0.015	0.002	10.45
51	04/11/94	0.285	−0.73	1.31	0.55	10.68	0.015	0.424	0.007	10.35
51	07/21/94	0.300	30.72	−1.92	1.60	7.97	0.019	0.091	0.006	20.94
51	10/12/94	0.587	10.19	7.08	0.92	10.20	0.015	0.015	0.002	10.45

Chesapeake Bay Pore Water Data

Sta-tion	Dist km	Date	NH$_4$ μM	PO$_4$ μM	Si μM	Cl μM	pH	SO$_4$ μM	Alk μM
Station	DIST	Date	NH4	PO4	SI	CL	PH	SO4	Alk
927RR	285.4	09/25/73	11.84	0.49	7.50	6.1	7.19	19.2	7.39
926S	283.5	06/18/73	5.62	0.75		5.8	7.30		5.08
922W	275.7	12/18/73	5.95	0.97	7.26	22.3	7.24	62.4	4.32
922Y	275.7	08/18/72	7.62	1.73	13.53	20.0	7.48	66.4	5.05
922Y	275.7	09/02/74		3.48		55.2	6.99		5.43
919T	269.9	09/25/73	6.09	0.91	5.57	151.7	7.06	566.4	3.80
915L	262.1	04/17/74	12.23	2.19	5.90	100.5	7.23		5.18
914C	260.2	12/19/73	5.07	1.93	7.24	68.5	7.42	245.6	3.65
914Q	260.2	12/19/73	5.61	2.25	6.55	66.8	7.31	233.6	3.97
914S	260.2	08/18/72	4.79	1.37	8.65	79.4	7.34	284.0	4.27
914S	260.2	12/19/73	18.88	1.32	7.72	160.7	7.12	423.2	8.50
914S	260.2	05/14/75	2.97	1.85	3.64	82.7	7.38	251.2	2.91
912K	256.3	04/17/74	10.95	0.96	5.20	75.8	7.28		0.61
904D	240.8	12/17/73	2.31	0.85	8.42	192.2	7.04	980.0	1.62
904N	240.8	08/08/72	19.58	1.62	7.15	183.7	7.39	456.0	14.10
904N	240.8	09/25/73	33.00	3.75	8.49	237.1	7.46	314.9	19.65
904N	240.8	12/18/73	13.29	5.00	8.50	197.8	7.45	468.5	11.04
904N	240.8	04/17/74	21.86	0.92	7.24	223.8	7.53		8.40
904N	240.8	09/01/74	30.54	6.44		216.1	7.30		19.40
904N1	240.8	12/18/72	52.88	2.55	10.98		7.68	364.8	
904N1	240.8	04/18/74	19.23	4.55	6.81	215.8	7.46		14.20
904N2	240.8	04/18/74	21.07	3.84	7.11	213.6	7.39		13.81
904N3	240.8	04/18/74	9.08	3.67	8.16	205.6	7.30		6.97
904N4	240.8	04/18/74	16.19	4.68	8.10	213.9	7.17		11.85
904N5	240.8	04/18/74	20.74	4.16	8.01	210.0	7.35		9.62
904P	240.8	12/19/73	3.97	2.70	10.36	188.0	7.37	736.0	4.80
858C	229.1	09/02/74		8.57		284.3	7.43		18.27
858C	229.1	05/14/75	19.76	9.09	11.60	275.8	7.58	668.0	14.23
858C	229.1	05/14/75		7.20	8.58	269.4	7.26	700.8	9.31
856-3	225.2	05/26/71	3.62	2.59	8.43				
856-4	225.2	05/26/71	0.91	0.70	6.82				
856-6	225.2	05/26/71	4.84	3.54	13.49				
856-7	225.2	05/26/71	6.46	0.09	1.83				
856-7	225.2	01/17/72	8.64	3.00	7.30	232.8	7.43		7.25
856B	225.2	01/19/72	3.72	2.51	8.17	258.4	6.98		3.44
856C	225.2	06/30/71	6.91	4.16	10.77	275.3	7.31	1105.6	5.37
856C	225.2	08/11/71	10.48	7.47	15.52	261.9	7.34	678.4	8.52
856C	225.2	11/24/71	6.82	3.64	10.54	276.8	7.21	1068.8	6.85
856C	225.2	01/19/72	3.61	2.13	6.39	274.6	7.38	1168.8	4.35
856C	225.2	03/29/72				270.9		834.2	
856C	225.2	04/28/72	7.62	4.98	9.69	259.6		1032.0	6.18
856C	225.2	07/19/72	9.31	1.66	6.36	157.5	7.31	470.4	4.85
856C	225.2	08/22/72	19.48	4.31	11.42	262.7	7.32	638.4	15.05
856C	225.2	06/19/73	28.53	6.09			7.17		13.77

Chesapeake Bay Pore Water Data

Sta-tion	Dist km	Date	NH$_4$ μM	PO$_4$ μM	Si μM	Cl μM	pH	SO$_4$ μM	Alk μM
856C	225.2	09/26/73	32.95	7.11	11.58	279.1	7.48	368.8	21.94
856C1	225.2	09/27/71			14.09	292.8	7.30	763.2	14.35
856C1	225.2	09/28/71	15.12	9.69	16.05	292.8	7.30	716.8	14.03
856C1	225.2	10/26/71	12.59	7.54	13.39	295.8	7.17	908.0	13.13
856C1	225.2	03/15/72	6.72	3.39	8.17	240.7	7.28	724.8	7.58
856C1	225.2	05/25/72	3.85	1.51	6.93	249.0	7.70	1140.0	3.47
856C1	225.2	06/27/72	6.81	3.41	8.17	231.6	7.29	683.2	7.07
856C1	225.2	11/22/72	13.69	4.58	3.42	275.7	7.62		9.70
856C1	225.2	12/19/72	52.97	2.57	11.08		7.68	371.2	
856C2	225.2	10/27/71			8.81	296.1		1225.6	5.30
856C2	225.2	03/15/72	8.21	3.98	7.19	293.4	7.08	977.6	8.48
856C2	225.2	05/26/72	3.14	2.29	6.87	242.7	7.40	1063.2	3.55
856C2	225.2	11/22/72	13.83	5.81	7.61	271.5	7.62	856.0	10.62
856C5	225.2	10/28/71			9.39	291.9	7.47		4.72
856C6	225.2	10/28/71			12.25	290.3	7.61		6.95
856C7	225.2	10/28/71			11.48	290.6	7.23		6.55
856E	225.2	05/26/71	1.39	0.50	9.87				
856E	225.2	06/30/71	6.73	4.53	10.77	234.5	7.59	954.2	6.00
856E	225.2	08/11/71	7.13	4.34	12.85	211.0	7.37	750.4	7.63
856E	225.2	09/25/71	9.27	3.81	11.32	241.3	7.07	717.6	8.58
856E	225.2	09/27/71	7.69	7.15	14.08	232.4	7.24	696.0	9.34
856E	225.2	01/16/72	9.89	5.65	7.37	243.9	7.38	756.0	9.03
856E	225.2	02/24/72	7.36	2.40	6.77	223.5	7.40	849.6	5.25
856E	225.2	03/29/72				209.9		929.6	
856E	225.2	04/27/72	3.27	0.99	6.75	208.1		926.4	3.25
856E	225.2	05/24/72	3.33	2.80	10.30	204.3	7.42	774.9	4.80
856E	225.2	06/27/72				117.4		272.8	3.44
856E	225.2	07/17/72	4.03	2.07	6.58	87.1	7.30	319.2	2.80
856E	225.2	08/15/72	7.28	1.93	11.35	183.5	7.22	494.4	5.74
8560	225.2	07/20/71	7.72	4.67	9.99	228.1	7.38	752.0	7.13
8560	225.2	11/22/71	8.72	5.29	9.85	285.1	7.46	746.4	7.90
8560	225.2	03/14/72	1.64	1.17	4.74	195.4		910.4	
8560	225.2	06/26/72	4.92	1.92	8.48	115.7	7.40	479.2	3.67
8560	225.2	05/26/71	4.38	0.87	7.88				
8560	225.2	11/23/71			11.78	234.4	7.50		8.70
8560	225.2	12/20/71			10.33	225.7	7.40	746.4	5.82
8560	225.2	03/14/72	3.09	2.42	10.46	198.3	7.36	854.4	2.47
8560	225.2	11/23/71			10.40	234.1	7.37	662.4	7.65
8560	225.2	11/23/71			11.02	232.3	7.34		6.70
856F	225.2	01/17/72	0.96	0.55	4.60	195.9	7.27		2.55
854C	221.4	04/19/74	8.82	3.89	7.80	205.7	7.25		6.22
854C	221.4	04/19/74	8.48	4.39	7.70	205.7	7.32		7.80
853C1	219.4	05/14/74	22.54	11.86	10.39	251.0	7.46		15.74
853C2	219.4	05/14/74	23.53	10.99	10.05	249.9	7.37		15.63
853C3	219.4	05/14/74	23.22	6.95	10.81	239.0	7.38		15.84

Chesapeake Bay Pore Water Data

Sta-tion	Dist km	Date	NH$_4$ μM	PO$_4$ μM	Si μM	Cl μM	pH	SO$_4$ μM	Alk μM
853C4	219.4	05/14/74	20.27	8.65	9.74	248.4	7.58		15.65
853C5	219.4	05/14/74	30.58	9.45	13.18	252.8	7.30		19.27
853E3	219.4	05/15/74	4.79	2.88	8.31	196.5	7.21		5.83
853E4	219.4	05/15/74	6.85	3.85	10.16	194.2	7.41		6.48
853E5	219.4	05/15/74	5.88	3.95	9.03	196.4	7.43		5.37
853E6	219.4	05/15/74	5.77	3.94	8.52	193.0	7.39		4.60
853E7	219.4	05/15/74	7.98	4.11	8.98	182.8	7.40		6.62
853G1	219.4	05/16/74	0.97	0.82	5.55	136.9	7.18		1.52
853G2	219.4	05/16/74	0.80	0.62	5.81	134.8	7.19		0.62
853G3	219.4	05/16/74	2.65	0.98	10.40	133.2	7.32		3.30
853G4	219.4	05/16/74	1.05	0.34	10.24	132.3	7.10		1.66
853R	219.4	08/21/72	7.75	10.71	11.79	222.0	7.45	976.8	7.00
848D	209.7	12/20/73	15.21	8.65	11.85	246.1	7.56	528.0	12.65
848E	209.7	09/28/73	7.36	11.41	15.18	284.2	7.07	396.0	29.44
848F	209.7	12/20/73	7.94	5.17	10.11	252.5	7.36	936.0	6.02
848I	209.7	12/20/73	1.74	1.04	5.94	182.1	7.33	912.0	1.73
845G	203.9	08/18/72	20.13	5.29	10.54	184.5	7.31		9.73
845G	203.9	08/21/72	11.20	4.94	13.71	289.3	7.49	976.0	8.40
834C	182.5	03/28/74	11.29	5.07	11.86	255.7	7.43		7.78
834F	182.5	03/26/74	25.63	12.01	19.99	284.0	7.53		16.01
834G	182.5	08/17/72	9.61	3.22	14.74	284.1	7.43	1081.6	16.17
834G	182.5	09/26/73	6.90	4.69	14.47	248.2	7.56	2496.8	7.91
834G	182.5	05/14/75		2.05	11.72	250.9	7.23	787.2	4.03
834G	182.5	07/17/82	2.73	2.59	8.93	179.0	7.46		5.42
834I	182.5	03/28/74	1.81	1.87	7.29	195.5	7.14		2.18
818P	151.5	07/18/72	17.22	9.96	15.74	231.7	7.42		10.23
818P	151.5	09/28/73	14.56	7.06	17.15	321.1	7.42	913.6	13.59
818S	151.5	08/17/72	17.28	9.07	19.83	318.0	7.36	759.4	12.64
813Z	141.7	03/28/74	10.61	3.67	10.83	209.0	7.30		8.51
812	139.8	03/26/74	2.31	1.48	11.49	202.9	7.02		2.08
812O	139.8	03/27/74	4.36	2.13	13.22	219.7	7.22		3.72
812R	139.8	03/26/74	12.62	3.25	15.05	291.5	7.48		7.44
812S	139.8	05/13/75	34.37	14.86	24.97	328.6	7.41	581.6	20.12
804C	124.3	08/17/72	19.88	10.08	23.60	307.9	7.48	647.2	6.63
804C	124.3	09/27/73	29.12	4.32	19.60	318.7	7.34	558.4	19.25
747A	91.3	05/13/75	10.04	5.26	25.40	336.2	7.15	1134.4	9.93
744A	85.4	08/16/72	19.97	11.93	24.09	323.7	7.31	687.2	13.87
744A	85.4	09/26/73	2.90	2.26	12.47	344.2	7.07	2058.2	2.16
744B	85.4	03/27/74	1.92	0.98	10.75	287.1	7.22		2.58
744NN	85.4	03/21/74		4.54	14.01	273.9	7.32		8.66
744U	85.4	03/27/74		0.85	8.95	281.6	7.27		2.31
724R	46.6	08/16/72	43.20	3.22	13.68	368.1	7.43	1640.8	3.06
724R	46.6	06/19/73	9.69	8.97			7.21		8.26
724R	46.6	09/27/73	21.01	12.28	18.01	417.5	7.20	1352.0	17.14

MERL Overlying Water Data

Treatment	Day	Mn_d	Mn_p	pH	O_2	Treatment	Day	Mn_d	Mn_p	pH	O_2
		— mg Mn/m³ —			mg/L			— mg Mn/m³ —			mg/L
CO	149	10.8			8.30	2X	644	11.7	14.5	7.74	5.50
CO	175	20.0		7.96	7.50	2X	700	2.5	4.1	7.88	8.20
CO	208	17.5		9.90	6.25	2X	778	6.9	3.5	8.27	12.50
CO	235	10.0		7.85	6.65	2X	826	11.1	4.0	8.50	11.35
CO	249	8.4	3.2	7.85	7.20	2X	853	11.1	4.4	8.50	12.24
CO	279	7.7	1.9	7.91	7.87	2X	885	10.3	11.4	8.52	12.28
CO	313	8.3		7.84	7.97	2X	915	5.1	12.3	8.28	9.31
CO	348	9.1		8.06	9.68	2X	945	3.5	10.3	8.06	8.58
CO	383	5.7		8.12		2X	974	7.3	8.9	7.96	7.55
CO	390	6.0	0.5	8.37	13.10	2X	995	4.2	9.0	7.81	6.14
CO	445	6.3		5.88	11.67	4X	142	14.7			8.30
CO	496	12.1	3.0	8.02	8.40	4X	176	8.0		8.28	11.90
CO	538	9.9	9.1	7.88	7.13	4X	225	25.9		7.45	5.65
CO	556	15.1	5.5	7.79	5.59	4X	256	31.5		7.62	4.70
CO	583	19.0	5.1	7.72	5.10	4X	279	14.3	2.5	7.80	7.21
CO	643	18.5	4.9	7.90	6.17	4X	314	13.9		7.80	7.39
CO	698	8.9	2.0	7.92	7.20	4X	348	11.3	2.5	7.93	9.35
CO	776	5.6	0.2	8.34	12.31	4X	379	5.7		8.37	14.40
CO	826	5.9	1.1	8.23	11.16	4X	390	4.7		8.80	17.93
CO	853	7.2	2.0	8.06	8.85	4X	448	14.7	6.6	9.04	18.20
CO	885	11.5	1.3	7.91	6.96	4X	497	15.5	0.8	8.41	7.99
CO	915	13.6	2.5	7.87	6.13	4X	538	7.1	18.4	8.20	10.78
CO	945	21.0	6.2	7.65	5.21	4X	558	16.5	13.1	8.06	8.25
CO	974	19.3	9.8	7.74	5.30	4X	584	27.1	7.1	7.74	5.55
CO	995	20.5	4.2	7.80	5.94	4X	645	19.8	11.2	7.68	5.40
C5	147	14.8			8.30	4X	701	6.5	7.1	7.83	7.40
C5	178	18.2		8.10	7.10	4X	777	5.3	3.4	8.41	13.50

MERL Overlying Water Data

Treat-ment	Day	Mn$_d$	Mn$_p$	pH	O$_2$
		— mg Mn/m^3 —			mg/L
C5	224	15.1		7.89	6.41
C5	267	11.2		7.85	6.54
C5	279	4.7	9.3	7.85	6.66
C5	309	3.5		7.88	7.90
C5	348	6.3	2.7	8.10	10.02
C5	385	1.3		8.53	17.00
C5	390	6.6	2.5	8.58	14.28
C5	440	8.4		8.38	11.77
C5	496	7.9	3.8	8.03	8.22
C5	538	9.3	7.4	7.94	7.40
C5	556	15.5	8.2	7.87	5.61
C5	583	14.5	7.9	7.83	5.40
C5	643	17.9	9.2	7.92	6.58
C5	699	11.4	3.7	7.98	7.71
C5	826	13.9	2.1	8.28	11.18
C5	853	14.2	3.6	8.26	10.28
C5	885	20.6	5.3	8.05	8.04
C5	915	16.4	7.3	7.94	6.77
C5	945	15.8	10.6	7.77	6.41
C5	974	16.3	16.5	7.85	6.41
C5	995	3.0	7.9	7.88	6.94
C8	147	10.4			8.67
C8	175	11.5		8.07	8.30
C8	237	10.5		7.84	6.00
C8	265	6.7		7.84	6.08
C8	279	7.7	4.7	7.85	6.42
C8	310	16.8		7.84	8.10
4X	826	7.2	6.0	8.78	17.23
4X	853	9.3	8.9	8.57	12.90
4X	885	7.5	13.8	8.31	10.30
4X	915	4.4	13.6	8.15	8.86
4X	945	4.3	9.4	8.22	9.61
4X	974	9.1	9.8	7.82	4.85
4X	995	13.0	1.9	7.63	4.39
8X	148	14.2			8.83
8X	209	29.4		8.12	6.58
8X	234	47.5		7.31	3.75
8X	248	36.6	4.6	7.34	4.94
8X	279	23.9	4.9	7.45	4.59
8X	314	13.9		7.62	6.12
8X	348	14.6	1.9	7.80	7.84
8X	384	10.2		7.94	9.88
8X	390	11.3	1.4	8.00	10.56
8X	442	11.9		8.83	17.50
8X	497	14.4	7.6	8.31	13.51
8X	538	34.1	3.8	7.30	6.06
8X	558	38.2	4.9	4.21	4.25
8X	584	35.8	4.3	7.13	2.74
8X	645	25.9	3.2	7.30	3.40
8X	702	11.3	3.2	7.67	6.10
8X	776	8.2	1.0	8.07	10.66
8X	826	7.6	3.0	8.59	14.73
8X	853	12.1	7.0	8.19	9.47
8X	885	22.6	9.0	7.84	6.16

MERL Overlying Water Data

Treat-ment	Day	Mn$_d$	Mn$_p$	pH	O$_2$	Treat-ment	Day	Mn$_d$	Mn$_p$	pH	O$_2$
		— mg Mn/m^3 —			mg/L			— mg Mn/m^3 —			mg/L
C8	348	7.4	1.6	8.10	10.11	8X	915	25.8	8.7	7.54	4.83
C8	377	4.8		8.19	11.16	8X	945	23.4	7.7	7.43	4.59
C8	390	2.7	3.6	8.53	14.47	8X	974	17.0	10.9	7.46	3.57
C8	441	9.8		8.48	12.80	8X	995	18.8	5.1	7.41	2.59
C8	496	11.9		8.01	8.36	16X	141	14.2			8.50
C8	538	5.5	6.0	7.92	7.04	16X	176	3.3		8.34	14.00
C8	555	9.3	12.0	7.94	6.17	16X	225	38.0		7.73	6.25
C8	583	23.5	4.1	7.87	5.29	16X	254	28.3		7.46	3.97
C8	642	19.0	13.6	7.88	6.36	16X	279	29.1	2.7	7.46	4.26
C8	699	12.5	2.8	7.93	7.98	16X	309	22.2		7.52	5.80
C8	826	7.7	2.0	8.17	10.16	16X	343	17.4	1.6	7.76	8.11
C8	853	9.7	1.7	8.22	8.96	16X	380	10.7		8.17	14.30
C8	885	13.1	1.9	7.96	7.40	16X	390	3.6	9.1	8.50	15.15
C8	915	14.6	2.3	7.88	6.24	16X	447	10.1		8.70	13.62
C8	945	18.8	4.6	7.68	6.02	16X	498	12.1	4.9	8.60	15.82
C8	974	23.3	4.8	7.80	6.08	16X	524	12.3	21.4	8.36	9.53
C8	995	29.7	5.5	7.81	6.23	16X	559	13.4	18.6	8.17	7.51
1X	148	14.2			8.10	16X	583	26.2	5.1	7.58	3.86
1X	182	23.7		7.84	6.80	16X	646	25.6	4.8	7.58	5.60
1X	225	18.2		7.93	7.35	16X	702	12.5	2.4	7.74	7.30
1X	255	14.3		7.96	8.33	16X	777	6.2	2.4	7.96	15.40
1X	279	14.8	6.6	7.82	6.81	16X	826	5.5	8.7	8.72	14.34
1X	308	12.6		7.96	9.37	16X	853	6.7	6.9	8.70	12.95
1X	348	11.5	2.5	8.02	9.47	16X	885	10.5	7.3	8.58	10.85
1X	378	7.9		8.11	10.80	16X	915	5.5	10.8	8.41	7.88
1X	390	3.0	3.0	8.51	15.20	16X	945	8.4	9.9	8.00	5.78
1X	440	9.3		8.78	17.31	16X	974	18.9	4.3	7.66	3.15

MERL Overlying Water Data

Treat-ment	Day	Mn$_d$ —mg Mn/m³—	Mn$_p$	pH	O$_2$ mg/L	Treat-ment	Day	Mn$_d$ —mg Mn/m³—	Mn$_p$	pH	O$_2$ mg/L
1X	497	17.9	1.6	8.34	11.05	16X	995	15.1	1.4	7.56	4.12
1X	538	19.8	17.3	7.93	8.13	32X	141	13.1			8.50
1X	556	23.0	12.5	7.89	7.22	32X	195	13.3		8.56	11.76
1X	584	24.0	8.6	7.82	6.03	32X	235	4.7	12.9	8.28	6.70
1X	644	26.7	8.2	7.72	5.10	32X	249	16.0	10.6	7.74	2.78
1X	700	13.7	1.3	7.86	7.30	32X	279	20.1	2.7	7.53	4.13
1X	826	6.2	2.2	8.33	11.68	32X	313			8.27	13.29
1X	853	8.9	3.6	8.15	9.47	32X	343	10.1	4.1	8.05	8.91
1X	885	15.9	3.0	7.94	7.53	32X	387	9.3		8.54	14.30
1X	915	16.5	3.4	7.79	6.23	32X	390	11.5	2.5	8.51	14.37
1X	945	20.8	5.7	7.69	6.02	32X	447	6.9		8.53	13.62
1X	974	29.9	13.5	7.68	4.93	32X	498	4.1	8.0	8.77	16.68
1X	995	39.3	6.9	7.59	4.29	32X	538	6.0	18.1	8.39	10.76
2X	140	14.3			8.50	32X	560	8.0	14.6	8.22	9.10
2X	196	8.2		8.21	10.20	32X	582	27.3	6.4	7.62	3.63
2X	238	19.0		7.77	6.40	32X	646	18.0	1.9	7.80	4.80
2X	255	15.7	3.3	7.75	6.30	32X	701	1.2	12.0	7.81	7.20
2X	279	11.8	10.4	7.78	6.70	32X	778	8.9	0.8	8.30	12.70
2X	310	17.8		7.68	7.50	32X	826	2.3	4.2	8.75	16.60
2X	348	10.4	2.7	7.99	9.40	32X	853	4.8	4.9	8.66	13.82
2X	387	5.4		8.35	13.20	32X	885	8.0	5.4	8.50	13.35
2X	390	5.2		8.40	13.84	32X	915	12.6	5.1	7.81	6.69
2X	441	14.4	2.7	8.75	15.50	32X	945	22.2	7.8	7.40	4.73
2X	497	15.5	0.8	8.12	8.15	32X	974	21.8	5.0	7.34	4.44
2X	538	18.7	10.4	8.09	8.56	32X	995	17.2	2.0	7.53	5.03
2X	557	22.3	17.9	7.91	6.64						
2X	584	26.6	12.9	7.64	4.54						

Lake Champlain Overlying Water Data

Station	Day	NH4 mg N/L	NO3 mg N/L	PO4 mg P/L	O2 mg O2/L
2	136	0.018	0.051	0.005	
2	144				8.06
2	153	0.018	0.024	0.005	8.10
2	166	0.013	0.048	0.006	7.60
2	179	0.085	0.119	0.005	
2	193	0.103	0.338	0.013	
2	193				2.70
2	208	0.075	0.387	0.042	
2	208				5.00
2	216				0.89
2	228				5.78
2	245	0.031	0.119		
2	264	0.018	0.120	0.008	7.80
2	277	0.051	0.251		
2	277				6.87
2	283	0.020	0.182	0.011	
7	130	0.025	0.330	0.012	
7	158				10.49
7	165				10.82
7	166				11.20
7	166	0.015	0.332	0.010	
7	179	0.015	0.305	0.002	
7	193	0.015	0.295	0.006	10.40
7	208				10.60
7	208	0.015	0.351	0.009	
7	208	0.015	0.169	0.002	
7	228				9.49
2	138	0.015	0.015	0.002	
2	147	0.048	0.029		11.29
2	147	0.029	0.089		
2	147				10.44
2	147				9.90
2	159				9.23
2	166	0.015	0.023		
2	166				8.54
2	178	0.020	0.016		
2	178				6.08
2	195	0.015	0.015		
2	195				5.69
2	208	0.015	0.015		
2	208				9.80
2	236				9.34
2	250				
2	262	0.015	0.015	0.003	
2	264	0.015	0.015		
2	264				8.37
2	290				10.48
7	131	0.015	0.424	0.007	
7	159	0.015	0.206	0.004	9.20
7	192	0.033	0.241	0.003	8.70
7	207	0.010	0.016	0.009	7.60
7	207	0.015	0.016		7.60
7	236	0.015	0.015	0.003	8.90
7	250				9.00

Lake Champlain Overlying Water Data

Station	Day	NH4 mg N/L	NO3 mg N/L	PO4 mg P/L	O2 mg O2/L
7	245	0.015	0.274	0.011	
7	264	0.015	0.302		9.60
7	277	0.020	0.260	0.006	9.03
7	278				
50	262	0.015	0.015	0.003	10.70
50	277	0.015	0.015	0.002	9.70
50	290				

Lake Champlain Pore Water Data

Station	Day	NH4 mg N/L	NO3 mg N/L	PO4 mg P/L
2	822	3.65	0.002	1.40
2	926	8.96	0.012	1.06
2	1018	8.52	0.005	1.23
7	822	1.51	0.002	1.05
7	926	1.00	0.026	1.62
7	1018	1.18	0.006	1.47
37	822	0.85	0.001	0.40
37	926	0.40	0.026	0.56
37	1018	1.19	0.009	0.91
41	822	0.74	0.002	0.42
41	926	3.44	0.017	1.77
41	1018	1.00	0.006	0.28
50	822	0.60	0.028	0.09
50	926	1.05	0.012	0.63
50	1018	1.12	0.004	0.30
51	822	1.29	0.003	0.86
51	926	2.45	0.014	0.62
51	1018	2.08	0.005	0.51

Lake Champlain Sediment Data

Station	Day	PIP mg P/g	POP mg P/g
2	926	0.77	0.17
7	926	1.27	0.25
37	926	1.29	0.31
41	926	0.87	0.22
50	926	0.75	0.26
51	926	1.14	0.25

Appendix B: Computer Program

The FORTRAN computer program presented below computes the steady state fluxes of nutrients and oxygen. The formulation and equations are presented in Chapter 16. It is the program that produces the results illustrated in Figs. 16.4 and 16.5 and the model responses in Figs. 16.6 through 16.11. For the latter computations the depositional flux of POC is specified directly. For the former calculations, the diagenesis and depositional fluxes are computed from the ammonia flux.

Lines of code are identified by the letters (A), (B), and so on, inserted after the !, which marks the beginning of a comment. The program reads the parameters from file 11 (A), the data from file 13 (B), and outputs to file 15. The program is set up to compute the steady state fluxes corresponding to an inputted ammonia flux using Eqs. (16.5). Ammonia diagenesis is estimated from the ammonia flux together with the overlying water ammonia and dissolved oxygen concentrations (C). The result is converted to the carbon diagenesis flux (D).

A number of necessary parameters are set and the reaction rates are evaluated at the appropriate temperature (E). The depositional flux jcin is computed from the diagenesis flux taking into account the fact that the G_3 component is conservative (F). Therefore its presence is not reflected in the diagenesis flux, but since it is lost by burial, it must be supplied in addition to the G_1 and G_2 components. Since the computation is iterative, and there are nonlinear terms in the ammonia and silica kinetics, the beginning concentrations $(t - 1)$ are set to zero (G), and the iteration loop starts (H).

The steady state diagenesis equations are solved and carbon J_C, nitrogen J_N, and phosphorus J_P diagenesis fluxes are computed (I). Then the flux equations are solved by a call to the root finding program zbrent, which is a part of the Numerical Recipes collection of subroutines (Press et al., 1989) and, therefore, is not included in the listing (J). It finds a root of the FORTRAN function sedf(sod1),which is listed below, bracketed by sodmin and sodmax to an accuracy specified by eps. The $t - 1$ variables are updated (K) and convergence is tested by comparing to the layer 1 and 2 concentrations from the previous iteration (L). If no convergence is achieved, the concentrations are saved (M) and the iteration is repeated. At convergence, the solid phase concentrations are computed, output is written, and the programs ends.

The function sedf(sod1)evaluates the sediment fluxes, for an inputted sod1 value of SOD (N). It sets up the equations for layers 1 and 2 in terms of the general equations in Chapter 5 and solves them with a call to sedsffnl, listed below. Ammonia, nitrate, sulfide, phosphate, and silica are computed sequentially using $s = $ sod1/$[O_2(0)]$ (O). The final computations (P) are the actual SOD that results (sod), and the difference between sod and sod1, which is the error, that is assigned to the function sedf. It is this error that zbrent attempts to minimize, since when sod equals sod1, the appropriate SOD has been found.

The subroutine sedsffnl sets up the layers 1 and 2 equations (Q). If the half-saturation constant kmc1 is nonzero then there is a Michaelis-Menton expression that is evaluated (R). The coefficients $a_{i,j}$ and forcing functions b_j of a 2×2 set of equations

$$\begin{bmatrix} a_{11} & a_{12} \\ a_{21} & a_{22} \end{bmatrix} \begin{bmatrix} c_{T1} \\ c_{T2} \end{bmatrix} = \begin{bmatrix} b_1 \\ b_2 \end{bmatrix} \tag{B.1}$$

are set up and solved

$$\begin{bmatrix} c_{T1} \\ c_{T2} \end{bmatrix} = \frac{1}{\Delta} \begin{bmatrix} a_{22} & -a_{12} \\ -a_{21} & a_{11} \end{bmatrix} \begin{bmatrix} b_1 \\ b_2 \end{bmatrix} \tag{B.2}$$

where

$$\Delta = a_{11}a_{22} - a_{12}a_{21} \tag{B.3}$$

An error is indicated if $\Delta = 0$ (S).

The "program" sedsfdim (T) is actually just the dimension and common statements that are inserted into each of the programs via the include statements. This simply insures that the same statements are included in each of the programs.

The listing labeled input file (U) is the file of parameters, with the values and the appropriate units, that is read in at the start of the program. The format statement 1001 skips two lines, the labels and the units, and reads the values. This is a convenient way to make the parameter file legible. The listing labeled Output file (V) is actually a short segment of the file that is transposed so that it can be listed. The units have been added to this file to clarify the results.

Model Equations The model equations used in the steady state model are listed below. The solutions are found by numerically integrating the equations to steady state.

Diagenesis

$$H_2 \frac{dG_{POC,i}}{dt} = -K_{POC,i}\theta_{POC,i}^{(T-20)} G_{POC,i} H_2 - w_2 G_{POC,i} + f_{POC,i} J_{POC} \quad \text{(B.4)}$$

$$J_C = \sum_{i=1}^{2} K_{POC,i}\theta_{POC,i}^{(T-20)} G_{POC,i} H_2 \quad \text{(B.5)}$$

and similar equations for G_{PON} and G_{POP}, J_N and J_P.

Mass balance equations

These are the general equations for C_{T1} and C_{T2}.

$$H_1 \frac{dC_{T1}}{dt} = -\frac{\kappa_1^2}{s} C_{T1} + s(f_{d0}C_{T0} - f_{d1}C_{T1})$$

$$+ w_{12}(f_{p2}C_{T2} - f_{p1}C_{T1}) + K_{L12}(f_{d2}C_{T2} - f_{d1}C_{T1}) - w_2 C_{T1} + J_{T1}$$

$$\text{(B.6)}$$

$$H_2 \frac{dC_{T2}}{dt} = -\kappa_2 C_{T2} - w_{12}(f_{p2}C_{T2} - f_{p1}C_{T1})$$

$$- K_{L12}(f_{d2}C_{T2} - f_{d1}C_{T1}) + w_2(C_{T1} - C_{T2}) + J_{T2} \quad \text{(B.7)}$$

$$f_{di} = \frac{1}{1 + m_i \pi_i} \quad i = 1, 2 \quad \text{(B.8)}$$

$$f_{pi} = 1 - f_{di} \quad \text{(B.9)}$$

$$w_{12} = \frac{D_p \theta_{D_p}^{(T-20)}}{H_2} \frac{G_{POC,1}}{G_{POC,R}} \frac{[O_2(0)]}{K_{M,D_p} + [O_2(0)]} \quad \text{(B.10)}$$

$$K_{L12} = \frac{D_d \theta_{D_d}^{(T-20)}}{H_2/2} \quad \text{(B.11)}$$

The kinetic and source terms for each solute are listed next. If the term is not listed, then it is zero.

Ammonia

$$\kappa_1^2 = \kappa_{\text{NH}_4,1}^2 \theta_{\text{NH}_4}^{(T-20)} \left(\frac{K_{\text{M,NH}_4} \theta_{K_{\text{M,NH}_4}}^{(T-20)}}{K_{\text{M,NH}_4} \theta_{K_{\text{M,NH}_4}}^{(T-20)} + [\text{NH}_4(1)]} \right) \left(\frac{[\text{O}_2(0)]}{2K_{\text{M,NH}_4,\text{O}_2} + [\text{O}_2(0)]} \right)$$

$$\text{(B.12)}$$

$$J_{\text{T2}} = J_{\text{N}} \tag{B.13}$$

Nitrate

$$\kappa_1^2 = \kappa_{\text{NO}_3,1}^2 \theta_{\text{NO}_3}^{(T-20)} \tag{B.14}$$

$$\kappa_2 = \kappa_{\text{NO}_3,2} \theta_{\text{NO}_3}^{(T-20)} \tag{B.15}$$

$$J_{\text{T1}} = J_{\text{N}} - J[\text{NH}_4] \tag{B.16}$$

Sulfide

$$\kappa_1^2 = \left(\kappa_{\text{H}_2\text{S,d1}}^2 f_{\text{d1}} + \kappa_{\text{H}_2\text{S,p1}}^2 f_{\text{p1}} \right) \theta_{\text{H}_2\text{S}}^{(T-20)} \frac{[\text{O}_2(0)]}{2K_{\text{M,H}_2\text{S,O}_2}} \tag{B.17}$$

$$J_{\text{T2}} = a_{\text{O}_2,\text{C}} J_{\text{C}} - a_{\text{O}_2,\text{NO}_3} \left(\frac{\kappa_{\text{NO}_3,1}^2}{s} [\text{NO}_3(1)] + \kappa_{\text{NO}_3,2}[\text{NO}_3(2)] \right) \tag{B.18}$$

Oxygen

$$\text{CSOD} = \frac{\left(\kappa_{\text{H}_2\text{S,d1}}^2 f_{\text{d1}} + \kappa_{\text{H}_2\text{S,p1}}^2 f_{\text{p1}} \right) \theta_{\text{H}_2\text{S}}^{(T-20)}}{s} \frac{[\text{O}_2(0)]}{2K_{\text{M,H}_2\text{S,O}_2}} [\Sigma \text{H}_2\text{S}(1)] \tag{B.19}$$

$$\text{NSOD} = a_{\text{O}_2,\text{NH}_4} \frac{\kappa_{\text{NH}_4,1}^2 \theta_{\text{NH}_4}^{(T-20)}}{s} \left(\frac{K_{\text{M,NH}_4} \theta_{K_{\text{M,NH}_4}}^{(T-20)}}{K_{\text{M,NH}_4} \theta_{K_{\text{M,NH}_4}}^{(T-20)} + [\text{NH}_4(1)]} \right)$$

$$\left(\frac{[\text{O}_2(0)]}{2K_{\text{M,NH}_4,\text{O}_2} + [\text{O}_2(0)]} \right) [\text{NH}_4(1)] \tag{B.20}$$

Phosphate

$$\pi_1 = \pi_2(\Delta\pi_{PO_4,1}) \qquad\qquad [O_2(0)] > [O_2(0)]_{crit,PO_4} \quad \text{(B.21)}$$

$$\pi_1 = \pi_2(\Delta\pi_{PO_4,1})^{\dfrac{[O_2(0)]}{[O_2(0)]_{crit,PO_4}}} \qquad [O_2(0)] \leqslant [O_2(0)]_{crit,PO_4} \quad \text{(B.22)}$$

$$J_{T2} = J_P \qquad\qquad\qquad \text{(B.23)}$$

Particulate Silica

$$H_2\frac{dP_{Si}}{dt} = -S_{Si}H_2 - w_2P_{Si} + J_{P_{Si}} + J_{DetrSi} \qquad \text{(B.24)}$$

$$S_{Si} = K_{Si}\theta_{Si}^{(T-20)}\frac{P_{Si}}{P_{Si} + K_{M,P_{Si}}}\big([Si]_{sat} - f_{d2}[Si(2)]\big) \qquad \text{(B.25)}$$

Silicate

$$\kappa_2 = K_{Si}\theta_{Si}^{(T-20)}\frac{P_{Si}}{K_{M,P_{Si}} + P_{Si}}f_{d2}H_2 \qquad \text{(B.26)}$$

$$\pi_1 = \pi_2(\Delta\pi_{Si,1}) \qquad\qquad [O_2(0)] > [O_2(0)]_{crit,Si} \quad \text{(B.27)}$$

$$\pi_1 = \pi_2(\Delta\pi_{Si,1})^{\dfrac{[O_2(0)]}{[O_2(0)]_{crit,Si}}} \qquad [O_2(0)] \leqslant [O_2(0)]_{crit,Si} \quad \text{(B.28)}$$

$$J_{T2} = K_{Si}\theta_{Si}^{(T-20)}\frac{P_{Si}}{K_{M,P_{Si}} + P_{Si}}[Si]_{sat}H_2 \qquad \text{(B.29)}$$

Model Parameters The parameters used in the steady state model computations are listed below.

Symbol	Description	Value	Units
$\kappa_{NH_4,1}$	Reaction velocity for nitrification	0.131	m/d
θ_{NH_4}	Temperature coefficient for nitrification	1.123	–
K_{M,NH_4}	Nitrification half saturation constant for NH_4	0.728	mg N/L
$\theta_{K_{M,NH_4}}$	Temperature coefficient for K_{M,NH_4}	1.125	–
K_{M,NH_4,O_2}	Nitrification half saturation constant for O_2	0.37	mg O_2/L
$\pi_{NH_4,1}$	Aerobic layer partition coefficient	1.0	L/kg
$\pi_{NH_4,2}$	Anaerobic layer partition coefficient	1.0	L/kg
κ_2	Anaerobic layer reaction velocity	0.0	m/d
J_{T1}	Aerobic layer ammonia source	0.0	mg N/m²-d
J_{T2}	Nitrogen diagenesis, J_N , Eq. (7.6)	–	mg N/m²-d
$\kappa_{NO_3,1}$	Aerobic layer denitrification reaction velocity	0.10	m/d
$\kappa_{NO_3,2}$	Anaerobic layer denitrification reaction velocity	0.25	m/d
θ_{NO_3}	Temperature coefficient for denitrification	1.08	–
$a_{O_2,C}$	Oxygen to carbon Redfield stoichiometry	2.667	g O_2^*/g C
a_{O_2,NO_3}	Diagenesis (in O_2 equivalents) consumed by denitrification	2.857	g O_2^*/g N
$\kappa_{H_2S,dl}$	Aerobic layer reaction velocity for dissolved sulfide oxidation	0.20	m/d
$\kappa_{H_2S,pl}$	Aerobic layer reaction velocity for particulate sulfide oxidation	0.40	m/d
θ_{H_2S}	Temperature coefficient for sulfide oxidation	1.08	–
K_{M,H_2S,O_2}	Sulfide oxidation normalization constant for oxygen	4.0.	mg O_2/L
$\pi_{H_2S,1}$	Aerobic layer sulfide partition coefficient	100.0	L/kg
$\pi_{H_2S,2}$	Anaerobic layer sulfide partition coefficient	100.0	L/kg
$[\Sigma H_2S(1)]$	Total aerobic layer sulfide concentration	–	g O_2^*/m³
a_{O_2,NH_4}	Oxygen consumed by nitrification	4.571	g O_2/g N
J_{T2}	Phosphorus diagenesis J_P computed from Eq. (7.8)	–	mg P/m²-d
$\pi_{PO_4,2}$	Aerobic layer phosphate partition coefficient	100.	L/kg
$\Delta\pi_{PO_4,1}$	Aerobic layer incremental PO_4 partition coefficient	300.	

$[O_2(0)]_{crit,PO_4}$	Overlying water O_2 conc. at which $\pi_{PO_4,2}$ starts to decrease	2.0	mg/L
K_{Si}	Biogenic silica dissolution rate constant	0.50	d^{-1}
θ_{Si}	Temperature coefficient for K_{Si}	1.10	–
$[Si]_{sat}$	Saturation concentration for pore water silica	40.0	mg Si/L
$K_{M,P_{Si}}$	Particulate biogenic silica half saturation constant for dissolution = 100 mg Si/g	5.0×10^4	mg Si/L
$\Delta\pi_{Si,1}$	Aerobic layer incremental silica partition coefficient	10.0	–
$\pi_{Si,2}$	Anaerobic layer silica partition coefficient	100.0	L/kg
J_{Si}	Flux of biogenic silica from the overlying water	-	mg Si/m²-d
$J_{Detr,Si}$	Flux of detrital silica from the overlying water	100.	mg Si/m²-d

```fortran
c
c    SEDIMENT MODEL - Steady State Version
c
     include 'sedsfdim.for'
c--  Input variables
     integer monzz,dayzz,yearzz
     real jcinzz,sodzz,jnh4zz,jno3zz,jpo4zz,jsizz,
    .o2zz,nh4zz,no3zz,po4zz,sizz,tempzz
c--  output variables
     real nh40z,nh41z,nh42z,nh4t2z,no30z,no31z,no32z,no3t2z
c--  function for zbrent root finder
     external sedf
c--  leave the file names blank - the file names come from the command line
     open(11,file=' ',form='formatted')      !Parameter input
     open(13,file=' ',form='formatted')      !Data input
     open(15,file=' ',form='formatted')      !Output file
     read(11,1001) m1,m2,Dp,w2,Dd,thtaDp,thtaDd               !(A)
     read(11,1001) kappnh4,pienh4,thtanh4,kmnh4,thtakmnh4,kmnh4o2
     read(11,1001) kappln03,k2no3,thtano3
     read(11,1001) kappd1,kapppl,pie1s,pie2s,thtapd1,kmhso2
     read(11,1001) ksi,csisat,dpie1si,pie2si
     read(11,1001) h2,thtasi,kmpsi,o2critsi
     read(11,1001) dpie1po4,pie2po4,o2crit,kmo2Dp
     read(11,1001) frpon1,kpon1,thtapon1
     read(11,1001) frpon2,kpon2,thtapon2
     read(11,1001) frpon3,kpon3,thtapon3
     read(11,1001) frpoc1,kpoc1,thtapoc1
     read(11,1001) frpoc2,kpoc2,thtapoc2
     read(11,1001) frpoc3,kpoc3,thtapoc3
     read(11,1001) frpop1,kpop1,thtapop1
     read(11,1001) frpop2,kpop2,thtapop2
     read(11,1001) frpop3,kpop3,thtapop3
     read(11,1001) ratiocn, ratiocp, ratiocsi
1001 format(//7f12.2)
     read(13,*) ninp                                          !(B)
c--  Main loop
     do 999 i=1,ninp
     read(13,*) monzz,dayzz,yearzz,sodzz,jnh4zz,jno3zz,jpo4zz,jsizz,
    .o2zz,nh4zz,no3zz,po4zz,sizz,tempzz
c--  if missing data do not process
     if((jnh4zz.eq.0.0).or.(sodzz.eq.0.0).or.
    .(o2zz.eq.0.0).or.(tempzz.eq.0.0).or.
    .(jnh4zz.eq.-999.0).or.(sodzz.eq.-999.0).or.
    .(o2zz.eq.-999.0).or.(tempzz.eq.-999.0)) then

     goto 999
     endif
c--  Compute jn                                               !(C)
     xknh4=kappnh4*thtanh4**(tempzz-20.)
c--  making the conversion factor for jn to sod a function of temperature
     ao2n=106./16.*12./14.*32./12.*1.068**(tempzz-20.)
     o2=o2zz
c--  convert from mg/m2-d to g/m2-d
     xjnh4=jnh4zz/1000.
     xnh4=nh4zz/1000.
c--  compute intermediate variables
     z1=xjnh4**2
     z2=xknh4**2
     z3=ao2n**2
     z4=1./z3
     z5=o2**2
     z5b=(-4.*ao2n*o2*xknh4**4*4*xnh4**3
    .  -z3*z1*z2*xnh4**2+18.*ao2n*o2*z1*z2*xnh4+27.*z5*z1*z2+4.*z3
    .  *xjnh4**4)
c--  check for negative root. If this occurs, assume that the olw nh4=0
     if(z5b.le.0.0) xnh4=0.0
     z5a=(z4*o2*xknh4*sqrt(-4.*ao2n*o2*xknh4**4*4*xnh4**3
    .  -z3*z1*z2*xnh4**2+18.*ao2n*o2*z1*z2*xnh4+27.*z5*z1*z2+4.*z3
    .  *xjnh4**4)/sqrt(3.)/6.0+z4*(9.*ao2n*o2*xjnh4*z2*xnh4
    .  +27.*z5*xjnh4**2
    .  +2.*z3*xjnh4**3)/54.0)
     z6=z5a/abs(z5a)*(abs(z5a))**(1.0/3.0)
     xjn=z6+(3.*o2*z2*xnh4+ao2n*z1)/(ao2n*z6)/9.0+xjnh4/3.0
     acn=106./16.*12./14.
     jcinzz=acn*xjn*1000.
c--  do not process missing data record
     if(jcinzz.le.0.0) goto 999                               !(D)
c--  Set maximum no of iterations and error                   !(E)
     isteady=20
     esteady=0.001
c--  set min and max and accuracy for root finder
     sodmin = 1.0e-6
     sodmax = 1.0e6
     eps = 1.0e-5
c--  set overlying water concentrations
     nh40=nh4zz
     no30=no3zz
```

```fortran
      o20=o2zz
      if(o20.eq.0.0) o20=1.0e-2
      po40=po4zz
      si0=sizz
      tempd=tempzz
      jcin=jcinzz/(frpoc1+frpoc2)                                    !(F)
c-- stoichiometric ratios for N,P and Si
      jnin=jcinzz/(frpon1+frpon2)/ratiocn
      jpin=jcinzz/(frpop1+frpop2)/ratiocp
      jsiin=jcin/ratiocsi
c-- evaluate the temperature dependent coefficients
      temp20=tempd-20.0
      temp202=temp20/2.0
      xapppnh4=kappnh4*thtanh4**temp202
      xappd1=kappd1*thtapd1**temp202
      xappp1=kappp1*thtapd1**temp202
      xapp1no3=kapp1no3*thtano3**temp202
      xk2no3=k2no3*thtano3**temp20*h2
      xksi=ksi*thtasi**temp20*h2
      kl12=Dd/(h2/2.0)*thtaDd**temp20
      w12=Dp/(h2/2.0)*(thtaDp**temp20)*poc1/1.0e5
      xkpon1=kpon1*thtapon1**temp20*h2
      xkpon2=kpon2*thtapon2**temp20*h2
      xkpon3=kpon3*thtapon3**temp20*h2
      xkpoc1=kpoc1*thtapoc1**temp20*h2
      xkpoc2=kpoc2*thtapoc2**temp20*h2
      xkpoc3=kpoc3*thtapoc3**temp20*h2
      xkpop1=kpop1*thtapop1**temp20*h2
      xkpop2=kpop2*thtapop2**temp20*h2
      xkpop3=kpop3*thtapop3**temp20*h2
      xksi=ksi*thtasi**temp20*h2
c-- set t minus 1 concentrations to anything                        !(G)
      nh4t2tm1=0.0
      sit2tm1=0.0
      psitm1=0.0
c-- Loop to converge
      do 1113 istst=1,isteady                                       !(H)
      iststsav=istst
c-- compute diagenesis
      pon1=(frpon1*jnin)/(xkpon1+w2)                                !(I)

      pon2=(frpon2*jnin)/(xkpon2+w2)
      pon3=(frpon3*jnin)/(xkpon3+w2)
      xjn=xkpon1*pon1+xkpon2*pon2+xkpon3*pon3
      jn=xkpon1*pon1+xkpon2*pon2+xkpon3*pon3
      poc1=(frpoc1*jcin)/(xkpoc1+w2)
      poc2=(frpoc2*jcin)/(xkpoc2+w2)
      poc3=(frpoc3*jcin)/(xkpoc3+w2)
      xjc=xkpoc1*poc1+xkpoc2*poc2+xkpoc3*poc3
      jc=xkpoc1*poc1+xkpoc2*poc2+xkpoc3*poc3
      pop1=(frpop1*jpin)/(xkpop1+w2)
      pop2=(frpop2*jpin)/(xkpop2+w2)
      pop3=(frpop3*jpin)/(xkpop3+w2)
      xjp=xkpop1*pop1+xkpop2*pop2+xkpop3*pop3
      jp=xkpop1*pop1+xkpop2*pop2+xkpop3*pop3
      fd2=1.0/(1.0+m2*pie2si)
      k3=xksi*(csisat-fd2*sit2tm1)/(psitm1+kmpsi)
      psi=jsiin/(k3+w2)                                             !(J)
c-- solve the nh4,no3,sod, po4 and si equations
      sod=zbrent(sedf,sodmin,sodmax,eps)                            !(K)
c-- replace the t minus 1 concentrations
      nh4tm1=nh41
      psitm1=psi
      sit2tm1=sit2
c-- compare to the previous iteration
      iagain=0                                                      !(L)
c-- do at least one iteration
      if(istst.eq.1) iagain=1
      if(abs(nh4t2sav-nh4t2).gt.abs(esteady*nh4t2sav)) iagain=1
      if(abs(no3t2sav-no3t2).gt.abs(esteady*no3t2sav)) iagain=1
      if(abs(hst2sav-hst2).gt.abs(esteady*hst2sav)) iagain=1
      if(abs(sit2sav-sit2).gt.abs(esteady*sit2sav)) iagain=1
      if(abs(po4t2sav-po4t2).gt.abs(esteady*po4t2sav)) iagain=1
      if(abs(pon1sav-pon1).gt.abs(esteady*pon1sav)) iagain=1
      if(abs(pon2sav-pon2).gt.abs(esteady*pon2sav)) iagain=1
      if(abs(pon3sav-pon3).gt.abs(esteady*pon3sav)) iagain=1
      if(abs(poc1sav-poc1).gt.abs(esteady*poc1sav)) iagain=1
      if(abs(poc2sav-poc2).gt.abs(esteady*poc2sav)) iagain=1
      if(abs(poc3sav-poc3).gt.abs(esteady*poc3sav)) iagain=1
      if(abs(pop1sav-pop1).gt.abs(esteady*pop1sav)) iagain=1
      if(abs(pop2sav-pop2).gt.abs(esteady*pop2sav)) iagain=1
      if(abs(pop3sav-pop3).gt.abs(esteady*pop3sav)) iagain=1
      if(abs(psisav-psi).gt.abs(esteady*psisav)) iagain=1
```

```fortran
c-  iagain=0 => converged; otherwise save the ending conditions and repeat
      if(iagain.eq.0) go to 1114
      if(istst.ne.isteady) then                                    !(M)
      nh4t2sav=nh4t2
      no3t2sav=no3t2
      hst2sav=hst2
      sit2sav=sit2
      po4t2sav=po4t2
      pon1sav=pon1
      pon2sav=pon2
      pon3sav=pon3
      poc1sav=poc1
      poc2sav=poc2
      poc3sav=poc3
      pop1sav=pop1
      pop2sav=pop2
      pop3sav=pop3
      psisav=psi
      end if
1113  continue
      print *,'Did not equilibrate in',isteady,' iterations'
1114  continue
c-  output solid phase concentrations in mg/g (O2) mg/kg (po4) mg/g (Si)
      hst2r=hst2/1000./m2
      po4t2r=po4t2/1000./m2
      sit2r=(psi+sit2)/1.0e6/m2
      psir=(psi)/1.0e6/m2
c-  output POC,PON,etc in mg/g
      pon1r=pon1/1.0e6/m2
      pon2r=pon2/1.0e6/m2
      pon3r=pon3/1.0e6/m2
      ponr=pon1r+pon2r+pon3r
      poc1r=poc1/1.0e6/m2
      poc2r=poc2/1.0e6/m2
      poc3r=poc3/1.0e6/m2
      pocr=poc1r+poc2r+poc3r
      pop1r=pop1/1.0e6/m2
      pop2r=pop2/1.0e6/m2
      pop3r=pop3/1.0e6/m2
      popr=pop1r+pop2r+pop3r
c-  output dissolved concentrations in mg/L
      nh40z=nh40/1000.
      nh41z=nh41/1000.
      nh42z=nh42/1000.
      nh4t2z=nh4t2/1000.
      no30z=no30/1000.
      no31z=no31/1000.
      no32z=no32/1000.
      no3t2z=no3t2/1000.
      si0z=si0/1000.
      si1z=si1/1000.
      si2z=si2/1000.
      sit2z=sit2/1000.
      po40z=po40/1000.
      po41z=po41/1000.
      po42z=po42/1000.
      po4t2z=po4t2/1000.
      hs1z=hs1
      hs2z=hs2
      hst2z=hst2
      if(i.eq.1)
     . write(15,1003)'mon,','day,','year,','sodzz,','jnh4zz,','jno3zz,',
     . 'jpo4zz,','jsizz,','o2zz,','nh4zz,','no3zz,','po4zz,','sizz,',
     . 'tempzz,','jnin,','jn,','jnh4,','nh40z,','nh41z,','nh42z,',
     . 'pon1r,','pon2r,','pon3r,','jno3,','no30z,','no31z,','no32z,',
     . 'jpin,','jp,','jpo4,','po40z,','po41z,','po42z,','pop1r,',
     . 'pop2r,','pop3r,','jsiin,','jsi,','si0z,','si1z,','si2z,',
     . 'sit2r,','jcin,','jc,','jhs,','hs1z,','hs2z,','hst2r,','poc1r,',
     . 'poc2r,','poc3r,','o20,','sod,','s,','h1,','csod'
1003  format(60a11)
      h1=k112*h2/s
      xmon=monzz
      xday=dayzz
      xyear=yearzz
      write(15,103) xmon,xday,xyear,sodzz,jnh4zz,jno3zz,jpo4zz,
     . jsizz,o2zz,nh4zz,no3zz,po4zz,sizz,tempzz,jnin,jn,jnh4,nh40z,
     . nh41z,nh42z,pon1r,pon2r,pon3r,jno3,no30z,no31z,no32z,jpin,jp,
     . jpo4,po40z,po41z,po42z,pop1r,pop2r,pop3r,jsiin,jsi,si0z,si1z,
     . si2z,sit2r,jcin,jc,jhs,hs1z,hs2z,hst2r,poc1r,poc2r,poc3r,
     . o20,sod,s,h1,csod
103   format(60(e10.3,','))
999   continue
      end
```

```fortran
      function sedf(sod1)                                          !(N)
      include 'sedsfdim.for'
c- Ammonia
      s = sod1/o20                                                 !(O)
      k0h1p=0.
      k1h1p=0.
      k2h2d=0.
      k2h2p=0.
      if(kmnh4.ne.0.0) then
      k0h1d=xappnh4**2/s*kmnh4*(o20/((kmnh4o2+o20))
      k1h1d=s
      else
      k1h1d=xappnh4**2/s*(o20/((kmnh4o2+o20))+s
      k0h1d=0.
      endif
      j1=s*nh40
      k3=0.0
      j2=jn
      pie1=pienh4
      pie2=pienh4
      kmc1=kmnh4
      call sedsffnl(nh41,nh42,nh4t1,nh4t2,nh41tm1)
      jnh4=s*(nh41-nh40)
c- Oxygen consumed by nitrification
c     a1 = 64/14 * 1/1000 - mole ratio and mg/m2-day to gm/m2-day
      a1 = 0.0045714
      if(kmnh4.ne.0.0) then
      jo2nh4=a1*k0h1d*nh41/(kmnh4+nh41tm1)
      else
      jo2nh4=a1*(k1h1d-s)*nh41
      endif
c- Nitrate
      k0h1d=0.
      k0h1p=0.
      kmc1=0.0
      k1h1d=xapp1no3**2/s+s
      k1h1p=0.
      k2h2d=xk2no3
      k2h2p=0.
      if(kmnh4.ne.0.0) then
      j1=s*no30+
     .  xappnh4**2/s*kmnh4*(o20/((kmnh4o2+o20))*nh41/(kmnh4+nh41tm1)
      else
      j1=s*no30+xappnh4**2/s*(o20/((kmnh4o2+o20))*nh41
      endif
      k3=0.0
      j2=0.0
      pie1=0.
      pie2=0.
      call sedsffnl(no31,no32,no3t1,no3t2,0.0)
      jno3=s*(no31-no30)
c- Sulfide
c     diagenesis consumed by denitrification
c       a2 =  10/8 * 32/14 * 1/1000
      a2 = 0.0028571
      xjcno3=a2*xapp1no3**2/s*no31 + a2*xk2no3*no32
c- Convert carbon diagenesis flux to o2* units and decrement jc
c       a0= 32./12./1000.
      a0=2.66666666e-3
      xjco2 = a0*jc
      xjc1 = amax1(xjco2-xjcno3,0.0)
203   continue
c-- Sulfide in O2 equivalents
      k0h1d=0.
      k0h1p=0.
      kmc1=0.
      k1h1d=xappd1**2/s*(o20/kmhso2) + s
      k1h1p=xappp1**2/s*(o20/kmhso2)
      k2h2d=0.
      k2h2p=0.
      j1=0.
      k3=0.0
      j2=xjc1
      pie1=pie1s
      pie2=pie2s
      call sedsffnl(hs1,hs2,hst1,hst2,0.0)
      jhs=s*hs1
      csod=(xappd1**2/s*fd1 + xappp1**2/s*fp1)*(o20/kmhso2)*hst1
c- Silica
      k0h1d=0.
      k0h1p=0.
      kmc1=0.0
      k1h1d=s
      k1h1p=0.
      k2h2d=0.
```

```fortran
      k2h2p=0.
      j1=s*si0
c-- Oxygen dependency of pie1
      if(o20.lt.o2crit) then
        pie1=pie2si*dpie1si**(o20/o2crit)
      else
        pie1=pie2si*dpie1si
      endif
      pie2=pie2si
c- Silica dissolution kinetics
      fd2=1./(1.+m2*pie2)
      k3=xksi*psi/(psi+kmpsi)*fd2
      j2=xksi*psi/(psi+kmpsi)*csisat
      call sedsffnl(si1,si2,sit1,sit2,0.0)
      jsi=s*(si1-si0)
c- Phosphate
      k0h1d=0.
      k0h1p=0.
      kmc1=0.0
      k1h1d=s
      k1h1p=0.
      k2h2d=0.
      k2h2p=0.
      j1=s*po40
      k3=0.0
      j2=jp
c-- Oxygen dependency of pie1
      if(o20.lt.o2crit) then
        pie1=pie2po4*dpie1po4**(o20/o2crit)
      else
        pie1=pie2po4*dpie1po4
      endif
      pie2=pie2po4
      call sedsffnl(po41,po42,po4t1,po4t2,0.0)
      jpo4=s*(po41-po40)
c- SOD - evaluate the error
      sod=csod + jo2nh4
      sedf = sod - sod1
      return
      end                                              !(P)
```

```fortran
      subroutine sedsffnl(c1,c2,ct1,ct2,c1tm1)         !(Q)
      include 'sedsfdim.for'
c- Notation
c   c1 = layer 1 dissolved conc.
c   c2 = layer 2 dissolved conc.
c   ct1 = layer 1 total conc.
c   ct2 = layer 2 total conc.
c   c1tm1 = c1 at time level t - 1

      fd1=1./(1.+m1*pie1)
      fp1=m1*pie1/(1.+m1*pie1)
      fd2=1./(1.+m2*pie2)
      fp2=m2*pie2/(1.+m2*pie2)
c- Transport and Decay terms
      f12 = w12*fp1 + kl12*fd1
      f21 = w12*fp2 + kl12*fd2
c   evaluate the MM term at time level t-1          !(R)
      if(kmc1.ne.0.0) then
        xk0 = (k0h1d*fd1 + k0h1p*fp1)/(kmc1+c1tm1)
      else
        xk0=0.0
      endif
      xk1 =  xk0 + k1h1d*fd1 + k1h1p*fp1
      xk2 = k2h2d*fd2 + k2h2p*fp2
c- Matrix and forcing function
      a11 = -f12 -xk1 - w2
      a21 =  f12 + w2
      a12 = f21
      b1 = -j1
      a22=-f21-xk2-w2-k3
      b2=-j2
c-- Solve the 2x2 set of linear equations, stop if determinant=0
      det=a11*a22-a12*a21                             !(S)
      if(det.eq.0.0) stop
      ct1=(b1*a22-b2*a12)/det
      ct2=(b2*a11-b1*a21)/det
      c1=fd1*ct1
      c2=fd2*ct2
      return
      end
```

sedsfdim.for !(T)

```
c- Input parameters
      real
     . m1,m2,kappnh4,kmnh4,kmnh4o2,kapp1no3,k2no3,kappd1,kappp1,
     . kmhso2,ksi,kmpsi,kmo2Dp,kmo2Dd,kpon1,kpon2,kpon3,kpoc1,
     . kpoc2,kpoc3,kpop1,kpop2,kpop3
      common /in/
     . m1,m2,Dp,w2,Dd,thtaDp,kappnh4,pienh4,thtanh4,kmnh4,
     . kmnh4o2,kapp1no3,k2no3,thtano3,kappd1,kappp1,piels,pie2s,
     . thtapd1,kmhso2,ksi,csisat,dpielsi,pie2si,h2,thtasi,kmpsi,
     . dpielpo4,pie2po4,o2crit,kmo2Dp,kmo2Dd,frpon1,kpon1,thtapon1,
     . frpon2,kpon2,thtapon2,frpon3,kpon3,thtapon3,frpoc1,kpoc1,
     . thtapoc1,frpoc2,kpoc2,thtapoc2,frpoc3,kpoc3,thtapoc3,frpop1,
     . kpop1,thtapop1,frpop2,kpop2,thtapop2,frpop3,kpop3,thtapop3.
     . ratiocn, ratiocp, ratiocsi
c- concentrations
      real
     . nh40,nh41,nh42,nh4t1,nh4t2,nh4t1m1,jnh4,nh4t2sav,no30,no31,no32,
     . no3t1,no3t2,jno3,no3t2sav,jhs,jsi,jpo4
      common /conc/
     . nh40,nh41,nh42,nh4t1,nh4t2,nh4t1m1,jnh4,nh4t2sav,no30,no31,no32,
     . no3t1,no3t2,jno3,no3t2sav,hs0,hs1,hs2,hst1,hst2,jhs,hst2sav,
     . si0,si1,si2,sit1,sit2,jsi,sit2sav,po40,po41,po42,po4t1,po4t2,
     . jpo4,po4t2sav
c- diagenesis
      real
     . jnin,xjn,jn,jcin,xjc,jc,jpin,xjp,jp,jsiin,jo2nh4
      common /diag/
     . jnin,xjn,jn,jcin,xjc,jc,jpin,xjp,jp,psi,psitm1,psisav,psiav,
     . jsiin,xjcno3,xjco2,xjc1,jo2nh4,xjco2av,xjc1av
c- sedsffnl storage
      real
     . k0h1d,k0h1p,k1h1d,k1h1p,k2h2d,k2h2p,k3,
     . pie1,pie2,j1,j2,kmc1,w12,kll2,tempd,o20
      common /pars/
     . k0h1d,k0h1p,k1h1d,k1h1p,k2h2d,k2h2p,k3,
     . pie1,pie2,j1,j2,kmc1,w12,kll2,tempd,o20
c- sedf storage
      real kll2nom
      common /store/
     . xapppnh4,xappd1,xappp1,xapp1no3,xk2no3,xksi,temp20,temp202,
     . fd1,fp1,fd2,fp2,sod,csod,s,kll2nom
```

Input file !(U)

Test Run
----

m1	m2	Dp	w2	Dd	thtaDp	thtaDd
.5	.5	0.6E-04	6.85E-06	5.0E-04	1.15	1.08
kg/L	kg/L	m2/d	m/d	m2/d	-	-

kappnh4	pienh4	thtanh4	kmnh4	thtakmnh4	kmnh4o2
1.313E-01	1.00	1.123	728.0	1.125	0.74
m/d	L/kg	-	ug/L	-	mg O2/L

kapp1no3	k2no3	thtano3
0.10	0.25	1.080
m/d	m/d	-

kappd1	kappp1	piels	pie2s	thtapd1	kmhso2
0.2	0.4	100.0	100.0	1.0800	4.0
m/d	m/d	L/kg	L/kg	-	mg O2/L

ksi	dpielsi	csisat	kmpsi	pie2si	o2critsi
0.5	10.0	40000.	5.0E+07	100.0	1.0
/d	-	ug Si/L	L/kg	L/kg	-

h2	thtasi	o2crit	kmo2Dp
0.1	1.100	2.0	4.0
/d	-	mg O2/L	mg O2/L

m	dpielpo4	pie2po4
-	300.0	100.0
ug Si/L	L/kg	L/kg

frpon1	kpon1	thtapon1
0.65	0.035	1.10
gN/gN	/d	-
frpon2	kpon2	thtapon2
0.25	0.0018	1.15
gN/gN	/d	-
frpon3	kpon3	thtapon3
0.10	0.0	1.17
gN/gN	/d	-

frpoc1	kpoc1	thtapoc1
0.65	0.035	1.10
gC/gC	/d	-
frpoc2	kpoc2	thtapoc2
0.20	0.0018	1.15
gC/gC	/d	-
frpoc3	kpoc3	thtapoc3
0.15	0.0	1.17
gC/gC	/d	-

frpop1	0.65	gP/gP	kpop1	0.035	/d	thtapop1	1.10	-
frpop2	0.20	gP/gP	kpop2	0.0018	/d	thtapop2	1.15	-
frpop3	0.15	gP/gP	kpop3	0.0	/d	thtapop3	1.17	-
ratiocn	5.68	gC/gN	ratiocp	41.	gC/gP	ratiocsi	2.5	gC/gSi

Output file - One line, transposed

					!(V)
sodzz	g/m2-d	1.17	jpo4	mg/m2-d	9.64
jnh4zz	mg/m2-d	62.8	po40z	mg/L	0.010
jno3zz	mg/m2-d	-14.1	po41z	mg/L	0.056
jpo4zz	mg/m2-d	0.00	po42z	mg/L	4.32
jsizz	mg/L	325.	pop1r	mg/g	0.007
o2zz	mg/L	5.57	pop2r	mg/g	0.050
nh4zz	mg/m3	125.0	pop3r	mg/g	0.585
no3zz	mg/m3	103.0	jsiin	mg/m2-d	219.
po4zz	mg/m3	9.60	jsi	mg/m2-d	94.2
sizz	mg/m3	385.	si0z	mg/L	0.385
tempzz	oC	15.8	si1z	mg/L	0.841
jnin	mg/m2-d	91.0	si2z	mg/L	27.7
jn	mg/m2-d	80.3	sit2r	mg/L	36.4
jnh4	mg/L	67.4	jcin	mg/m2-d	547.0
nh40z	mg/L	0.125	jc	mg/m2-d	457.0
nh41z	mg/L	0.451	jhs	mg/m2-d	0.006
nh42z	mg/g	22.6	hs1z	mg/L	0.028
pon1r	mg/g	0.050	hs2z	mg/L	303.
pon2r	mg/g	0.426	hst2r	mg/g	31.00
pon3r	mg/m2-d	2.66	poc1r	mg/g	0.303
jno3	mg/L	7.33	poc2r	mg/g	2.05
no30z	mg/L	0.103	poc3r	mg/g	24.0
no31z	mg/L	0.139	o20	mg/L	5.57
no32z	mg/m2-d	13.4	sod	g/m2-d	1.15
jpin	mg/m2-d	11.2	s	m/d	0.206
jp			h1	m	0.002
			csod	g/m2-d	1.09

Nomenclature

A	Cross sectional area (**19.9**, *457*).*
a_{CN}	Redfield carbon to nitrogen stoichiometric coefficient (**1.10**, *9*).
a_{C,N_2}	Carbon required to denitrify nitrate to nitrogen to nitrogen gas (**9.17**, *190*).
a_{CP}	Redfield carbon to phosphorus stoichiometric coefficient (**1.10**, *9*).
a_{CSi}	Carbon to phosphorus stoichiometric coefficient (Table 1.1, *10*).
$[A_i]$	Concentration of chemical species A_i (**8.15**, *170*).
a_{ik}	Stoichiometric coefficient of the kth component in the ith species (**8.19**, *171*).
$[Alk]$	Dissolved alkalinity concentration (**17.3**, *397*).
$Alk(0)$	Total alkalinity concentration in layer 0 (**17.25a**, *402*).
Alk_0	Dissolved alkalinity concentration in layer 0 (**18.62a**, *440*).
$Alk(1)$	Total alkalinity concentration in layer 1 (**17.25a**, *402*).
Alk_1	Dissolved alkalinity concentration in layer 1 (**18.62a**, *440*).
$Alk(2)$	Total alkalinity concentration in layer 2 (**17.25a**, *402*).
Alk_2	Dissolved alkalinity concentration in layer 2 (**18.62a**, *440*).
Alk_d	Dissolved alkalinity (**17.26a**, *402*).
Alk_T	Total alkalinity concentration (**17.11**, *398*).
a_{O_2C}	Redfield oxygen to carbon stoichiometric coefficient (**16.4**, *368*).

---

*(**Equation number**, *page number*).

a_{O_2,NH_4}	Stoichiometric ratio of oxygen consumed to ammonia oxidized (**10.25**, *201*).
$a_{O_2,NH_4 \to N_2}$	Stoichiometric ratio for the oxygen consumed by ammonia oxidation and denitrification to nitrogen gas (**10.40**, *205*).
$a_{O_2,NH_4 \to NO_3}$	Stoichiometric ratio for the oxygen consumed by ammonia oxidation to nitrate (**10.45**, *206*).
$a_{O_2^*,N}$	Redfield ratio of oxygen equivalents to nitrogen (**10.44**, *206*).
$a_{SO_4,Alk}$	Alkalinity produced per mole of sulfate reduced (**12.23**, *258*).
$a_{SO_4,C}$	Sulfate reduced per mole of carbon oxidized (**11.1**, *222*).
β	Parameter group $w_2/(sf_{d1}r_{12})$ (**18.32**, *420*).
β_{Mn}	Ratio of overlying water dissolved oxygen concentration to critical oxygen concentration for aerobic layer manganese partitioning (**18.45b**, *428*).
β_{PO_4}	Ratio of overlying water dissolved oxygen concentration to critical oxygen concentration for aerobic layer phosphate partitioning (**6.21**, *141*).
β_{Si}	Ratio of overlying water dissolved oxygen concentration to critical oxygen concentration for aerobic layer silica partitioning (**7.19**, *156*).
$[B_k]$	Concentration of chemical component B_k (**8.19**, *171*).
$[Ca]$	Molar concentration of dissolved calcium (**17.1**, *397*).
$Ca(0)$	Total calcium concentration in layer 0 (**17.25c**, *402*).
$Ca(1)$	Total calcium concentration in layer 1 (**17.25c**, *402*).
$Ca(2)$	Total calcium concentration in layer 2 (**17.25c**, *402*).
$CaCO_3(2)$	Concentration of $CaCO_3(s)$ in layer 2 (**17.25**, *402*).
Ca_d	Dissolved calcium concentration $Ca_T - CaCO_3$ (**17.13**, *399*).
Ca_T	Total calcium concentration (**17.10**, *398*).
Ca_x	Excess calcium concentration $Ca_T - 1/2Alk_T$ (**17.15**, *399*).
$[CH_4(0)]$	Methane concentration in layer 0 in oxygen equivalents (**11.38a**, *235*).
$[CH_4(1)]$	Methane concentration in layer 1 in oxygen equivalents (**11.38a**, *235*).
$[CH_4(2)]$	Methane concentration in layer 2 in oxygen equivalents (**11.38a**, *235*).
$[CH_4]_{sat}$	Methane concentration in equilibrium methane gas in oxygen equivalents (**10.51**, *208*).
$[CH_4(sat)]$	Methane concentration in equilibrium methane gas in oxygen equivalents (**11.37**, *234*).
CSOD	Carbonaceous sediment oxygen demand. The flux of oxygen to the sediment due to the oxidation of the carbon component of organic matter (**9.11**, *189*).
$CSOD_{max}$	The CSOD that would result from the complete oxidation of all the dissolved methane transported to the sediment-water interface (**10.28**, *203*).
C_{T1}	Total (dissolved + particulate) concentration in layer 1 (**5.1a**, *121*).

C_{T2}	Total (dissolved + particulate) concentration in layer 2 (**5.1b**, *121*).
D_d	Diffusion coefficient of dissolved solutes (**21.6**, *514*).
$\Delta\pi_{Mn,1}$	Factor that increases the aerobic layer partition coefficient $\pi_{Mn,1} = \Delta\pi_{Mn,1}\pi_{Mn,2}$ relative to the anaerobic layer partition coefficient $\pi_{Mn,2}$ (**7.18**, *156*).
$\Delta\pi_{PO_4,1}$	Factor that increases the aerobic layer partition coefficient $\pi_{PO_4,1} = \Delta\pi_{PO_4,1}\pi_{PO_4,2}$ relative to the anaerobic layer partition coefficient $\pi_{PO_4,2}$ (**6.19**, *141*).
$\Delta\pi_{Si,1}$	Factor that increases the aerobic layer partition coefficient $\pi_{Si,1} = \Delta\pi_{Si,1}\pi_{Si,2}$ relative to the anaerobic layer partition coefficient $\pi_{Si,2}$ (**7.18**, *156*).
D_p	Diffusion coefficient of particulate solutes due to particle mixing (bioturbation) (**13.1**, *276*).
$E_{0,-1}$	Vertical mixing coefficient between the bottom water column layer and the layer immediately above (**20.10**, *484*).
$f_{colloidal}$	Colloidal fraction of the dissolved water column manganese (**19.33**, *473*).
f_{d0}	Dissolved fraction of the total concentration in layer 0 (**5.2a**, *122*).
f_{d1}	Dissolved fraction of the total concentration in layer 1 (**5.2a**, *122*).
f_{d2}	Dissolved fraction of the total concentration in layer 2 (**5.2a**, *122*).
$f_{d,FeS}$	Fraction FeS dissolved (**21.26**, *518*).
$f_{p,FeS}$	Fraction FeS particulate (**21.26**, *518*).
f_{d,Cd_T}	Fraction Cd_T dissolved (**21.30**, *519*).
f_{p,Cd_T}	Fraction Cd_T particulate (**21.30**, *519*).
Fe(0)	Total iron (II) concentration in layer 0 (**20.9**, *482*).
Fe(1)	Total iron (II) concentration in layer 1 (**20.9**, *482*).
Fe(2)	Total iron (II) concentration in layer 2 (**20.9**, *482*).
FeOOH(0)	Particulate iron (III) concentration in layer 0 (**20.9**, *482*).
FeOOH(1)	Particulate iron (III) concentration in layer 1 (**20.9**, *482*).
FeOOH(2)	Particulate iron (III) concentration in layer 2 (**20.9**, *482*).
f_ℓ	Labile fraction of the total particulate COD (**8.33**, *174*).
$f_{O_2^*}$	Oxidized fraction of the dissolved COD transported to the sediment-water inferface (**8.14**, *166*).
f_{p1}	Particulate fraction of the total concentration in layer 1 (**5.2b**, *122*).
f_{p2}	Particulate fraction of the total concentration in layer 2 (**5.2b**, *122*).
$f_{POC,1}$	Fraction of particulate organic carbon POC in the G_1 reactivity class (**12.1**, *252*).
$f_{POC,2}$	Fraction of particulate organic carbon POC in the G_2 reactivity class (**12.1**, *252*).
$f_{POC,3}$	Fraction of particulate organic carbon POC in the G_3 reactivity class (**12.1**, *252*).
$f_{PON,1}$	Fraction of particulate organic nitrogen PON in the G_1 reactivity class (**12.1**, *252*).

$f_{PON,2}$ Fraction of particulate organic nitrogen PON in the G_2 reactivity class (**12.1**, *252*).

$f_{PON,3}$ Fraction of particulate organic nitrogen PON in the G_3 reactivity class (**12.1**, *252*).

$f_{POP,1}$ Fraction of particulate organic phosphorus POP in the G_1 reactivity class (**12.1**, *252*).

$f_{POP,2}$ Fraction of particulate organic phosphorus POP in the G_2 reactivity class (**12.1**, *252*).

$f_{POP,3}$ Fraction of particulate organic phosphorus POP in the G_3 reactivity class (**12.1**, *252*).

f_r Refractory PCOD component fraction of the algal carbon (**8.39**, *177*).

H_0 Depth of layer 0 (**10.51**, *208*).

H_0 Concentration of H^+ in layer 0 (**18.69**, *441*).

H_1 Depth of layer 1 (**3.15**,*69*).

H_1 Concentration of H^+ in layer 1 (**18.70**, *442*).

$\dot{H_1}$ Time derivative of the aerobic layer depth $dH_1 dt$ (**13.18a**, *287*).

$\dot{H_1^-}$ Negative of the time derivative of the aerobic layer depth $-dH_1 dt$ limited to $dH_1/dt \leqslant 0$ (**13.14b**, *286*).

$\dot{H_1^+}$ Time derivative of the aerobic layer depth $dH_1 dt$ limited to $dH_1/dt \geqslant 0$ (**13.14a**, *286*).

H_2 Depth of the anaerobic layer (Fig. 3.1,*442*).

$\dot{H_2}$ Time derivative of the anaerobic layer $dH_2 dt$ (**13.18b**, *287*).

H_{sat} Depth at which the dissolved methane concentration reaches its solubility limit $[CH_4]_{sat}$ (**10A.5**, *217*).

H_{SO_4} Depth of the zone of sulfate reduction (**11.14**, *228*).

H_T Total depth of the active sediment layer $H_1 + H_2$ (**13.22**, *287*).

J_{Alk} Source of alkalinity in the anaerobic layer due to sulfate reduction (**17.25b**, *402*).

$J[Alk]$ Flux of alkalinity from the sediment to layer 0 (**17.29**, *404*).

J_{aq} Flux of solute to layer 0 (**5.20a**, *125*).

J_{br} Flux of solute that is buried (**5.20c**, *125*).

J_C Carbon diagenesis flux (**12.2**, *253*).

$J[Ca]$ Flux of calcium to layer 0 (**17.30**, *404*).

$J_{CaCO_3(s)}$ Burial flux of calcium carbonate (**17.37**, *405*).

J_{CH_4} Methane source due to carbon diagenesis (**10.34**, *204*).

$J[CH_4(aq)]$ Dissolved methane flux to layer 0 (**10.13**, *199*).

$J[CH_4(g)]$ Flux of methane gas to layer 0 (**10.14**, *199*).

J_{DetrSi} Silica source due to clay mineral weathering (**16A.21**, *391*).

J_{Gas} Volumetric gas flux to layer 0 (**10.56**, *210*).

$J[Mn]$ Flux of dissolved manganese (II) to layer 0 (**18.26**, *419*).

J_{MnO_2} Flux of particulate Mn(IV) to the sediment (**18.21c**, *418*).

J_N	Source of ammonia due to nitrogen diagenesis as a flux (**12.5**, *253*).
$J[N_2(g)]$	Flux of nitrogen gas to layer 0 (**4.55**, *114*).
$J_N(i)$	Nitrogen diagenesis for station i as a flux (**3.37**, *76*).
$J_N(i, t_j)$	Nitrogen diagenesis for station i at time t_j as a flux (**3.37**, *76*).
J_{NO_3}	Nitrate source flux due to nitrification as a flux $S[NO_3]H_1$ (**4.3**, *95*).
$J[NO_3]$	Nitrate flux to layer 0 (**4.4**, *95*).
$J_N(T)$	Nitrogen diagenesis at temperature T (**4.33**, *105*).
J_P	Source of phosphorus due to phosphorus diagenesis as a flux (**12.6**,*253*).
J_{PCOD}	Flux of particulate chemical oxygen demand COD to the sediment (**8.14**, *166*).
J_{POC}	Depositional flux of particulate organic carbon POC to the sediment (**12.46**, *269*).
J_{PON}	Depositional flux of particulate organic nitrogen PON to the sediment (**12.1**, *252*).
J_{POP}	Depositional flux of particulate organic phosphorus POP to the sediment (**12.47**, *269*).
J_{PSi}	Depositional flux of particulate silica PSi to the sediment (**7.15**, *154*).
$J_{re,2}$	Solute reacted in anaerobic layer as a flux (**5.20b**, *125*).
$J[Si]$	Silica flux to layer 0 (**7.8**, *153*).
$J[SO_4]$	Sulfate flux to layer 0 (**11.15**, *228*).
J_{SS}	Suspended solids flux (**1.5**, *5*).
J_{T1}	Source of solute to the aerobic layer (**5.1a**, *121*).
J_{T2}	Source of solute to the anaerobic layer (**5.1b**, *121*).
K_1	First order removal reaction rate constant in layer 1 (**5.1a**, *121*), (**18.66a**, *441*).
K_2	First order removal reaction rate constant in layer 2 (**5.1b**, *121*), (**18.66b**, *441*).
κ_1	Reaction velocity for first order removal reaction rate constant in layer 1 (**5.15**, *125*).
κ_2	Reaction velocity for first order removal reaction rate constant in layer 2 (**5.16**, *125*).
κ_D	Diffusion mass transfer coefficient for dissolved methane (**10.17**, *199*).
$\kappa_{Fe,1}$	Aerobic layer Fe(II) oxidation reaction velocity (**20.9**, *482*).
$\kappa_{Fe,2}$	Anaerobic layer FeOOH reduction reaction velocity (**20.9**, *482*).
$\kappa_{H_2S,d1}$	Aerobic layer dissolved H_2S oxidation reaction velocity (**9.3**, *185*).
$\kappa_{H_2S,p1}$	Aerobic layer particulate H_2S oxidation reaction velocity (**9.4**, *185*).
$\kappa_{Mn,0}$	Overlying water aerobic layer Mn(II) oxidation reaction velocity (**19.7**, *457*).
$\kappa_{Mn,1}$	Aerobic layer Mn(II) oxidation reaction velocity (**18.17**, *417*).
$\kappa_{MnO_2,2}$	Anaerobic layer $MnO_2(s)$ reduction reaction velocity (**18.19**, *417*).
$\kappa_{NH_4,1}$	Aerobic layer ammonia oxidation reaction velocity (**3.17**, *69*).
$\kappa_{NO_3,1}$	Aerobic layer nitrate reduction reaction velocity (**4.8**, *96*).

$\kappa_{NO_3,2}$	Anaerobic layer nitrate reduction reaction velocity (**4.9**, *96*).
$\kappa_{NO_3,2}^*$	Effective anaerobic layer nitrate reduction reaction velocity-mass transfer coefficient (**4.10**, *96*).
$\kappa_{Si,2}$	Particulate silica dissolution reaction velocity (**7.7**, *152*).
k_1	First order decay rate constant in layer 1 (**5.1a**, *121*).
k_2	First order decay rate constant in layer 2 (**5.1b**, *121*).
K_2	Equilibrium constant for the reaction $HCO_3^- \leftrightarrow H^+ + CO_3^{2-}$ (**17.5**, *397*).
K_{CaAlk}	Apparent solubility product for $CaCO_3$ in terms of total dissolved calcium and dissolved alkalinity (**17.9**, *398*).
k_{CdS}	Oxidation rate of CdS (**21.28**, *519*).
$k_{CH_4,1}$	First-order rate constant for dissolved methane oxidation (**10.20**, *200*).
$k_C(T)$	Chemical SOD at temperature T (**8.13**, *165*).
k_{Disp}	Rate of the displacement reaction.
$k_{Fe,OH,0}'$	Overlying water Fe(II) oxidation rate constant (**20.10**, *484*).
$k_{FeOOH,0}$	Overlying water FeOOH(s) reduction rate constant (**20.10**, *484*).
$k_{Fe,OH,1}$	Aerobic layer Fe(II) oxidation rate constant (**20.10**, *484*).
$k_{FeOOH,2}$	Anaerobic layer FeOOH(s) reduction rate constant (**20.10**, *484*).
k_{FeS}	Oxidation rate of FeS (**21.22**, *517*).
k_{FeS_d}	Oxidation rate of dissolved FeS (**21.23**, *518*).
k_{FeS_p}	Oxidation rate of particulate FeS (**21.21**, *517*).
$k_{H_2S,d1}$	Aerobic layer dissolved H_2S oxidation reaction constant (**9.2**, *185*).
$k_{H_2S,p1}$	Aerobic layer particulate H_2S oxidation reaction constant (**9.2**, *185*).
K_{L01}	Surface mass transfer coefficient between overlying water and aerobic layer 1 (**3.1a**, *65*).
K_{L01,O_2}	Surface mass transfer coefficient for O_2 (**3.12**, *68*).
K_{L12}	Mass transfer coefficient between aerobic and anaerobic layers 1 and 2 (**13.6**, *281*).
K_{L12,SO_4}	Mass transfer coefficient for sulfate between layers 1 and 2 (**11.19**, *229*).
K_{L12,CH_4}	Mass transfer coefficient for methane between layers 1 and 2 (**11.36b**, *234*).
K_{L01,NH_4}	Mass transfer coefficient for ammonia between layers 1 and 2 (**3.10d**, *67*).
K_{M,O_2}	O_2 half-saturation constant for oxidation of POC (**21.11**, *515*).
K_{M,D_p}	Half-saturation constant for dissolved oxygen for particle mixing (**13.2**, *277*).
K_{M,H_2S,O_2}	Scales layer 0 oxygen concentration for the second order sulfide oxidation reaction (**9.2**, 185).
$k_{Mn,0}'$	Oxidation of Mn(II) first-order rate constant in layer 0 (**19.7**, *457*).
$k_{Mn,0}$	Oxidation of Mn(II) second-order rate constant in layer 0 (**19.7**, *457*).
$k_{Mn,OH,0}$	Overlying water aerobic layer Mn(II) oxidation pH dependent reaction rate constant (**19.36**, *473*).

$k_{Mn,OH,MnO_2,0}$ Overlying water aerobic layer Mn(II) oxidation pH dependent auto-catalytic reaction rate constant (**19.35**, *473*).

$k_{MnO_2,0}$ Reduction of MnO_2 first-order rate constant in the overlying water (**20.21**, *504*).

$k'_{Mn,1}$ Oxidation of Mn(II) first-order rate constant in the aerobic layer (**18.13**, *416*).

$k_{Mn,1}$ Oxidation of Mn(II) second-order rate constant in the aerobic layer (**18.46a**, *428*).

$k'_{Mn,OH,1}$ Overlying water aerobic layer Mn(II) oxidation pH dependent first order reaction rate constant (**20.21**, *504*).

$k_{Mn,OH,1}$ Overlying water aerobic layer Mn(II) oxidation pH dependent reaction rate constant (**18.79**, *446*).

$k_{MnO_2,2}$ Reduction of MnO_2 first-order rate constant in the anaerobic layer (**18.18**, *417*).

$K_{M,FeOOH,0}$ Oxygen half-saturation constant for iron oxidation in layer 0 (**20.11**, *485*).

$K_{M,MnO_2,0}$ Oxygen half-saturation constant for manganese oxidation in layer 0 (**20.12**, *485*).

K_{M,NH_4} Ammonia half-saturation constant for nitrification (**3.29**, *72*).

K_{M,O_2} Oxygen half-saturation constant for POC oxidation (**21.11**, *515*).

K_{M,O_2,NH_4} Oxygen half-saturation constant for nitrification (**3.45**, *81*).

K_{MOH} Equilibrium constant for metal hydrolysis (**19.3**, *454*).

$K_{M,P_{Si}}$ Half-saturation constant for particulate silica mineralization (**7.17**, *156*).

$k_{NH_4,1}$ First-order rate constant for nitrification (**3.29**, *72*).

$k_{NO_3,1}$ First-order rate constant for denitrification in layer 1 (**4.5a**, *96*).

$k_{NO_3,2}$ First-order rate constant for denitrification in layer 2 (**4.5b**, *96*).

k_{PCOD} First-order rate constant for PCOD mineralization (**8.26**, *173*).

$k_{POC,1}$ First-order rate constant for G_1 POC mineralization (**12.1**, *252*).

$k_{POC,1,SO_4}$ First-order rate constant for G_1 POC mineralization using SO_4 (**21.13**, *515*).

$k_{POC,2}$ First-order rate constant for G_2 POC mineralization (**12.1**, *252*).

$k_{POC,2,SO_4}$ First-order rate constant for G_2 POC mineralization using SO_4 (**21.16**, *516*).

$k_{PON,1}$ First-order rate constant for G_1 PON mineralization (**12.3**, *253*).

$k_{PON,2}$ First-order rate constant for G_2 PON mineralization (**12.3**, *253*).

$k_{POP,1}$ First-order rate constant for G_1 POP mineralization (**12.4**, *253*).

$k_{POP,2}$ First-order rate constant for G_2 POP mineralization (**12.4**, *253*).

k_S First-order rate constant coefficient for accumulated stress (**13.3**, *280*).

K_{s1} Solubility product for $CaCO_3(s)$ (**17.1**, *397*).

k'_{Si} Specific reaction rate for silica dissolution (**7.1**, *150*).

k_{Si} First-order rate constant for silica dissolution (**7.2**, *150*).

k_{T1}	Overall first-order rate constant for reaction and mass transfer in layer 1 (**5.5**, *123*).
m_1	Solids concentration in layer 1 (**1.4**, *4*).
m_2	Solids concentration in layer 2 (**1.4**, *4*).
$Mn(0)$	Total Mn(II) concentration in layer 0 (**18.21a**, *418*).
$Mn(0)_i$	Total Mn(II) concentration in the inflow to layer 0 (**19.12**, *459*).
$Mn(1)$	Total Mn(II) concentration in layer 1 (**18.21a**, *418*).
$Mn(2)$	Total Mn(II) concentration in layer 2 (**18.21a**, *418*).
$MnO_2(0)$	MnO_2 concentration in layer 0 (**18.21c**, *418*).
$MnO_2(0)_i$	MnO_2 concentration in the inflow to layer 0 (**19.12**, *459*).
$MnO_2(1)$	MnO_2 concentration in layer 1 (**18.21c**, *418*).
$MnO_2(2)$	MnO_2 concentration in layer 2 (**18.21c**, *418*).
$\mu(T)$	Maximum biological SOD at temperature T (**8.13**, *165*).
$n_1(z)$	Ammonia concentration in layer 1 at depth z (**3.20**, *70*).
$n_2(z)$	Ammonia concentration in layer 2 at depth z (**3.20**, *70*).
$[NH_4(0)]$	Ammonia concentration in layer 0 (**3.1a**, *65*).
$[NH_4(1)]$	Ammonia concentration in layer 1 (**3.1a**, *65*).
$[NH_4(2)]$	Ammonia concentration in layer 2 (**3.1b**, *65*).
$[NO_3(z)]$	Nitrate concentration at depth z (**4.1**, *95*).
NSOD	Nitrogenous sediment oxygen demand due to ammonia oxidation (**9.14**, *190*).
η	Dimensionless reaction rate-diffusion coefficient-sedimentation velocity number (**8.32**, *173*).
ν_{ji}	Reaction stoichiometry, the quantity of A_i produced by the jth reaction (**8.15**, *170*).
$[O_2(0)]$	Oxygen concentration in layer 0 (**8.5**, *163*).
$[O_2(1)]$	Oxygen concentration in layer 1 (**3.33**, *73*).
$[O_2(H_1)]$	Oxygen concentration at the bottom of layer 1 (**8.6**, *164*).
$[O_2(0)]_{crit,Fe}$	Oxygen half-saturation constant for aerobic layer iron partition coefficient (Table 20.3, *490*).
$[O_2(0)]_{crit,Mn}$	Oxygen half-saturation constant for aerobic layer manganese partition coefficient (**18.45a**, *428*)
$[O_2]_{crit,PO_4}$	Oxygen half-saturation constant for aerobic layer phosphate partition coefficient (**6.20**, *141*).
$[O_2]_{crit,Si}$	Oxygen half-saturation constant for aerobic layer silica partition coefficient (**7.18**, *156*).
$[O_2^*(z)]$	Oxygen equivalents concentration at depth z (**8.28**, *173*).
Ω_{PO_4}	Equivalent burial rate of particulate phosphate (**6.15**, *138*).
Ω_{Si}	Equivalent burial rate of particulate silica (**7.14**, *153*).
$[PCOD(z)]$	Concentration of particulate chemical oxygen demand at depth z (**8.27**, *173*).
pH_0	pH in layer 0 (**18.69**, *441*).
pH_1	pH in layer 1 (**18.71**, *442*).
ϕ	Porosity (**1.1**, *3*).

$\pi_{Cd,FeOOH}$	Cd partition coefficient to FeOOH (**21A.5f**, *537*).
$\pi_{Cd,POC}$	Cd partition coefficient to POC (**21A.5e**, *537*).
π_{FeS}	FeS partition coefficient (**21A.2**, *536*).
$\pi_{Fe,2}$	Iron (II) anaerobic layer partition coefficient (Table 20.3, *490*).
$\pi_{H_2S,1}$	Total sulfide partition coefficient in layer 1 (Table 9.1, *187*).
$\pi_{H_2S,2}$	Total sulfide partition coefficient in layer 1 (Table 9.1, *187*).
$\pi_{Mn,1}$	Mn(II) partition coefficient in layer 1 (**18.12**, *415*).
$\pi_{Mn,2}$	Mn(II) partition coefficient in the aerobic layer 2 (**18.12**, *415*).
$\pi_{Mn,OC}$	Mn(II) partition coefficient to particulate organic carbon (**19.5**, *455*).
$\pi_{NH_4,1}$	Ammonia partition coefficient in layer 1 (Appendix B).
$\pi_{PO_4,2}$	Phosphate partition coefficient in layer 2 (**6.19**, *141*).
$\pi_{Si,1}$	Silica partition coefficient in layer 1 (**7.18**, *156*).
$\pi_{Si,2}$	Silica partition coefficient in layer 2 (**7.18**, *156*).
$[PO_4(0)]$	Phosphate concentration in layer 0 (**6.1a**, *133*).
$[PO_4(1)]_T$	Phosphate concentration in layer 1 (**6.1a**, *133*).
$[PO_4(2)]_T$	Phosphate concentration in layer 2 (**6.1b**, *133*).
POC_i	Concentration of the G_i component of POC (**12.1**, *252*).
POC_1	Concentration of the G_1 component of POC (**12.1**, *252*).
POC_2	Concentration of the G_2 component of POC (**12.1**, *252*).
POC_3	Concentration of the G_3 component of POC (**12.1**, *252*).
$POC_{1,R}$	Reference G_1 POC concentration for bioturbation (**13.1**, *276*).
PON_i	Concentration of the G_i component of PON (**12.3**, *253*).
PON_1	Concentration of the G_1 component of PON (**12.3**, *253*).
PON_2	Concentration of the G_2 component of PON (**12.3**, *253*).
PON_3	Concentration of the G_3 component of PON (**12.3**, *253*).
POP_i	Concentration of the G_i component of POP (**12.4**, *253*).
POP_1	Concentration of the G_1 component of POP (**12.4**, *253*).
POP_2	Concentration of the G_2 component of POP (**12.4**, *253*).
POP_3	Concentration of the G_3 component of POP (**12.4**, *253*).
P_{Si}	Particulate silica concentration (**7.1**, *150*).
Q	Flow through layer 0 layer (**19.9**, *457*).
q_0	Hydraulic detention time in layer 0 (**19.9**, *457*).
R_0	Zero-order consumption rate of oxygen (**8.3**, *163*).
r_{12}	Ratio of total concentration in layer 1 to layer 2 (**5.7**, *123*).
r_{21}	Ratio of total concentration in layer 2 to layer 1 (**5.10**, *124*).
R_C	Zero-order consumption rate of oxygen (**8.12**, *165*).
ρ_b	Bulk density (**1.2**, *4*).
ρ_s	Solids density (**1.2**, *4*).
ρ_w	Density of water (**1.2**, *4*).
R_j	Difference between the backward and forward reaction rates of the jth fast reversible reaction (**8.15**, *170*).
R_N	Nitrification rate in the aerobic layer (**8.12**, *165*).
$R_{NH_4,1}$	Rate of nitrification in layer 1 (**3.29**, *72*).
$R_{N,max}$	Maximum nitrification rate $R_N \leqslant R_{N,max}$ (**8.12**, *165*).

S	Accumulated benthic stress (**13.3**, *280*).
s	Ratio of SOD to overlying water dissolved oxygen (**3.14**, *68*).
s_{20}	s at $T = 20°C$ (**4.31**, *104*).
S_{CH_4}	Methane source in the anaerobic layer (**10.3**, *197*).
S_i	Net source of A_i due to the slow reactions (**8.15**, *170*).
$[Si(0)]$	Dissolved silica concentration in layer 0 (**7.5a**, *152*).
$[Si(1)]_T$	Total silica concentration in layer 1 (**7.5a**, *152*).
$[Si(2)]_T$	Total silica concentration in layer 2 (**7.5b**, *152*).
$\sigma_{Cd,FeOOH}$	Cd sorption capacity of FeOOH (**21A.7**, *538*).
$\sigma_{Cd,POC}$	Cd sorption capacity of POC (**21A.6**, *538*).
$\Sigma H_2S(1)]$	Total sulfide concentration in layer 1 (**16A.16**, *390*).
$[Si]_{sat}$	Saturated concentration of silica (**7.1**, *150*).
$S_{Mn(II)}$	Rate of Mn(II) production in layer 2 per unit area (**18.18**, *417*).
S_{MnO_2}	Rate of MnO$_2$(s) production in layer 1 per unit area (**18.13**, *416*).
S_N	Rate of ammonia diagenesis (**3.20**, *70*).
$S[NO_3]$	Rate of nitrate production in the aerobic layer (**3.45**, *81*).
$[SO_4(0)]$	Sulfate concentration in layer 0 (**11.2a**, *223*).
$[SO_4(1)]$	Sulfate concentration in layer 1 (**11.2a**, *223*).
$[SO_4(2)]$	Sulfate concentration in layer 1 (**11.2a**, *223*).
SOD	Sediment oxygen demand (**9.19**, *190*).
SOD_{max}	SOD which results from the complete oxidation of dissolved methane and ammonia in the aerobic layer (**10.50**, *207*).
S_{Si}	Rate of production of dissolved silica (**7.1**, *150*).
θ_{CdS}	Temperature coefficient for k_{CdS} (**21.28**, *519*).
θ_{D_d}	Temperature coefficient for D_d (**21.6**, *514*).
θ_{D_p}	Temperature coefficient for D_p (**13.1**, *276*).
$\theta_{Fe,0}$	Temperature coefficient for $k_{Fe,0}$ (**20.10**, *484*).
$\theta_{FeOOH,0}$	Temperature coefficient for $k_{FeOOH,0}$ (**20.10**, *484*).
θ_{FeS}	Temperature coefficient for k_{FeS} (**21.22**, *517*).
$\theta_{Fe,1}$	Temperature coefficient for $k_{Fe,1}$ (**20.3**, *490*).
$\theta_{FeOOH,2}$	Temperature coefficient for $k_{FeOOH,2}$ (**20.3**, *490*).
θ_{H_2S}	Temperature coefficient for $k_{H_2S,d1}$ and $k_{H_2S,p1}$ (**9.2**, *185*).
$\theta_{K_{M,NH_4},1}$	Temperature coefficient for K_{M,NH_4} (**3.29**, *72*).
$\theta_{Mn,0}$	Temperature coefficient for $k_{Mn,0}$ (**19.7**, *457*).
$\theta_{Mn,1}$	Temperature coefficient for $\kappa_{Mn,1}$ (**18.17**, *417*).
$\theta_{MnO2,2}$	Temperature coefficient for $\kappa_{MnO_2,2}$ (**18.19**, *417*).
θ_N	Temperature coefficient for $J_N(T)$ (**4.33**, *105*).
$\theta_{NH_4,1}$	Temperature coefficient for $\kappa_{NH_4,1}$ (**3.31**, *72*).
θ_{NO_3}	Temperature coefficient for $\kappa_{NO_3,1}$ (**4.8**, *96*).
$\theta_{POC,1}$	Temperature coefficient for $k_{POC,1}$ (**12.1**, *252*).
$\theta_{POC,2}$	Temperature coefficient for $k_{POC,2}$ (**12.1**, *252*).
$\theta_{PON,1}$	Temperature coefficient for $k_{PON,1}$ (**12.3**, *253*).
$\theta_{PON,2}$	Temperature coefficient for $k_{PON,2}$ (**12.3**, *253*).
$\theta_{POP,1}$	Temperature coefficient for $k_{POP,1}$ (**12.4**, *253*).

$\theta_{POP,2}$	Temperature coefficient for $k_{POP,2}$ (**12.4**, *253*).
θ_s	Temperature coefficient for s_{20} (**4.31**, *104*).
θ_{Si}	Temperature coefficient for k_{Si} (**7.1**, *150*).
$t_{D_p 1/4}$	Time for biomass buildup to 1/4 max (**21.5**, *514*).
$t_{D_p 1/2}$	Time for biomass buildup to 1/4 max (**21.5**, *514*).
TIC_0	Total inorganic carbon in layer 0 (**18.62b**, *440*).
TIC_1	Total inorganic carbon in layer 1 (**18.62b**, *440*).
TIC_2	Total inorganic carbon in layer 2 (**18.62b**, *440*).
w_{12}	Particle mixing velocity between layers 1 and 2 without benthic stress (**13.5**, *281*).
w_{12}^*	Particle mixing velocity between layers 1 and 2 (**13.2**, *277*).
w_2	Sedimentation velocity (**2.37**, *36*).
z_{D_p}	e-folding depth of particle mixing (**21.4**, *514*).

Bibliography

Abodollahi H. and Nedwell D.B. (1979). Seasonal temperature as a factor influencing bacterial sulfate reduction in a saltmarsh sediment. *Micro. Ecol.*, 5:73–79.

Abramowitz M. and Stegun I.A. (1954). *Handbook of Mathematical Functions.* National Bureau of Standards. Applied Mathematics Series 55. U.S Government Printing Office, Wash. DC.

Aitchison J. and Brown J.A.C. (1957). *The Lognormal Distribution.* Cambridge University Press, Cambridge, England.

Allen H., Gongmin F., and Deng B. (1993). Analysis of acid volatile sulfide (AVS) and simultaneously extracted metals (SEM) for estimation of potential toxicity in aquatic sediments. *Environ. Toxicol. Chem.*, 12:1441–1453.

Aller R. (1980a). Diagenetic processes near the sediment-water interface of Long Island Sound. I. Decomposition and Nutrient Element Geochemistry (S, N, P). In B. Saltzman, editor, *Estuarine Physics and Chemistry: Studies in Long Island Sound. Advances in Geophysics*, pp. 237–350. Academic Press, New York.

Aller R. (1980b). Diagenetic processes near the sediment-water interface of Long Island Sound. II. Fe and Mn. In B. Saltzman, editor, *Estuarine Physics and Chemistry: Studies in Long Island Sound. Advances in Geophysics*, pp. 351–415. Academic Press, New York.

Aller R. (1994). The sedimentary Mn cycle in Long Island Sound: Its role as intermediate oxidant and the influence of bioturbation, O_2, and C_{org} flux on diagenetic reaction balances. *J. Mar. Res.*, 52:259–295.

Aller R. and Aller J. (1998). The effect of biogenic irrigation intensity and solute exchange on diagenetic reaction rates in marine sediments. *J. Mar. Res.*, 56:905–936.

Aller R., Benninger L., and Cochran J. (1980). Tracking particle-associated processes in nearshore environments by use of $^{234}Th/^{238}U$ disequilibrium. *Earth and Planetary Sci.*, 47:161–175.

Aller R., Mackin J., Ullman W., Chen-Hou W., Shing-Min T., Jian-Cai J., Yong-Nian S., and Jia-Zhen H. (1985). Early chemical diagenesis, sediment-water solute exchange, and storage of reactive organic matter near the mouth of the Changjiang, East China Sea. *Continental Shelf Res.*, 4:227–251.

Aller R.C. (1982). The effects of macrobenthos on chemical properties of marine sediment and overlying water. In P. McCall and M. Tevesz, editors, *Animal-Sediment Relations. The Biogenic Alteration of Sediments*, pp. 53–102. Plenum Press, New York.

Aller R.C. and Benninger L.K. (1981). Spatial and temporal patterns of dissolved ammonium, manganese, and silica fluxes from bottom sediments of Long Island Sound, USA. *J. Mar. Res.*, 39:295–314.

Aller R.C. and Yingst J.Y. (1980). Relationships between microbial distributions and the anaerobic decomposition of organic matter in surface sediment of Long Island Sound, USA. *Mar. Biol.*, 56:29–42.

Almegren T. and Hagstrom I. (1974). The oxidation rate of sulfide in sea water. *Wat. Res*, 8:395–400.

Andersen J.M. (1978). Importance of the denitrification process for the rate of degradation of organic matter. In H.I. Golterman and W. Junk, editors, *Lake Sediments in Interactions between Sediments and Fresh Water*. B.V. Publishers, The Hague.

Anikouchine W.A. (1967). Dissolved chemical substances in compacting marine sediments. *J. Geophysic. Res.*, 72:505–509.

Antoniou P., Hamilton J., Koopman B., Jain R., Holloway B., Lyberatos G., and Svoronos S.A. (1990). Effect of temperature and pH on the effective maximum specific growth rate of nitrifying bacteria. *Wat. Res*, 24(1):97–101.

Argaman Y. and Miller E. (1979). Modeling recycled systems for biological nitrification and denitrification. *J. Water Pollution Control Federation*, 51(4):749–758.

Baccini P. (1985). Phosphate interactions at the sediment-water interface. In W. Stumm, editor, *Chemical Processes in Lakes*, pp. 189–205. John Wiley & Sons, New York.

Baes C.F. and Mesmer R.E. (1976). *The Hydrolysis of Cations*. John Wiley & Sons, New York.

Baity H. (1938). Some factors affecting the aerobic decomposition of sewage sludge deposits. *Sewage Works J.*, 10:539–568.

Balzer W. (1996). Particle mixing processes of Chernobyl fallout in deep Norwegian Sea sediments: Evidence for seasonal effects. *Geochim. Cosmochim. Acta*, 60:3425–3433.

Barnes R.O. and Goldberg E.D. (1976). Methane production and consumption in anoxic marine sediments. *Geology*, 4:297–300.

Barrow N.J. (1983). A mechanistic model for describing the sorption and desorption of phosphate by soil. *Journal of Soil Science*, 34:733–750.

Berg P., Risgaard-Petersen N., and Rysgaard S. (1998). Interpretation of measured concentration profiles in sediment pore water. *Limnol. Oceanogr.*, 43:1500–1510.

Berner R.A. (1971). *Principles of Chemical Sedimentology*. McGraw-Hill, New York.

Berner R.A. (1974). Kinetic models for the early diagenesis of nitrogen, sulfur, phosphorus and silicon in anoxic marine sediments. In E. Goldberg, editor, *The Sea*, vol. 5, pp. 427–450. John Wiley & Sons, New York.

Berner R.A. (1977). Stoichiometric models for nutrient regeneration in anoxic sediments. *Limnol. Oceanogr.*, 22(5):781–786.

Berner R.A. (1980a). *Early Diagenesis. A Theoretical Approach*. Princeton University Press, Princeton, NJ.

Berner R.A. (1980b). A rate model for organic matter decomposition during bacterial sulfate reduction in marine sediments. In *Biogeochemistry of Organic Matter at the Sediment-*

Water Interface, pp. 35–44. Comm. Natl. Recherche Scientific, France.

Berry W., Hansen D., Mahony J., Robson D., Di Toro D., Shipley B., Rogers B., Corbin J., and Boothman W. (1996). Predicting the toxicity of metals-spiked laboratory sediments using acid volatile sulfide and interstitial water normalizations. *Environ. Toxicol. Chem.*, 15(12):2067–2079.

Billen G. (1978). A budget of nitrogen recycling in North Sea sediments off the Belgian coast. *Est. Coast. Mar. Sci.*, 7:127–146.

Billen G. (1982). An idealized model of nitrogen recycling in marine sediments. *Am. J. Sci.*, 282:512–541.

Billen G. (1988). Modelling benthic nitrogen cycling in temperate coastal ecosystems. In T.H. Blackburn and J. Sorensen, editors, *Nitrogen Cycling in Coastal Marine Environments*, pp. 341–378. John Wiley & Sons, New York.

Billen G., Dessery S., Lancelot C., and Maybeck M. (1989). Seasonal and inter-annual variations of nitrogen diagenesis in the sediments of a recently impounded basin. *Biogeochem.*, 8:73–100.

Bird R., Stewart W., and Lightfoot E. (1960). *Transport Phenomena*. John Wiley & Sons, New York.

Blackburn T. (1994). Simulation model of the coupling between nitrification and denitrification in a freshwater sediment. *Applied and Environmental Microbiology*, 60:3089–3095.

Blackburn T.H. (1990). Denitrification model for marine sediment. In N. Revsbech and J. Sorensen, editors, *Denitrification in Soil and Sediment*, pp. 323–337. Plenum Press, New York.

Bostrom B., Andersen J.M., Fleischer S., and Jansson M. (1988). Exchange of phosphorus across the sediment-water interface. *Hydrobiol.*, 170:229–244.

Boudreau B.P. (1990). Asymptotic forms and solutions of the model for silica-opal diagenesis in bioturbated sediments. *J. Geophysic. Res.*, 95(C5):7367–7379.

Boudreau B.P. (1996). *Diagenetic Models and Their Implementation. Modelling Transport and Reactions in Aquatic Sediments.* Springer-Verlag, Berlin.

Boudreau B.P. (1998). Mean mixed depth of sediments: The wherefore and the why. *Limnol. Oceanogr.*, 43:524–526.

Boudreau B.P. and Canfield D. (1993). A comparison of closed- and open-system models for porewater pH and calcite-saturation state. *Geochim. Cosmochim. Acta*, 57:317–334.

Boudreau B.P. (1986a). Mathematics of tracer mixing in sediments: I. Spatially-dependent diffusive mixing. *Am. J. Sci.*, 286:161–198.

Boudreau B.P. (1986b). Mathematics of tracer mixing in sediments: II. Nonlocal mixing and biological conveyor-belt phenomena. *Am. J. Sci.*, 286:199–238.

Boudreau B.P. (1991). Modelling the sulfide-oxygen reaction and associated pH gradients in porewaters. *Geochim. Cosmochim. Acta*, 55:145–159.

Bouldin D. (1968). Models for describing the diffusion of oxygen and other mobile constituents across the mud-water interface. *J. Ecol.*, 56(77):77–87.

Boynton W., Kemp W., and Barnes J. (1985a). *Ecosystem Processes Component. Data Rept. No. 2.* Chesapeake Biological Laboratory, University of Maryland System, Solomons, MD. UMCEES CBL No. 85–121.

Boynton W., Kemp W., Garber J., and Barnes J. (1986). *Ecosystem Processes Component Level I Data Interim Rept.* Chesapeake Biological Laboratory, University of Maryland System, Solomons, MD. UMCEES CBL Ref No. 86–56.

Boynton W., Kemp W., Garber J., Barnes J., Robertson L., and Watts J. (1988a). *Ecosystem Process Component. Maryland Chesapeake Bay Water Quality Monitoring Program.* Maryland Department of the Environment. Level I. Rept. No.5 [UMCEES] CBL Ref. No.

88–126.

Boynton W., Kemp W., Garber J., Barnes J., Robertson L., and Watts J. (1988b). *Ecosystem Process Component. Maryland Chesapeake Bay Water Quality Monitoring Program. Maryland Department of the Environment. Level I. Rept. No.5.* Chesapeake Biological Laboratory, University of Maryland System, Solomons, MD. UMCEES CBL Ref. 88–69.

Boynton W., Kemp W., Lubbers L., Wood K., and Keefe C. (1985b). *Ecosystems Processes Component. Data Rept. No. 1.* Chesapeake Biological Laboratory, University of Maryland System, Solomons, MD. UMCEES CBL Ref. No. 84–109.

Boynton W., Kemp W., Lubbers L., Wood K., Keefe C., and Barnes J. (1985c). *Ecosystems Process Component. Study Plan.* Chesapeake Biological Laboratory, University of Maryland System, Solomons, MD. UMCEES CBL Ref. No. 85–16.

Boynton W., Kemp W., Osborne C., Kaymeyer K., and Jenkins M. (1981). Influence of water circulation rate on in situ measurements of benthic community respiration. *Marine Biol.*, 65:185–190.

Boynton W., Kemp W.M., Lubbers L., Wood K., and Keefe C. (1985d). *Ecosystem Processes Component, Pilot Study for Maryland Office of Environmental Programs.* Chesapeake Biological Laboratory, University of Maryland System, Solomons, MD. UMCEES CBL Ref. No. 85–3.

Boynton W.R. and Kemp W.M. (1985). *Chesapeake Bay Water Quality Monitoring Program. Ecosystem Processes Component.* Chesapeake Biological Laboratory, University of Maryland System, Solomons, MD.

Boynton W.R., Kemp W.M., Barnes J.M., Cowan J.L.W., Stammerjohn S.E., Matteson L.L., Rohland F.M., and Garber J.H. (1990). Long-term characteristics and trends of benthic oxygen and nutrient fluxes in the Maryland portion of Chesapeake Bay. In *New Perspectives in the Chesapeake System: A research and management partnership.*, pp. 339–354. Chesapeake Res. Consortium #137, Baltimore, MD.

Boynton W.R., Sampou P., Barnes J., Weaver B., and Magdeburger L. (1995). *Sediment-water fluxes and sediment analyses in Chesapeake Bay: Tidal fresh Potomac River and Maryland Mainstem.* Chesapeake Biological Laboratory, University of Maryland System, Solomons, MD. UMCEES-CBL-Ref 95–112.

Brandes J. and Devol A. (1995). Simultaneous nitrate and oxygen respiration in coastal sediments: Evidence for discrete diagenesis. *J. Mar. Res.*, 53:771–797.

Bray J. (1973). *The Behavior of Phosphate in the Interstitial Waters of Chesapeake Bay Sediments.* Ph.D. thesis, The Johns Hopkins University, Baltimore, MD.

Bray J., Bricker O., and Troup B. (1973). Phosphate in interstitial waters of anoxic sediments: Oxidation effects during sampling procedure. *Sci.*, 180:1362–1364.

Brezonik P. (1994). *Chemical Kinetics and Process Dynamics in Aquatic Systems.* CRC Press, Boca Raton, FL.

Brezonik P.L. (1977). Denitrification in natural waters. *Prog. Wat. Technol.*, 8:373–392.

Bricker O., Matisoff G., and Holdren G. (1977). *Interstitial water chemistry of Chesapeake Bay sediments. Basic Data Report No.9.* Maryland Geological Survey Technical report.

Bricker O. and Troup B. (1975). Sediment-water exchange in Chesapeake Bay. In L. Cronin, editor, *Estuarine Research*, pp. 3–27. Academic Press, New York.

Bufflap S. and Allen H. (1995). Comparison of pore water sampling techniques for trace metals. *Wat. Res*, 29(9):2051–2054.

Buffle J., De Vitre R., Perret D., and Leppard G. (1988). Combining field measurements for speciation in non perturbable water samples. In J. Kramer and H. Allen, editors, *Metal Speciation: Theory, Analysis and Application.* Lewis Publisher, Chelsea, MI.

Burdige D. (1989). *Sediment Monitoring Program for Southern Chesapeake Bay*. Dept. of Oceanography, Old Dominion University, Norfolk VA, ODRUF technical report 89-6.

Burdige D. (1991). The kinetics of organic matter mineralization in anoxic marine sediments. *J. Mar. Res.*, 49:727–761.

Butler J. (1991). *Carbon Dioxide Equilibria and Their Applications*. Lewis Publishers, Chelsea, MI.

Canfield D.E. (1989). Reactive iron in marine sediments. *Geochim. Cosmochim. Acta*, 53:619–632.

Capone D. and Kiene R. (1988). Comparison of microbial dynamics in marine and freshwater sediments: Contrasts in anaerobic catabolism. *Limnol. Oceanogr.*, 33:725–749.

Caraco N., Cole J., and Likens G. (1993). Sulfate control of phosphorus availability in lakes. In P.C.M. Boers, T.E. Cappenberg, and W. van Raaphorst, editors, *Hydrobiologia*, volume 253, pp. 275–280. Kluwer.

Carbanaro R. (1999). *Modeling metal sulfide fluxes from sediments*. MS thesis, Environmental Engineering Dept. Manhattan College, Riverdale NY.

Carignan R., Rapin F., and Tessier A. (1985). Sediment porewater sampling for metal analysis: A comparison of techniques. *Geochim. Cosmochim. Acta*, 49:2493–2497.

Carnahan B., Luther H., and Wilkes J. (1969). *Applied Numerical Methods*. John Wiley & Sons, New York.

Cerco C. (1988). Sediment nutrient fluxes in a tidal freshwater embayment. *Wat. Resources Bull.*, 24(3):255–260.

Cerco C. (1995a). Response of Chesapeake Bay to nutrient load reductions. *J. Environ. Engr. ASCE*, 121(8):549–557.

Cerco C. (1995b). Simulation of long-term trends in Chesapeake Bay eutrophication. *J. Environ. Engr. ASCE*, 121(4):298–310.

Cerco C. and Cole T. (1993). Three-dimensional eutrophication model of Chesapeake Bay. *J. Environ. Engr. ASCE*, 119(6):1006–1025.

Cerco C. and Cole T. (1994). *Three-dimensional Eutrophication Model of Chesapeake Bay*. U.S. Army Corps of Engineers. Waterways Experiment Station. Vicksburg MI. Report El–94–4.

Cerco C. and Cole T. (1995). *User's guide to the CE-QUAL-ICM Three-dimensional Eutrophication Model*. U.S. Army Corps of Engineers. Waterways Experiment Station. Vicksburg MI. Report El–95–15.

Cerco C.F. (1985). *Effect of Temperature and Dissolved Oxygen on Sediment-Water Nutrient Fluxes*. Virginia Institute of Marine Science, Gloucester Pt. VA. Technical report.

Chambers R.M. and Odum W.E. (1990). Porewater oxidation, dissolved phosphate and the iron curtain. *Biogeochem.*, 10:37–52.

Chapra S. (1997). *Surface Water-Quality Modeling*. McGraw-Hill, New York.

Cline J. and Richards F. (1969). Oxygenation of hydrogen sulfide in seawater at constant salinity, temperature, and pH. *Environ. Sci. Technol.*, 3(9):838.

Cochran J.K. and Aller R.C. (1979). Particle reworking in sediments from the New York Bight Apex: Evidence from ^{234}Th/^{238}U disequilibrium. *Est. Coast. Mar. Sci.*, 9:739–747.

Conley D.J. and Schelske C.L. (1989). Processes controlling the benthic regeneration and sedimentary accumulation of biogenic silica in Lake Michigan. *Arch. Hydrobiol.*, 116(1):23–43.

Conley D.J., Schelske C.L., Dempsey B.G., and Campbell C.D. (1986). Distribution of biogenic silica in the surficial sediments of Lake Michigan. *Can. J. Earth Sci.*, 23:1442–1449.

Cooke J.G. and White R.E. (1988). Nitrate enhancement of nitrification depth in sediment/water microcosms. *Environ. Geol. Wat. Sci.*, 11(1):85–94.

Copp J. and Dold P. (1998). Confirming the nitrate-to-oxygen conversion factor for denitrification. *Water Research*, 32:1296–1304.

Cornwell J. (1993). Preliminary data for Onondaga Lake sediment process study. University of Maryland Center for Environmental Studies. Horn Point Laboratory, Cambridge MD. University of Maryland Center for Environmental Studies. Horn Point Laboratory, Cambridge MD. *Personal communication.*

Cornwell J. and Owens M. (1998a). *Benthic phosphorus cycling in Lake Champlain.* University of Maryland Center for Environmental Studies. Horn Point Laboratory, Cambridge MD. University of Maryland Center for Environmental Studies. Horn Point Laboratory, Cambridge MD. Technical report TS–168–98.

Cornwell J. and Owens M. (1998b). *Sediment biogeochemical processes in Jamaica Bay, New York.* University of Maryland Center for Environmental Studies. Horn Point Laboratory, Cambridge MD. University of Maryland Center for Environmental Studies. Horn Point Laboratory, Cambridge MD. Technical report.

Cowan J. and Boynton W. (1996). Sediment-water oxygen and nutrient exchanges along the longitudinal axis of Chesapeake Bay: Seasonal patterns, controlling factors and ecological significance. *Estuaries*, 19:562–580.

Crank J. (1975). *The Mathematics of Diffusion.* p. 39. Oxford Univ. Press, London.

Davies S. and DeWiest R. (1966). *Hydrogeology.* Wiley-Interscience, New York.

Davies S. and Morgan J. (1989). Mn(II) oxidation kinetics. *J. Colloid Interface Sci.*, 129:63–77.

Davison W. (1985). Conceptual models for transport at a redox boundary. In W. Stumm, editor, *Chemical Processes in Lakes*, pp. 31–53. John Wiley & Sons, New York.

de Beer D., Glud A., Epping E., and Kuhl M. (1997). A fast-responding CO_2 microelectrode for profiling sediments, microbial mats, and biofilms. *Limnol. Oceanogr.*, 42:1590–1600.

DeLand E.C. (1967). *Chemist-The RAND Chemical Equilibrium Program* Technical report RM–5404–PR. RAND Corp., Santa Monica, CA.

deRooij N. (1991). *Mathematical simulation of biochemical processes in natural waters by the model CHARON.* Delft Hydraulics Institute, Delft, The Netherlands. Report R1310–10.

DeWitt T., Swartz R., Hansen D., McGovern D., and Berry W. (1996). Bioavailability and chronic toxicity of cadmium in sediment to the estuarine amphipod *Leptocheirus plumulosus*. *Environ. Toxicol. Chem.*, 15(12):2095–2101.

Di Toro D.M. and Fitzpatrick J. (1993). *Chesapeake Bay Sediment Flux Model.* HydroQual, Inc. Mahwah, NJ. Prepared for U.S. Army Engineer Waterways Experiment Station, Vicksburg, MS. Contract Report EL–93–2.

Di Toro D.M., Fitzpatrick J., and Isleib R. (1998). *A Model of Manganese and Iron Fluxes from Sediments.* HydroQual, Inc. Mahwah, NJ. Prepared for U.S. Army Engineer Waterways Experiment Station, Vicksburg, MS. Contract Report W–98–1.

Di Toro D.M., Fitzpatrick J., and Thomann R. (1981). *Documentation for water quality analysis simulation program (WASP) and model verification program (MVP).* HydroQual, Inc. Mahwah, NJ. Prepared for U.S. EPA. Report EPA–600–3–81–044.

Di Toro D.M., Lowe S., and Fitzpatrick J. (2001). Application of a water column-sediment eutrophication model to a mesocosm experiment. I. Calibration. *J. Environ. Engr. ASCE.*

Di Toro D.M., Mahony J., and Gonzalez A. (1996a). Particle oxidation model of synthetic FeS and sediment AVS. *Environ. Toxicol. Chem.*, 15(12):2156–2167.

Di Toro D.M., Mahony J., Hansen D., and Berry W. (1996b). A model of the oxidation of iron and cadmium sulfide in sediments. *Environ. Toxicol. Chem.*, 15(12):2168–2186.

Di Toro D.M. and Paquin P. (1984). Time variable model of the fate of DDE and Lindane in a quarry. *Environ. Toxicol. Chem.*, 3:335–353.

Di Toro D.M. (1976). Combining chemical equilibrium and phytoplankton models—a general methodology. In R.P. Canale, editor, *Modeling Biochemical Processes in Aquatic Ecosystems*, pp. 233–256. Ann Arbor Science Press, Ann Arbor, MI.

Di Toro D.M. (1980a). Applicability of cellular equilibrium and Monod theory to phytoplankton growth kinetics. *Ecol. Modelling*, 8:201–218.

Di Toro D.M. (1980b). The effect of phosphorus loadings on dissolved oxygen in Lake Erie. In R.C. Loehr, C.S. Martin, and W. Rast, editors, *Phosphorus Management Strategies for Lakes*, pp. 191–205. Ann Arbor Science, Ann Arbor, MI.

Di Toro D.M. (1986). A diagenetic oxygen equivalents model of sediment oxygen demand. In K. Hatcher, editor, *Sediment Oxygen Demand. Processes, Modeling, and Measurement*, pp. 171–208. University of Georgia, Athens, GA.

Di Toro D.M., Mahony J.D., Hansen D.J., Scott K.J., Carlson A.R., and Ankley G.T. (1991a). Acid volatile sulfide predicts the acute toxicity of cadmium and nickel in sediments. *Environ. Sci. Tech.*, 26(1):96–101.

Di Toro D.M., Mahony J.D., Hansen D.J., Scott K.J., Hicks M.B., Mayr S.M., and Redmond M.S. (1990a). Toxicity of cadmium in sediments: The role of acid volatile sulfide. *Environ. Toxicol. Chem.*, 9:1487–1502.

Di Toro D.M. and Matystik W.F. (1980). *Mathematical Models of Water Quality in Large Lakes Part I: Lake Huron and Saginaw Bay*. Report EPA–600/3–80-056. U.S. EPA Environmental Resarch Lab, Duluth, MN.

Di Toro D.M., O'Connor D.J., Thomann R.V., and St. John J. (1982). Simplified model of the fate of partitioning chemicals in lakes and streams. In A.W. Maki, K.L. Dickson and J. Cairns, editors, *Modeling the Fate of Chemicals in the Environment*, pp. 165–190. Ann Arbor Science Publications, Ann Arbor, MI.

Di Toro D.M., Paquin P.R., Subburamu K., and Gruber D.A. (1990b). Sediment oxygen demand model: Methane and ammonia oxidation. *J. Environ. Engr. ASCE*, 116(5):945–986.

Di Toro D.M., Zarba C.S., Hansen D.J., Berry W.J., Swartz R.C., Cowan C.E., Pavlou S.P., Allen H.E., Thomas N.A., and Paquin P.R. (1991b). Technical basis for the equilibrium partitioning method for establishing sediment quality criteria. *Environ. Toxicol. Chem.*, 11(12):1541–1583.

Diamond M., Mackay D., Cornett R., and Chant L. (1990). A model of the exchange of inorganic chemicals between water and sediments. *Environ. Sci. Technol.*, 24:713–722.

Dortch M. and Hamlin-Tillman D. (1995). Disappearance of reduced manganese in reservoir tailwaters. *J. Environ. Engr.*, 121(4):287–297.

Droop M. (1973). Some thoughts on nutrient limitation in algae. *J. Phycology*, 9:495–506.

Dzombak D.A. and Morel F.M.M. (1990). *Surface Complexation Modeling. Hydrous Ferric Oxide*. John Wiley & Sons, New York, NY.

Edwards R.W. and Rolley H.L.J. (1965). Oxygen consumption of river muds. *J. Ecol.*, 53(1):1–19.

Effler S. and Harnett G. (1996). Background. In S. Effler, editor, *Limnological and Engineering Analysis of a Polluted Urban Lake*. Springer-Verlag, New York.

Effler S. and Whitehead K. (1996a). Particle composition. In S. Effler, editor, *Limnological and Engineering Analysis of a Polluted Urban Lake*. Springer-Verlag, New York.

Effler S. and Whitehead K. (1996b). Tributaries and Discharges. In S. Effler, editor, *Limnological and Engineering Analysis of a Polluted Urban Lake*. Springer-Verlag, New York.

Emerson S., Jacobs L., and Tebo B. (1983). The behavior of trace metals in marine anoxic waters: Solubilities at the oxygen-hydrogen sulfide interface. In K.W. Bruland, C.S. Wong, E. Boyle and J. Burton, editors, *Trace Metals in Sea Water*, pp. 579–608. Plenum Press,

New York.

Emmemegger L., King D., Sigg L., and Sulzberger B. (1998). Oxidation kinetics of Fe(II) in a eutrophic Swiss lake. *Environ. Sci. Technol.*, 32:2990–2996.

Fair G.M., Moore E.W., and Thomas Jr. H. (1941). The natural purification of river muds and pollutional sediments. *Sewage Works J.*, 13(2):270–307.

Fallon R.D. and Brock T.D. (1979). Decomposition of blue-green algal (Cyanobacteria) blooms in Lake Mendota, Wisconsin. *Appl. Environ. Microbiol.*, 37:820–830.

Fillos J. and Molof A. (1972). Effect of benthal deposits on oxygen and nutrient economy of flowing waters. *J. Water Pollut. Control Federation*, 44(4):644–662.

Forssberg K. (1985). *Flotation of Sulphide Minerals*. Elsevier Science, New York, NY.

Frithsen J.B., Keller A.A., and Pilson M. (1985a). *Effects of inorganic nutrient additions in coastal areas: A mesocosm experiment data report. Volume 1.* Marine Ecosystems Research Laboratory, Graduate School of Oceanography, University of Rhode Island, Kingston, RI. MERL Series Report No. 3.

Frithsen J.B., Keller A.A., and Pilson M. (1985b). *Effects of inorganic nutrient additions in coastal areas: A mesocosm experiment data report. Volume 3.* Marine Ecosystems Research Laboratory, Graduate School of Oceanography, University of Rhode Island, Kingston, RI. MERL Series Report No. 5.

Frithsen J.B., Lane P., Keller A.A., and Pilson M. (1985c). *Effects of inorganic nutrient additions in coastal areas: A mesocosm experiment data report. Volume 2.* Marine Ecosystems Research Laboratory, Graduate School of Oceanography, University of Rhode Island, Kingston, RI. MERL Series Report No. 4.

Gachter R., Meyer J., and Mares A. (1988). Contribution of bacteria to release and fixation of phosphorus in lake sediments. *Limnol. Oceanogr.*, 33(6):1542–1558.

Gachter R. and Meyers J. (1993). The role of microorganisms in mobilization and fixation of phosphorus in sediments. In P.C.M. Boers, T.E. Cappenberg, and W. van Raaphorst, editors, *Hydrobiologia*, volume 253, pp. 103–121. Kluwer.

Galant S. and Appleton J.P. (1973). *The Rate-Controlled Method of Constrained Equilibrium Applied to Chemical Reactive Open Systems.* Fluid Mechanics Laboratory, MIT, Cambridge MA.

Garber J., Boynton W., et al. (1988). *Ecosystem Processes Component and Benthic Exchange and Sediment Transformations.* Combined Rept. No.1. Maryland Department of the Environment.

Gardiner R., Auer M., and Canale R. (1984). Sediment Oxygen Demand in Green Bay (Lake Michigan). In M. Pirbazari and J. Devinny, editors, *Environmental Engineering. Proceedings of the 1984 Specialty Conference*, pp. 514–519. Am. Soc. Civil Engr., NY.

Garrells R.M. and Christ C.L. (1965). *Solutions, Minerals and Equilibrium.* Harper, New York.

Gee C.S., Suidan M.T., and Pfeffer J.T. (1990). Modeling of nitrification under substrate-inhibiting conditions. *J. Environ. Engr. ASCE*, 116(1):18–31.

Gelda R., Auer M., and Effler S. (1995). Determination of sediment oxygen demand by direct measurement and by inference from reduced species accumulation. *Mar. Freshwater Res.*, 46:81–88.

Gerino M., Aller R., Lee C., Cochran J., Aller J., Green M., and Hirschberg D. (1998). Comparison of different tracers and methods used to quantify bioturbation during a spring bloom: 234-thorium, luminophores and chlorophyll *a*. *Estuarine Coastal and Shelf Sci.*, 46:531–547.

Giblin A., Hopkinson C., and Tucker J. (1997). Benthic metabolism and nutrient cycling in Boston Harbor, Massachusetts. *Estuaries*, 20:346–364.

Giblin A. and Wieder R. (1992). Sulfur cycling in marine and freshwater wetlands. In R.W. Howarth, J.W.B. Stewart, and M. Ivanov, editors, *Sulfur Cycling on the Continents*. John Wiley & Sons, New York.

Goldberg E.D. and Koide M. (1963). Rates of sediment accumulation in the Indian Ocean. In *Earth Science and Meteoritics*, pp. 90–102. North-Holland, Amsterdam.

Goldhaber M.B., Aller R.C., Cochran J.K., Rosenfeld J.K., Martens C.S., and Berner R.A. (1977). Sulfate reduction diffusion and bioturbation in Long Island Sound sediments: Reports of the FOAM group. *Am. J. Sci.*, 277:193–237.

Goloway F. and Bender M. (1982). Diagenetic models of interstitial nitrate profiles in deep sea suboxic sediments. *Limnol. Oceanogr.*, 27(4):624–638.

Gratton Y., Edenborn H., Silverberg N., and Sundby B. (1990). A mathematical model for manganese diagenesis in bioturbated sediments. *Am. J. Sci.*, 290:246–262.

Green M. and Aller R. (1998). Seasonal patterns of carbonate diagenesis in nearshore terrigenous muds: Relation to spring phytoplankton bloom and temperature. *J. Mar. Res.*, 56:1097–1123.

Green M., Aller R., and Aller J. (1993). Carbonate dissolution and temporal abundances of Foraminifera in Long Island Sound sediments. *Limnol. Oceanogr.*, 38:331–345.

Grill E.V. and Richards F.A. (1964). Nutrient regeneration from phytoplankton decomposing in seawater. *J. Mar. Res.*, 22:51–59.

Gruber D., Bauer K., Di Toro D., Magruder C., Paquin P., and Grant R. (1987). *The Milwaukee Harbor Estuary Study, An Overview*. Milwaukee Metropolitan Sewerage District, Milwaukee WI. Technical report.

Gschwend P.M. and Wu S. (1985). On the constancy of sediment-water partition coefficients of hydrophobic organic pollutants. *Environ. Sci. Technol.*, 19(1):90.

Hall P., Anderson L., van der Loef M., Sundby B., and Westerlund S. (1989). Oxygen uptake kinetics in the benthic boundary layer. *Limnol. Oceanogr.*, 34:734–746.

Hamming R. (1962). *Numerical Methods for Scientists and Engineers*. McGraw-Hill, New York.

Hansen D., Berry W., Mahony J., Boothman W., Di Toro D., Robson D., Ankley G., Ma D., Yan Q., and Pesch C. (1996a). Predicting the toxicity of metals-contaminated field sediments using interstitial concentrations of metal and acid volatile sulfide normalizations. *Environ. Toxicol. Chem.*, 15(12):2080–2094.

Hansen D., Mahony J., Berry W., Benyi S., Corbin J., Pratt S., Di Toro D., and Able M. (1996b). Chronic effect of cadmium in sediments on colonization by benthic marine organisms: An evaluation of the role of interstitial cadmium and acid volatile sulfide in biological availability. *Environ. Toxicol. Chem.*, 15(12):2126–2137.

Hare L., Carignan R., and Huerta-Diaz M. (1994). A field study of metal toxicity and accumulation by benthic invertebrates; implication for the acid-volatile sulfide (AVS) model. *Limnol. Oceanogr.*, 39(7):1653–1668.

Hargrave B. (1984). Sinking of particulate matter from the surface water of the ocean. In J. Hobbie and P. Leb, editors, *Heterotrophic Activity in the Sea*. Plenum Press, New York.

Harper M., Davison W., and Tych W. (1999). One-dimensional views of three-dimensional sediments. *Environ. Sci. Technol.*, 33:2611–2616.

Hatcher K. (1986). Introduction of Sediment Oxygen Demand Modeling. In K. Hatcher, editor, *Sediment Oxygen Demand. Processes, Modeling and Measurement*, pp. 113–138. Institute of Natural Resources, University of Georgia, Athens, GA.

Henriksen K. and Kemp W. (1988). Nitrification in estuarine and coastal marine sediments. In T. Blackburn and J. Sorensen, editors, *Nitrogen Cycling in Coastal Marine Environments*. John Wiley, NY.

Hirst D. (1962). The geochemistry of modern sediment from the Gulf of Paria II. The location and distribution of trace elements. *Geochim. Cosmochim. Acta*, 26:1147–1187.

Holdren Jr. G. (1977). *Distribution and Behavior of Manganese in the Interstitial Waters of Chesapeake Bay Sediments during Early Diagenesis*. Ph.D. thesis, The Johns Hopkins University, Baltimore, Maryland.

Holdren Jr. G., Bricker O., and Matisoff G. (1975). A model for the control of dissolved manganese in the interstitial waters of Chesapeake Bay. In T. Church, editor, *Marine Chemistry in the Coastal Environment*, pp. 364–381. ACS Symposium Series 18. American Chemical Society, Wash. DC.

Howarth R.W. and Jorgensen B.B. (1984). Formation of ^{35}S-labelled elemental sulfur and pyrite in coastal marine sediments (Limfjorden and Kysing Fjord, Denmark) during short-term $^{35}SO_4$ reduction experiments. *Geochim. Cosmochim. Acta*, 48:1807–1818.

Hunt C. and Kelly J. (1988). Manganese cycling in coastal regions: Response to eutrophication. *Est. Coast. Shelf Sci.*, 26:527–558.

Hurd D. (1973). Interactions of biogenic opal, sediment, and seawater in the central equatorial Pacific. *Geochim. Cosmochim. Acta*, 37:2257–2282.

Hutchinson G. (1957). *A Treatise on Limnology, Volume 1, Part 2 - Chemistry of Lakes*. John Wiley & Sons, New York.

HydroQual (1995). *A water quality model for Massachusetts Bay and Cape Cod Bay. The bays eutrophication model calibration*. HydroQual, Inc. Mahwah, NJ. Prepared for the Massachusetts Water Resources Authority, Boston MA. Report NAIC0103.

HydroQual (1996). *Water quality modeling analysis of hypoxia in Long Island Sound using LIS3.0*. HydroQual, Inc. Mahwah, NJ. Prepared for the U.S. EPA. Report NENG0035.

HydroQual I. (1992). *User's guide for RCA. Release 2.0*. HydroQual, Inc. Mahwah, NJ. Technical report.

Imoboden D.M. (1975). Interstitial transport of solutes in non-steady state accumulations and compacting sediments. *Earth. Planet. Sci. Letters*, 27:221–228.

Ishikawa M. and Nishimura H. (1989). Mathematical model of phosphate release rate from sediments considering the effect of dissolved oxygen in overlying water. *Wat. Res*, 23(3):351–359.

Jahnke R., Craven D., and Gaillard J.F. (1994). The influence of organic matter diagenesis on $CaCO_3$ dissolution at the deep-sea floor. *Geochim. Cosmochim. Acta*, 58(13):2799–2809.

Jahnke R.A., Emerson S.R., and Murray J.W. (1982). A model of oxygen reduction, denitrification, and organic matter mineralization in marine sediments. *Limnol. Oceanogr.*, 27(4):610–623.

Jaquet J., Nembrini G., Garcia P., and Vernet J. (1982). The manganese cycle in Lac Leman, Switzerland: The role of Metallogenium. *Hydrobiol.*, 91:323–340.

Jenkins M. and Kemp W. (1984). The coupling of nitrification and denitrification in two estuarine sediments. *Limnol. Oceanogr.*, 29(3):609–610.

Jenne E. (1968). Controls on Mn, Fe, Co, Ni, Cu, and Zn concentration in soils and water: The significant role of hydrous Mn and Fe oxides. In R. Baker, editor, *Trace Inorganics in Water. Advances in Chemistry Series 73*. American Chemical Society, Wash. DC.

Jewell W.J. and McCarty P.L. (1971). Aerobic decomposition of algae. *Environ. Sci. Technol.*, 5:1023–1031.

Jorgensen B. and Revsbech N. (1985). Diffusive boundary layers and the oxygen uptake of sediments and detritus. *Limnol. Oceanogr.*, 30(1):111–122.

Jorgensen B.B. (1977a). Bacterial sulfate reduction within reduced microniches of oxidized marine sediments. *Mar. Biol.*, 41:7–17.

Jorgensen B.B. (1977b). The sulfur cycle of a coastal marine sediment (Limfjorden, Denmark). *Limnol. Oceanogr.*, 22:814–832.

Jorgensen B.B. (1982). Ecology of the bacteria of the sulphur cycle with special reference to anoxic-oxic interface environments. *Phil. Trans. R. Soc. Lond. B*, 298:543–561.

Jorgensen B.B. and Revsbech N.P. (1990). Oxygen uptake, bacterial distribution, and carbon-nitrogen-sulfur cycling in sediments from the Baltic Sea - North Sea transition. *Ophelia*, 31(1):29–49.

Jorgensen B.B., Revsbech N.P., and Cohen Y. (1983). Photosynthesis and structure of benthic microbial mats: microelectrode and SEM studies of four cyanobacterial communities. *Limnol. Oceanogr.*, 28:1075–1093.

Jorgensen S.E., Kamp-Nielsen L., and Jacobsen O.S. (1975). A submodel for anaerobic mud-water exchange of phosphate. *Ecol. Modelling*, 1:133–146.

Kamp-Nielsen L. (1975). A kinetic approach to the aerobic sediment-water exchange of phosphorus in Lake Estrom. *Ecol. Modelling*, 1:153–160.

Kamp-Nielsen L., Mejer H., and Jorgensen S.E. (1982). Modelling the influence of bioturbation on the vertical distribution of sedimentary phosphorus in L. Esrom. *Hydrobiol.*, 91:197–206.

Kana T., Sullivan M., Cornwell J., and Groszkowski K. (1998). Denitrification in estuarine sediments determined by membrane inlet mass spectrometry. *Limnol. Oceanogr.*, 43:334–339.

Kaplan I.R. (1974). *Natural Gases in Marine Sediments.* Plenum Press, New York.

Kaplan I.R. and Rittenberg S.C. (1964). Microbiological fractionation of sulfur isotopes. *J. Gen. Microbiol.*, 34:195–212.

Kelly J., Berounsky V., Nixon S., and Oviatt C. (1985). Benthic-pelagic coupling and nutrient cycling across an experimental eutrophication gradient. *Mar. Ecol. Prog. Ser.*, 26:207–219.

Kemp W.M., Sampou P., Caffrey J., Mayer M., Henriksen K., and Boynton W.R. (1990). Ammonium recycling versus denitrification in Chesapeake Bay sediments. *Limnol. Oceanogr.*, 35(7):1545–1563.

King D. (1998). Role of carbonate speciation on the oxidation rate of Fe(II) in aquatic systems. *Environ. Sci.l Technol.*, 32:2997–3003.

Klapwijk A. and Snodgrass W.J. (1982). Biofilm model for nitrification, denitrification, and sediment oxygen demand in Hamilton Harbor. *55th Annual Water Pollution Control Federation Conf.*

Klapwijk A. and Snodgrass W.J. (1986). Biofilm Model for Nitrification, Denitrification, and Sediment Oxygen Demand in Hamilton Harbor. In K. Hatcher, editor, *Sediment Oxygen Demand. Processes, Modeling and Measurement*, pp. 75–97. Institute of Natural Resources, University of Georgia, Athens, GA.

Klump J.V. (1980). *Benthic Nutrient Regeneration and the Mechanisms of Chemical Sediment-Water Exchange in an Organic-Rich Coastal Marine Sediment.* Ph.D. thesis, University of North Carolina, Chapel Hill, NC.

Klump J.V. and Martens C.S. (1981). Biogeochemical cycling in an organic rich coastal marine basin. II. Nutrient sediment-water exchange processes. *Geochim. Cosmochim. Acta*, 45:101–121.

Klump J.V. and Martens C.S. (1983). Benthic Nitrogen Regeneration. E.J. Carpenter, D.G. Capone, editors, in *Nitrogen in the Marine Environment*, pp. 411–457. Academic Press, New York.

Klump J.V. and Martens C.S. (1989). The seasonality of nutrient regeneration in an organic-rich sediment: Kinetic modeling of changing pore-water nutrient and sulfate distributions. *Limnol. Oceanogr.*, 34(3):559–577.

Kramer J.R. (1964). Theoretical model of the chemical composition of fresh water with application to the Great Lakes. *Pub. No. 11, Great Lakes Research Division, Univ. of Mich.*, p. 147.

Kreyszig E. (1972). *Advanced Engineering Mathematics*. John Wiley & Sons, New York.

Krishnaswami S. and Lal D. (1978). Radionuclide Limnochronology. In A. Lerman, editor, *Lakes. Chemistry, Geology, Physics*, pp. 153–177. Springer-Verlag, New York.

Lauria J. and Goodman A. (1983). Mass Flux Measurement of Sediment Oxygen Demand. In K. Hatcher, editor, *Sediment Oxygen Demand. Processes, Modeling and Measurement*, pp. 75–97. Institute of Natural Resources, University of Georgia, Athens, GA.

Lawson D.S., Hurd D.C., and Pankratz H.S. (1978). Silica dissolution rates of decomposing phytoplankton assemblages at various temperatures. *Am. J. Sci.*, 278:1373–1393.

Lerman A. (1975). Maintenance of steady state in oceanic sediments. *Am. J. Sci.*, 275:609–635.

Lerman A. and Taniguchi H. (1972). Strontium 90-diffusional transport in sediments of the Great Lakes. *J. Geophysic. Res.*, 77(3):474.

Lewandoswki Z. (1982). Temperature dependency of biological denitrification with organic materials addition. *Wat. Res*, 16:19–22.

Liber K., Call D., Markee T., Schmude K., Balcer M., Whiteman F., and Ankley G. (1996). Effects of acid volatile sulfide on zinc bioavailability and toxicity to benthic macroinvertebrates: A spiked-sediment field experiments. *Environ. Toxicol. Chem.*, 15(12):2113–2125.

Lijklema L. (1980). Interaction of ortho-phosphate with iron(III) and aluminum hydroxides. *Environ. Sci. Technol.*, 14:537–541.

Lindsay W.L. (1979). *Chemical Equilibria in Soils*. John Wiley & Sons, New York, NY.

Loewenthal R. and Marais G. (1976). *Carbonate Chemistry of Aquatic Systems: Theory and Application*. Ann Arbor Science, Ann Arbor, MI.

Lowe S. and Di Toro D. (2001). Application of a water column-sediment eutrophication model to a mesocosm experiment. II. Analysis. *J. Environ. Engr. ASCE*.

Luther III G., Brendel P., Lewis B., Sundby B., Lefrancois L., Silverberg N., and Nuzzio D. (1998). Simultaneous measurement of O_2, Mn, Fe, I−, and S(−II) in marine pore waters with a solid-state voltammetric microelectrode. *Limnol. Oceanogr.*, 43:325–333.

Mackenthun A. and Stefan H. (1998). Effect of flow velocity on sediment oxygen demand: Experiments. *J. Environ. Engr. ASCE*, 124:222–230.

Mackin J., Aller R., Vigil H., and Rude P. (1991). *Nutrient and dissolved oxygen fluxes across the sediment-water interface, in Long Island Sound Study. Final Report, Sediment Geochemistry and Biology* Marine Sciences Research Center, SUNY at Stony Brook, NY. U.S. EPA Contract CE 002870026.

Macsyma (1993). *Macsyma*. Macsyma, Inc. 20 Academy St. Arlington, MA.

Mahony J., Di Toro D., Gonzalez A., Curto M., Dilg M., De Rosa L., and Sparrow L. (1996). Partitioning of metals to sediment organic carbon. *Environ. Toxicol. Chem.*, 15(12):2187–2197.

Mancini J. (1983). A method for calculating effects on aquatic organisms of time varying concentrations. *Wat. Res*, 17:1355–1362.

Marinelli R. and Boudreau B. (1996). An experimental and modeling study of pH and related solutes in an irrigated anoxic coastal sediment. *J. Marine Research*, 54:939–966.

Marquardt D.W. (1963). An algorithm for least-squares estimation of non-linear parameters. *J. Soc. Indust. Appl. Math.*, 11(2):431–441.

Martens C.S. and Berner R.A. (1977). Interstitial water chemistry of anoxic Long Island Sound sediments. I. Dissolved gases. *Limnol. and Oceanogr.*, 22(1):10–25.

Matisoff G. (1977). *Early Diagenesis of Chesapeake Bay Sediments: A Time Series Study of Temperature, Chloride and Silica*. Ph.D. thesis, The Johns Hopkins University, Baltimore, MD.

Matisoff G. (1982). Mathematical Models of Bioturbation. In P. McCall and M. Tevesz, editors, *Animal-Sediment Relations. The Biogenic Alteration of Sediments*, pp. 289–330. Plenum Press, New York.

Matisoff G., Bricker O., Holdren Jr. G., and Kaerk P. (1975). Spatial and temporal variations in the interstitial water chemistry of Chesapeake Bay sediments. In T. Church, editor, *Marine Chemistry in the Coastal Environment*, pp. 343–363. ACS Symposium ser. 18, American Chemical Society, Wash., DC.

Matisoff G. and Wang X. (1998). Solute transport in sediments by freshwater infaunal bioirrigators. *Limnol. Oceanogr.*, 43:1487–1499.

Matisoff G., Wang X., and McCall P. (1999). Biological redistribution of lake sediments by tubificid oligochaetes: *Branchiura sowerbyi* and *Limnodrilus hoffmeisteri/Tubifex tubifex*. *J. Great Lakes Res.*, 25:205–219.

Maughan J. (1986). *Relationship Between Macrobenthic Infauna and Organic Carbon*. Ph.D. thesis, Graduate School of Oceanography, University of Rhode Island, Kingston, RI.

McCall P. and Tevesz M. (1982). The effects of benthos on physical properties of freshwater sediments. In P. McCall and M. Tevesz, editors, *Animal-Sediment Relations. The Biogenic Alteration of Sediments*, pp. 105–176. Plenum Press, New York.

Messer J.J. and Brezonik P.L. (1984). Laboratory evaluation of kinetic parameters for lake sediment denitrification models. *Ecol. Modelling*, 21:277–286.

Metcalf and Eddy, Inc., Hazen and Sawyer, PC (1995). *The Croton water treatment plant at Jerome Park Reservoir. Appendix E. Volume II.* Report prepared for City of New York, Department of Environmental Protection.

Metropolitan Washington Council of Governments (1990). *Anacostia River.* Technical report. Washington DC.

Meyer G. and Gruendling G. (1979). Limnology of Lake Champlain. New England River Basins Commision, Burlington, NY.

Millero F. (1990). Effect of ionic interactions on the oxidation rates of metals in natural waters. In D. Melchior and R. Bassett, editors, *Chemical Modeling of Aqueous Systems II*, pp. 447–460. Am. Chem. Soc., Washington, DC.

Millero F.J. (1986). The thermodynamics and kinetics of the hydrogen sulfide system in natural waters. *Mar. Chem.*, 18:381–391.

Millero F.J. (1991). The oxidation of H_2S in the Chesapeake Bay. *Est. Coast. Shelf Sci.*, 33:521–527.

Millero F.J., Hubinger S., Fernandez M., and Garnett S. (1987). Oxidation of H_2S in seawater as a function of temperature, pH, and ionic strength. *Environ. Sci. Technol*, 21:439–443.

Morel F. and Morgan J.J. (1972). A numerical technique for computing equilibria in aqueous chemical systems. *Environ. Sci. Technol.*, 6:58.

Morel F.M.M. (1983). *Principles of Aquatic Chemistry*. John Wiley & Sons, New York, NY.

Morgan J. (1967). Chemical equilibria and kinetic properties of manganese in natural water. In S. Faust and J. Hunter, editors, *Principles and Applications of Water Chemistry*. John Wiley & Sons, New York, NY.

Morse J. and Mackenzie F. (1990). *Geochemistry of Sedimentary Carbonates*. Elsevier, New York.

Morse J.W., Millero F.J., Cornwell J.C., and Rickard D. (1987). The chemistry of the hydrogen sulfide and iron sulfide systems in natural waters. *Earth Sci. Rev.*, 24:1–42.

Mortimer C.H. (1941). The exchange of dissolved substances between mud and water. I and II. *J. Ecol*, 29:280–329.

Mortimer C.H. (1942). The exchange of dissolved substances between mud and water in lakes. III and IV. *J. Ecol*, 30:147–201.

Mortimer C.H. (1971). Chemical exchanges between sediments and water in the Great Lakes–Speculations on probable regulatory mechanisms. *Limnol. Oceanogr.*, 16(2):387–404.

Mudroch A. and MacKnight S.D. (1991). *CRC Handbook of Techniques for Aquatic Sediments Sampling*. CRC Press, Boca Raton, FL.

Nakajima M., Hayamizu T., and Nishimura H. (1984). Effect of oxygen concentration on the rates of denitrafication and denitrification in the sediments of an eutrophic lake. *Wat. Res*, 18(3):335–338.

Nakamura Y. and Stefan H. (1994). Effect of flow velocity on sediment oxygen demand: Theory. *J. Environ. Engr. ASCE*, 120:996–1016.

Nedwell D.B. and Floodgate G.D. (1972). Temperature-induced changes in the formation of sulfide in a marine sediment. *Mar. Biol.*, 14:18–24.

Nelson M.B. (1978). *Kinetics and Mechanisms of the Oxidation of Ferrous Sulfide* Ph.D. thesis, Stanford University, Palo Alto, CA.

Nixon S., Oviatt C., Frithsen J., and Sullivan B. (1986). Nutrients and the productivity of estuarine and coastal marine ecosystems. *J. Limnol. Soc. South Africa*, 12(1/2):43–71.

Nixon S. and Pilson M. (1983). Nitrogen in Estuarine and Coastal Marine Ecosystems. In E. Carpenter and D. Capone, editors, *Nitrogen in the Marine Environment*. Academic Press, New York.

Nixon S.W., Oviatt C.A., and Hale S.S. (1976). Nitrogen regeneration and the metabolism of coastal marine bottom communities. In *The Role of Terrestrial and Aquatic Organisms in Decomposition Processes*, pp. 269–283. Blackwell, Oxford.

Nowicki B. and Oviatt C. (1990). Are estuaries traps for anthropogenic nutrients? Evidence from estuarine mesocosms. *Mar. Ecol. Prog. Ser.*, 66:131–146.

Nurnberg G.K. (1988). Prediction of phosphorus release rates from total and reductant-soluble phosphorus in anoxic lake sediments. *Can. J. Fish. Aquat. Sci.*, 45:453–462.

Nyffeler U., Santschi P., and Li Y.H. (1986). The relevance of scavenging kinetics to modeling of sediment-water ineractions in natural waters. *Limnol. Oceanogr.*, 31(2):277–292.

O'Brien D.J. and Birkner F.B. (1977). Kinetics of oxygenation of reduced sulfur species in aqueous solution. *Environ. Sci. Technol.*, 11(12):1114–1120.

O'Connor D. (1983). Wind effects on gas-liquid transfer coefficients. *J. Environ. Engr.*, 109(3).

O'Connor D. and Dobbins W. (1958). Mechanism of reaeration in natural streams. *Trans. ASCE*, 123:641–684.

O'Connor D.J. and Connolly J. (1980). The effect of concentration of adsorbing solids on the partition coefficient. *Water Resources*, 14:1517–1523.

Officer C. and Lynch D. (1989). Bioturbation, sedimentation and sediment-water exchange. *Estuarine, Coastal and Shelf Sci.*, 28:1–12.

Otsuki A. and Hanya T. (1972). Production of dissolved organic matter from dead green algal cells. 1. Aerobic microbial decomposition. *Limnol. Oceanogr.*, 17:248–257.

Oviatt C.A., Keller A., Sampou P., and Beatty L. (1986). Patterns of productivity during eutrophication: a mesocosm experiment. *Marine Ecology – Progress Series*, 28:69–80.

Oviatt C.A., Doering P., and Nowicki B. (1993). Net system production in coastal waters as a function of eutrophication, seasonality and benthic macrofaunal abundance. *Estuaries*, 16(2):247–254.

Oviatt C.A., Doering P., Nowicki B., Reed L., Cole J., and Frithsen J. (1995). An ecosystem level experiment on nutrient limitation in temperate coastal marine environments. *Mar. Ecol. Prog. Ser.*, 116:171–179.

Oviatt C.A., Pilson M., Nixon S., Frithsen J., Rudnick D., Kelly J., Grassle J., and Grassle J. (1984). Recovery of a polluted estuarine system: A mesocosm experiment. *Mar. Ecol. Prog. Ser.*, 16:203–217.

Oviatt C.A., Quinn J., Maughan J., Ellis J., Sullivan B., Gearing J., Gearing P., Hunt C., Sampou P., and Latimer J. (1987). Fate and effects of sewage sludge in the coastal marine environment: A mesocosm experiment. *Mar. Ecol. Prog. Ser.*, 41:187–203.

Painter H.A. and Loveless J.E. (1983). Effect of temperature and pH value on the growth-rate constants of nitrifying bacteria in the activated-sludge process. *Wat. Res*, 17(3):237–248.

Park S. and Jaffe P. (1996). Development of a sediment redox potential model for the assessment of postdepositional metal mobility. *Ecol. Modeling*, 91:169–181.

Phelps E. (1944). *Stream Sanitation*. John Wiley & Sons, Inc., New York.

Pilson M., Oviatt C., and Nixon S. (1980). Annual nutrient cycles in a marine microcosm. In J. Giesy, editor, *Microcosms in Ecological Research. Conf-781101*. Technical Information Center, US Dept. Energy., Wash., DC.

Porcella D., Mills W., and Bowie G. (1986). A review of modeling formulations for sediment oxygen demand. In K. Hatcher, editor, *Sediment Oxygen Demand. Processes, Modeling and Measurement*, pp. 121–138. Institute of Natural Resources, University of Georgia, Athens, GA.

Postma D. and Jakobsen R. (1996). Redox zonation: Equilibrium constraints on the Fe(III)/SO_4-reduction interface. *Geochim. Cosmochim. Acta*, 60:3169–3175.

Press W., Flannery B., Teukolsky S., and Vetterling W. (1989). *Numerical Recipes. The Art of Scientific Computing. (FORTRAN Version)*. Cambridge University Press, New York, NY.

Rabouille C. and Gaillard J. (1990). The validity of steady-state flux calculations in early diagenesis: A computer simulation of deep-sea silica diagenesis. *Deep Sea Res.*, 37(4):625–646.

Rabouille C. and Gaillard J. (1991). Toward the EDGE: Early diagenesis global explanation. A model depicting the early diagenesis of organic matter, O_2, NO_3, Mn, and PO_4. *Geochim. Cosmochim. Acta*, 55:2511–2525.

Redfield A.C., Ketchum B.H., and Richards F.A. (1963). The influence of organisms on the composition of seawater. M.N. Hill, editor, in *The Sea*, Vol.2, pp. 26–77. Wiley-Interscience, New York.

Reeburgh W.S. (1969). Observations of gases in Chesapeake Bay sediments. *Limnol. Oceanogr.*, 14:368.

Reeburgh W.S. and Heggie D.T. (1974). Depth distributions of gases in shallow water sediments. In *Natural Gases in Marine Sediments*, pp. 27–46. Plenum Press, New York.

Revsbach N.P., Sorensen J., and Blackburn T.H. (1980). Distribution of oxygen in marine sediments measured with microelectrode. *Limnol. Oceanogr.*, 25(3):403–411.

Richards F.A. (1965). Anoxic Basins and Fjords. J.P. Riley, G. Skirrow, editors, in *Chemical Oceanography*, Vol.1 pp. 611–644. Academic Press, New York.

Robbins J., Edgington D., and Kemp A. (1978). Comparative ^{210}Pb, ^{137}Cs, and pollen geochronologies of sediments from Lakes Ontario and Erie. *Quaternary Res.*, 10.

Robbins J., Husby-Coupland K., and White D. (1984). Precise radiotracer measurement of the rate of sediment reworking by *Stylodrilus heringianus* and the effects of variable dissolved oxygen concentrations. *J. Great Lakes Res.*, 10(4):335–347.

Robbins J., Keilty T., White D., and Edginton D. (1989). Relationships between tubificid abundances, sediment composition and accumulation rates in Lake Erie. *Canadian J. Fish.*

Aquatic Sci., 46(2):223–231.

Robbins J., McCall P., Fisher J., and Krezoski J. (1979). Effect of deposit feeders on migration of cesium-137 in lake sediments. *Earth Plant. Sci. Lett.*, 42:277–287.

Robbins J.A. (1986). A model for particle-selective transport of tracers in sediments with conveyor belt deposit feeders. *J. Geophysic. Res.*, 91(c7):8542–8558.

Robbins J.A. and Edgington D.N. (1975). Determination of recent sedimentation rates in Lake Michigan using Pb-210 and Cs-137. *Geochim. Cosmochim. Acta*, 39:285–304.

Roden E. and Tuttle J. (1995). Carbon cycling in mesohaline Chesapeake Bay sediments 2: Kinetics of particulate and dissolved organic carbon turnover. *J. Mar. Res.*, 54:343–383.

Roden E., Tuttle J., Boynton W., and Kemp W. (1995). Carbon cycling in mesohaline Chesapeake Bay sediments 1: POC deposition rates and mineralization pathways. *J. Mar. Res.*, 53:799–819.

Rolley H. and Owens M. (1967). Oxygen consumption rates and some chemical properties of river muds. *Wat. Res.*, 1:759–766.

Rudd J.W.M. and Taylor C.D. (1980). Methane cycling in aquatic environments. In *Advances in Aquatic Microbology.*, volume 2, pp. 77–150. Academic Press, New York.

Rudolph J., Frenzel P., and Pfennig N. (1991). Acetylene inhibition technique underestimates in situ denitrification rates in intact cores of freshwater sediment. *FEMS Microbiol. Ecol.*, 85:101–106.

Rysgaard S., Thastum P., Dalsgaard T., Christensen P., and Sloth N. (1999). Effects of salinity on NH_4^+ adsorption capacity, nitrification, and denitrification in Danish estuarine sediments. *Estuaries*, 22:21–30.

Salomons W. and Forstner U. (1984). *Metals in the Hydrocycle.* Springer-Verlag, New York.

Sampou P. (1990). *Sediment-water exchanges and diagenesis of Anacostia River sediments.* University of Maryland Center for Environmental and Estuarine Studies. Horn Point Laboratory, Cambridge, MD. Report 90–040–PE–4845.

Sampou P., Kemp W.M., Cornwell J., Rosman L., and Owens M. (1989). *Chesapeake Bay sediment data collection program 1988/1989: Nitrogen cycling and anaerobic processes.* CEES, Univ. of Maryland, Cambridge, MD.

Sampou P. and Oviatt C. (1991). A carbon budget for a eutrophic marine ecosystem and the role of sulfur metabolism in sedimentary carbon, oxygen and energy dynamics. *J. Mar. Res.*, 49:825–844.

Santschi P., Amdurer M., Adler D., O'Hara P., Li Y.H., and Doering P. (1987). Relative mobility of radioactive trace elements across the sediment-water interface in the MERL model ecosystems of Narragansett Bay. *J. Mar. Res*, 45:1007–1048.

Schaffner L., Dellapenna T., Hinchey E., Friedrichs C., Neubauer M., Smith M., and Kuehl S. (1999). Physical energy regimes, seabed dynamics and organism-sediment interactions along an estuarine gradient. In J. Aller and R. Aller, editors, *Organism-Sediment Interactions.* University of South Carolina Press, Columbia, SC.

Schink D.R. and Guinasso Jr. N.L. (1980). Processes affecting silica at the abyssal sediment-water interface. In *Biogeochimie de la Matiere Organique a l'Interface Eau-Sediment Marin*, volume 293, pp. 81–92. Colloq. Int. du CNRS.

Schink D.R., Guinasso Jr. N.L., and Fanning K.A. (1975). Processes affecting the concentration of silica at the sediment-water interface of the Atlantic Ocean. *J. Geophysical Res.*, 80:3013–3031.

Schluter M., Sauter E., Hansen H., and Suess E. (2000). Seasonal variations of bioirrigation in coastal sediments: Modeling of field data. *Geochim. Cosmochim. Acta*, 64:821–834.

Schmidt J., Deming J., Jumars P., and Keil R. (1998). Constancy of bacterial abundance in surficial marine sediments. *Limnol. Oceanogr.*, 43:976–982.

Schnoor J. (1996). *Environmental Modeling*. John Wiley & Sons, New York, NY.

Schubel J. and Hirschberg D. (1977). Pb^{210} determined sedimentation rate and accumulation of metals in sediments at a station in Chesapeake Bay. *Chesapeake Science*, 18:379–382.

Seitzinger S., Gardner W., and Spratt A. (1991). The effect of salinity on ammonium sorption in aquatic sediments: Implications for benthic nutrient recycling. *Estuaries*, 14(2):167–174.

Shapiro N. (1962). *Analysis by Migration in the Presence of Chemical Reaction*. RAND Corp., Santa Monica, CA.

Sigg L. and Stumm W. (1980). The interaction of anions and weak acids with the hydrous goethite (a-FeOOH) surface. *Colloids and Surf.*, 2:101–117.

Slomp C., Malschaert J., and Van Raaphorst W. (1998). The role of adsorption in sediment-water exchange of phosphate in North Sea continental margin sediments. *Limnol. Oceanogr.*, 43:832–846.

Smith S. and Jaffe P. (1998). Modeling the transport and reaction of trace metals in water-saturated solis and sediments. *Water Resources Res.*, 34(11):3135–3147.

Smits J. and van der Molen D. (1993). Application of SWITCH, a model for sediment-water exchange of nutrients, to Lake Veluwe in the Netherlands. In P.C.M. Boers, T.E. Cappenberg, and W. van Raaphorst, editors, *Hydrobiologia*, volume 253, pp. 281–300. Kluwer.

Soetaert K., Herman P., and Middelburg J. (1996). Dynamic response of deep-sea sediments to seasonal variations: A model. *Limnol. Oceanogr.*, 41:1651–1668.

Sorensen J. and Revsbech N.P. (1990). Denitrification in stream biofilm and sediment: in situ variation and control factors. N.P. Revsbech and J. Sorensen, editors, in *Denitrification in Soil and Sediment*, pp. 277–289. Plenum Press, New York.

Stenstrom M.K. and Poduska R.A. (1980). The effect of dissolved oxygen concentration on nitrification. *Wat. Res*, 14:643–649.

Stephens M., Kadko D., Smith C., and Latasa M. (1997). Chlorophyll-a and pheopigments as tracers of labile organic carbon at the central equatorial Pacific seafloor. *Geochim. Cosmochim. Acta*, 61:4605–4619.

Stevens D.K., Berthouex P.M., and Chapman T.W. (1989). Dynamic model of nitrification in fluidized bed. *J Environ. Engr. Div. ASCE*, 115(5):910–929.

Stone A. (1987). Microbial metabolites and the reductive dissolution of manganese oxides: Oxalate and pyruvate. *Geochim. Cosmochim. Acta*, 51:919–925.

Streeter H. (1935). Measures of natural oxidation in polluted streams. I. The oxygen demand factor. *Sewage Works Journal*, 7(2):251–279.

Stumm W. (1966). Redox potential as an environmental parameter: Conceptual significance and operational limitation. In *3rd International Conference on Water Pollution Research*, pp. 283–308. Munich, Germany.

Stumm W. and Lee G. (1961). Oxygenation of ferrous iron. *Indust. Eng. Chem.*, 53:143–146.

Stumm W. and Morgan J.J. (1970). *Aquatic Chemistry*. John Wiley & Sons, New York.

Stumm W. and Morgan J. (1981). *Aquatic Chemistry*. John Wiley & Sons, New York.

Stumm W. and Morgan J. (1996). *Aquatic Chemistry. Chemical Equilibria and Rates in Natural Waters*. John Wiley & Sons, New York.

Sun M., Aller R.C., and Lee C. (1991). Early diagenesis of chlorophyll-a in Long Island Sound sediments: A measure of carbon flux and particle reworking. *J. Mar. Res.*, 49:379–401.

Sun M., Aller R.C., and Lee C. (1994). Spatial and temporal distributions of sedimentary chloropigments as indicators of benthic processes in Long Island Sound. *J. Mar. Res.*, 52:149–176.

Sundby B., Anderson L., Hall P., Iverfeldt A., van der Loeff M., and Westerlund S. (1986). The effect of oxygen on release and uptake of cobalt, manganese, iron and phosphate at the sediment-water interface. *Geochim. Cosmochim. Acta*, 50:1281–1288.

SWRPC (1987). *A Water Resources Management Plan for the Milwaukee Harbor Estuary.* SE Wisconsin Regional Planning Commission, Waukesha, WI. 53187.

Tessier A., Campbell P.G.C., and Bisson M. (1979). Sequential extraction procedure for the speciation of particulate trace metals. *Anal. Chem.*, 51(7):844–851.

Tessier A., Couillard Y., Campbell P., and Auclair J. (1993). Modeling Cd partitioning in oxic lake sediments and Cd concentrations in the freshwater bivalve: *Anodonta grandis* (Mollusca, Pelecypoda). *Limnol. Oceanogr.*, 38(1):1–17.

Thomann R. and Mueller J. (1987). *Principles of Surface Water Quality Modeling and Control.* Harper, New York.

Thoms S., Matisoff G., McCall P., and Wang X. (1995). *Models for alteration of sediments by benthic organisms.* Water Environment Research Foundation. Alexandria VA. Report 92-NPS-2.

Thorstenson D.C. (1970). Equilibrium distribution of small organic molecules in natural waters. *Geochim. Cosmochim. Acta.*, 34:745–770.

Toth D.J. and Lerman A. (1977). Organic matter reactivity and sedimentation rates in the ocean. *Am. J. Sci.*, 277:465–485.

Troup B. (1974). *The Interaction of Iron with Phosphate, Carbonate and Sulfide in Chesapeake Bay Interstitial Waters: A Thermodynamic Interpretation.* Ph.D. thesis, The Johns Hopkins University, Baltimore, MD.

Troup B. and Bricker O. (1975). Processes affecting the transport of materials from continents to oceans. In T. Church, editor, *Marine Chemistry in the Coastal Environment*, pp. 133–149. ACS Symposium Ser. 18, American Chemical Society, Wash. DC.

Troup B., Bricker O., and Bray J. (1974). Oxidation effect on the analysis of iron in the interstitial water of recent anoxic sediments. *Nature*, 249(5454):237–239.

Turekian K.K., Benoit G.J., and Benninger L.K. (1980). The mean residence time of plankton-derived carbon in a Long Island Sound sediment core: A correction. *Est. Coast. Mar. Sci.*, 11:583–587.

Ullman W.J. and Aller R.C. (1989). Nutrient release rates from the sediments of Saginaw Bay, Lake Huron. *Hydrobiol.*, 171:127–140.

Valiela I. (1995). *Marine Ecological Processes.* Springer-Verlag, New York.

van Cappellen P. and Berner R. (1988). A mathematical model for the early diagenesis of phosphorus and fluorine in marine sediments: Apatite precipitation. *Am. J. Sci.*, 288:289–333.

van Cappellen P. and Gaillard J. (1996). Biogeochemical dynamics in aquatic sediments. In P.C. Lichtner, C.I. Steefel, and E.H. Oelkers, editors, *Reactive Transport in Porous Media. Reviews in Mineralogy 34.*, pp. 335–376. Mineralogical Society of America, Wash. DC.

van Cappellen P. and Wang Y. (1995). Metal cycling in surface sediments: Modeling the interplay of transport and reaction. In H. Allen, editor, *Metal Contaminated Aquatic Sediments*, pp. 21–64. Ann Arbor Press, Ann Arbor, MI.

van Cappellen P. and Wang Y. (1996). Cycling of iron and manganese in surface sediments: A general theory for the coupled transport and reaction of carbon, oxygen, nitrogen, sulfur, iron, and manganese. *Am. J. Sci.*, 296:197–243.

Van der Molen D. (1991). A simple, dynamic model for the simulation of the release of phosphorus from sediments in shallow, eutrophic systems. *Wat. Res*, 25(6):737–744.

Vanderborght J. and Billen G. (1975). Vertical distribution of nitrate concentration in interstitial water of marine sediments with nitrification and denitrification. *Limnol. Oceanogr.*, 20:953–961.

Vanderborght J.P., Wollast R., and Billen G. (1977). Kinetic models of diagenesis in disturbed sediments. I. Mass transfer properties and silica diagenesis. *Limnol. Oceanogr.*, 22:787–793.

Vaughan D. and Craig J. (1978). *Mineral Chemistry of Metal Sulfides.* Cambridge University Press, Cambridge, UK.

Versar Inc. (1990). *Impact assessment of Craighill Channel Dredged Material Placement in the Deep Trough.* Versar, Inc., ESM Operations, 9200 Rumsery Rd., Columbia, MD. Technical report.

von Langen P., Johnson K., Coale K., and Elrod V. (1997). Oxidation kinetics of manganese (II) in seawater at nanomolar concentrations. *Geochim. Cosmochim. Acta*, 61:4945–4954.

Vosjan J.H. (1974). Sulfate in water and sediment of the Dutch Wadden Sea. *J. Mar. Res.*, 1:15–21.

Walker R. and Snodgrass W. (1986). Model for sediment oxygen demand in lakes. *J. Environ. Engr. ASCE*, 112(1):25–43.

Walker W. (1991). *Compilation and review of Onondaga Lake water quality data.* Submitted to Onondaga County Department of Drainage and Sanitation, Syracuse, NY. Technical report.

Wang Y. and van Cappellen P. (1996). A multicomponent reactive transport model of early diagenesis: Application to redox cycling in marine sediments. *Geochim. Cosmochim. Acta*, 60:2993–3014.

Warwick J.J. (1986). Diel variation of in-stream nitrification. *Wat. Res*, 20(10):1325–1332.

Wehrli B., Friedl G., and Manceau A. (1995). Reaction rates and products of manganese oxidation at the sediment-water interface. In C.P. Huang, C.R. O'Melia, and J.J. Morgan, editors, *Aquatic Chemistry. Interfacial and Interspecies Processes*, Advances in Chemistry Series 255. American Chemical Society, Wash. DC.

Wersin P., Hohener P., Giovanoli R., and Stumm W. (1991). Early diagenetic influences on iron transformations in a freshwater lake sediment. *Chem. Geol.*, 90:223–252.

Westall J. and Hohl H. (1980). A comparison of electrostatic models for the oxide/solution interface. *Adv. Colloid Interface Sci.*, 12:265–294.

Westrich J.T. (1983). *The Consequences and Controls of Bacterial Sulfate Reduction in Marine Sediments Ph.D. Thesis.* Yale University, New Haven.

Westrich J.T. and Berner R.A. (1984). The role of sedimentary organic matter in bacterial sulfate reduction: The G model tested. *Limnol. Oceanogr.*, 29(2):236–249.

Westrich J.T. and Berner R.A. (1988). The effect of temperature on rates of sulfate reduction in marine sediments. *Geomicrobiol. J.*, 6:99–117.

Wheatcroft R. and Martin W. (1996). Spatial variation in short-term (^{234}Th) sediment bioturbation intensity along an organic-carbon gradient. *J. Marine Research*, 54:763–792.

Wheatland A.B. (1954). Factors affecting the formation and oxidation of sulphides in a polluted estuary. *J. Hyg.*, 52:194–210.

Whittemore R. (1986). The significance of interfacial water velocity on the measurement of sediment oxygen demand. In K. Hatcher, editor, *Sediment Oxygen Demand. Processes, Modeling and Measurement.*, pp. 63–74. Institute of Natural Resources, University of Georgia, Athens, GA.

Wilkinson L. (1990). *SYSTAT: The System for Statistics.* SYSTAT, Inc., Evanston, IL.

Wilmot P.D., Cadee K., Katinic J.J., and Kavanagh B.V. (1988). Kinetics of sulfide oxidation by dissolved oxygen. *J. Wat. Pollut. Control Fed.*, 60(7):1264–1270.

Wong G.T.F. and Grosch C.E. (1978). A mathematical model of the distribution of dissolved silicon in interstitial waters-an analytical approach. *J. Mar. Res.*, 36:735–750.

Yamamoto S., Alcauskas J., and Crozier T. (1976). Solubility of methane in distilled water and seawater. *J. Chem. Engr. Data*, 21(1):78–80.

Yao W. and Millero F. (1995). Oxidation of hydrogen sulfide by Mn(IV) and Fe(III) (hydr)oxides in seawater. In M. Vairavamurthy and M. Schoonen, editors, *Geochemical Transformations of Sedimentary Sulfur*. Am. Chem. Soc., Wash. DC.

Yin C. and Johnson D. (1984). Sedimentation and particle class balances in Onondaga Lake. *Limnol. Oceanogr.*, 29:1193–1201.

Yoshida T. (1981). Mathematical model of phosphorus release from lake sediment. *Verh. Internat. Verein. Limnol.*, 21:268–274.

Young J.C. and Thompson L.O. (1979). Control strategy for biological nitrification systems. *J Wat. Pollut. Cont. Fed.*, 51(7):1824–1840.

Zhang J.Z. and Millero F.J. (1991). The rate of sulfite oxidation in seawater. *Geochim. Cosmochim. Acta*, 55:677–685.

Index